MECHANICAL SYSTEM COMPONENTS

This book is part of the

ALLYN AND BACON SERIES IN ENGINEERING

CONSULTING EDITOR: FRANK KREITH
UNIVERSITY OF COLORADO

MECHANICAL SYSTEM COMPONENTS

JAMES F. THORPE
University of Cincinnati

ALLYN AND BACON
Boston London Sydney Toronto

A division of Simon & Schuster
160 Gould Street
Needham Heights, Massachusetts 02194

Library of Congress Cataloging-in-Publication Data

Thorpe, James F., 1926–
Mechanical system components / James F. Thorpe.
p. cm. — (Allyn and Bacon series in engineering)
Includes bibliographies and index.
ISBN 0-205-11713-9
1. Machinery—Design. 2. Power transmission. I. Title.
II. Series.
TJ230.T48 1989
621.8′15—dc19 88-30464
CIP

Printed in the United States of America

10 9 8 7 6 5 4 3 2 1 93 92 91 90 89

Contents

5 Hydrostatic Drives (Transmissions) 246

6 Hydrodynamic Drives 284

7 Hydroviscous Drives 320

Preface

The impetus to write this book has come from the concern of engineering educators and engineers in industry who have been re-awakened to the importance of design experiences in the education of engineers. For some years, there was an aberration in engineering education in which the analytic (or scientific) training of engineers was overemphasized at the expense of the synthetic (or artisan) experiences. The reasons for this are many, but the primary cause was the proliferation of graduate and research programs developed in engineering colleges in response to the development of new technologies in areas such as nuclear energy, electronics, computers, and space exploration.

The rationale for introducing new scientifically oriented courses, which displaced traditional courses in graphics, shop practice, laboratory, design, and manufacture, was that the traditional breed of engineer was incapable of dealing with new technologies and that the proper role of engineers is to develop these new technologies rather than to be involved in the more routine tasks of designing and manufacturing traditional products such as appliances, machine tools, automobiles, and earth-moving and agricultural equipment.

The movement of engineering education toward a more scientifically oriented curriculum produced many benefits for both students and faculty, but it went too far in submerging design as an important engineering discipline. Some highly respected engineers and engineering educators have stated that this emphasis, which failed to support more general industrial needs, has contributed to a decline in the ability of the United States to compete in the international arena of industrial technology.

In an effort to insure that engineering design experiences are restored to engineering curricula, the Program Criteria for Engineering Programs, as formulated by the Accreditation Board for Engineering and Technology (ABET), have been quite specific about requirements in the area of engineering design. ABET requires that each upper-division student, and particularly seniors, participate in an integrative design experience which includes such elements as creativity, design methodology, specifications, alternative solutions, feasibility, and detailed design under realistic constraints (e.g., economic factors, safety, reliability, aesthetics, ethics, and social impact). In other criteria, this required integrative experience has simply been referred to as a capstone design experience, meaning that there shall be a course or courses in which the student has an opportunity to tie together (and extend) the many separate disciplinary courses of the curriculum.

Although there is more than one way to implement an integrative design experience (which might depend on the particular field of engineering), I believe strongly that the most natural and cost-effective way, for mechanical engineering students, is by means of design projects involving teams of students. This approach requires that students function well as team members, as team leaders, and as engineers committed to accomplishing an engineering design task within the constraints of specifications, timetables, documentation, and presentation. All of this should be done in an entrepreneurial and competitive atmosphere. There is no denying that this approach presents the instructor with a formidable task which will require more work and preparation than will a simple analysis course of the same number of credit hours.

The difficulties for the instructor in managing a capstone design course go beyond the problem of integrating the individual disciplinary courses the student has had prior to the senior year. Once the instructor has identified a suitable project for achieving the ABET requirements (indeed, this may be the most difficult task of all), the project will inevitably also require additional knowledge and information that students (and perhaps the instructor) do not have in areas such as mechanical components, electrical components, electrical controls, pneumatics, fluid power technology, power transmission devices, and special machine concepts. A technical thesaurus covering these areas would be of immense help in filling this void. If the design project is to be successful, this information cannot be learned concurrently with the capstone design project; it must precede it.

The intent of the capstone design project is to go beyond the traditional machine design project which focuses primarily on stress analysis and the synthesis of mechanical elements such as springs, fasteners, bearings, gears, shafts, and so forth. The capstone project will, ideally, emphasize the systems approach which will require (in addition to some detailed design) synthesis of a variety of off-the-shelf components including engines, motors, motor starters, actuators, clutches, gearboxes, fluid power components, transmissions, pneumatic components, control systems (electrical, electronic, microelectronic, and hydraulic), computers, programmable controllers, microprocessors, and so forth. This synthesis is to proceed in a milieu of human factors, product safety and liability, engineering economy, and project management. This is a tall order for both student and instructor!

One difficulty in delivering such a capstone design experience to senior mechanical engineering students is that mechanical engineering curricula are, by and large, not presently structured properly in the design area. A vacuum exists between the machine design course (whatever you may call it) and the capstone design experience desired by ABET; a transitional course is badly needed, and the purpose of this book is to provide a resource for such a course.

A course work conceptual scheme for the mechanical engineering design area intended to remedy this situation is sketched on page xv. Courses and resources for block III are absent from almost all mechanical engineering curricula. This is the fundamental cause of difficulties in implementing the ABET Program Criteria in the area of mechanical engineering design.

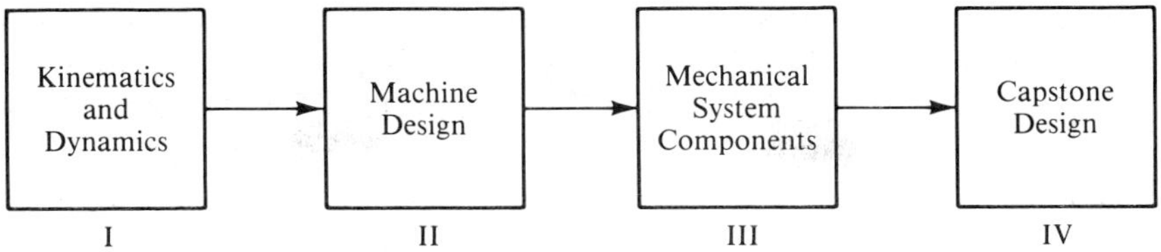

As an instructor in the mechanical engineering design area, assigned the responsibility of developing capstone design experiences for senior mechanical engineering students, I developed the present thesaurus of mechanical system components from notes over a period of several years. Teaching a mechanical system components course using these notes has been the single most important contribution to the outstanding success our senior mechanical engineering design teams have had at the University of Cincinnati in biting off design projects almost too big to chew. Six examples of these projects are available, through the publisher, to instructors who adopt this book. Of course, success also requires outstanding students and student project leaders as well as design projects of proper scope and dimensions.

In teaching a course in mechanical system components using this thesaurus, it is also important to provide students with industrial catalogs particularly in the areas of fluid power technology and mechanical power transmission and to develop a library accessible to students in which there are catalogs covering a broad range of components (from *A*ctuators to *Z*ero-max drives). Synthesis of off-the-shelf (commercially available) components into a workable machine requires that students know how to use catalog and engineering data and how to communicate with manufacturers' representatives.

This book was written in response to the needs of students matriculated in mechanical engineering programs. However, it is also perfectly suitable as a textbook for students matriculated in mechanical engineering technology programs who may be engaged in senior design projects. And, no doubt, a number of practicing engineers will find this book useful if for no other reason than it contains information about a broad range of mechanical system components (and concepts) in a single reference.

The resources used in compiling the information in this book have been cited at the end of the chapters. One reference, however, deserves special mention. This is the little jewel of a book titled *The Characteristics of Mechanical Engineering Systems*, by R. Holmes, Pergamon Press (unfortunately out of print).

In design projects associated with the use of this book, I have made heavy demands of time and energy from student design teams that culminated their design experiences under my direction. The outstanding work done by these design teams has been my most gratifying professional experience during a long career as an engineer and engineering educator. It is fitting, therefore, that this book be dedicated to those students who met a very high expectation of professional engineering performance.

The author extends thanks to the following reviewers whose contributions have enriched the text:

Thomas R. Chase
University of Minnesota

Darrell A. Frohrib
University of Minnesota

Henry O. Fuchs
Stanford University

David H. Laanen
Arizona State University

Jami Shah
Arizona State University

And, finally, I express my appreciation to Rhonda Christman for her invaluable help and assistance in creating this book.

J.F.T.

MECHANICAL SYSTEM COMPONENTS

1

Introduction

> The arc of a balance-wheel
> Flows like a curved rush of swallows, come
> over a hill . . .
> Things lost come again in sudden new beauty.
> Look long on an engine. It is sweet to the
> eyes.
>
> *MacKnight Black*

Two basic kinds of machinery design problems face the mechanical engineer concerned with mechanical systems. One type of problem involves the detailed design of some part in a machine in which the stress, deformation, fatigue, wear, corrosion, reliability, and physical shape must be determined from fundamental disciplines of mechanical engineering (e.g., strength of materials). This type of problem is exemplified by the detailed design procedures a mechanical designer would have to carry out in specifying a shaft (and its bearings) to be used in transmitting power in some machine. Traditional machine design courses in mechanical engineering curricula deal with this detailed kind of design problem.

The second type of problem is broader in design scope and deals with synthesizing a mechanical system from a variety of off-the-shelf components such as motors, clutches, gearboxes, and brakes. In this type of problem, the designer is not concerned with the design details delineated above. Instead, the component is treated as a black box. The engineer simply wants to know the properties of the black box that will permit it to be synthesized with other components in a compatible (if not optimum) way. One such property is the component transfer function which must be known in order to analyze the transient performance of the mechanical system and to design its controls.

Prerequisite to the determination of the transient performance of a mechanical system is knowledge of its steady-state operating performance. Here we need to know the characteristic power parameters of the black box component that it creates at its output or requires at its input. For brevity we will simply call these the output and input steady-state performance characteristics (or just the characteristics).

A block diagram illustrating an elementary mechanical system synthesis is sketched in Fig. 1-1. The system shown has its components in series. This arrangement is for convenience only; components in a real machine can be in both series and parallel. The important things to notice about this simple system are (a) that power is transmitted through the machine from the power source (the prime mover) to some power sink (the load) and (b) that the output of a component is the input to the adjacent component. Two links are shown between adjacent components because the transmission of power requires that both an intensive (potential) variable and an extensive (flow) variable be transmitted.

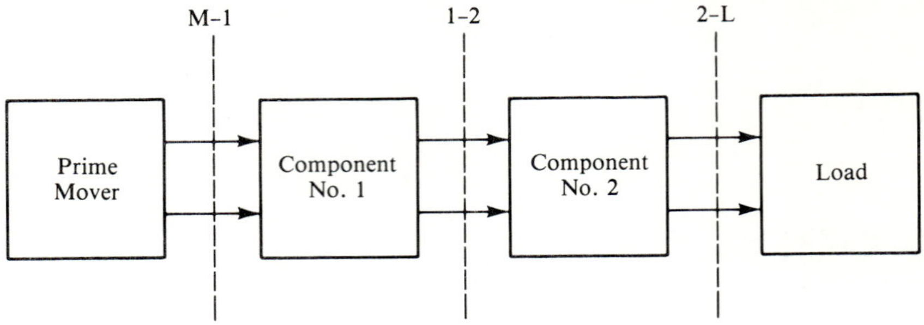

Figure 1-1. Elementary mechanical system.

TABLE 1-1. Characteristic Power Parameters.

Type of Power	Potential Parameter	Flow Parameter
Mechanical (linear)	Force, F	Linear velocity, V
Mechanical (rotary)	Torque, T	Angular velocity, ω
Hydraulic	Head, P	Discharge, Q
Electrical	Voltage, V	Current, I
Magnetic	Magnetomotive force, F	Magnetic flux, Φ
Thermal	Temperature, T	Heat flux, q

Characteristic power variables (the characteristics) of various kinds of power are tabulated in Table 1-1.

1-1 STEADY-STATE PERFORMANCE

The sequence of components in series shown in Fig. 1-1 is called a power train. Each interface between components involves a component output and a component input. At the particular interface 1-2, let us suppose that the potential variable is torque, T, and that the flow variable is angular velocity, ω. Assume further that the components are directly coupled, as shown in Fig. 1-2, so that their angular velocities are identical (assume them to be cw when viewed from the left).

If the power flow $P_{1\text{-}2}$ is from component no. 1 to component no. 2 through interface 1-2, then component no. 1 will be the driving member and component no. 2 will be the driven member. And since the two components are directly coupled,

$$\omega_1 = \omega_2 \qquad \text{(both cw)} \tag{1-1}$$

Under the steady-state operation, both of the angular velocities remain constant and

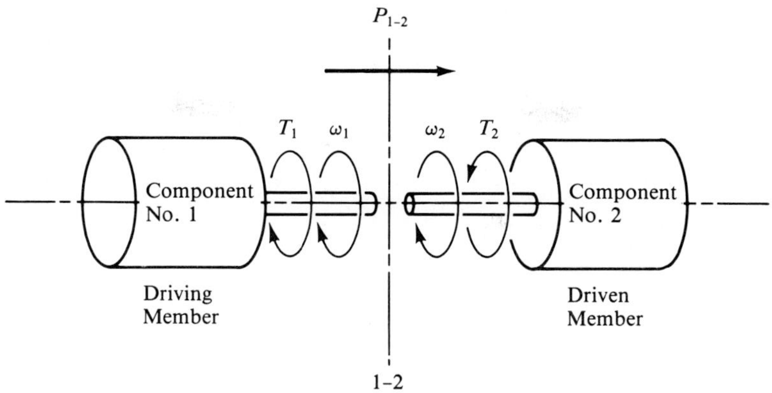

Figure 1-2. Interface conditions between components.

$$\frac{d\omega}{dt} = \alpha = 0 \qquad \textbf{(1-2)}$$

The equation of motion then requires that

$$\sum T = T_{\text{net}} = T_1 + T_2 = 0 \qquad \textbf{(1-3)}$$

or

$$T_1 = -T_2 \qquad \textbf{(1-4)}$$

Since component no. 2 is the driven member, it is acting as a load insofar as the system is concerned, which means that T_2 must be ccw if ω_2 is cw (loads are always represented by a force, or torque, in opposition to their velocity, or angular velocity, when the system is in a steady-state mode of operation). Equation (1-4) (the law of action and reaction) then requires that T_1 be cw.

So we see that the power output from a driving member consists of a torque and angular speed having the same angular sense while the input of a driven member consists of a torque and speed having opposite angular senses. Or, looking at the situation slightly differently, we can say that every driven member (load) requires a torque and speed, of the same angular sense, to drive it. Typically, component no. 1 could be an electric motor and component no. 2 could be a gearbox.

We could have made a similar analysis for the flow of electrical power through interface 1-2. In that case, the increase in electrical potential (voltage increase) would play the same role as torque and electrical current would play the role of angular speed. We would find that in an electrical load the direction of the current flow I is in opposition to voltage increase V and that to drive an electrical load we need a supply in which the directions of current flow and voltage increase are the same.

A situation is frequently encountered in mechanical systems in which the role of driven and driving members can become reversed. As an illustration, consider the system of components shown in Fig. 1-3. Here we have a prime

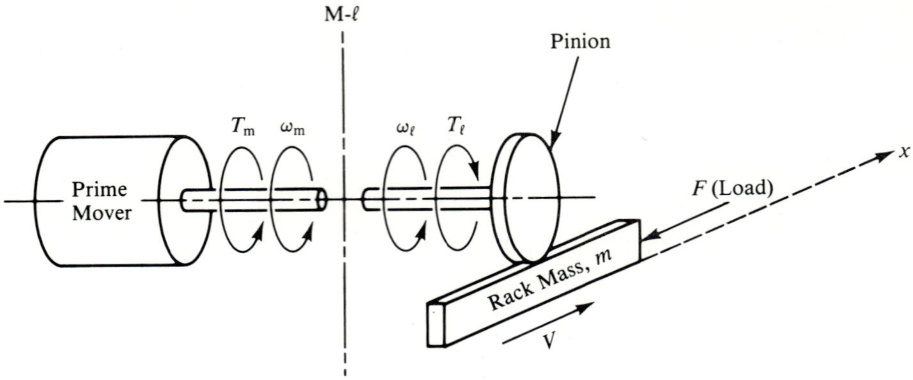

Figure 1-3. Prime mover driving rack and pinion load.

mover driving a pinion which, in turn, drives a rack at constant velocity V against a resisting force F. Then the power flow is from the prime mover, through the pinion, to the rack.

Now suppose that the load force F is suddenly removed and that the rack has a considerable mass m. Its inertia will cause the rack to tend to continue moving in the x-direction and the moment of its momentum will be in the same sense as the angular velocity ω (cw). This is now called an over-running (or overhauling) load. It can create a hazardous situation in a machine and must almost always be braked. Note that the same situation can be created if the direction of force F is simply changed, suddenly.

We will assume that the prime mover of Fig. 1-3 is an electric dynamo. When load F is present (as shown in the figure), the dynamo acts as a motor because the dynamo torque and speed both have the same cw angular direction (viewed from the left). When load F is suddenly removed, the moment of momentum of the over-running load will require a ccw torque from the dynamo if the over-running load is to be braked (the directions of torque and speed are always in opposition in a brake).

When the torque and speed of a dynamo are in opposite directions, the dynamo is acting as a load (or brake) which also means the dynamo is acting as a generator. When load F is removed, we have completely reversed the roles of the dynamo and the rack in our effort to brake the over-running load (the expression "over-running load" is really a misnomer in that a load always should be a force in opposition to the motion). Now the rack becomes the prime mover (by virtue of the kinetic energy it possesses) and the prime mover (dynamo as a generator) becomes the load.

There are two basic ways in which the generator can provide the necessary braking torque. The first method places an electrical resistance across the terminals of the dynamo so that the electrical power generated by the over-running load is dissipated in the resistance as an I^2R loss. This method is called dynamic braking. In the second method, called regenerative braking, the generator simply supplies electrical power back into the electrical supply lines from which it derived its power when it acted as a motor to drive load *F*.

Another example of the reversal of roles of prime mover and load is the automobile which when operating uphill (a hauling load) requires power flow from the engine to the wheels. When the automobile is going downhill (overhauling load), its momentum makes it the prime mover, requiring braking either by the compression of the engine (now acting as a load) or by a system of conventional friction brakes (or both). Unfortunately, no attempt is presently being made to brake regeneratively in the automobile; the energy is simply dissipated.

The above illustrations can be summarized nicely, for a dynamo, in terms of a four-quadrant operation diagram such as that shown in Fig. 1-4. In the first quadrant, we have a dynamo acting as a motor in hauling a load upward. The motor electromagnetic torque must be in the same ccw direction as the rotor angular velocity in order to balance the cw load torque. In the fourth quadrant, the load is descending (overhauling) and the dynamo must act as a brake (which it

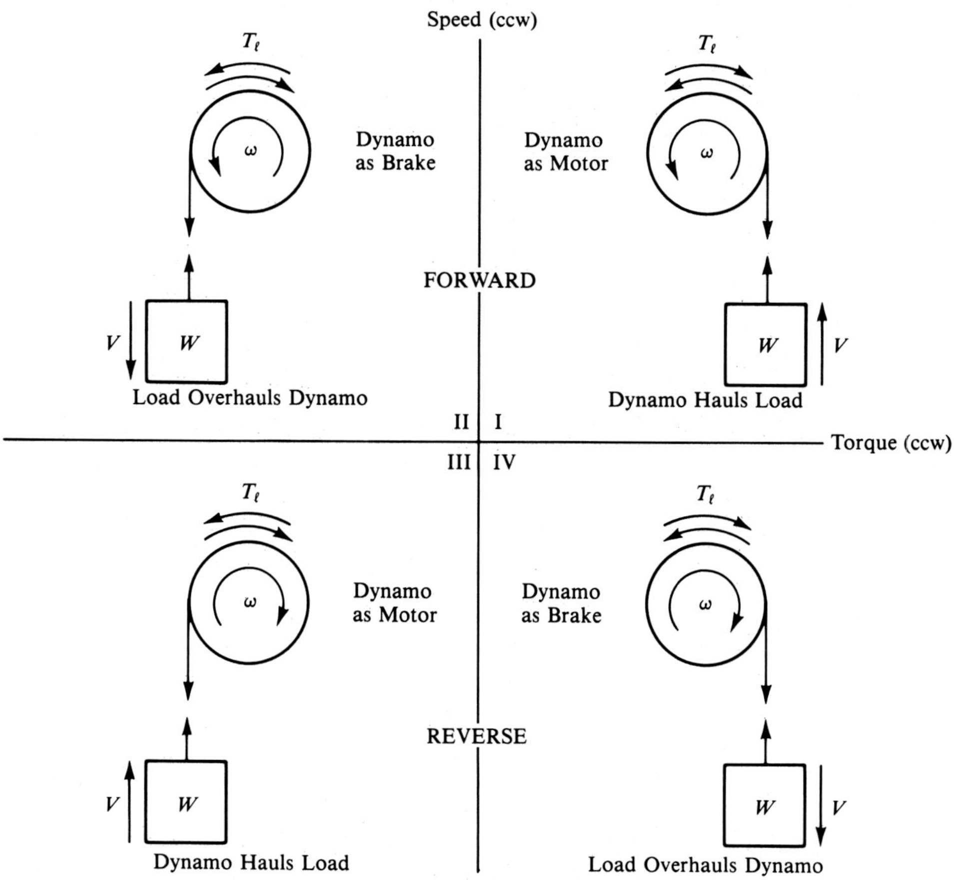

Figure 1-4. Four-quadrant dynamo operation.

can do effectively by acting as a generator). The generator electromagnetic torque must now be in the opposite ccw direction to the rotor angular speed in order to balance the cw load torque. The second and third quadrants show the same types of operations when the dynamo rotational speeds are reversed.

1-2 TYPES OF COMPONENTS

In terms of the parameters of Table 1-1 and Fig. 1-1, we can identify two basic kinds of components in addition to the prime mover and the load: power converters (or dynamos) and power transmitters (or transmissions).

1-2-1 Power Converters

We define a power converter (dynamo) to be a component that changes the power at its input to a different kind of power at its output. An example is an electric motor which has voltage and current as input electrical variables and torque and angular velocity as output mechanical variables. Other power converters are electrical generators, internal combustion engines, gas and hydraulic turbines, axial flow and centrifugal pumps and fans, compressors and blowers, and heat engines.

For the purpose of synthesizing mechanical systems, one of the most important engineering properties of a power converter is its output characteristic (sometimes we will also be interested in the converter input characteristic). A power converter output characteristic (simply called the characteristic) is the relationship between output potential and flow parameters which represents intrinsic behavior under steady-state conditions. This information is provided by the manufacturer of the power converter.

An illustrative output characteristic curve is shown in Fig. 1-5. The power converter whose characteristic is shown here is an AC induction motor. Its output variables are torque and speed. Some of the important points on the characteristic are indicated. Note that the characteristic corresponds to a particular applied input voltage, V_1. If the input voltage is changed, the torque-speed characteristic will also change.

The induction motor is generally considered to be a constant-speed power converter because the speed changes are only a small percentage of the synchronous speed over the operating torque range around rated torque. This is another way of saying that the induction motor has good speed regulation.

Each point of the characteristic curve of Fig. 1-5 is an equilibrium (steady-state) operating point determined experimentally by changing the load torque on the motor and then determining the speed at which the motor comes to equilibrium under this load. Some typical characteristics of other important power converters are displayed in Fig. 1-6.

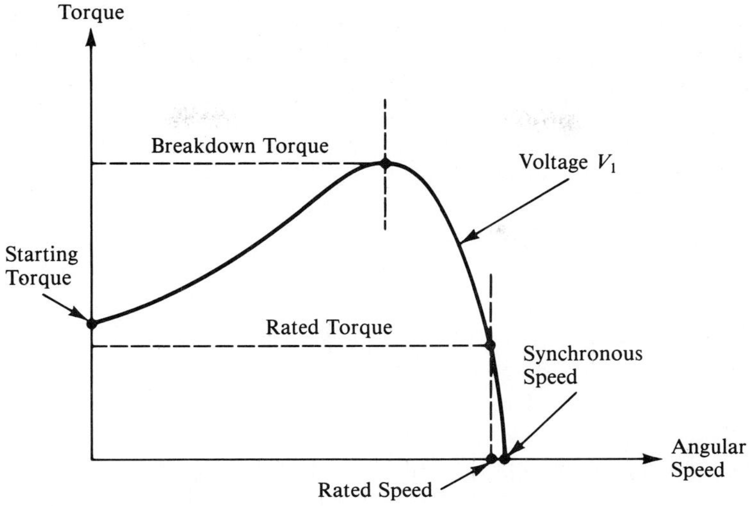

Figure 1-5. Induction motor output characteristic.

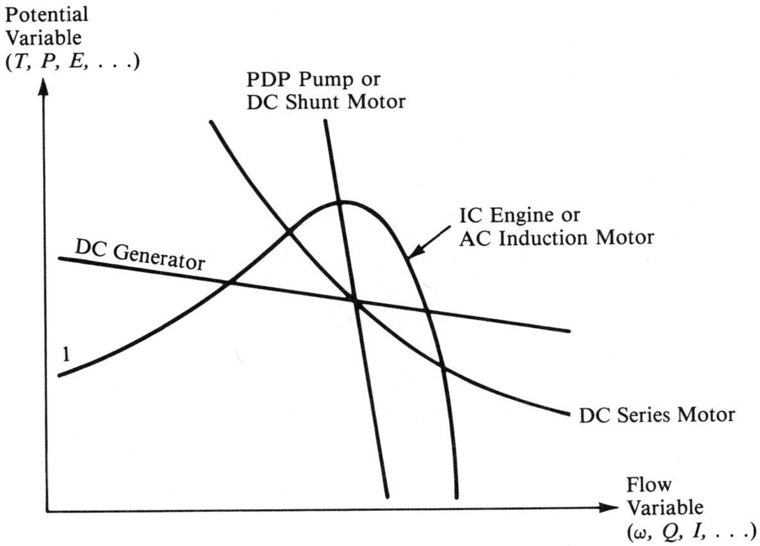

Figure 1-6. Typical converter characteristics.

1-2-2 Power Transmitters

A power transmitter is a component having the same kind of potential and flow parameters at both input and output. The four basic kinds of power transmitters are couplings, clutches, speed changers (or ratioing components), and back-to-back converters.

A speed reducer (gearbox) is an example of a power transmitter in which mechanical energy at a given torque and speed at the input is changed to mechanical energy at a different torque and speed at the output. In the ideal reducer, there are no frictional effects to cause dissipation of the mechanical energy. In this case, the torque-speed relationship will be the same form as that of a simple frictionless lever (Fig. 1-7).

Since we have assumed no power dissipation in the gearbox, the power output must be equal to the power input. Therefore,

$$\omega_o T_o = \omega_i T_i \tag{1-5}$$

Equation (1-5) is a realistic relationship for spur or helical gearboxes where the speed reduction is not large. For worm gear reducers, however, as much as 50 percent (sometimes more) of the input power can be dissipated, so Eq. (1-5) would not realistically represent the speed-torque relationship.

In a geared speed reducer, there may be backlash or shaft torsional deflection, but there will be no slip between input and output shafts. A gearbox is a positive drive component in this sense and is often called a ratioing element because its speed reduction is a constant number (ratio), defined by

$$r = \frac{\omega_i}{\omega_o} \tag{1-6}$$

A coupling is simply a ratioing component for which $r = 1$.

An electrical transformer is the electrical analog of the gearbox and is also a ratioing device if we neglect electrical power dissipation. The input and output powers then obey the relationship

$$E_o I_o = E_i I_i \tag{1-7}$$

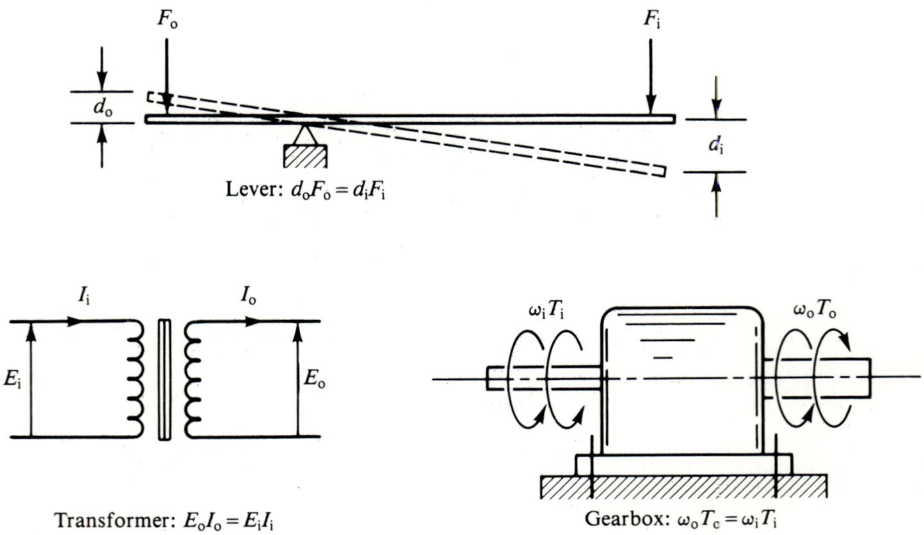

Figure 1-7. Ratioing components.

In a stepdown transformer we are reducing the voltage, so the transformer ratio will be

$$r = \frac{E_i}{E_o} \tag{1-8}$$

Clutches and torque converters are also power transmitters but are not simple ratioing components because of slip between input and output halves of the component. A schematic diagram for a simple clutch is shown in Fig. 1-8.

A clutch consists of an input and an output which are not positively connected. We might call a clutch a slip coupling because the output and input speeds are not necessarily equal as they are in a simple positively connected coupling.

The slip s of a clutch is defined as

$$s = \frac{\omega_i - \omega_o}{\omega_i} \tag{1-9}$$

or

$$\frac{\omega_o}{\omega_i} = 1 - s \tag{1-10}$$

When the slip is equal to zero, we have $\omega_o = \omega_i$, which is the same as for a simple coupling. Some clutches are designed to be locked up so that $s = 0$ can be obtained.

More generally, the slip is not zero. When $s = 1$, we have an output speed of zero. The output shaft is said to be stalled.

Clutches must always have some means of controlling the slip between input and output shafts. In a simple frictional shoe or disk clutch, the control mechanism is generally a mechanical linkage which may include a spring or springs. In an electrical clutch, the slip is controlled by changing the electrical excitation.

As we shall see later in the book, clutches can be mechanical, electrical, hydrostatic, hydrodynamic, and hydroviscous. Each of these devices becomes a brake if one-half of the clutch is held fixed. In other words, brakes and clutches are the same kinds of components.

If we place two dynamos back to back, we will have a device whose input variables are the same kind as the output variables. The overall device is

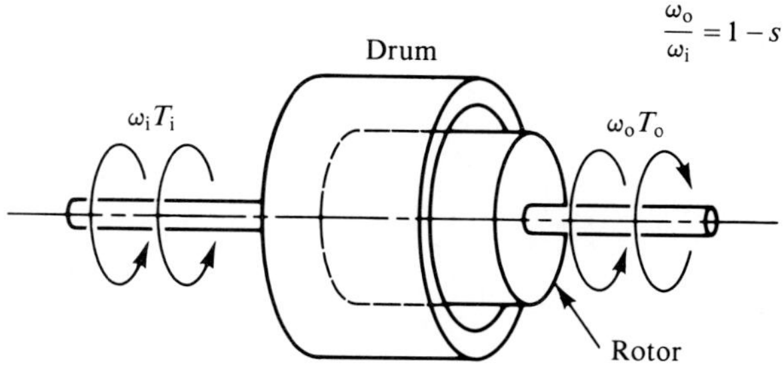

Figure 1-8. Schematic diagram of a clutch.

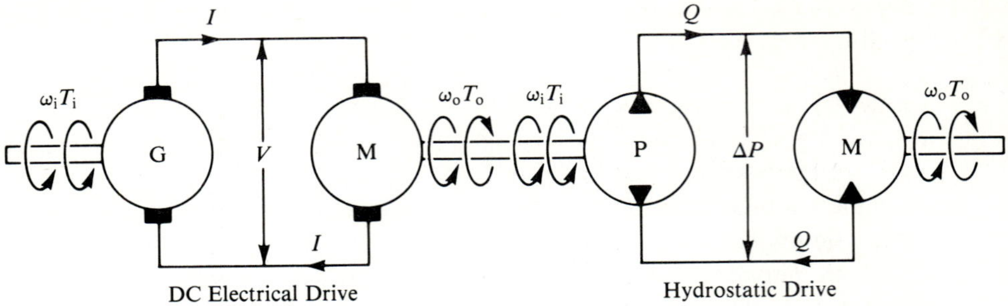

Figure 1-9. Typical drives (or transmissions).

therefore a power transmitter, but we generally call it a drive or transmission. Two examples are shown in Fig. 1-9.

The transmission on the left is called the Ward-Leonard drive. It consists of a DC electrical generator connected directly to a DC motor. A torque is required to drive the generator and the motor produces a torque, so the overall system is simply a transmitter of power even though the individual components are power converters. The virtue of such a transmission is that a broad and flexible range of control can be obtained by means of the independent excitations of the generator and motor fields. A typical control characteristic made possible by this kind of transmission is sketched in Fig. 1-10. It is called traction speed control because it

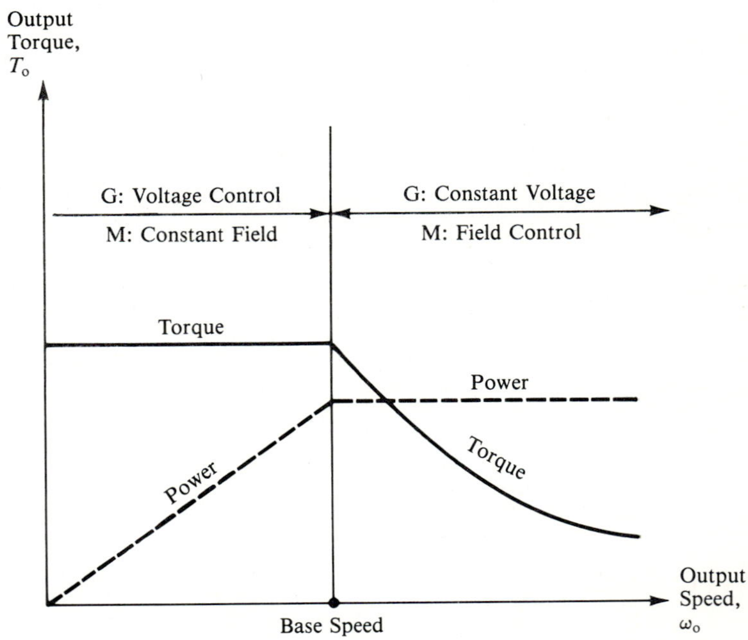

Figure 1-10. Traction control characteristics.

is a scheme for controlling electric motors used in traction applications such as rolling mills, winding machines, and vehicles.

In the low-speed range, constant torque control is obtained by keeping the motor field excitation constant and varying the armature voltage. In the high-speed range, constant horsepower control is obtained by keeping the armature voltage constant and varying the motor field control.

An analogous type of transmission can be obtained by using hydraulic dynamos rather than DC dynamos. Such a scheme is shown on the right side of Fig. 1-9. Here we have a positive-displacement hydraulic pump connected directly to a positive-displacement hydraulic motor. Again, a traction control characteristic can be obtained with this hydrostatic drive by judiciously changing the stroke of the pump and/or motor.

1-2-3 Loads

We have already defined a load as a power sink in which the load force (or torque) is in the opposite sense to the load velocity (or angular velocity). Strictly speaking, this is a resistive load, but since we are limiting our discussion here to steady-state operation, we cannot have any other kind (e.g., an inertia load).

To drive a load at steady speed, we must supply, say, a torque and angular speed having the same sense (as would be provided at the output of a component). The ultimate objective in designing machines is to meet the requirements of a load, or demand, characteristic. The nature of the load characteristic

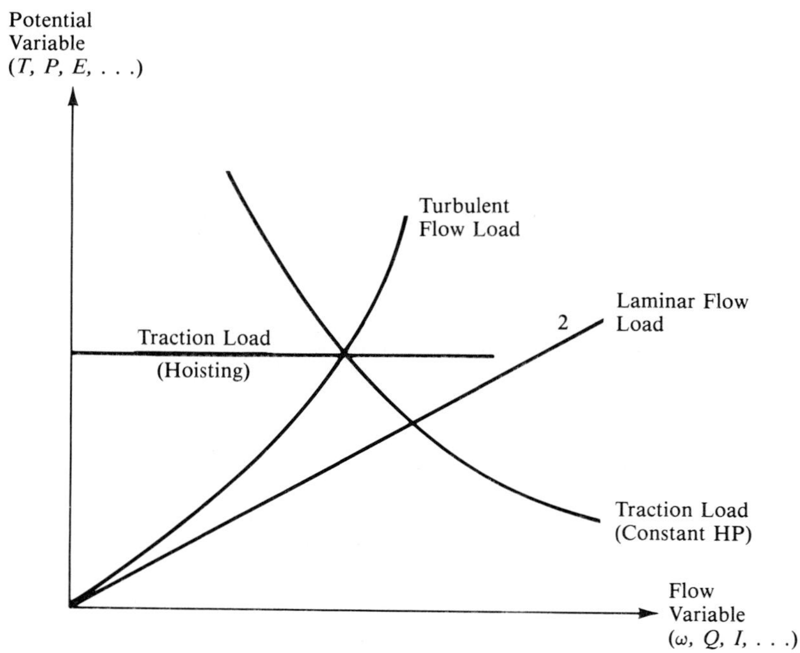

Figure 1-11. Typical load characteristics.

may be instrumental in our choice of components for driving and controlling the load.

In the steady state, loads have characteristics that can be displayed in a manner similar to that used to display converter characteristics. The display is a curve in the potential versus flow variable plane. Some typical load characteristics are sketched in Fig. 1-11.

The laminar flow load characteristic is the kind of load associated with machine tools in which the guideways utilize hydrostatic bearings. The turbulent flow type of load characteristic is important in pipe or duct flow systems to be powered by centrifugal or axial flow pumps (or blowers). And, finally, we have already discussed the traction type of load characteristic.

Each point of the load characteristic curves of Fig. 1-11 is an equilibrium operating point determined either by calculation or by experiment. For example, the turbulent flow load characteristic curve simply shows that as the flow in a piping system increases, the pressure drop (potential variable) required to cause this flow increases in an approximately quadratic fashion.

1-3 EQUILIBRIUM OPERATING POINT OF A SYSTEM

Suppose we start up an elementary mechanical system such as that sketched in Fig. 1-1. If the system has a simple resistive load characteristic, as discussed above, it will eventually come to an equilibrium operating condition in which a compatibility among the components is achieved. We call this steady-state operation.

The equilibrium operating conditions at each interface between components can be determined by a process of matching the characteristics of driving and driven members. We will illustrate this for the simple system shown in Fig. 1-2.

Suppose component no. 1 of Fig. 1-2 is an AC induction motor having the output characteristic curve labeled 1 in Fig. 1-6. Let component no. 2 be a load characterized by the curve labeled 2 in Fig. 1-11. If we plot both curves on the same torque-speed plane, the result will be as shown in Fig. 1-12.

The intersection of the two curves in Fig. 1-12 defines the equilibrium, or operating point of the simple system of Fig. 1-2. This is the point where the conditions of Section 1-1 are satisfied, namely,

$$\omega_1 = \omega_2 , \qquad T_1 = -T_2 \tag{1-11}$$

The operating point of the system defined by the intersection of the AC motor and load curves in Fig. 1-12 may, or may not, be an optimum one. It must be remembered that the mechanical efficiency of the motor (not shown) will be a function of the motor speed and will generally have a maximum value. So, if we want the system to operate at maximum efficiency, we will have to ensure that the operating point occurs at the motor speed where the efficiency is a maximum. This is called "matching" a converter characteristic to the load characteristic.

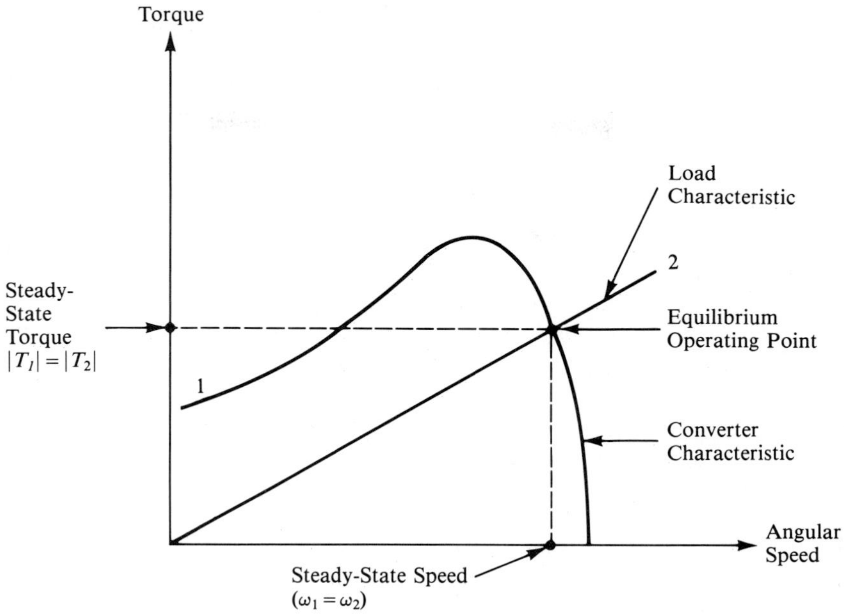

Figure 1-12. Matching converter and load characteristics.

Instead of having the system operate at maximum converter efficiency, it may be desirable to have it operate at a speed where the converter (in this case, the induction motor) develops its maximum power output. This would be another criterion by which a converter is matched to its load.

EXAMPLE 1-1

An AC induction motor has a synchronous speed of 3600 rpm and a torque-speed output characteristic that is linear in the speed range between 3240 rpm and synchronous speed. We will assume that the motor delivers 5 hp at a slip of 0.03 (3 percent).

Suppose the motor is directly coupled to a load whose torque-speed characteristic is

$$T = An$$

where T is the load torque in ft-lb and n is the load speed in rpm. Determine the constant A such that the motor will drive the load at maximum power in the speed range between 3240 rpm and 3600 rpm.

Solution

The motor characteristic curve is sketched in Fig. 1-13. The linear curve is an approximation of the induction motor characteristic shown in Fig. 1-5 in the vicinity of rated torque. This curve has the general linear form

$$T_m = an + b$$

where T_m is motor torque in ft-lb and n is motor speed in rpm.

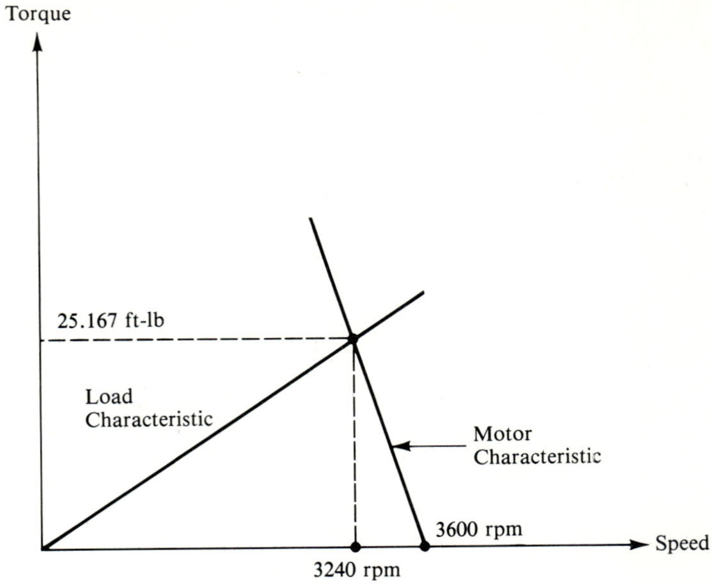

Figure 1-13. Matching load and motor characteristics.

The motor speed at a slip of 3 percent is obtained from

$$\frac{3600 - n}{3600} = 0.03$$

$$n = 3492 \text{ rpm}$$

The motor torque at this speed is then determined from the given power data:

$$T_m = \frac{33{,}000(5)}{2\pi(3492)} = 7.52 \text{ ft-lb}$$

Now the constants a and b of the motor characteristic curve can be obtained from the two points

$$n = 3600 \text{ rpm},\ T_m = 0 \qquad n = 3492 \text{ rpm},\ T_m = 7.52$$

resulting in

$$a = -0.0696 \qquad b = 250.67$$

So the motor output characteristic in the specified interval is

$$T_m = -0.0696n + 250.67$$

The motor power in this interval is then

$$P_m = n \cdot T_m = -0.0696n^2 + 250.67n$$

To determine the speed for maximum power, we differentiate the above expression to obtain

$$\frac{dP_m}{dn} = 0 = -0.1392n + 250.67$$

The maximum power occurs at the speed

$$n = \frac{250.67}{0.1392} = 1800 \text{ rpm}$$

This speed lies outside the specified interval, however, and it is apparent that maximum power in the interval occurs at the speed of 3240 rpm. The motor torque at this speed is then given by

$$T_m = -0.0696(3240) + 250.67 = 25.167 \text{ ft-lb}$$

The motor and load torques must be matched at this operating point of $n = 3240$ rpm, $T_n = 25.167$ ft-lb. Therefore, the constant A must be given by

$$A = \frac{25.167}{3240} = 7.77 \times 10^{-3}$$

The intersection of this load curve with the motor curve is shown in Fig. 1-13.

We note that the horsepower requirement to drive this load at a speed of 3240 rpm is

$$\text{HP} = \frac{2\pi(3240)(25.167)}{33{,}000} = 15.5 \text{ hp}$$

Therefore, a 20 hp motor would have to be used. ■

1-4 STABILITY OF AN EQUILIBRIUM OPERATING POINT

We have been focusing on the steady-state (equilibrium or quasistatic) behavior of mechanical systems. This approach has permitted us to expose the operating point of the system at which the system will come to equilibrium as dictated by the point of intersection of converter and load characteristics. However, we generally want to know more than just the operating point. An important question is whether or not this equilibrium operating point is stable.

There are several ways of approaching this problem. The detailed approach requires an intimate knowledge of control theory, of transfer functions, and of the concept of stability of a feedback control system. We discuss this approach in a cursory fashion later. Here, our approach will be to look at the stability of the operating point in terms of the geometry of the characteristic converter and load characteristics in the vicinity of their point of intersection (the operating point). This is a rather simple approach but it has limitations in that the curves being considered are really valid only for equilibrium conditions whereas our analysis will be for transient conditions.

Of course we are implying that equilibrium and stability are not the same (which can be confusing). We can clarify this in terms of the simple system sketched in Fig. 1-14.

On the left, a pencil is shown standing on its point (which is slightly flattened). After some effort, we can balance the pencil so that it will remain in equilibrium. This equilibrium is highly unstable, however, in that a slight disturbance will cause the pencil to lose its equilibrium and fall down. We say that the

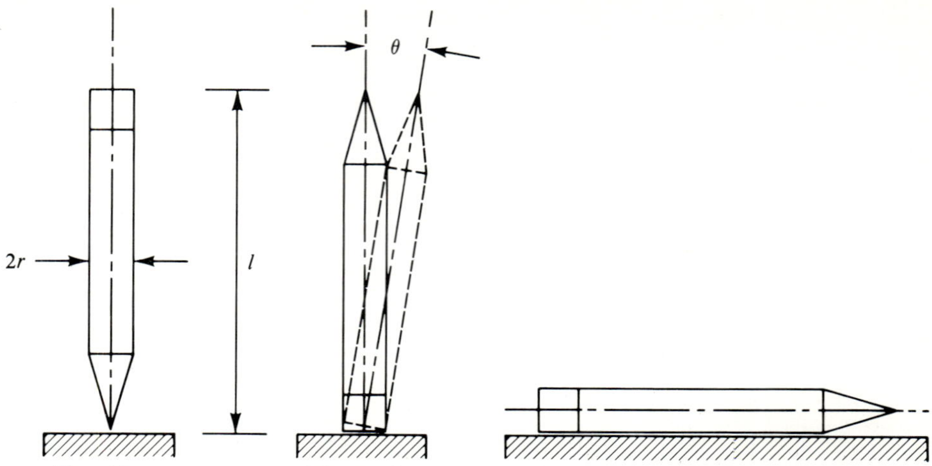

Figure 1-14. Stability of equilibria illustrated.

equilibrium of a pencil balanced on its point is an unstable equilibrium. The equilibrium is unstable even to infinitesimal disturbances.

In the middle sketch, the pencil is shown balanced on its (new) eraser. Now the equilibrium of the pencil will be stable to infinitesimal disturbances but unstable with respect to finite disturbances which cause the pencil to tip at angles greater than $\theta = 2r/l$. This equilibrium is called metastable equilibrium.

On the right, we see that the equilibrium of the pencil is stable to all disturbances in the sense that once the disturbance is removed, the pencil will again assume the equilibrium position of lying flat on its side. This state is called stable equilibrium.

We note that the key concept in stability analysis is the response of the system (pencil) to disturbances that displace it from its equilibrium position. This is the same approach we will use to determine the stability of the equilibrium operating point of a mechanical system.

Consider the characteristic torque-speed curves of a prime mover and its load as sketched in Fig. 1-15. We have chosen torque and speed for convenience. The analysis that follows will also be valid for other system characteristic curves (e.g., head versus discharge in a piping and pump hydraulic system).

Two possible operating points, A and B, are shown in Fig. 1-15. We want to investigate the stability of these operating points. It is important to understand that we are going to define stability of the operating point in terms of the response to a small disturbance (or perturbation) in speed.

Let the *perturbation* be a *disturbance* in prime mover speed, $\Delta\omega$. We want to determine if this perturbation decays, or grows, after the disturbance is removed. Of course, associated with the perturbation in speed are the perturbations ΔT_{m} and ΔT_{ℓ} in prime mover and load torques respectively. During this transient period, these perturbations are governed by the fundamental equation

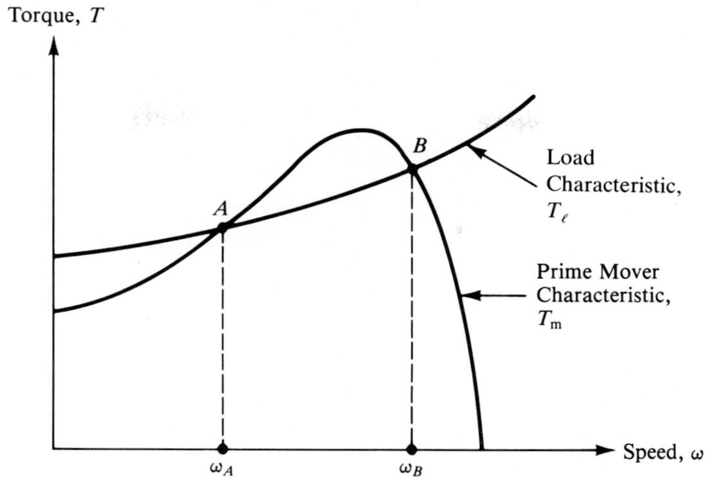

Figure 1-15. Stability of operating points A and B.

$$T_{\mathrm{m}} - T_\ell = I\,\frac{d\omega}{dt} \tag{1-12}$$

Assuming the disturbances are applied to point A, we have

$$T_{\mathrm{m}} = T_{\mathrm{m}(A)} + \Delta T_{\mathrm{m}} \tag{1-13}$$

$$T_\ell = T_{\ell(A)} + \Delta T_\ell \tag{1-14}$$

$$\omega = \omega_A + \Delta\omega \tag{1-15}$$

Combining Eqs. (1-12) through (1-15) produces

$$T_{\mathrm{m}(A)} + \Delta T_{\mathrm{m}} - T_{\ell(A)} - \Delta T_\ell = I\,\frac{d(\omega_A + \Delta\omega)}{dt} \tag{1-16}$$

But the equilibrium relationship at point A is

$$T_{\mathrm{m}(A)} = T_{\ell(A)} \tag{1-17}$$

$$\frac{d\omega_A}{dt} = 0 \tag{1-18}$$

Therefore,

$$\Delta T_{\mathrm{m}} - \Delta T_\ell = I\,\frac{d(\Delta\omega)}{dt} \tag{1-19}$$

Perturbations ΔT_ℓ and ΔT_{m} can be estimated from the slopes of the characteristic curves in the vicinity of point A. The estimates are

$$\Delta T_{\mathrm{m}} = \left(\frac{\partial T_{\mathrm{m}}}{\partial\omega}\right)_A \cdot \Delta\omega \tag{1-20}$$

$$\Delta T_\ell = \left(\frac{\partial T_\ell}{\partial \omega}\right)_A \cdot \Delta\omega \tag{1-21}$$

Substituting Eqs. (1-20) and (1-21) into Eq. (1-19) gives

$$\left[\left(\frac{\partial T_{\mathrm{m}}}{\partial \omega}\right)_A - \left(\frac{\partial T_\ell}{\partial \omega}\right)_A\right]\Delta\omega = I\,\frac{d(\Delta\omega)}{dt} \tag{1-22}$$

The constant coefficient of $\Delta\omega$ on the left side of Eq. (1-22) involves the slopes of the characteristic curves through point A. With this coefficient defined as C_A, Eq. (1-22) becomes

$$\frac{d(\Delta\omega)}{\Delta\omega} = \left(\frac{C_A}{I}\right) dt \tag{1-23}$$

This expression can be integrated to obtain

$$\Delta\omega = K \exp\left[\frac{C_A}{I}\,t\right] \tag{1-24}$$

Now we can see that $\Delta\omega$ will either decay or grow depending on the sign of C_A. For point A of Fig. 1-15, we see that

$$\left(\frac{\partial T_{\mathrm{m}}}{\partial \omega}\right)_A > \left(\frac{\partial T_\ell}{\partial \omega}\right)_A \quad \text{and} \quad C_A > 0 \tag{1-25}$$

Point A is therefore an unstable operating point because a small disturbance will produce a perturbation growing exponentially. Once point A is disturbed, the system tends to seek a new (more stable) operating point.

Point B, on the other hand, is a stable operating point because

$$\left(\frac{\partial T_{\mathrm{m}}}{\partial \omega}\right)_B < \left(\frac{\partial T_\ell}{\partial \omega}\right)_B \quad \text{and} \quad C_B < 0 \tag{1-26}$$

A small perturbation $\Delta\omega$ will decay away once the disturbance is removed, and the operating point tends to be restored to point B.

Another way of explaining this situation is as follows. Consider a positive fluctuation $\Delta\omega$ at operating point A. The fluctuations ΔT_{m} and ΔT_ℓ corresponding to $\Delta\omega$ are such that the prime mover torque T_{m} exceeds the load torque T_ℓ. The excess torque will tend to accelerate the load so ω tends to move away from ω_A in the same direction as $\Delta\omega$. At point B, just the opposite occurs. For a positive fluctuation $\Delta\omega$ at point B, the associated torques T_{m} and T_ℓ are such that T_ℓ exceeds T_{m}. Torque deficit tends to decelerate the load so ω tends to move back toward the equilibrium speed ω_B (opposite to $\Delta\omega$).

The real danger of an operating point such as point A will become manifest if another prime mover is operated in parallel with the prime mover of Fig. 1-15. Since point A is unstable, pulsations (or surging) can occur between the two prime movers as the operating point of the unstable configuration moves away from, and back toward, point A.

Finally, it should be noted that many prime movers cannot develop an appreciable torque at low speeds (as, for example, in an internal combustion engine). Necessarily, then, the prime mover characteristic curve must be an ascending one during the low-speed region of the characteristic curve until a maximum value is reached beyond which the device may exhibit a descending (called drooping) characteristic. The low-speed ascending region is not generally where an operating point should be established; the operating point should be established on the drooping portion of the characteristic curve of the prime mover.

1-5 SYSTEM TRANSIENT PERFORMANCE

Thus far we have considered the properties of components and systems under the assumption that they are operating in a steady-state (equilibrium) mode. Although this assumption permits us to expose some fundamental concepts and properties of the components themselves, which are necessary in order to be able to choose components and to synthesize them into a working system, it is limiting in several ways.

The assumption of steady-state operation is tantamount to having a system subjected only to steady, resistive loads. Although this assumption may be realistic for many systems most of the time, it cannot be true during start-up and shut-down operations which necessarily involve unsteady-state (or transient) conditions. An example of such a system is a belt conveyor delivering bulk materials, such as gravel or coal, at a steady rate. The conveyor must be started (perhaps under load), and then stopped at some point in its operation.

Then there are many kinds of systems in which steady-state conditions of operation will never be achieved. Such systems are constantly in a state of either being cycled or otherwise operating in an unsteady mode (which may even be unpredictable). In this case, the system will be subjected to inertia and spring loading in addition to resistive loads. An example is an earth-moving machine such as a backhoe digger.

Designing mechanical systems involves more than just the synthesis of power converters, transmissions, and loads as dictated by their steady-state parameters. Even the belt conveyor carrying bulk materials will require a control system albeit a simple one. And, more complex systems subjected to both external and internal transient loads may require sophisticated control systems.

So our view of a mechanical system must be expanded beyond that of just the components in the power train; it must include control components. A schematic diagram indicating this broader concept for a typical system is presented in Fig. 1-16.

Note that the controller may be controlling several components of the power train. The prime mover could be an electric motor and the transmission components might include an electrically controlled clutch. The setpoint command block is the point in the system where inputs are provided. It is where the system is told, for example, at what speed it is to operate. In this instance, speed would be a controlled variable.

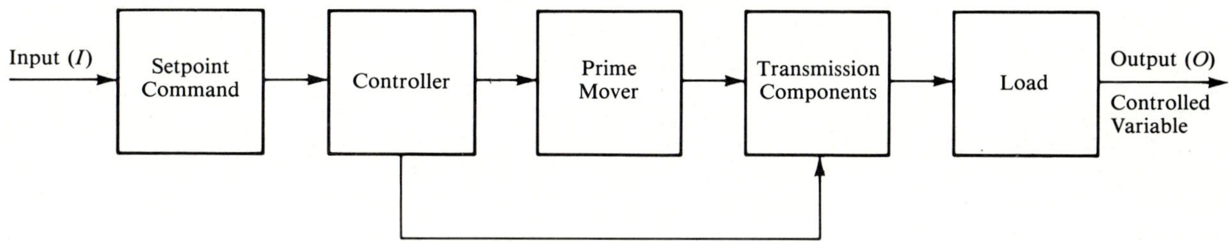

Figure 1-16. Power train and control system.

1-5-1 Open-Loop Control

Figure 1-16 can be simplified schematically by lumping the power train components into a single block labeled "system feedforward components" and by lumping the setpoint command device and controller into a block labeled "controller feedforward components." The result is shown in Fig. 1-17.

The system shown in Fig. 1-17 is called an open-loop or feedforward-only control system. All of the control flow is from left to right (or feedforward). This type of system does not have the capability of providing precise control of the output variable and can be subject to drift. Some of the characteristics of open-loop control are as follows:

- **(a)** It is a simple control scheme compared to others we might use.
- **(b)** Fluctuations, or disturbances, of the load or in the system cannot be controlled by the controller per se because no communications link exists between the controller and the variable being controlled. Disturbances can be modulated only by the intrinsic regulation properties of the system.
- **(c)** The relationship between the response of the controlled variable and the setpoint command must be determined analytically from prior knowledge of the system dynamics or by calibration.
- **(d)** As components in the system age, the relationship between the controlled variable and the setpoint command will drift from an initial calibration which means that open-loop control systems may have to be recalibrated (perhaps frequently).
- **(e)** Open-loop control systems are generally stable.

The open-loop system shown in Fig. 1-17 is not subject to any disturbances. When this situation prevails, the output variable (labeled O) is a function only of the input variable (labeled I). We will designate the ratio of output to input variables using the symbol G: Then

$$\frac{O}{I} = G \tag{1-27}$$

The simple relationship of Eq. (1-27) is merely an operational form. Control theory is an engineering discipline in which the nature of this operational relationship is explored using the differential equations governing the transient

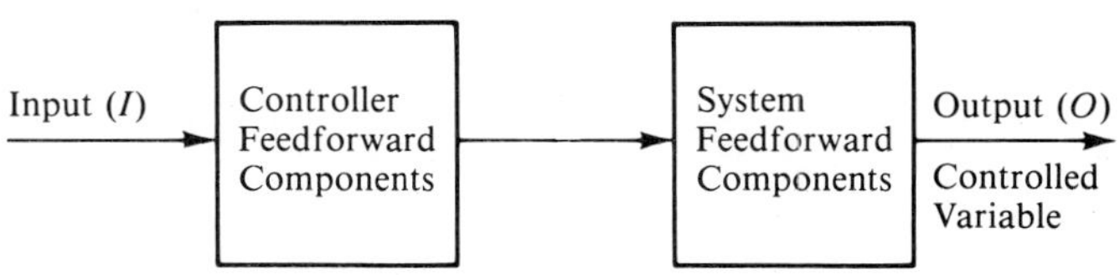

Figure 1-17. Open-loop control scheme.

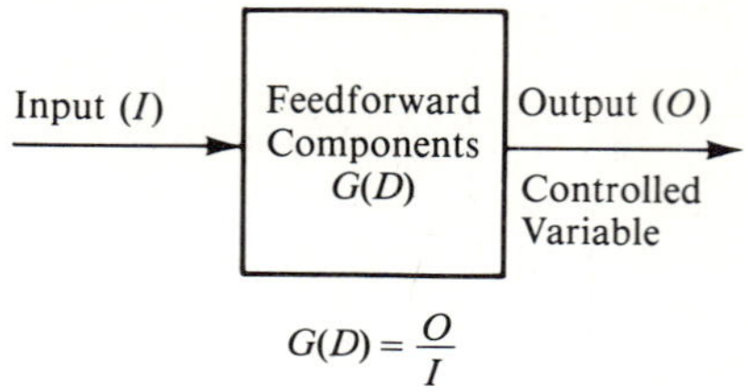

Figure 1-18. Open-loop transfer function.

behavior of both system and control components. Control theory analysis reveals that ratio G is a function only of system constants and the differential operator $D = d/dt$ when all system relationships can be represented by linear differential equations. When this is true, ratio G is expressed as

$$G = G(D) \tag{1-28}$$

The function $G(D)$ is called the system transfer function. Operationally, then, Fig. 1-17 can be reduced to Fig. 1-18. In the open-loop system sketched in Fig. 1-18, the transfer function $G(D)$ is called the feedforward transfer function.

EXAMPLE 1-2

A simple mechanical system consisting of a rotating mass mounted on a torsional shaft (spring) is subjected to an applied torque T_a as shown in Fig. 1-19. The rotational moment of inertia of the system is I and the spring constant of the torsional shaft is K.

The equation of motion of the system is

$$\sum T = T_{net} = I\alpha = I \frac{d^2\theta}{dt^2} \tag{1-29}$$

The net external torque is the difference between the applied torque T_a and the spring torque of the torsion bar T_s, where

$$T_s = K\theta \tag{1-30}$$

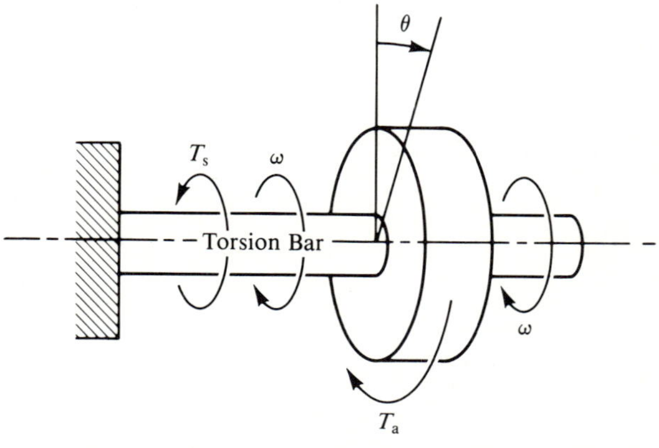

Figure 1-19. Simple dynamic system.

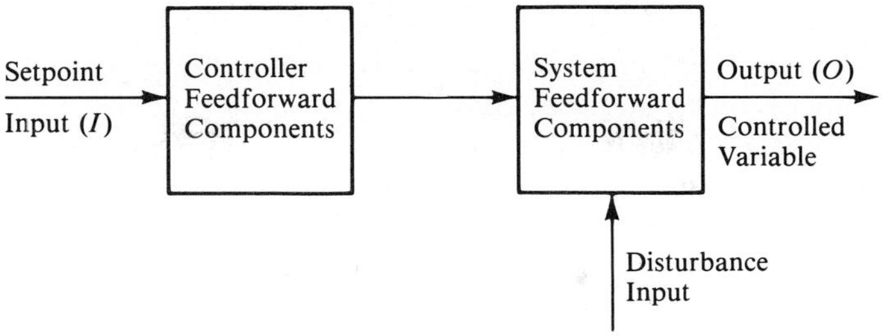

Figure 1-20. More general open-loop system.

So the governing differential equation becomes

$$T_a - K\theta = I\frac{d^2\theta}{dt^2} \tag{1-31}$$

In operational form, Eq. (1-31) is the same as

$$T_a = (ID^2 + K)\theta \tag{1-32}$$

If θ is regarded as the output response to an input T_a, then

$$\frac{O}{I} = \frac{\theta}{T_a} = G(D) = \frac{1}{ID^2 + K} \tag{1-33}$$

So we see that the transfer function of this very simple system is the reciprocal of $ID^2 + K$. Note that in the steady state (θ = a constant) the transfer function becomes simply a constant:

$$G(D) = \frac{1}{K} \tag{1-34}$$

■

The simple system of our example adequately conveys the idea of a system transfer function but does not contain any control element. Remember that one of the basic functions of a control system is to provide an amplification capability whereby we can control high-level machine power using only low-level control signals. A good example is the 12 V rheostat which might be used to control a 440 V motor.

Although actual power transmission systems will be much more complicated than our simple example, the concept of the transfer function remains unchanged. The transfer function will simply be a much more complicated algebraic expression of all the inertias, springs, and dampers in the system. Also, in developing these basic ideas we have chosen to omit potential disturbances which will act as additional inputs to the system. A more general open-loop system is sketched in Fig. 1-20.

1-5-2 Closed-Loop Control

The operation of a mechanical system in an open-loop control mode is based on trust or hope. Let us suppose that the system has been run and calibrated to determine how the setpoint dial is to be graduated based on measurements taken at the output variable. The calibration is likely to consist of a group of tests run under various equilibrium conditions.

Suppose the output variable is speed. Once the system has been calibrated, all further attempts to measure the output speed are abandoned and we simply trust that if the setpoint dial is set, say, at 100 feet per second (fps), then the output speed will be 100 fps. Such a control scheme is adequate for many mechanical engineering systems that stand alone (do not have to be synchronized with other systems) or in which precise control of the output is not crucial to the execution of the system function.

If a system controlled in the open-loop mode is subjected to disturbances, the speed of the output may change despite the fact that the setpoint command signal has not changed. The response of the speed to load disturbances will be determined by the intrinsic behavior of the system which we call its speed regulation. Again, if the speed regulation of the system is good (small changes in speed caused by large disturbances), the open-loop system may be adequate.

The limitations of an open-loop control system can be overcome by using a simple, but ingenious, idea. This idea involves the continuous monitoring of the output variable to be controlled and continuous comparison of the measured output to the value commanded at the input by the setpoint. If there is a difference (called the error signal e), we use this difference to change the system in such a way as to liquidate the error.

A scheme for implementing this idea is sketched in Fig. 1-21. The control scheme of Fig. 1-21 is called a closed-loop or feedback control system. In addition to the components of an open-loop system, a closed-loop system requires

(a) An output variable sensor (labeled c)
(b) A feedback path and feedback transfer function, $H(D)$

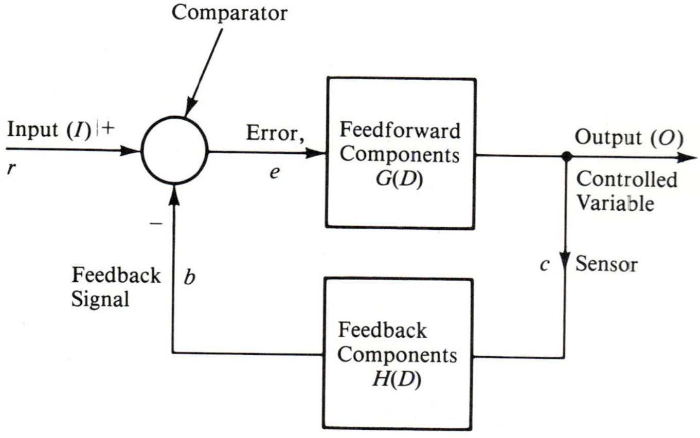

Figure 1-21. Closed-loop control scheme.

(c) A comparator for generating the error e between the setpoint command r and feedback signal b (the block labeled $H(D)$ changes signal c to signal b)

Of course, Fig. 1-21 is a simple block diagram intended to illustrate the main idea rather than to obscure it by showing all the detailed components required to actually implement it. Keep in mind that the originating command could be via a knob on a graduated dial while the output variable could be, perhaps, temperature. There will have to be components (not shown) in this system that will make it possible for the comparator to work with commensurable input and feedback signals in order to generate the error signal e. And the error signal might be at a very low level of power, in which case it must be amplified and transduced as it is fed forward in the system.

To illustrate how a closed-loop control system operates, consider the system of Fig. 1-21 to be driving a mechanical load requiring load torque T at load speed ω, where speed ω is to be the controlled variable (the speed is being measured by a tachometer at the output whose signal is labeled c). Suppose that the system has been commanded (via setpoint r) to operate at a load speed of 100 rpm and that, in fact, the load is operating at 100 rpm while carrying a torque at 1000 in-lb. If the tachometer is a small electrical generator, signal c is a voltage. The transfer function $H(D)$ changes the tachometer signal c to b so it is commensurable with the setpoint reference signal labeled r.

The error signal at the output of the comparator will be zero because the comparator equation is

$$e = r - b = 100 - 100 = 0 \tag{1-35}$$

Since no error signal is being provided to the controller (which is part of the feedforward components of Fig. 1-21), no corrective actions are fed forward to the system power transmission components and the system will be operating (as commanded) in the steady state at a speed of 100 rpm and a torque of 1000 in-lb.

Now suppose the torque demanded by the load suddenly increases to 1200 in-lb. As dictated by the intrinsic dynamic behavior of the power transmission system, torque increase will tend to slow down the output of the mechanical system. In an open-loop system, the output speed would slow down until a new equilibrium is established at the new torque demand of 1200 in-lb. Suppose the new equilibrium speed is 98 rpm. We would say that the open-loop system has good speed regulation because the speed decrease was only 2 percent for a 20 percent increase in load torque.

In many systems a speed swing of only 2 percent would be of no consequence and open-loop control would be adequate. But in other systems the speed would have to remain precisely at 100 rpm regardless of changes in load torque. This kind of precision speed control can be provided only by a closed-loop control system and would be accomplished in the following way.

The slowing down of the controlled variable ω will decrease signal c and hence signal b. Since the setpoint command will not have changed, an error signal is created at the comparator. The error signal is fed forward and the system will be designed to respond in such a way that the increased torque required is supplied by changing a power supply variable rather than by simply slowing down (as in the open-loop system). The power supply variable will continue to change so long as an error signal is present. This change will cause a restoration of speed to 100 rpm at which point the error signal vanishes and the system is again in an equilibrium state at the new torque of 1200 in-lb.

It is easy to show, using the block diagram of Fig. 1-21, that the overall transfer function of a closed-loop system is

$$\frac{O}{I} = \frac{G}{1 + HG} \tag{1-36}$$

There are two things we might note about the relationship of Eq. (1-36). First, if the feedback path is broken ($H = 0$), we again have the open-loop system relationship Eq. (1-27). Second, if the product HG could assume the value -1, the denominator of Eq. (1-36) would become zero which implies infinite output O for infinitesimal input I. In other words, in going to the benefits of the closed-loop control scheme we take a risk that the resulting system might be unstable. Much of the study in the formal discipline of control theory is devoted to an analysis of the stability of closed-loop systems (i.e., the circumstances whereby $HG = -1$).

Our discussion of some of the elementary concepts of control theory in this chapter cannot substitute for a formal course in this important subject. Our objective has been only to present some basic ideas that will permit us, in subsequent chapters, to explain how certain mechanical system components can be controlled.

A more complete exposé of control theory would include further discussion of the role of transducers and amplifiers in both analog and digital control systems. A transducer is merely a component that changes one type of signal to another kind. It is nothing more than a power converter at a low level of power. We might use a transducer to change rotational speed into a voltage, a temperature into a sequence of current pulses, or acceleration into pressure.

An amplifier is nothing more than a power transmitter, as we discussed earlier. It is used to obtain what is called gain in a control system in order that the interfaces between components are compatibly matched. The feedforward and feedback components of Fig. 1-21 will inevitably contain a number of these amplifiers and transducers.

A more complete block diagram of a mechanical system and its closed-loop control scheme is shown in Fig. 1-22. Here we have shown the system input disturbances that will exist in real systems in addition to the setpoint input. Each block of Fig. 1-22 is characterized by its own transfer function containing intrinsic dynamic information about the transient behavior of the system.

The relationship between output and input in this more general system is

$$\frac{O}{I} = \frac{AG_1G_2 + (BG_2)D/I}{1 + G_1G_2H} \tag{1-37}$$

Some of the characteristics of such a closed-loop control system are as follows:

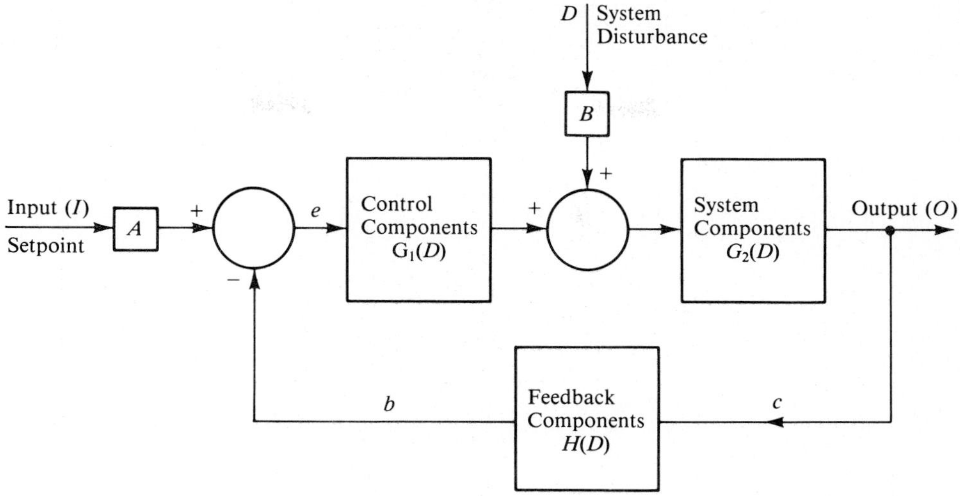

Figure 1-22. General block diagram of feedback-controlled system.

(a) It is a more complex system to analyze than an open-loop system.
(b) It does not require frequent calibration and can be implemented without the necessity of knowing the analytical relationships governing the behavior of components.
(c) It controls by changing certain system variables rather than just by regulating them.
(d) It requires communication (feedback) between the controlled variable(s) and the controller. This feedback requires the use of sensors and transducers.

1-5-3 Characteristics of Open-Loop versus Closed-Loop Control

Suppose we have a prime mover (power converter) directly coupled to, and driving, a load as illustrated by Fig. 1-2. The prime mover characteristic curve corresponds to the curve(s) that droop from left to right in the sketches of Fig. 1-23.

In both of the sketches of Fig. 1-23, we show the system in an initial equilibrium state with the operating point as dictated by the intersection of the prime mover curve with the (ascending) load curve at point 1. Open-loop control is shown at the left and closed-loop control is shown at the right.

We assume that the load suddenly increases from torque level T_1 to level T_2. In the open-loop system, no attempt is made to change the prime mover characteristic in response to the unsensed decrease in load speed ($\Delta\omega$) caused by the increase in load torque (ΔT). Consequently, the operating point must move along this original prime mover characteristic curve from point 1 to point 2 in

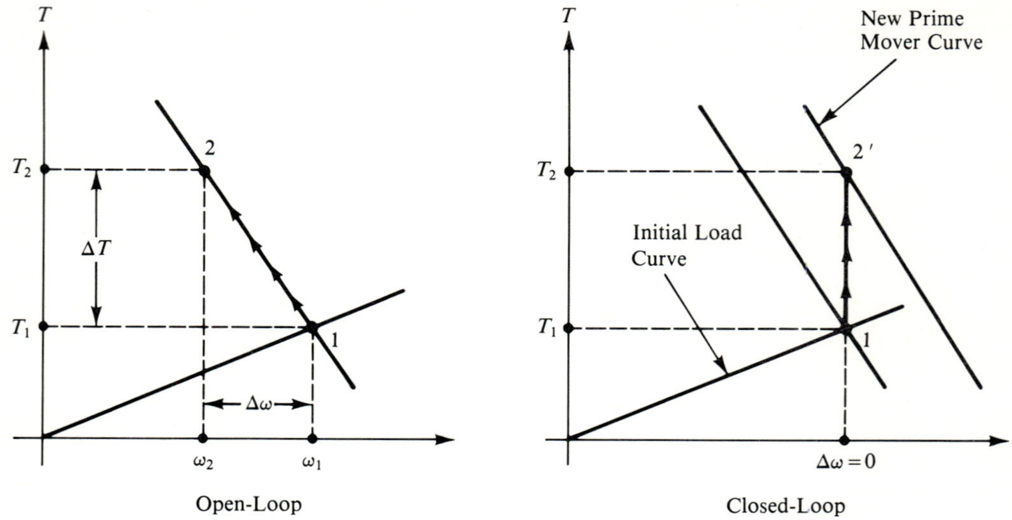

Figure 1-23. Open-loop versus closed-loop control.

order to get to the new load level T_2. The speed is, at the same time, diminished to the new load speed ω_2.

The ratio $\Delta\omega/\Delta T$ is a measure of the speed regulation of the open-loop system shown at the left of Fig. 1-23. If this measure is small, we say that the speed regulation is good and we may be able to tolerate the relatively small fluctuation in speed ($\Delta\omega$) without the necessity for any control action. We might call this passive control.

If the ratio $\Delta\omega/\Delta T$ is such that poor speed regulation results, the decrease in speed may be unacceptable (in some systems, even a small decrease in speed will be unacceptable as, for example, in a web tensioning control scheme). Then we need dynamic control which is achieved with the closed-loop control scheme shown at the right of Fig. 1-23.

In the closed-loop control scheme, the incipient decrease in speed (caused by the increase in load) is sensed with a tachometer generator. Through the feedback loop and the comparator (Fig. 1-21), an error signal is created which will ultimately cause the prime mover curve to change to the new position intersecting point 2′ of Fig. 1-23. For example, if the prime mover is an electric motor, we might increase the supply voltage to the motor thus shifting its torque-speed characteristic upward. Now, in moving from state point 1 to state point 2′, we do not move along the original prime mover curve but, instead, move from the initial curve to a completely new prime mover curve. This movement will provide a new load torque T_2 with no change in load speed. In other words, we control rather than regulate.

In a more detailed study of the control of a system such as the one we have very simply described, the precision with which we can accomplish the changes in the closed-loop control scheme is analyzed in terms of proportional, integral, and derivative (PID) techniques. The study of this subject requires a complete textbook devoted to it.

1-6 SUMMARY

In this chapter we have tried to develop some generic concepts common to all mechanical systems. The basic concepts include component matching, system operating point, stability of an operating point, transfer function, open-loop control, and closed-loop control. Additionally, components have been classified into three generic types: power converters, power transmitters, and loads.

The concept of matching a load to a prime mover converter was discussed and shown to be a simple steady-state operating concept when the load is directly coupled to the prime mover. As we will show in Chapter 2, this problem becomes more complicated if a ratioing component is placed between the prime mover and the load.

1-7 REFERENCES

1-1. R. Holmes, *The Characteristics of Mechanical Engineering Systems*, Pergamon, Oxford, UK, 1977.

1-2. R. L. Jaeschke, *Controlling Power Transmission Systems*, Penton/IPC, Cleveland, 1978.

PROBLEMS

1-1. Imagine a closed-loop system in a steady-state mode of operation with an output speed of 100 rpm as commanded by a setpoint dial setting of 100 rpm. It is desired to increase the speed of the output (while the load remains constant), and the setpoint dial value is increased to 110 rpm. Explain, in words, how the system will respond to this change commanded at its input.

1-2. Show that the overall transfer function of the closed-loop system of Fig. 1-21 is

$$\frac{O}{I} = F(D) = \frac{G(D)}{1 + G(D)H(D)}$$

1-3. Show that the relationship among O, I, and D of Fig. 1-22 is

$$\frac{O}{I} = \frac{AG_1G_2 + (BG_2D)/I}{1 + G_1G_2H}$$

1-4. A centrifugal pump drives a load whose characteristic is

$$P = 1.2Q \text{ kN/m}^2 \qquad (Q \sim \text{m}^3/\text{min})$$

The pump output characteristic is

$$P = \frac{N-6}{2} - \frac{Q}{5} \text{ kN/m}^2 \qquad (N \sim \text{rev/sec})$$

The pump has a mechanical efficiency of $\eta_p = 0.75$. It is driven by a 110 V DC shunt motor having an armature resistance of 0.1 Ω and a field resistance of 110 Ω. The motor efficiency is $\eta_m = 0.95$.

(a) At what speed must the pump be driven such that the system discharge will be $5\ m^3/min$?

(b) What input current will the motor draw at this discharge?

(c) Display the equilibrium operating point of this system on the *P-Q* plane as the intersection of the load and converter curves.

1-5. Determine the load characteristics of a hydraulic system in which water at 60°F is to be pumped from one reservoir to another one which is 150 feet higher in elevation. The pipe used is 1000 feet long and 12 inches in diameter (ID). Assume the friction factor to be 0.025 and neglect entrance and exit losses. Plot the load characteristic curve in units of feet of water (head, P) and gallons per minute (discharge, Q). (*Answer*: $P = 150 + 3.1 \times 10^{-6} Q^2$)

1-6. The rotor of an electric motor is subjected to a load torque T_ℓ, a frictional torque T_f exerted by the bearings, and an electromagnetic torque T_e developed by the armature (the motive torque). The relationship of frictional torque T_f to rotor speed can be assumed to be of the form

$$T_f = B\omega = B\frac{d\theta}{dt}$$

where B is a frictional coefficient. If the rotational moment of inertia is I, show that the ratio of rotor displacement θ (the output) to the net driving torque $T_e - T_\ell$ (the input) is given by

$$\frac{\theta}{T_e - T_\ell} = \frac{1}{ID^2 + BD} \qquad \text{(transfer function)}$$

where

$$D = \frac{d}{dt}$$

Part A

MECHANICAL COMPONENTS

The crank-throws give the double-bass, the
  feed-pump sobs an' heaves,
An' now the main eccentrics start their quarrel
  on the sheaves:
Her time, her own appointed time, the rocking
  link-head bides,
Till—hear that note?—the rod's return whings
  glimmerin' through the guides.
They're all awa! True beat, full power, the
  clangin' chorus goes
Clear to the tunnel where they sit, my purrin'
  dynamoes.

Rudyard Kipling

As we saw in Chapter 1, a power transmission system consists of a prime mover, intermediate transmission components, a load, and a control system. For the mechanical systems of interest to us here, prime movers will be turbines, internal combustion engines, and electric motors. These power sources generally operate at high speeds because of both efficiency and simplicity of design which high speed makes possible. Synchronous speeds of electric motors are commonly 3600 rpm and 1800 rpm while internal combustion engines may operate in the range 2000 to 3600 rpm. Turbines operate at even higher speeds.

In contrast to the high speed of prime movers, many of the loads supplied by power transmission systems operate at low speeds. A classic example is the belt conveyor system whose drive pulley may operate at, say, 100 rpm while its prime mover operates at 1750 rpm. Obviously, the transmission components between prime mover and load will have to implement this change in speed.

A reduction in speed can be obtained basically in only two ways. In the first method, a positive reduction in speed is obtained by using toothed linkages such as sprockets, chains, or gearing. This method has the advantage that registration can be maintained between input and output of the speed reduction mechanism. In other words; if the speed reduction is an integer r (revolutions of the input to revolutions of the output), then a mark scribed on the output shaft will return to an original angular position, after one revolution, at the same time that a corresponding mark scribed on the input shaft will return to its original angular position, after r revolutions. We say that a registration exists between the shafts which is forever preserved in a positive transmission device. In many machines, this registration is essential to the synchronization of the motions of various components.

Another advantage of positive drives is that they are usually more efficient than nonpositive drives (worm gear drives excepted) and hence dissipate less energy. But this advantage comes at the expense of speed control. If, at some point in a positive drive transmission, a change in speed ratio is required, it is generally feasible only in an incremental fashion by shifting some of the gears along their splined axes. A good example is the five-speed geared transmission of today's automobiles. If continuous changes in the speed ratio are required, this generally cannot be accomplished with a positive drive by itself.

The second method of obtaining a reduction in speed is through the mechanism of slip. Devices for accomplishing this are called clutches or slip drives. These components make possible a continuous change in speed ratio (they are called adjustable-speed or infinitely variable drives) but they necessarily involve dissipation of energy because of slippage in the transmission. We will discuss some of these slip drives in later chapters.

There is a third way of effecting a speed reduction between input and output of a transmission which is neither positive nor slip in nature. The drives involved are called traction drives. Ideally they are intended to be positive drives, but in order to transmit power they require creep to exist between mating members. Creep is a very small unintended slip such as might occur in a V-belt drive. Although creep may be small in such a transmission, it destroys the registration between input and output rotations.

2

Geared Transmissions

Where a positive drive is required in a machine it can be obtained mechanically by using a chain drive or by using a geared transmission. Where synchronization or precise registration among machine members is required this can be provided by a differential or planetary geared transmission.

Geared drives can be made very efficient through the use of precisely machined helical gears and can be combined with other types of transmission components in an ingenious variety of ways. Where a low-cost, positive reduction in speed is needed, the worm gear reducer is an effective transmission albeit at the expense, perhaps, of mechanical efficiency.

In this chapter we will discuss the integrated gearing systems used for reducing speed (in rare cases to increase speed) that are called gear reducers (or speed reducers). First we will discuss simple two-terminal gearboxes and how to apply them. Then, three- and four-terminal epicyclic gear trains will be considered from both the kinematic and power viewpoints.

From a mechanical systems viewpoint, geared transmissions are a subset of a generic type of component called a ratioing element. In contrast to continuously variable speed transmissions, however, geared transmissions operate at distinct, fixed ratios.

2-1 TWO-TERMINAL SPEED (GEARED) REDUCERS

Machine design courses go into considerable detail about the geometry, kinematics, stress, and strength of gears and gear teeth. The information given is generally not sufficient to enable one to become a gear designer but is adequate for teaching how to choose a set of gears from a catalog of gears in meeting some power transmission requirement.

Yet for every engineer likely to become a gear designer, there are probably a hundred engineers who simply are choosing a set of gears from a catalog to be used in some power transmission application. And for every set of individual gears selected from a catalog by a machine designer, probably a hundred speed reducers are being chosen as components of mechanical systems by project and applications engineers.

A gear reducer (or simply a reducer) is an assemblage of gears, shafts, and bearings integrated within a housing. These mechanical components have been used in systems transmitting only fractions of a horsepower and in others having severe power and torque requirements (up to 50,000 hp and more).

The term *reducer* refers to a reduction of angular speed; that is, the output shaft of the reducer possesses an angular speed less than that of the input shaft. The ratio r of a speed reducer refers to the input shaft rpm (or rad/sec) divided by the output shaft rpm:

$$r = \frac{\omega_i}{\omega_o} = \frac{1}{e} \tag{2-1}$$

Parameter e is called the train value of the gear train; it is the ratio of output speed to input speed.

Although some gear reducers can be run backward to operate as speed increasers, most cannot. If a speed increase is required, then generally a special gearbox must be designed. Most worm gear reducers are designed to be operated only in the speed reduction mode. They may be self-locking which will not permit them to be run backward.

A typical speed reducer is shown in Fig. 2-1. The reducer shown is a helical gear speed reducer. In this particular device, two sets of helical gears are used to effect a speed reduction in such a way that both input and output shafts are co-axial. This type of reducer is called a double-reduction, concentric (or in-line) speed reducer.

A gear reducer might just as well be called a gear amplifier because at the same time that speed is being reduced, the torque is being amplified. If there were no frictional losses in a gear reducer, then the ratio r would also represent the output torque divided by the input torque (this is why the output shaft of a reducer always can be identified as the larger shaft). In other words, a speed reducer is analogous to a simple lever.

In the absence of frictional losses, the output power must equal the input power. Therefore,

$$HP_o = \frac{2\pi n_o T_o}{33{,}000} = \frac{2\pi n_i T_i}{33{,}000} = HP_i \tag{2-2}$$

or

$$\frac{T_o}{T_i} = \frac{n_i}{n_o} = r \tag{2-3}$$

If there are frictional losses in a gear reducer, the output torque will be somewhat less than that predicted by Eq. (2-3) with the deficit representing power dissipation.

The mechanical efficiency η of a speed reducer is defined as the ratio of the

Figure 2-1. Concentric, helical gear, speed reducer. (Courtesy of Foote-Jones Gear Division, Dresser Industries.)

output power to the input power:

$$\eta = \frac{HP_o}{HP_i} = \frac{n_o T_o}{n_i T_i} = \frac{\omega_o T_o}{\omega_i T_i} \tag{2-4}$$

Equation (2-4) can also be written as

$$\eta = \left(\frac{1}{r}\right)\frac{T_o}{T_i} \tag{2-5}$$

If we denote the ideal output torque, as represented by Eq. (2-3) in the absence of frictional losses, as T'_o, then Eqs. (2-3) and (2-5) will show that

$$\eta = \frac{T_o}{T'_o} \tag{2-6}$$

As an example, the catalog data for one manufacturer's triple-reduction helical gear reducer shows an output torque of 850 in-lb and an output speed of 17 rpm when the input torque is 8.4 in-lb at an input speed of 1800 rpm (the ratio of the reducer is $r = 105.9$). The theoretical output torque (neglecting losses) as calculated using Eq. (2-3) would be 890 in-lb, so the actual output torque is about 95.5 percent of the ideal value. From Eq. (2-6), the mechanical efficiency of this reducer is $\eta = 95.5$ percent, which is quite high despite the fact that three sets of gears are used (the efficiency of each set of helical gears averages to around 98.5 percent which includes bearing friction and oil-pumping losses). This efficiency is typical of a helical speed reducer where the losses will be relatively small because we have minimal sliding friction with this type of gear tooth.

A contrasting example is provided by a triple-reduction worm gear reducer having a ratio of 100,000:1. Catalog data indicate that for a particular reducer the output torque is 2715 in-lb at 0.018 rpm when the input torque is 5.32 in-lb at 1800 rpm. The frictionless output torque, as determined from Eq. (2-3), would be 532,000 in-lb, so we see that the actual output torque is only about 0.5 percent of the ideal value. Almost all of the input horsepower (about 0.15 hp) is dissipated in the form of heat. Dissipation is a consequence of the inherently inefficient mode of torque transmission in a worm and worm-wheel where there is appreciable sliding friction. In our example, the inherent inefficiency of the worm gear set is multiplied by 3 because the reducer is a triple-reduction unit (has three worm gear sets).

2-2 TYPES OF SPEED REDUCERS

Geared speed reducers are available in almost every conceivable combination of gear types and shaft arrangements. We find that bevel gear pairs, helical (and spur) gear pairs, and worm gear pairs are combined in a diverse variety of arrangements designed to give just the desired combination of shaft arrangement, speed reduction, power capability, mechanical efficiency, and smooth operation. One such hybrid reducer is shown in Fig. 2-2. This reducer incorporates a helical gear pair and a worm gear pair to give a shaft configuration at right angles.

In selecting a speed reducer from the large variety of types available, consideration must usually be given to the following factors:

(a) *Shaft configuration*. The input and output shaft centerlines might be parallel, at right angles, or concentric. If shafts are to be at right angles, then at least one bevel gear or worm gear set will be required.

(b) *Speed reduction*. For moderate speed reduction (say up to $r = 200$), spur gear or helical gear reducers will be adequate. For the higher reductions, either worm gear reducers or planetary reducers will have to be used. Here the trade-off is simple. The worm gear reducer is low cost but inefficient whereas the planetary unit is just the reverse; the planetary unit will also permit co-axial input and output shafts which the worm gear unit cannot provide.

(c) *Noise, shock, and vibration*. For smooth, quiet operation, helical gear reducers are preferable to spur gear reducers. Where very high power is to be transmitted, the helical drive (perhaps incorporating bevel gears also) is highly preferable to the worm gear drive because of the cost of power dissipation. On the other hand, the worm gear drive is cheap and has good capability of handling shock loads.

Figure 2-2. Worm and helical gear speed reducer. (Courtesy of PT Components, Inc.)

2-2-1 Worm Gear Reducers

Worm gear reducers are advantageous in designs calling for a high ratio in a small space at low cost. They are also available in a wide range of sizes from fractional horsepower and low torque output to output torques of thousands of in-lb. And, compared to spur or helical gearing, worm gear reducers have a higher tolerance to shock loads.

Worm gear reducers are not generally designed to be run backward. In fact, under certain conditions it may be virtually impossible to back-drive a worm gear reducer. The impossibility of running backward is called self-locking, which can be very desirable in some machine design applications. In self-locking, the reducer cannot be operated by applying torque to the low-speed shaft. This condition involves the lead angle of the worm being less than the friction angle of the gear materials. However, it should be remembered that if the reducer is subjected to a source of vibration, a slight motion may be induced in the gears which could reduce the friction angle below the worm lead angle. Then, what otherwise would have been a self-locked gear pair can become a reducer running backward if a load is available to cause this. Self-locking, in other words, is not a substitute for a brake in a mechanical system.

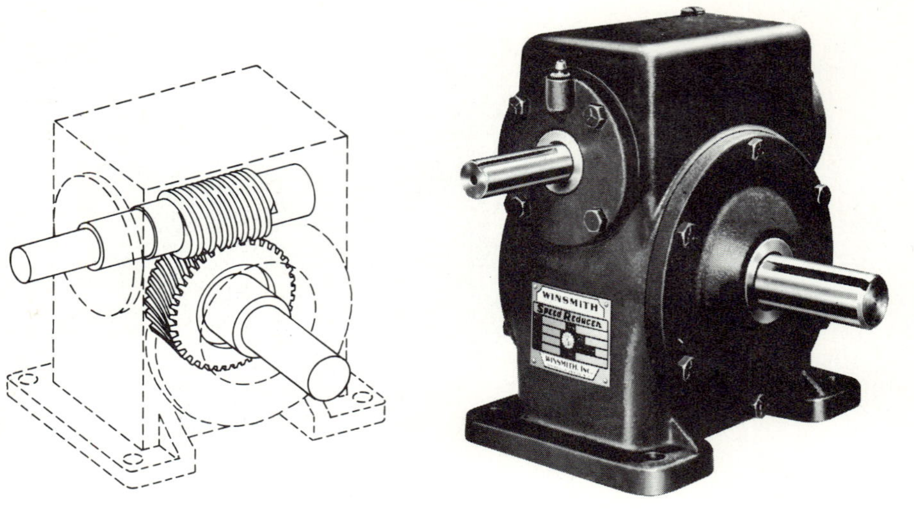

Figure 2-3. Single-reduction worm gear reducer. (Courtesy of Winsmith, Peerless-Winsmith, Inc.)

A single-reduction worm gear reducer is shown in Fig. 2-3. Single-reduction worm gear drives are available in ratios of up to around 70:1. Note that the shaft configuration is necessarily one at right angles. These components are available with input horsepowers ranging up to several hundred hp and output torques ranging up to hundreds of thousands of in-lb.

Double-reduction worm gear reducers are used where higher ratios are required. Ratios of up to around 4900:1 can be obtained simply by using two 70:1 worm gear pairs in series. Such a reducer is shown in Fig. 2-4.

Note that in the double-reduction worm gear reducer the input and output shafts are parallel (but not concentric). These units are usually confined to input horsepower ratings of 100 hp or less in the off-the-shelf variety.

Where a very high ratio is required, a triple-reduction worm gear unit such as that sketched in Fig. 2-5 can be used. Of course, the power dissipation in such units becomes a large fraction of the input horsepower when ratios ranging up to 200,000:1 are used. Such a ratio is easily achievable with a triple-reduction unit.

Because of the relatively high dissipation of energy in worm gear speed reducers (particularly in the higher ratios), selection of a reducer from application speed and torque (or horsepower) values may be dictated by thermal limitations of the reducer rather than by mechanical strength limitations. Thermally selected reducers must be adequately cooled regardless of their strength capabilities. Consequently, worm gear reducers are generally rated both by thermal and by mechanical horsepower capacities.

In the speed ratio range where the high ratios of the single-reduction worm gear reducer overlap the low ratios of the double-reduction unit, a helical gear pair often can be combined with a single worm gear pair to produce a more

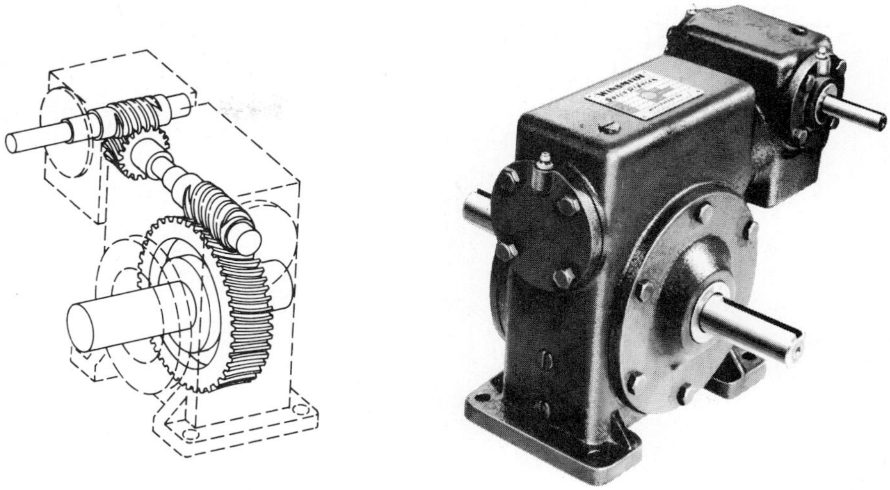

Figure 2-4. Double-reduction worm gear reducer. (Courtesy of Winsmith, Peerless-Winsmith, Inc.)

economical reducer such as that shown in Fig. 2-2. One manufacturer provides reducers of this type covering speed ratios in the range 14:1 to 460:1.

Worm gear speed reducers can be found in a wide variety of mechanical systems. They are, by far, the most commonly used speed reducers in the lower horsepower applications where power dissipation costs are not considered to be as important as capital costs.

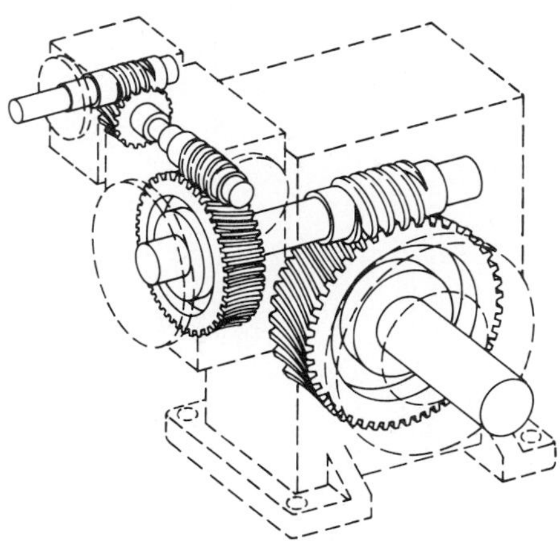

Figure 2-5. Triple-reduction worm gear reducer. (Courtesy of Winsmith, Peerless-Winsmith, Inc.)

2-2-2 Parallel-Shaft Helical Gear Reducers

In the higher horsepower applications where thousands of hp are to be transmitted in a mechanical system, the mechanical efficiency of the speed reducer becomes a critical factor. High efficiency dictates the use of helical gearing (and perhaps bevel gearing if shafts must be at right angles).

In applications requiring high horsepower transmission, the most commonly used speed reducer is the parallel-shaft reducer. A double-reduction parallel-shaft reducer is shown in Fig. 2-6.

Speed reduction units, such as are indicated in Fig. 2-6, are available as off-the-shelf components in single-reduction, double-reduction, and triple-reduction versions. Housings may be either cast iron or weldments.

One disadvantage of the conventional parallel-shaft speed reducer is that the gearbox tends to be long because of the way in which the gear pairs have been distributed to optimize bearing performance. A more compact arrangement can be obtained with the concentric configuration shown in Fig. 2-1. However, these concentric-shaft gear reducers are not available, off-the-shelf, in the very high horsepower capacities.

Figure 2-6. Double-reduction helical gear reducer. (Courtesy of PT Components, Inc.)

Figure 2-7. Bevel and helical gear reducer. (Courtesy of PT Components, Inc.)

In applications where shafts must be at right angles, a bevel gear pair can be incorporated into a parallel-shaft gearbox to produce a highly efficient unit possessing the requisite horsepower capacity. Such a unit is shown in Fig. 2-7.

EXAMPLE 2-1

In cases where more than one type of speed reducer might be available in meeting the application mechanical requirements of speed and torque, the ultimate decision must be based on cost considerations. Although the sophisticated method for making such an economic decision will involve the time value of money (interest rates) and taxes, here we will assume that there are only two major cost factors: capital cost and operating cost.

Suppose we have a conveyor application requiring a head shaft speed of 58 rpm and 75 hp. We plan to use an 1150 rpm induction motor as the prime mover. The conveyor is to operate 24 hours a day and 365 days a year.

TABLE 2-1. Cost Comparison of Two Reducers.

	Worm Gear Reducer	Parallel-Shaft Reducer
Capital cost	\$6000	\$9000
Efficiency	0.93	0.96
Annual operating cost	\$26,350	\$25,527
Differential annual cost	\$823	\$0

Upon examining various catalogs that provide off-the-shelf reducers, we determine that both a single-reduction worm gear reducer and a parallel-shaft reducer (with bevel gear input) can satisfy the load and ratio (19.8:1) requirements.

The worm gear unit costs \$6000 and its mechanical efficiency is 93 percent. The parallel shaft reducer costs \$9000 and its efficiency is 96 percent. Assuming that the induction motor is 100 percent efficient and that the cost of electricity is 5¢/kWh we are able to determine that the capital costs and yearly operating costs are as presented in Table 2-1.

We see that the difference in annual operating costs is \$823 in favor of the more expensive parallel-shaft reducer. Consequently, it will take only 3.64 years to recover the \$3000 difference in capital cost between the worm gear unit and the parallel-shaft unit. This may well justify purchasing the more expensive parallel-shaft reducer. ■

2-2-3 Shaft-Mounted Speed Reducers

The shaft-mounted speed reducer represents an evolution of shaft and gear pair configurations resulting from the huge market for speed reducer applications involving particular mechanical systems such as conveyor systems. In a large number of conveyor systems, where a head shaft is to be driven and the horsepower requirements are not too large, thoughtful engineers noted that the flexible coupling required to connect a conventional speed reducer to the head shaft could be eliminated. This is accomplished by making the output shaft of the reducer hollow and hanging the reducer directly on the head shaft as indicated in Fig. 2-8.

Shaft-mounted speed reducers are easy to install, do not require a motor base, are lightweight, and do not need a coupling for the output shaft. They come in capacities of up to around 150 hp.

In the installation shown in Fig. 2-8, the reducer is mounted directly on the head shaft and is being powered by a variable-speed drive input. The belt pulley on the motor has a variable pitch diameter (it is a spring-loaded split pulley) sheave whose pitch diameter can be adjusted by sliding the motor base by means of take-up screws.

With the reducer mounted (hung) on the conveyor head shaft, the reducer

Figure 2-8. Shaft-mounted speed reducer (safety guards have been removed for photographic purposes). (Courtesy of Dodge/Reliance Electric, Greenville, SC.)

must be restrained from tending to rotate about the head shaft. This is done by means of the torque arm shown extending to the left in Fig. 2-8.

The restraining torque arm required with a shaft-mounted reducer can be used advantageously by equipping it with an overload release limit switch as illustrated in Fig. 2-9. On the left of Fig. 2-9, the belt drive is tight, the reducer is in its normal operating position, and the overload release device has not been actuated. If the reducer encounters an overload while in this configuration, the increased torque will cause an increased force in the torque arm which will cause the overload release mechanism to be tripped which, in effect, shortens the torque arm. In turn, the belt will be loosened, the power will be cut off, and (perhaps) an alarm sounded. This condition is shown at the right of Fig. 2-9.

Shaft-mounted speed reducers can be purchased in a variety of combinations integrated with other components. Figure 2-10 shows one such unit in which the input shaft of the reducer is powered directly with an integrated hydraulic motor. This design is useful in applications where electrical power is unavailable or where the reducer location is remote from the system's ultimate prime mover.

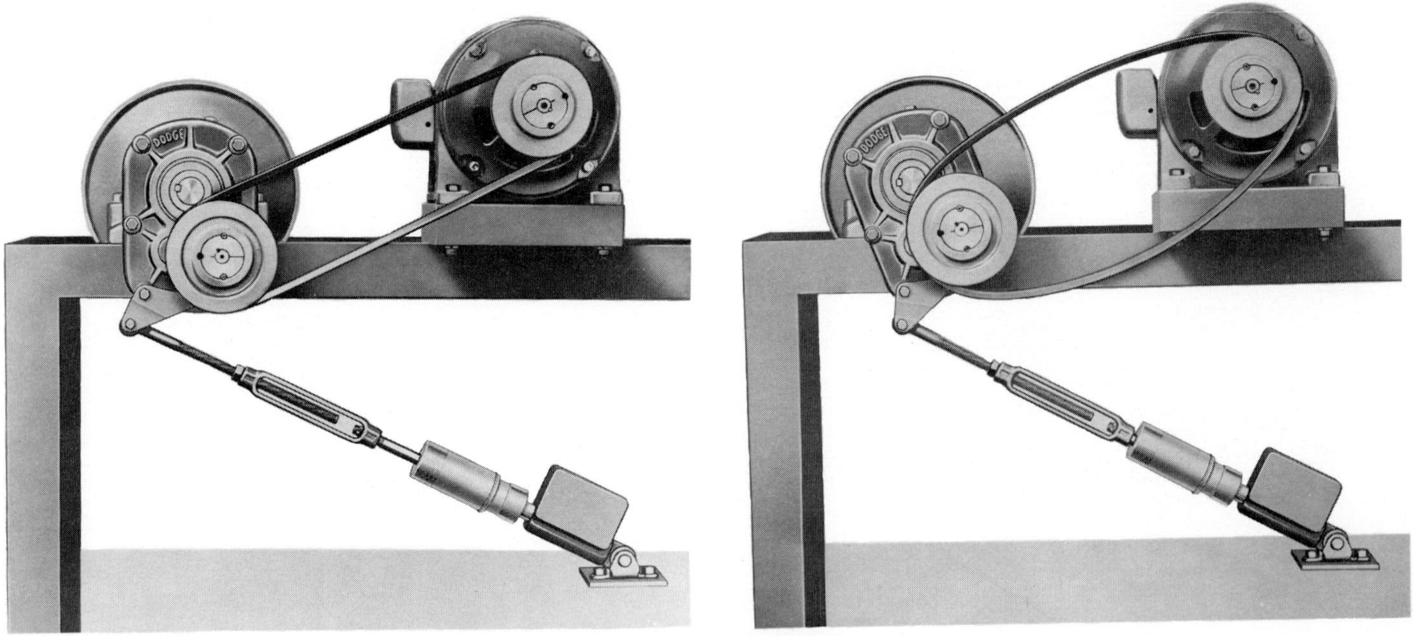

Figure 2-9. Torque arm overload release (safety guards have been removed for photographic purposes). (Courtesy of Dodge/Reliance Electric, Greenville, SC.)

Figure 2-10. Hydraulically powered shaft-mounted reducer. (Courtesy of Dodge/Reliance Electric, Greenville, SC.)

EXAMPLE 2-2

A conveyor head shaft requires 100 hp and rotates at 50 rpm. Power is transmitted by a double-reduction shaft-mounted reducer whose ratio is 24:1. As shown in Fig. 2-11, there is a 10 inch pitch diameter V-belt sheave mounted on the reducer input shaft which is driven by another V-belt sheave mounted on a 1750 rpm induction motor. Assume that the tension F_s in the slack side of the V-belt drive is 20 percent of the tension F_t in the taut side. Determine

(a) The pitch diameter of the motor sheave
(b) The tension F_t on the taut side of the V-belt
(c) The input shaft overhung load, F_i
(d) The input shaft torque, T_i
(e) The output shaft torque, T_o
(f) The load that must be preset into the torque arm overload release if it is to be tripped by a 25 percent overload

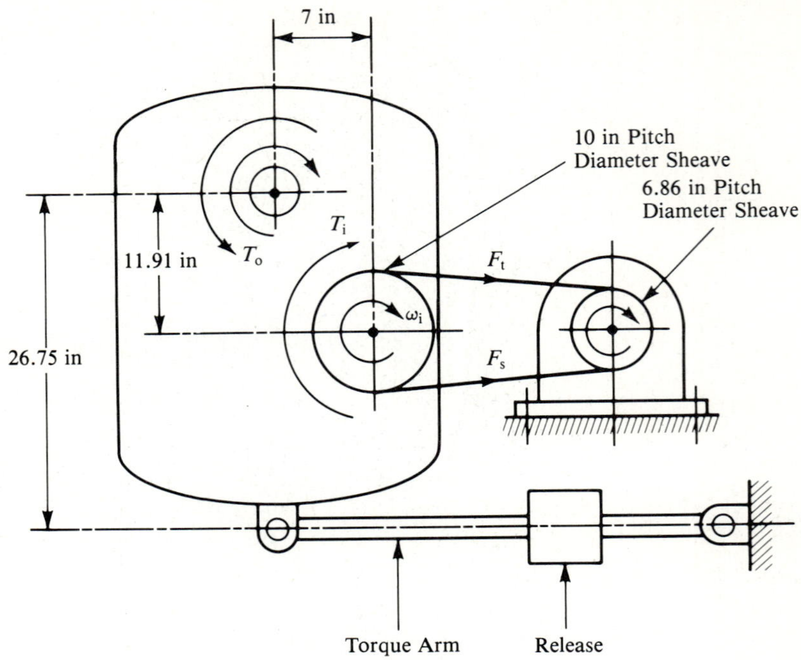

Figure 2-11. Schematic diagram for Example 2-2.

Solution

The reducer input speed is just 50(24) = 1200 rpm. Therefore, the motor sheave diameter is

$$PD_m = \frac{1200}{1750}(10) = 6.86 \text{ inches}$$

The reducer input torque is given by

$$T_i = \frac{33{,}000(100)(12)}{2\pi(1200)} = 5252 \text{ in-lb}$$

Therefore, the output torque is just

$$T_o = 24(5252) = 126{,}050 \text{ in-lb}$$

Since this is a double-reduction gearbox, both the input and output angular speeds are (say) cw which means that the input torque is cw while the output torque is ccw.

The input torque can also be expressed in terms of the V-belt tensions, that is,

$$(F_t - F_s)(5) = T_i = 5252$$

where

$$F_s = 0.2F_t$$

Solving these equations gives

$$F_t = 1313 \text{ lb} \qquad F_s = 263 \text{ lb}$$

The overhung load is approximately (neglecting the small angles in the V-belt drive)

$$F_i = F_s + F_t = 1576 \text{ lb}$$

Now, if we take moments about the output shaft of the reducer, the free-body diagram of the reducer will produce the moment equation

$$M_0 = 126{,}050 + 1576(11.91) - 5252 - F_a(26.75) = 0$$

Solving this equation, we find that the force in the torque arm is $F_a = 5217$ lb. Therefore, the overload release mechanism must be preset to the value 1.25(5217) = 6521 lb. ■

2-2-4 Gearmotors

Gear reducers can be purchased with electric drive motors attached integrally to the reducer by means of an adapter flange. Figure 2-12 shows one such design in which a single-reduction worm gear reducer is integrated, in one housing, with an electric motor. This arrangement is called a gearmotor.

Gearmotors are available in a wide variety of types and combinations including single-, double-, and triple-reduction worm gear units. They are also available with other types of gear reducers such as the planetary reducer to be discussed later.

Figure 2-12. Gearmotor. (Courtesy of Winsmith, Peerless-Winsmith, Inc.)

2-3 APPLICATIONS

In the task of selecting a speed reducer for a given machinery application, the engineer serves merely as an interface between the speed, torque, or horsepower requirements of the load and the catalog data of the manufacturer which define reducer model numbers as a function of speed, torque, and horsepower ratings. Catalog data for speed reducer characteristics are almost always tabulated in terms of

- **(a)** Speed reducer model number
- **(b)** The speed ratio, r
- **(c)** Input horsepower
- **(d)** Output torque
- **(e)** Overhung load
- **(f)** Thrust load
- **(g)** Reducer rotating inertia referred to the high-speed shaft

It is important to note that input horsepower and output torque are used to define the reducer's capability. From these values, reducer efficiency can be determined if the ratio and input speed of the reducer are also specified. The efficiency is

$$\eta = \frac{2\pi}{33{,}000}\left(\frac{T_o}{P_i}\right)\left(\frac{1}{r}n_i\right) \tag{2-7}$$

In Eq. (2-7), power P_i is in hp, torque T_o is in ft-lb, and input speed n_i is in rpm.

A typical page from a speed reducer catalog is reproduced in Fig. 2-13. These data are for a single-reduction worm gear speed reducer.

2-3-1 Service Factors

In application, gear reducers are subject to fluctuating loads, potential overloads, and potential shock loads. Service factors associated with such loadings have been compiled by the American Gear Manufacturers Association (AGMA) from long historical experience. A service factor is a multiplier used to increase the nominal load (or rated load) on the reducer for the purpose of selecting (sizing) a reducer that will satisfy not only the steady load applied to the reducer but also the potential overload or shock conditions. The rated load of a reducer is just the gearbox rating for a service factor of 1.0.

The procedure involved in selecting a service factor generally requires that two tables be used. The first table is an applications table in which various kinds of machinery are categorized in terms of whether they represent applications involving uniform loads (U), moderate shock loads (M), or heavy shock loads (H). These categories are called load classes and the table is called the load class table (see Table 2-2).

After the load class (U, M, or H) has been established, a second table is used in which the numerical value of the service factor is chosen, depending on (a) load class (U, M, H), (b) service duty (duration of operation), and (c) type of

Input Horsepower, Output Torque, Overhung Loads and Thrust Loads for Universal Floor Mount Single Reduction Worm Gear Speed Reducers (Based on 1.0 Service Factor)

Table No. 1

1750 RPM DRIVER — 5 to 1 RATIO — 350 RPM OUTPUT

REDUCER★ REF. NO.	INPUT HORSEPOWER Therm.	INPUT HORSEPOWER Mech.	OUTPUT TORQUE INCH-POUNDS Therm.	OUTPUT TORQUE INCH-POUNDS Mech.	MAXIMUM OVERHUNG LOADS-LBS. Input Shaft	MAXIMUM OVERHUNG LOADS-LBS. Output Shaft	MAX. THRUST LOAD-LBS. Output Shaft
133U1-5E	1.20	1.20	177	177	55	160	290
154U1-5E	1.56	1.56	245	245	125	285	395

1750 RPM DRIVER — 10 to 1 RATIO — 175 RPM OUTPUT

REDUCER★ REF. NO.	INPUT HORSEPOWER Therm.	INPUT HORSEPOWER Mech.	OUTPUT TORQUE INCH-POUNDS Therm.	OUTPUT TORQUE INCH-POUNDS Mech.	MAXIMUM OVERHUNG LOADS-LBS. Input Shaft	MAXIMUM OVERHUNG LOADS-LBS. Output Shaft	MAX. THRUST LOAD-LBS. Output Shaft
133U1-10E	.78	.78	216	216	55	160	300
154U1-10E	1.09	1.09	306	306	125	285	475
175U1-10E	1.34	1.43	385	414	125	285	475
206U1-10E	1.81	2.02	540	605	150	500	660
237U1-10E	2.16	2.72	660	830	175	525	675
262U1-10E	2.95	3.45	880	1030	175	725	880
300U1-10E	3.72	5.54*	1180	1750*	335	900	950
300U1-10FE	5.54	5.54	1750	1750	260	900	950
325U1-10E	4.25	6.55*	1340	2075*	350	1150	1050
325U1-10FE	6.55	6.55	2075	2075	325	1150	1050
375U1-10FE	8.75	8.90	2800	2850	400	1240	1160
450U1-10FE	12.80	13.20	4200	4310	500	2085	1540
516U1-10FE	16.75	17.80	5490	5840	331	2434	1610

1750 RPM DRIVER — 15 to 1 RATIO — 116.6 RPM OUTPUT

REDUCER★ REF. NO.	INPUT HORSEPOWER Therm.	INPUT HORSEPOWER Mech.	OUTPUT TORQUE INCH-POUNDS Therm.	OUTPUT TORQUE INCH-POUNDS Mech.	MAXIMUM OVERHUNG LOADS-LBS. Input Shaft	MAXIMUM OVERHUNG LOADS-LBS. Output Shaft	MAX. THRUST LOAD-LBS. Output Shaft
133U1-15E	.625	.625	233	233	55	160	300
154U1-15E	.83	.83	334	334	125	285	475
175U1-15E	1.02	1.05	430	442	125	285	475
206U1-15E	1.35	1.51	580	650	150	500	660
237U1-15E	1.68	2.02	740	895	175	535	770
262U1-15E	2.18	2.62	940	1130	175	725	880
300U1-15E	2.76	4.20*	1250	1900*	345	900	950
300U1-15FE	4.20	4.20	1900	1900	180	900	950
325U1-15E	3.14	4.85*	1460	2250*	400	1150	1050
325U1-15FE	4.67	4.85	2170	2250	119	1150	1050
375U1-15FE	6.66	6.66	3100	3100	400	1370	1270
450U1-15FE	9.80	10.00	4550	4660	465	2300	1540
516U1-15FE	12.69	13.85	5870	6440	257	2700	1900

1750 RPM DRIVER — 20 to 1 RATIO — 87.5 RPM OUTPUT

REDUCER★ REF. NO.	INPUT HORSEPOWER Therm.	INPUT HORSEPOWER Mech.	OUTPUT TORQUE INCH-POUNDS Therm.	OUTPUT TORQUE INCH-POUNDS Mech.	MAXIMUM OVERHUNG LOADS-LBS. Input Shaft	MAXIMUM OVERHUNG LOADS-LBS. Output Shaft	MAX. THRUST LOAD-LBS. Output Shaft
133U1-20E	.520	.520	236	236	60	160	300
154U1-20E	.70	.70	334	334	125	285	475
175U1-20E	.85	.89	435	455	125	285	475
206U1-20E	1.13	1.20	620	660	150	500	660
237U1-20E	1.35	1.62	765	920	175	535	770
262U1-20E	1.74	2.00	1000	1150	175	725	880
300U1-20E	2.26	3.42*	1260	1950*	350	900	950
300U1-20FE	3.42	3.42	1950	1950	175	900	950
325U1-20E	2.54	3.98*	1480	2315*	400	1150	1050
325U1-20FE	3.82	3.98	2230	2315	97	1150	1050
375U1-20FE	5.35	5.35	3200	3200	400	1370	1270
450U1-20FE	8.00	8.00	4800	4800	430	2300	1540
516U1-20FE	10.36	10.90	6190	6510	202	2700	1900

1750 RPM DRIVER — 30 to 1 RATIO — 58.3 RPM OUTPUT

REDUCER★ REF. NO.	INPUT HORSEPOWER Therm.	INPUT HORSEPOWER Mech.	OUTPUT TORQUE INCH-POUNDS Therm.	OUTPUT TORQUE INCH-POUNDS Mech.	MAXIMUM OVERHUNG LOADS-LBS. Input Shaft	MAXIMUM OVERHUNG LOADS-LBS. Output Shaft	MAX. THRUST LOAD-LBS. Output Shaft
133U1-30E	.359	.359	240	240	60	160	300
154U1-30E	.520	.520	345	345	125	285	475
175U1-30E	.605	.648	425	455	125	285	475
206U1-30E	.85	.91	625	670	150	500	660
237U1-30E	1.06	1.24	800	935	175	535	770
262U1-30E	1.33	1.54	1020	1175	175	725	880
300U1-30E	1.72	2.49*	1330	1930*	345	900	950
300U1-30FE	2.49	2.49	1930	1930	175	900	950
325U1-30E	1.76	2.76*	1480	2320*	370	1150	1050
325U1-30FE	2.63	2.76	2210	2320	67	1150	1050
375U1-30FE	3.83	3.83	3220	3220	395	1370	1270
450U1-30FE	5.70	5.70	4880	4880	500	2300	1540
516U1-30FE	7.47	7.71	6380	6600	143	2700	1900

1750 RPM DRIVER — 40 to 1 RATIO — 43.7 RPM OUTPUT

REDUCER★ REF. NO.	INPUT HORSEPOWER Therm.	INPUT HORSEPOWER Mech.	OUTPUT TORQUE INCH-POUNDS Therm.	OUTPUT TORQUE INCH-POUNDS Mech.	MAXIMUM OVERHUNG LOADS-LBS. Input Shaft	MAXIMUM OVERHUNG LOADS-LBS. Output Shaft	MAX. THRUST LOAD-LBS. Output Shaft
133U1-40E	.312	.312	233	233	60	160	300
154U1-40E	.450	.450	335	335	125	285	475
175U1-40E	.528	.562	425	454	125	285	475
206U1-40E	.71	.76	620	665	150	500	660
237U1-40E	.83	1.00	780	925	175	535	770
262U1-40E	1.03	1.19	995	1150	175	725	880
300U1-40E	1.36	2.00	1330	1960*	355	900	950
300U1-40FE	2.00	2.00	1960	1960	190	900	950
325U1-40E	1.53	2.33*	1540	2340*	370	1150	1050
325U1-40FE	2.24	2.33	2260	2340	57	1150	1050
375U1-40FE	3.14	3.14	3250	3250	400	1370	1270
450U1-40FE	4.73	4.73	5000	5000	420	2300	1540
516U1-40FE	6.05	6.05	6550	6550	264	2700	1900

1750 RPM DRIVER — 50 to 1 RATIO — 35 RPM OUTPUT

REDUCER★ REF. NO.	INPUT HORSEPOWER Therm.	INPUT HORSEPOWER Mech.	OUTPUT TORQUE INCH-POUNDS Therm.	OUTPUT TORQUE INCH-POUNDS Mech.	MAXIMUM OVERHUNG LOADS-LBS. Input Shaft	MAXIMUM OVERHUNG LOADS-LBS. Output Shaft	MAX. THRUST LOAD-LBS. Output Shaft
133U1-50E	.262	.262	226	226	65	160	300
154U1-50E	.370	.370	319	319	125	285	475
175U1-50E	.445	.445	424	424	120	285	475
206U1-50E	.59	.62	610	640	150	500	660
237U1-50E	.72	.83	765	885	175	535	770
262U1-50E	.92	1.04	970	1110	175	725	880
300U1-50E	1.24	1.66*	1410	1880*	350	900	950
300U1-50FE	1.66	1.66	1880	1880	295	900	950
325U1-50E	1.28	1.96*	1560	2400*	370	1150	1050
325U1-50FE	1.86	1.96	2280	2400	48	1150	1050
375U1-50FE	2.70	2.70	3320	3320	400	1370	1270
450U1-50FE	3.86	3.86	4860	4860	460	2300	1540
516U1-50FE	4.96	4.96	6330	6330	242	2700	1900

★Reducer Reference Number indicates Basic Reducer Size and Ratio. From the Assembly Layouts on Page M-4, determine the reducer configuration desired and complete Part Numbers from instructions on that page.

Above ratings are not applicable when reducer shafts are subjected to combined overhung and thrust loads.
Find ratings for input speeds not shown by straight line interpolation.
Maximum overhung loads are at center of keyseats and on one end of output shaft only. Overhung loads applied closer to the reducer housing are desirable, but overhung loads farther out on the shaft and overhung loads on both ends of output shaft should be referred to the Browning Engineers.
Contact the Browning Engineers for the following:

1. **High starting torques exceeding 200% of the reducer mechanical rating.**
2. **Frequent starting or repetitive shock applications.**
3. **Applications where high energy loads must be absorbed as when stalling.**
4. **Input shaft vertical.**

***These ratings cannot be used for continuous duty when service factor exceeds 1.25; use fan-cooled reducer.**

Figure 2-13. Catalog data for a single-reduction worm gear reducer. (Courtesy of Browning Mfg. Div., Emerson Electric Co.)

TABLE 2-2. Load Classes. (Courtesy of PT Components, Inc.)

Driven machine	Load class ■
Agitators	
pure liquids	U
liquids and solids	MS
liquids—variable density	MS
Blowers	
centrifugal	U
lobe	MS
vane	U
Brewing and distilling	
bottling machinery	U
brew kettles—continuous duty	U
cookers—continuous duty	U
mash tubs—continuous duty	U
scale hopper—frequent starts	MS
Can filling machines	U
Cane knives⊙	MS
Car dumpers	HS
Car pullers	MS
Clarifiers	U
Classifiers	MS
Clay working machinery	
brick press	HS
briquette machine	HS
clay working machinery	MS
pug mill	MS
Compressors	
centrifugal	U
lobe	MS
reciprocating	
multi-cylinder	ṀS
single-cylinder	HS
Conveyors—uniformly loaded or fed	
apron	U
assembly	U

Driven machine	Load class ■
jig drives	HS
maneuvering winches	MS
pumps	MS
screen drive	HS
stackers	MS
utility winches	MS
Dry dock cranes	
main hoist	△
auxiliary hoist	△
boom, luffing	△
rotating, swing or slew	□
tracking, drive wheels	★
Elevators	
bucket—uniform load	U
bucket—heavy load	MS
bucket—continuous	U
centrifugal discharge	U
escalators	U•
freight	MS•
gravity discharge	U
man lifts	▲
passenger	▲
Extruders (Plastic)	
film⊙	U
sheet⊙	U
coating⊙	U
rods⊙	U
pipe⊙	U
tubing⊙	U
blow molders⊙	MS
pre-plasticizers⊙	MS
Fans	
centrifugal	U
cooling towers induced draft	▲

Driven machine	Load class ■
main drive	♁
carriage drive	▲
conveyors	
burner	‡
main or heavy duty	★
main log	♁
re-saw	▲
merry-go-round	‡
slab	♁
transfer	‡
chains	
floor	★
green	◆
cut-off saws	
chain	◆
drag	◆
debarking drums	♁
feeds	
edger	‡
gang	♁
trimmer	‡
log deck	♁
log hauls—incline, well type	♁
log turning devices	♁
planer feed	‡
planer tilting hoists	★
rolls—live off bearing, roll cases	♁
sorting table	‡
tipple hoist	‡
transfers	
chain	◆
craneway	◆
tray drives	‡
veneer lathe drives	▲

Application	Load
belt	U
bucket	U
chain	U
flight	U
oven	U
screw	U
Conveyors—heavy duty	
not uniformly fed	
apron	MS
assembly	MS
belt	MS
bucket	MS
chain	MS
flight	MS
live roll	▲
oven	MS
reciprocating	HS
screw	MS
shaker	HS
Cranes	
main hoists	U
bridge travel	▲
trolley travel	▲
Crusher	
ore	HS
stone	HS
sugar⊙	MS
Dredges	
cable reels	MS
conveyors	MS
cutter head drives	HS

Application	Load
forced draft⊙	U
induced draft	MS
large, mine, etc.	MS
large, industrial	MS
light, small diameter	U
Feeders	
apron	MS
belt	MS
disc	U
reciprocating	HS
screw	MS
Food industry	
beet slicer	MS
cereal cooker	U
dough mixer	MS
meat grinders	MS
Generators—not welding	U
Hammer mills	HS
Hoists	
heavy duty	HS
medium duty	MS
skip hoist	MS
Laundry washers, reversing	MS
Laundry tumblers	MS
Line shafts	
driving processing equipment	MS
light	U
other line shafts	U
Lumber industry	
barkers	
spindle feed	‡

Application	Load
Machine tools	
bending roll	MS
punch press—gear driven	HS
notching press—belt driven	▲
plate planers	HS
tapping machine	HS
other machine tools	
main drives	MS
auxiliary drives	U
Metal mills	
draw bench carriage and main drive	MS
pinch, dryer and scrubber rolls—reversing	▲
slitters	MS
table conveyors	
non-reversing	
group drives	MS
individual drives	HS
reversing	▲
wire drawing and flattening machine	MS
wire winding machine	MS
Mills—rotary type	
ball⊙	MS
cement kilns⊙	MS
dryers and coolers⊙	MS
kilns, other than cement	MS
pebble⊙	MS
rod	
plain⊙	MS

■ U . . . . . Uniform load
MS . . . . Moderate shock load
HS . . . . Heavy shock load

⊙ 24 hours per day service factor only

▲ Consult Link-Belt Drive Division

△ Use service factor 1.00

□ Use service factor 1.25

★ Use service factor 1.50

♁ Use service factor 1.75

‡ 1.25 for 10, 1.50 for 24 hours

◆ 1.50 for 10, 1.75 for 24 hours

TABLE 2-2 (continued)

Driven machine	Load class ■
wedge bar⊙	MS
tumbling barrels	HS
Mixers	
concrete mixers—continuous	MS
concrete mixers—intermittent	MS
constant density	U
variable density	MS
Oil industry	
chillers	MS
oil well pumping	▲
paraffin filter press	MS
rotary kilns	MS
Paper mills	
agitators (mixers)	MS
barker—auxiliaries—hydraulic	MS
barker—mechanical	MS
barking drum	HS
beater and pulper	MS
bleacher	U
calenders	MS
calenders—super	HS
converting machine, except cutters, platers	MS
conveyors	U
couch	MS
cutters—platers	HS
cylinders	MS
dryers	MS
felt stretcher	MS
felt whipper	HS
jordans	HS
log haul	HS
presses	U
pulp machine reel	MS
stock chest	MS
suction roll	U

Driven machine	Load class ■
washers and thickeners	MS
winders	U
Printing presses	▲
Pullers	
barge haul	HS
Pumps	
centrifugal	U
proportioning	MS
reciprocating	
single acting; 3 or more cylinders	MS
double acting; 2 or more cylinders	MS
single acting; 1 or 2 cylinders	▲
double acting; single cylinder	▲
rotary	
gear type	U
lobe, vane	U
Rubber and plastics industries	
crackers⊙	HS
laboratory equipment	MS
mixing mills⊙	HS
refiners⊙	MS
rubber calenders⊙	MS
rubber mill—2 on line⊙	MS
rubber mill—3 on line⊙	U
sheeter	MS
tire building machines	▲
tire and tube press openers	▲
tubers and strainers⊙	MS
warming mills⊙	MS
Sand muller	MS
Sewage disposal equipment	
bar screens	U
chemical feeders	U
collectors	U

Driven machine	Load class ■
dewatering screws	MS
scum breakers	MS
slow or rapid mixers	MS
thickeners	MS
vacuum filters	MS
Screens	
air washing	U
rotary—stone or gravel	MS
traveling water intake	U
Slab pushers	MS
Steering gear	▲
Stokers	U
Sugar industry	
cane knives⊙	MS
crushers⊙	MS
mills⊙	MS
Textile industry	
batchers	MS
calenders	MS
cards	MS
dry cans	MS
dryers	MS
dyeing machinery	MS
knitting machines	▲
looms	MS
mangles	MS
nappers	MS
pads	MS
range drives	▲
slashers	MS
soapers	MS
spinners	MS
tenter frames	MS
washers	MS
winders	MS
Windlass	▲

•Caution: Relative to applications involving the handling of people, i.e. elevators, escalators, etc., Link-Belt Drive Division must be consulted prior to drive selection.

TABLE 2-3. Service Factors for Worm Gear Reducers. (Courtesy of PT Components, Inc.)

Load class	Prime mover†											
	Electric motor				Multicylinder internal combustion engine				Single cylinder internal combustion engine			
	Half hour per day	2 hours per day	up to 10 hours per day	24 hours per day	Half hour per day	2 hours per day	up to 10 hours per day	24 hours per day	Half hour per day	2 hours per day	up to 10 hours per day	24 hours per day
Uniform	0.80	0.90	1.00	1.25	0.90	1.00	1.25	1.50	1.00	1.25	1.50	1.75
Moderate shock	0.90	1.00	1.25	1.50	1.00	1.25	1.50	1.75	1.25	1.50	1.75	2.00
Heavy shock	1.00	1.25	1.50	1.75	1.25	1.50	1.75	2.00	1.50	1.75	2.00	2.25

▲Applications involving more than 10 starts and stops per hour, or where high energy loads must be absorbed are not covered. Consult Link-Belt Drive Division

†When a fluid coupling is used between the prime mover and speed reducer, service factors for moderate shock or heavy shock may be modified Link-Belt Drive Division Maximum momentary or starting load must not exceed 300% of speed reducer rating at service factor of 1

TABLE 2-4. Service Factors for Helical Offset Reducers. (Courtesy of PT Components, Inc.)

Load class	Service factor											
	Prime mover†											
	Electric motor				Multicylinder internal combustion engine					Single cylinder internal combustion engine		
	Half hour per day⊗	3 hours per day⊗	up to 10 hours per day	24 hours per day	Half hour per day⊗	3 hours per day⊗	up to 10 hours per day	24 hours per day	Half hour per day⊗	3 hours per day⊗	up to 10 hours per day	24 hours per day
Uniform	0.50	0.80	1.00	1.25	0.80	1.00	1.25	1.50	1.00	1.25	1.50	1.75
Moderate shock	0.80	1.00	1.25	1.50	1.00	1.25	1.50	1.75	1.25	1.50	1.75	2.00
Heavy shock	1.25	1.50	1.75	2.00	1.50	1.75	2.00	2.25	1.75	2.00	2.25	2.50

† When a fluid coupling is used between the prime mover and speed reducer, service factors for moderate shock or heavy shock may be modified, consult Link-Belt Drive Division.

⊗ Check starting and peak loads description on page F-105.

prime mover driving the speed reducer. One such service factor table, for worm gear reducers, is represented in Table 2-3.

It is important to know that different service factor tables apply to different types of reducers. Table 2-4, for example, gives service factors for helical offest reducers, and we can see that the numerical values differ, in some cases, from those in Table 2-3. The fact that some of the service factors are less than 1.0 means that gear reducers are designed to have a safety margin of two to three times their rated loads. Thus, a reducer will be able to withstand momentary overloads.

Sometimes the service factors may be incorporated directly into the load class table so that only one table is necessary. Also, gearmotors are given numerical load classes and are cataloged directly in terms of these classes. The actual procedure used to select a reducer seems to depend on the type of reducer, and the user generally will have to follow a procedure directly spelled out in the manufacturer's catalog.

2-3-2 Thermal Ratings

Some speed reducers (notably worm gear reducers) are given catalog ratings both for mechanical strength and for thermal capacity. If the mechanical efficiency of the reducer is low, there may be an appreciable dissipation of energy in the reducer which will cause a rise in temperature of mechanical parts and lubricating oil. In such cases, it may be that thermal limitations supersede mechanical ones and the reducer must be selected accordingly. So, where both ratings are given, the design engineer choosing the reducer must be careful to consider both.

EXAMPLE 2-3

A 10:1 worm gear speed reducer is connected between a 1750 rpm drive motor and a conveyor head shaft rotating at 175 rpm. The conveyor requires a torque of 1434 in-lb. The load class has been determined to be M (moderate shock) with a duty cycle of 8 to 10 hours per day. Select a speed reducer suitable for this application.

Solution

From Table 2-3, we find the service factor to be 1.25. Therefore, the reducer output torque rating must be

$$T_o = 1.25(1434) = 1792 \text{ in-lb}$$

Referring to the table of Fig. 2-13, we see that reducers 325U1-10E and 325U1-10FE both have a mechanical rating of 2075 in-lb. However, the thermal rating of reducer 325U1-10E is only 1340 in-lb whereas the thermal rating of reducer 325U1-10FE is 2075 in-lb. Therefore, the 325U1-10FE model must be chosen.

The input horsepower rating of the chosen reducer is seen to be 6.55 (Fig. 2-13). The actual average operating horsepower will, however, be only

$$\text{HP} = \frac{1434}{2075}(6.55) = 4.52 \text{ hp}$$

Consequently, a 5 hp motor can be used which will be capable of handling the moderate shock overloads. ■

2-3-3 Overhung Loads

In addition to being subjected to loads of a somewhat uncertain nature dictated by the load class tabulation, gear reducers are also often subjected to overhung loads at either the input or output shaft (or both). An overhung load (OHL) is a loading configuration at the input or output shaft that causes additional loads on the bearings in the reducer (i.e., loads in addition to those induced by the gears in the reducer). Where direct, flexible couplings are used to connect a reducer shaft either to load or prime mover, there will be no overhung load. But where belt pulleys, chain sprockets, or gears are used at the input or output shaft, an overhung load will occur.

An example is shown in Fig. 2-14, where the output shaft of a gear reducer is being powered by a chain and sprocket. The overhung load is the resultant force acting on the reducer shaft as a consequence of the type of component mounted on it. With a chain drive that has no tension in the slack side of the chain, the overhung load is simply the tension on the taut side of the chain. This tension force can be determined from either the shaft torque or horsepower for the shaft in question. That is,

$$\text{OHL} = \frac{T}{d/2} = \frac{126{,}000 \cdot \text{HP}}{n \cdot d} \tag{2-8}$$

where d is the sprocket diameter in inches, n is the shaft rpm, T is the torque in in-lb, and HP is the horsepower. They apply to both input and output shafts.

When the overhung load is caused by a chain sprocket coupling having no tension on the slack side of the chain, Eq. (2-8) applies. If, however, Fig. 2-14 represents a belt running over a pulley mounted on the reducer shaft, there could be a tension in both sides of the belt. The overhung load will then be the sum of these two tensions and cannot, therefore, be simply determined from the torque or horsepower. Consequently, an overhung load factor (called the K-factor by the AGMA) is applied to Eq. (2-8) in order to estimate the overhung load. Equation (2-8) is rewritten as

$$\text{OHL} = \frac{KT}{d/2} = \frac{126{,}000 \cdot K \cdot \text{HP}}{n \cdot d} \tag{2-9}$$

where K is the overhung load factor to be chosen from Table 2-5.

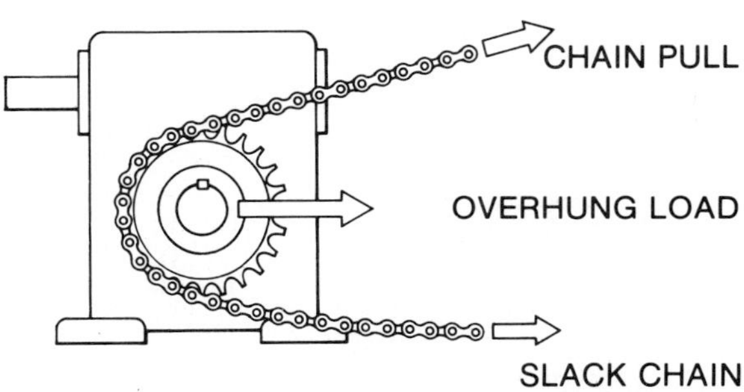

Figure 2-14. Overhung load from a chain and sprocket input. (Courtesy of Winsmith, Peerless-Winsmith, Inc.)

TABLE 2-5. Overhung Load Factors.

Chain and sprocket	1.00
Gear	1.25
V-belt	1.50
Flat belt	2.50
Variable-pitch belt	3.50

In applying Eq. (2-9), the values T and HP should already have been modified from their nominal values by applying the service factor mentioned earlier, and tabulated in tables such as Table 2-3 and Table 2-4.

EXAMPLE 2-4

Suppose we have the drive configuration shown in Fig. 2-15. Here we have a direct coupling on the input shaft, so there is no overhung load on the input to the reducer. At the output of this reducer is mounted a 21 tooth, No. 80, roller-chain sprocket which drives a 60 tooth sprocket mounted on the conveyor head shaft.

For illustrative purposes, we make the following assumptions:

(a) The conveyor operates approximately 10 hours per day.
(b) There are moderate shock loads (M).
(c) The torque and speed requirements of the conveyor head shaft are 9500 in-lb and 30 rpm respectively.
(d) The drive motor operates at 1750 rpm under load.

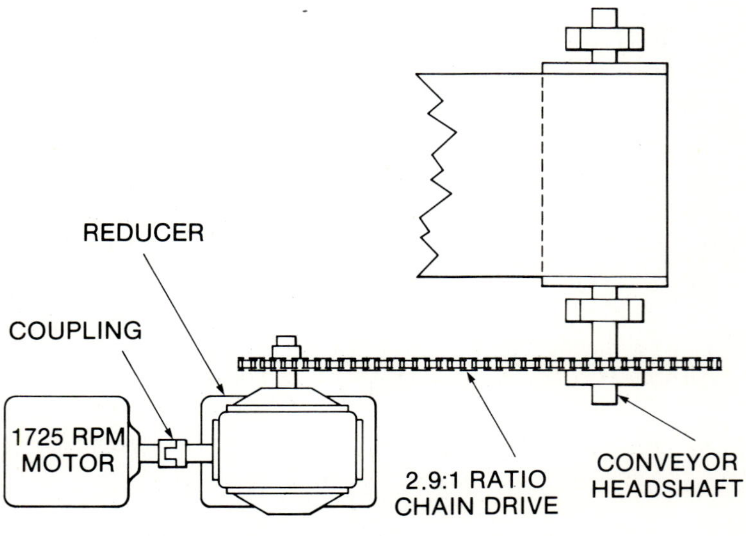

Figure 2-15. Schematic diagram for Example 2-4. (Courtesy of Winsmith, Peerless-Winsmith, Inc.)

Since the overall ratio of the drive is 1750/30 = 58.33 and the chain drive represents a ratio of 60/21 = 2.86, the gear reducer must provide a ratio of 58.33/2.86 = 20.4. So we see that a single-reduction worm gear reducer with ratio $r = 20:1$ will approximately satisfy both the speed reduction requirement and the shaft configuration at right angles.

Now, the nominal torque at the output of the reducer is obtained by reducing the head shaft torque requirement of 9500 in-lb by the chain drive ratio of 2.86. So the nominal reducer output torque is 9500/2.86 = 3322 in-lb. This value must now be modified by the service factor which, from Table 2-3, we find to be 1.25. So, in selecting the reducer we must require an output torque rating of 1.25(3322) = 4152 in-lb.

With the output torque rating of 4152 in-lb, input speed of 1750 rpm, and ratio of 20:1, we can now go to a reducer catalog and tentatively select a single-reduction worm gear reducer. One manufacturer's catalog data are represented in Fig. 2-16. We can see that a model size 120 will be required which provides an output torque of 6147 in-lb and has an input horsepower capacity of 9.57. Actually, this reducer is somewhat oversized for this application and at this point we may want to refer to other manufacturers' products as well (we can find another manufacturer who produces a reducer having a 4981 in-lb output torque rating and input horsepower rating of about 8).

The process of selecting this reducer is not yet complete because we must check the overhung load on the output shaft. The value is determined from Eq. (2-9) with $K = 1$ (see Table 2-5). Then

$$\text{OHL} = \frac{1(4152)}{3.35} = 1239 \text{ lb}$$

The diameter of the 21 tooth, No. 80, roller chain sprocket is 6.7 inches.

We now refer to the overhung load table for this size of reducer, also shown in Fig. 2-16. We see that the size 120 reducer has an overhung load capacity of some 3200 lb at the output speed of 1750/20 = 87.5 rpm (we did an interpolation in the table). So the overhung load capability of this reducer is quite adequate.

To complete this selection process, we would also have to know the thermal horsepower rating of this reducer. This particular reducer has a thermal rating exceeding its mechanical rating, so there will be no reducer heating problem.

The third task in designing this drive would be to select the motor to drive the reducer. We note that the mechanical efficiency of the size 120 reducer is 0.87 which means that the nominal input torque will be 3322/20(0.87) = 191 in-lb. The nominal input horsepower is then

$$\text{HP}_i = \frac{2\pi(1750)(191)}{33{,}000(12)} = 5.3$$

Therefore, we would have to select a 7.5 hp motor (since 6 hp motors are not manufactured). This motor will have enough capacity to handle the moderate shock loads imposed on it.

The final task in sizing the components of this application would be to determine if the RC-80 roller chain is adequate. This will be found to be the case. ∎

Rating Tables

Type S Single Worm
1750 RPM

Nominal ratio ★	Output speed RPM	Rating description ■	74	85	100	120	145	175	210	250	1200 1400
7½	233	Input Horsepower	6.52	9.16	12.76	20.73	31.96	46.58	72.03	104.2	LARGE WORM RATINGS—SEE PAGE F-209
		Output Torque, lb.-in.	1598	2270	3164	5055	7792	11795	18235	26672	
		Efficiency, %	92	93	93	94	94	95	95	96	
		Exact Ratio	7.4	7.4	7.4	7.2	7.2	7.4	7.4	7.4	
10	175	Input Horsepower	5.55	7.58	10.64	17.28	25.77	39.19	59.64	88.54	
		Output Torque, lb.-in.	1753	2421	3438	5500	8417	12937	19894	29542	
		Efficiency, %	90	91	92	93	93	94	95	95	
		Exact Ratio	9.75	9.75	9.75	9.50	9.75	9.75	9.75	9.75	
15	117	Input Horsepower	4.15	5.61	7.85	12.97	19.00	28.47	44.48	65.18	
		Output Torque, lb.-in.	1908	2636	3690	6027	9138	13843	21847	32034	
		Efficiency, %	87	89	89	90	91	92	93	93	
		Exact Ratio	14.67	14.67	14.67	14.33	14.67	14.67	14.67	14.67	
20	87.5	Input Horsepower	3.22	4.34	5.89	9.57	14.68	21.94	34.02	49.82	
		Output Torque, lb.-in.	1971	2725	3742	6147	9539	14418	22600	33480	
		Efficiency, %	83	85	86	87	88	89	90	91	
		Exact Ratio	20.5	20.5	20.5	20.5	20.5	20.5	20.5	20.5	
25	70.0	Input Horsepower	2.73	3.77	5.23	8.37	12.69	18.97	29.11	42.73	
		Output Torque, lb.-in.	1986	2730	3832	6149	9632	14564	22603	33563	
		Efficiency, %	81	82	83	85	86	87	88	89	
		Exact Ratio	24.5	24.5	24.5	24.0	24.5	24.5	24.5	24.5	
SEE TYPE H RATINGS† 30	58.3	Input Horsepower	2.41	3.27	4.45	7.10	10.98	16.13	24.80	36.33	
		Output Torque, lb.-in.	1996	2734	3834	6155	9637	14573	22656	33587	
		Efficiency, %	78	80	81	83	84	85	86	87	
		Exact Ratio	29.5	29.0	29.5	29.0	29.0	29.5	29.5	29.5	
40	43.8	Input Horsepower	1.87	2.50	3.42	5.40	8.25	12.17	18.83	27.37	
		Output Torque, lb.-in.	1969	2701	3750	6073	9514	14378	22507	33134	
		Efficiency, %	73	75	76	78	80	82	83	84	
		Exact Ratio	40.0	40.0	40.0	40.0	40.0	40.0	40.0	40.0	
50	35.0	Input Horsepower	1.54	2.04	2.85	4.49	6.71	9.89	15.17	22.05	
		Output Torque, lb.-in.	1910	2610	3691	5870	9187	13889	21857	32172	
		Efficiency, %	69	71	72	74	76	78	80	81	
		Exact Ratio	50.0	50.0	50.0	49.0	50.0	50.0	50.0	50.0	
60	29.2	Input Horsepower	1.31	1.70	2.37	3.67	5.60	8.10	12.31	18.03	
		Output Torque, lb.-in.	1810	2468	3490	5545	8686	13127	20475	30402	
		Efficiency, %	64	67	68	71	73	75	77	78	
		Exact Ratio	60.0	59.0	60.0	59.0	59.0	60.0	60.0	60.0	
70	25.0	Input Horsepower	1.06	1.41	1.95	3.08	4.68	6.60	10.18	14.87	
		Output Torque, lb.-in.	1659	2281	3192	5128	8028	12145	18985	28121	
		Efficiency, %	61	64	65	68	70	73	74	75	
		Exact Ratio	70.0	69.0	70.0	68.0	68.0	70.0	70.0	70.0	

★ Bold face indicates non-stock ratios. Consult Drive Division for delivery schedule.

■ Ratings based on service factor of 1. Complete list of service factors on pages F-165 and F-166. When reducer is not direct connected refer to overhung load ratings on page F-162.

† Before making final selection, consult Type H Ratings Tables.

Overhung load and thrust ratings

Type S—single worm

Speed Reducer Size	Maximum overhung load, pounds ■ High Speed Shaft R.P.M. ○							Low Speed Shaft R.P.M.									
	1750	1450	1150	870	720	580	300	250	200	150	100	75	50	25	10	5	1
74	529	498	486	402	348	295	154	885	928	998	1096	1266	1480	1818	2340	2340	2340
85	396	352	315	220	245	285	250	1256	1543	1550	1552	1990	2237	2900	2900	2900	2900
100	762	700	642	598	617	552	420	1984	2100	2259	2483	2691	3013	3300	3300	3300	3300
120	976	934	862	784	660	715	600	2537	2677	2879	3111	3376	3700	4830	5850	5850	5850
145	1618	1607	1541	1363	1200	1044	912	2750	2905	3100	3289	3567	3826	5022	7173	8400	8400
175	1587	1546	1463	1369	1203	1087	685	4994	5293	5699	6183	6686	7200	9278	9575	9575	9575
210	1724	1749	1787	1905	2006	1963	2212	6148	6467	6963	7633	8215	8637	11000	13490	13490	13490
250	2477	2528	2612	2762	2829	2800	1000	6747	7080	7599	8244	8898	9154	11722	16640	17130	17130
1200	1478	1546	1555	1598	1593	1671	1689	8350	8850	9310	9550	11450	13050	20000	20000	20000	20000
1400	3307	3346	3414	3499	3581	3648	3685	11900	12700	13400	14180	16250	18575	21100	21100	21100	21100

Figure 2-16. Catalog data for a single-reduction worm gear reducer. (Courtesy of PT Components, Inc.)

2-4 MULTI-TERMINAL GEARBOXES

Up to this point, we have been concerned with rather straightforward two-terminal gearboxes (speed reducers). These reducers have had a single input shaft and a single output shaft.

There are many mechanical power transmission applications in which it would be advantageous to have three-terminal (or even four-terminal) gearboxes with, say, two inputs and one output. As an example, suppose two prime movers are available to drive a power shaft. With a three-terminal gearbox, we could connect the two prime movers to the two input shafts and let the output shaft drive the power shaft. In doing this, though, we would have to take special precautions to insure that the gearbox is designed in such a way as to make possible the balanced acceptance of these inputs.

Another application of a three-terminal gearbox is shown in Fig. 2-17. Here we use the three-terminal gearbox as part of a variable-speed transmission in order, effectively, to increase the power capability of a simple variable-speed drive unit. We can also, by this means, often increase the speed ratio of the overall transmission beyond what it would be for the variable-speed drive itself (i.e., the device labeled PIV).

In the device shown in Fig. 2-17, some of the input power is bypassed (shunted) around the variable-speed drive (the PIV) and proceeds directly into the three-terminal gearbox through input shaft *A* while the rest of the input power goes through the PIV to input shaft *B*. We will discuss some of these so-called shunt transmissions in more detail later.

Another important feature of the multi-terminal gearbox is the control possibilities it provides. In a three-terminal gearbox, for example, we could use two of the terminals primarily for the transmission of power while exercising some degree of control by how the third terminal is actuated. Such gearboxes are often called differential gearboxes.

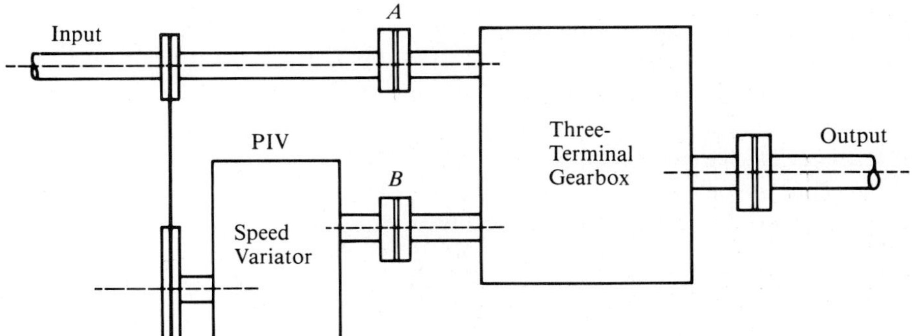

Figure 2-17. Shunt, variable-speed transmission.

2-5 EPICYCLIC GEARING

One of the most important kinds of multi-terminal gearboxes is the epicyclic gear train. In an epicyclic gear train, the axis of at least one gear is moving relative to some other gear axis, which is fixed. The gear mounted on the moving axis is called a planet gear; a point on the surface (tooth) of this gear traces out an epicycloidal path in space as the axis of the planet moves. It is from this characteristic that the epicyclic gear train gets its name. The moving frame in which the planet gear axis is mounted is called the arm or carrier. Figure 2-18 shows a typical epicyclic train.

Figure 2-18. Epicyclic gear train. (Courtesy of Andantex, Inc.)

The large number of useful epicyclic gear train configurations that have been developed for use in various kinds of mechanical systems is a tribute to the ingenuity of engineers, tinkerers, and inventors. Of all of these diverse types of epicyclic trains, the simple single-stage planetary gear train is the most elementary.

2-6 THE SIMPLE SINGLE-STAGE PLANETARY

A single-stage planetary gear train is sketched in Fig. 2-19. This simple epicyclic gear train gets its name from its analogy to the solar system. It consists of three

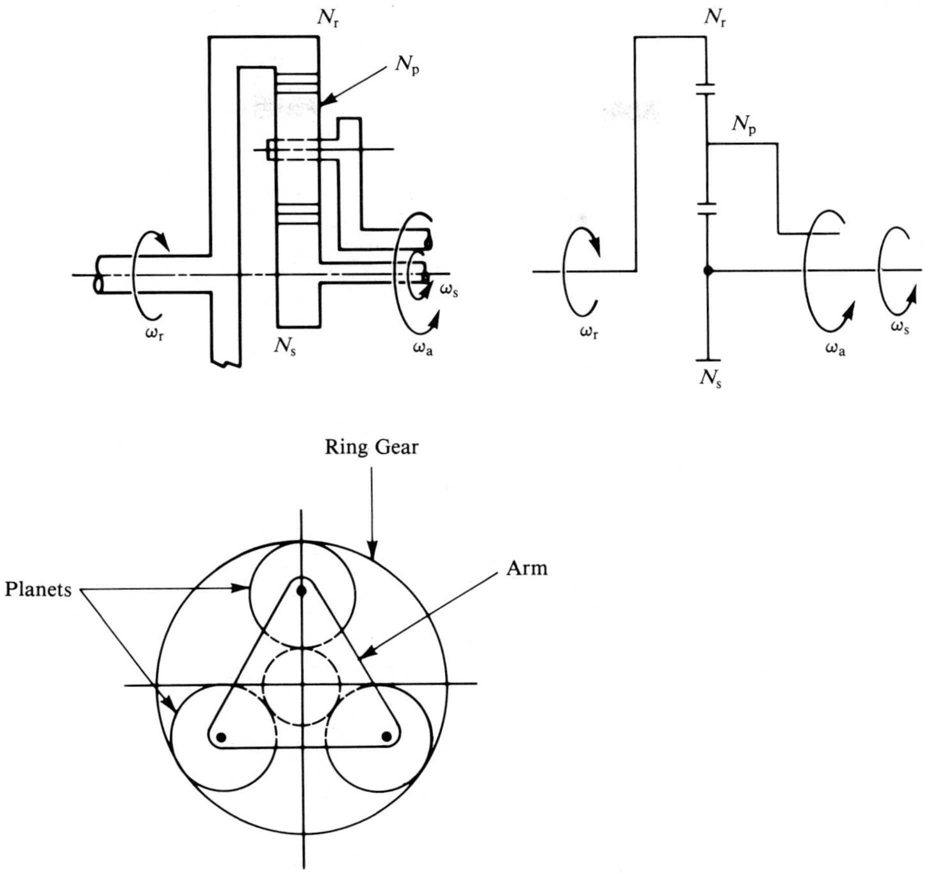

Figure 2-19. Simple planetary gear train.

gears: the sun gear, the planet gear, and the ring gear. The planet gear axis is mounted in a carrier (the arm) which, in turn, is connected to one of the terminals of the gear train. The ring gear is an internal gear in which the teeth are cut into the inside periphery of the ring. Note that the planet shaft, itself, is not an external terminal to the device. The terminals are connected to the sun gear, the ring gear, and the arm.

The schematic diagram used to indicate this planetary train symbolically is shown at the right of Fig. 2-19. Also, it must be remembered that a single planet, by itself, would only rarely be used in designing such a gear train. To balance the gear train, planets are generally provided in twos, threes, or fours. The carrier arrangement for three planets is indicated at the lower left of Fig. 2-19.

The kinematic relationship for the various angular velocities of the single-stage planetary train can be derived from the linear velocities shown in Fig. 2-20. The angular velocities shown in Fig. 2-20 are the same as those of Fig. 2-19 when either figure is viewed from the right. We will assign a positive value to all ccw

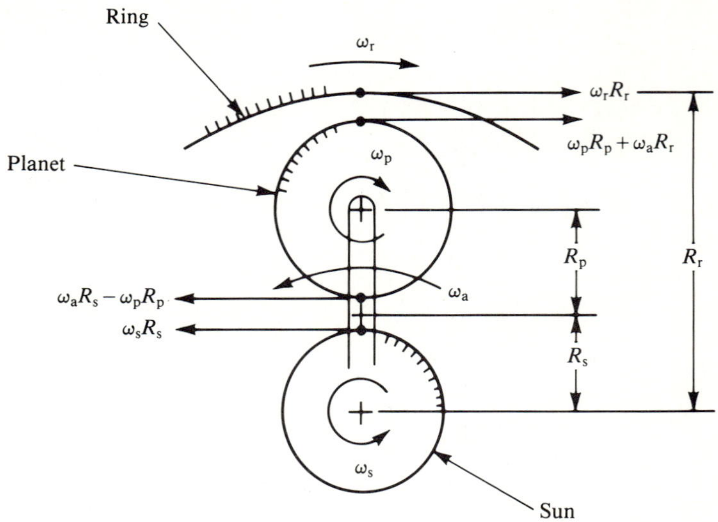

Figure 2-20. Kinematics of a single-stage planetary.

angular velocities and a negative value to all cw angular velocities when the gear trains are viewed from the right.

With angular speeds ω_s and ω_a taken as ccw and angular speed ω_r taken as cw, the planet angular speed ω_p will be cw. Angular velocities ω_s, ω_a, and ω_r are absolute velocities. However, the angular speed ω_p of the planet, as defined here, is not an absolute velocity. We will define it to be the angular speed of the planet relative to the arm. We do this to simplify the kinematic analysis to follow, in which ω_p will vanish from our final answer. It is important, however, to note that angular speed ω_p is generally taken to be the absolute value (the value relative to the planet axis rather than to the arm).

The tooth contact points of two different gears in contact must have equal linear velocities. Consequently, from Fig. 2-20 we see that

$$\omega_s R_s = \omega_a R_s - \omega_p R_p \tag{2-10}$$

$$\omega_r R_r = \omega_p R_p + \omega_a R_r \tag{2-11}$$

Note how the right sides of Eqs. (2-10) and (2-11) represent absolute velocities of points on the arm plus the velocities (of the planet) relative to the arm.

Eliminating the planet relative velocity ω_p, we obtain

$$\frac{\omega_s - \omega_a}{\omega_r - \omega_a} = -\frac{R_r}{R_s} = -\alpha \tag{2-12}$$

This can be put in a slightly different form:

$$\frac{n_s - n_a}{n_r - n_a} = -\frac{N_r}{N_s} = -\alpha \tag{2-13}$$

where n represents angular speed in rpm and N represents number of teeth on the gear in question.

2-6-1 Single-Stage Planetary as a Two-Terminal Device

With one of the terminals fixed, the single-stage planetary gearbox becomes a simple two-terminal gearbox of the kind we studied in Section 2-1. The speed ratio $\omega_i/\omega_o = r = n_i/n_o$ for various combinations of input, output, and fixed terminals is tabulated in Table 2-6. There are six possible combinations in which ratio r can be expressed simply as a function of parameter α when one terminal is fixed.

TABLE 2-6. Simple Planetary as a Two-Terminal Gearbox.

Input Terminal	Fixed Terminal	Output Terminal	Speed Ratio, r	Speed Ratio for $\alpha = 3$
Sun	Arm	Ring	$-\alpha$	-3
Ring	Arm	Sun	$-1/\alpha$	$-1/3$
Arm	Sun	Ring	$\alpha/(1+\alpha)$	$3/4$
Ring	Sun	Arm	$(1+\alpha)/\alpha$	$4/3$
Sun	Ring	Arm	$1+\alpha$	4
Arm	Ring	Sun	$1/(1+\alpha)$	$1/4$

The numerical values shown for r in Table 2-6 are for the special case in which sun gear and planet gear have the same diameter. Note that the speed ratio for such a planetary transmission can be varied from a low of 0.25 to a high of 4.

The highest ratio is provided when the ring gear is fixed and the arm is the output terminal. The highest practical ratio achievable for this case is about $r = 10$ because, as the sun gear is made smaller, it becomes strength limited. The alternative would be to increase the ring gear diameter, but then the virtues of the planetary gear train begin to disappear because the gearbox grows in size.

EXAMPLE 2-5

Consider a single-stage planetary gear train in which the pitch diameters of the sun and planet gears are 4 inches. Then the pitch diameter of the ring gear will have to be 12 inches. The parameter α will be $\alpha = 3$. Suppose the ring gear is fixed and the sun gear is the input terminal rotating at 3600 rpm (ccw). Determine

(a) The output (arm) speed
(b) The rpm of the planet relative to the arm
(c) The absolute angular speed of the planet
(d) The rpm of the planet if the planet axis is fixed (with the ring gear absent)

Solution

From Table 2-6, we see that

$$n_a = \frac{3600}{4} = 900 \text{ rpm (ccw)}$$

Then, from Eq. (2-10), we find that

$$n_s R_s = n_a R_s - n_p R_p$$

or

$$3600(2) = 900(2) - n_p(2)$$

Solving for n_p gives

$$n_p = -2700 \text{ rpm (cw) relative to the arm!}$$

The absolute velocity of the planet is obtained by compounding the relative velocity of the planet with the absolute angular velocity of the arm. Therefore,

$$n'_p = -2700 + 900 = -1800 \text{ rpm (cw)}$$

The angular velocities just calculated are equivalent to one rotation of the arm (ccw), three rotations of the planet relative to the arm (cw), two rotations of the planet with respect to the planet axis (cw), for every four rotations of the sun (ccw). Note that if the planet were to be rigidly connected to the arm, it would experience one rotation (ccw) as the arm carries it around once (ccw). This is why there is the one-revolution difference between the relative and absolute angular velocities of the planet.

With the planet axis fixed and the ring gear absent, the planet will experience four revolutions (cw) for every four revolutions of the sun gear (ccw). So we see that the effect of the planetary action is to reduce the rpm of the planet. ■

2-6-2 Single-Stage Planetary as a Differential

When all three terminals of the single-stage planetary are free to rotate, the gear train is called a differential system. Now power can be transmitted at all three terminals. One typical configuration has one input terminal driving two outputs; the alternative is to have two inputs driving a single output. Table 2-7 contains the various three-terminal configurations along with the kinematic conditions associated with each one.

It is important to note the directional senses assigned to the various terminals of Table 2-7. The directions are not totally arbitrary. They are dictated, in part, by the requirement that the ring gear and sun gear torques always have the same directional sense! And since we can view each case of Table 2-7 by viewing the gearbox either from the right or from the left, it is obvious that all cases will remain valid if the directional sense of each terminal is reversed.

When the planetary gear train operates at constant speed (no acceleration), the planet is merely an idler gear; it can experience no net torque. This leads to the conclusion that the tangential tooth forces on all of the gears have the same magnitude, as is shown in Fig. 2-21.

To balance the two tooth forces acting on the planet, the arm must exert a force of $2W_t$ on the planet axis as shown in Fig. 2-21. Then the internal torques, due to these internal tooth forces, acting on the sun, arm, and ring gears, can be

TABLE 2-7. Simple Planetary as a Differential.

	Input Terminal(s)			Output Terminal(s)		
Case	Gear	Speed	Torque	Gear	Speed	Torque
1	Sun	ccw	ccw	Arm	ccw	cw
				Ring	cw	ccw
2	Ring	ccw	ccw	Arm	ccw	cw
				Sun	cw	ccw
3	Arm	ccw	ccw	Ring	ccw	cw
				Sun	ccw	cw
4	Arm	ccw	ccw	Sun	ccw	cw
	Ring	cw	cw			
5	Arm	ccw	ccw	Ring	ccw	cw
	Sun	cw	cw			
6	Ring	ccw	ccw	Arm	ccw	cw
	Sun	ccw	ccw			

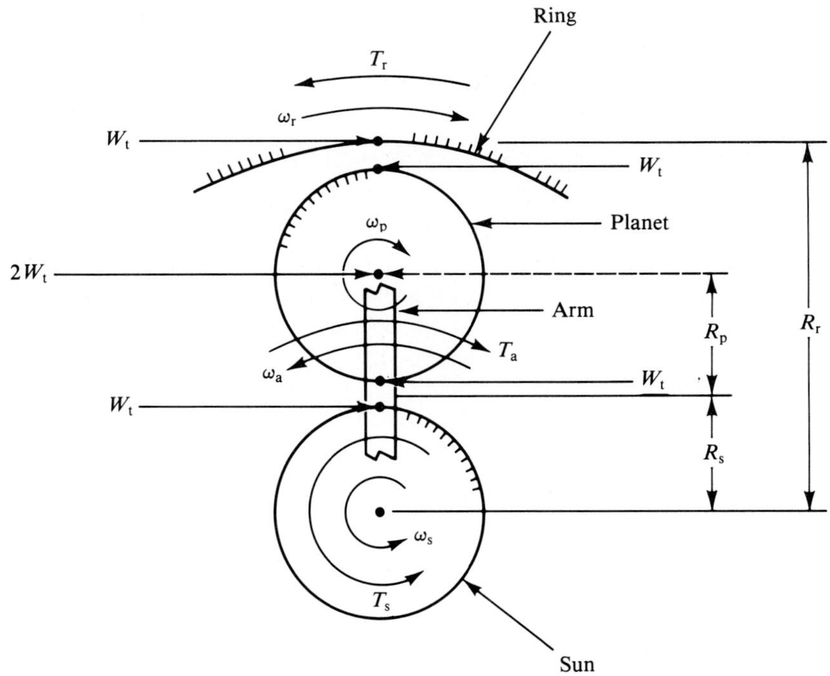

Figure 2-21. Planetary tangential tooth forces (with sun gear as input).

TABLE 2-8. Simple Planetary Internal Torques and Horsepower (with Sun Gear as Input).

Gear	Internal Torque	Speed	Internal Horsepower
Sun (input)	$-R_s W_t$ (cw)	$+n_s$ (ccw)	$\dfrac{-2\pi R_s W_t n_s}{33{,}000}$
Arm (output)	$\dfrac{+R_r + R_s}{2}(2W_t)$ (ccw)	$+n_a$ (ccw)	$\dfrac{+2\pi(R_r + R_s)W_t n_a}{33{,}000}$
Ring (output)	$-R_r W_t$ (cw)	$-n_r$ (cw)	$\dfrac{+2\pi R_r W_t n_r}{33{,}000}$

tabulated as shown in Table 2-8. This configuration corresponds to case 1 of Table 2-7. It is important to note that the directional sense of sun and ring torques will always be the same and opposite to the directional sense of the arm torque.

EXAMPLE 2-6

Using the dimensional data of Example 2-5, consider a single-stage planetary in which 100 hp is input to the arm at an arm speed of 900 rpm (ccw). Suppose this input drives a primary power output requiring 90 hp at the sun gear while the ring gear is used as an auxiliary power output consuming 10 hp. This is case 3 of Table 2-7. Determine

(a) The tangential tooth force in the gear train
(b) The speeds of the sun and ring gears

Solution

The input torque of the arm is given by

$$T_i = T_a = \frac{33{,}000(100)}{2\pi(900)} = 583.57 \text{ ft-lb (ccw)}$$

From the formulas in Table 2-8, we can calculate the tooth force:

$$(R_r + R_s)W_t = 583.57$$

$$\left(\frac{1}{2} + \frac{1}{6}\right)W_t = 583.57$$

$$W_t = 875.35 \text{ lb}$$

Then, from the power formulas of Table 2-8, we find

$$n_s = \frac{90(33{,}000)}{2\pi(1/6)(875.35)} = 3240 \text{ rpm (ccw)} \qquad \text{(see Table 2-7)}$$

$$n_r = \frac{10(33{,}000)}{2\pi(1/2)(875.35)} = 120 \text{ rpm (ccw)} \qquad \text{(see Table 2-7)}$$

Note that these output speeds were not arbitrary; they were dictated by the given conditions of horsepower split between the two specified outputs along with the specification that $n_a = 900$ rpm (ccw).

As a check, these three angular speeds must also satisfy Eq. (2-13). Substituting, we find

$$\frac{n_s - n_a}{n_r - n_a} = \frac{+3240 - (+900)}{+120 - (+900)} = -3 = -\alpha$$

This is precisely the same as the value for α calculated from

$$-\alpha = -\frac{N_r}{N_s} = -\frac{1/2}{1/6} = -3 \qquad ■$$

The lesson to be learned from this example is that one cannot go about arbitrarily specifying values for terminal speeds, torques, and powers for a planetary gear train. In this example we specified only the input speed and the two output horsepowers at the two output terminals. This was enough information to uniquely determine both magnitude and directional sense of the angular speeds of the output terminals. As we will see later, there is a general rule for determining when a planetary gear train problem is determinate.

In subsequent discussions, we will use the symbol N for dual purposes: (a) to refer to the number of teeth on a gear and (b) to name the gear. For example, gear N_6 refers to both the gear labeled number 6 and to the number of teeth on this gear.

2-7 COMPOUNDED EPICYCLIC GEAR TRAINS

We will define a compounded epicyclic gear train to be a gear train having at least two distinct (sets of) planets. Often the two (sets of) planets will be compound gear pairs; that is, they are rigidly fixed to the same planet shaft. An example is shown in Fig. 2-22.

Planets N_5 and N_6 are compounded together. They are rigidly fixed to the same shaft and will rotate at the same angular speed. In this case, the compound planet shaft is being carried by the arm which appears as terminal 1.

Gears N_3 and N_4 are also compounded together in this gear train. However, it is important to realize that gears N_2 and N_3 are not part of the epicyclic portion of the gear train! In the epicyclic portion, the sun gear is gear N_4, the ring gear is gear N_7, and the compounded planets are gears N_5 and N_6.

Several methods can be used to develop the angular speed relationships for more complex epicyclic gear trains. One way would be to analyze the internal linear velocities as we did in developing Eqs. (2-12) and (2-13). Another alternative is to use the tabular superposition method which can be found in most kinematics textbooks. Here, we will use the formula method outlined below.

The formula method is based on the fundamental idea that the absolute motion of one body can be determined by compounding absolute and relative motions of the body with the motions of other bodies. In epicyclic trains, this idea is used by referring speeds to the motion of the arm (or carrier). This procedure is

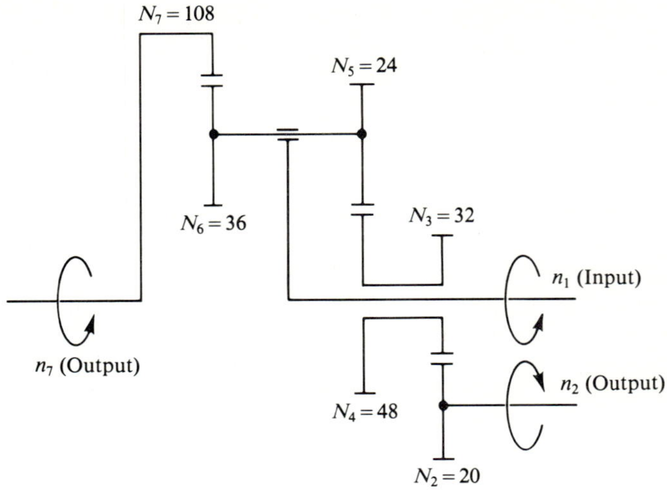

Figure 2-22. Compounded epicyclic gear train.

equivalent to reducing the arm speed to zero as is done in the tabular methods discussed in kinematics texts. Then, with the arm speed zero, the relative speeds are determined simply from the numbers of teeth on the gears. The formula expressing this idea will then correctly relate the motion of the various terminals of the epicyclic train.

The basic epicyclic speed formula is (without proof)

$$\frac{n_\ell - n_\mathrm{a}}{n_\mathrm{f} - n_\mathrm{a}} = \pm \frac{N_\mathrm{f} \ldots}{\ldots N_\ell} = \pm e \tag{2-14}$$

where n_ℓ = the absolute angular speed (rpm) of the last gear in the epicyclic train

n_f = the absolute angular speed (rpm) of the first gear in the epicyclic train

n_a = the absolute angular speed (rpm) of the arm

N_f = the number of teeth of the first gear of the epicyclic train

N_ℓ = the number of teeth of the last gear of the epicyclic train

e = the train value of the epicyclic train when the arm is fixed

On the right side of Eq. (2-14), the train value is the value with the arm fixed! It is determined in the same manner as for a conventional gear train in which all axes are fixed. The train value is then determined by considering numbers of teeth in proceeding through the gear train from the first gear to the last one. The sign used depends on the relative rotation sense between the first and last gears of the train when the arm is fixed.

We can see that Eq. (2-14) contains, as a special case, the kinematic relationship for the simple single-stage planetary, Eq. (2-13). In Eq. (2-13), the last gear would be the sun gear and the first gear would be the ring gear.

Parameter α is simply the train value e for the case in which the arm is fixed; for the simple planetary train, this is a negative value.

If we apply Eq. (2-14) to the configuration of Fig. 2-22, the result is

$$\frac{n_7 - n_1}{n_4 - n_1} = -\left(\frac{48}{24}\right)\left(\frac{36}{108}\right) = -0.666$$

Here we have arbitrarily defined the ring gear to be the last gear in the epicyclic portion and the sun gear to be the first. Gears N_2 and N_3 are not included in the formula because they are not part of the epicyclic train.

Now suppose that terminals 1 and 2 are rotating at speeds of 120 rpm (ccw) and 360 rpm (cw) respectively. The rpm of sun gear N_4 is then

$$n_4 = \frac{20}{32}(360) = 225 \text{ rpm (ccw)}$$

Equation (2-14) becomes

$$\frac{n_7 - (+120)}{225 - (+120)} = -0.666$$

$$n_7 = 50 \text{ rpm (ccw)}$$

We see that this device works in such a way that sun, arm, and ring gears all operate ccw under the conditions assumed. This configuration corresponds to either case 3 or case 6 of Table 2-7. Let us choose case 3, which means that the arm will be an input while ring gear and sun gear are outputs. The speed reduction between terminal 2 and terminal 7 is 7.2:1, and between terminals 2 and 1 it is 3:1.

2-8 EXTERNAL TORQUE AND POWER RELATIONSHIPS

In addition to the speed relationships for epicyclic gearboxes, we also wish to determine the relationships for division of external torque and power among the gearbox terminals. Here we are interested not in the detailed forces and torques inside the gearbox but only in the external values. Of course, speed, torque, and power relationships are not totally independent because power is proportional to the product of torque and speed.

Epicyclic gearboxes are usually designed using either spur, helical, or bevel gearing so although there will necessarily be some internal dissipation, the dissipation will be relatively small. For purposes of establishing torque and power division, then, we assume the gearboxes to be ideal; all of the input power must appear as output power.

If we identify the terminals of a three-terminal epicyclic train numerically as 1, 2, and 3, then the idealized power relationship is expressed as

$$T_1 n_1 + T_2 n_2 + T_3 n_3 = 0 \qquad \textbf{(2-15)}$$

We are not concerned with units at this point. Input power will be taken as positive, while output power is negative. If, for example, we were to have power inputs at terminals 1 and 2, Eq. (2-15) would show that $T_3 n_3$ must be negative, indicating that power must flow out of terminal 3.

The additional rule needed to establish torque and power division is the torque relationship

$$T_1 + T_2 + T_3 = 0 \tag{2-16}$$

In Eq. (2-16), ccw torques will be inserted as positive and cw torques will be inserted as negative (when the gear train is viewed from the right). And, we reiterate that the shaft rotation and torque will have the same directional sense at all power inputs while at power outputs they will have opposing directional senses.

Equation (2-16) results from a free-body diagram of the epicyclic portion of the gearbox under steady-state conditions which assume no accelerations to exist. Equation (2-16) simply states that the concentric external torques on the epicyclic gear train must be balanced. Where there are additional gears, not part of the epicyclic train, Eq. (2-16) will not be valid when referred to a terminal of such an additional gear!

The following example illustrates how Eqs. (2-14), (2-15), and (2-16) are used.

EXAMPLE 2-7

Consider the epicyclic configuration shown in Fig. 2-23. We specify that gear N_3 be an output transmitting 25 hp out of the gearbox. But note that gear N_3 is not a part of the epicyclic train. It cannot be entered directly into any of the epicyclic relationships established previously.

Using Eq. (2-14), we find the speed relationship for the epicyclic portion to be

$$\frac{n_2 - n_5}{n_1 - n_5} = (-)\left(\frac{N_1}{N_4}\right)\left(\frac{N_4}{N_2}\right) = (-)\,\frac{N_1}{N_2} \tag{2-17}$$

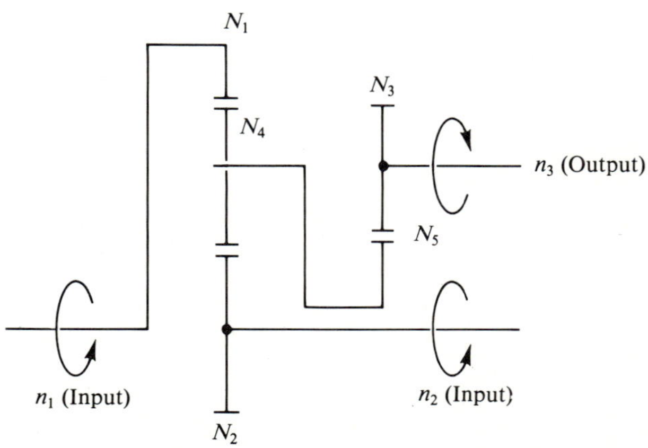

Figure 2-23. Schematic diagram for Example 2-7.

Let us assume the following specific numerical values: $N_2/N_1 = 1/7$, $N_3/N_5 = 1/2$, $n_1 = 700$ rpm (ccw), $n_3 = -1300$ rpm (cw). We wish to determine the speed n_2 corresponding to the conditions just defined.

The arm speed n_5 is obtained in the conventional manner for a simple gear pair:

$$n_5 = -n_3\left(\frac{N_3}{N_5}\right) = -(-1300)\left(\frac{1}{2}\right) = 650 \text{ rpm (ccw)}$$

Substituting into Eq. (2-17), we find that

$$\frac{n_2 - (+650)}{700 - (+650)} = -7$$

$$n_2 = 300 \text{ rpm (ccw)}$$

Next, we would like to determine how the torque and power are split among the terminals. Writing the torque and power equations, we obtain

$$T_1 + T_2 + T_5 = 0 \tag{2-18}$$

$$T_1 n_1 + T_2 n_2 + T_5 n_5 = 0 \tag{2-19}$$

It is important to note that it would be incorrect to write $T_1 + T_2 + T_3 = 0$! Since $T_3 \neq T_5$, there will have to be a reaction torque exerted by the frame between the shafts of gears N_3 and N_5.

The output power at terminal 3 (and hence at the arm) has been specified. Therefore, torque T_5 can be calculated:

$$T_5 = \frac{-25(33{,}000)}{2\pi(650)} = -202 \text{ ft-lb (cw)}$$

The output horsepower is negative. This negativeness produces a cw output torque T_5 consistent with the ccw speed n_5.

Substituting values into Eqs. (2-18) and (2-19), we obtain

$$T_1 + T_2 - 202 = 0$$

$$T_1(700) + T_2(300) - 202(650) = 0$$

Solving these equations, we find that

$$T_1 = 177 \text{ ft-lb (ccw)} \qquad T_2 = 25 \text{ ft-lb (ccw)}$$

The powers at terminals 1 and 2 are determined from

$$P_1 = \frac{2\pi(+700)(+177)}{33{,}000} = +23.6 \text{ hp}$$

$$P_2 = \frac{2\pi(+300)(+25)}{33{,}000} = +1.4 \text{ hp}$$

Since both of these values are positive, we see that both P_1 and P_2 must be input powers! Again we note that by specifying n_1, n_3, and P_3 we have uniquely fixed all other values at the other terminals. In fact, we have verified that this epicyclic gear train satisfies the conditions for case 6 of Table 2-7. ■

We are now in a position to formulate a statement of the conditions that determine a planetary configuration. For a three-terminal device, we have nine unknowns in speed, torque, and power (n_1, n_2, n_3, T_1, T_2, T_3, P_1, P_2, P_3), of which only six are independent. There are three independent governing equations—Eqs. (2-14) through (2-16). Consequently, by specifying three independent values of the n, T, or P values (arbitrarily), we will have a completely determinate problem.

2-9 RECIRCULATING POWER

Some rather unusual circumstances can occur in some epicyclic gear trains where gears are compounded together. It is sometimes possible to design an epicyclic train in which a circulation of power within the gearbox exceeds in magnitude the power being transmitted through the gearbox. One such design is sketched in Fig. 2-24.

In the epicyclic of Fig. 2-24, we see that ring gears N_3 and N_4 are compounded as are the sun gears N_1 and N_6. This compoundness means that

$$n_3 = n_4 \quad \text{and} \quad n_1 = n_6 \tag{2-20}$$

If we apply Eq. (2-14) to each side of the gear train of Fig. 2-24, we obtain the two equations

$$\frac{n_6 - n_a}{n_4 - n_a} = -\frac{N_4}{N_6} \tag{2-21}$$

$$\frac{n_1 - 0}{n_3 - 0} = -\frac{N_3}{N_1} \tag{2-22}$$

Combining Eqs. (2-20) through (2-22) gives the speed ratio r of this epicyclic train:

$$\frac{n_1}{n_a} = r = \frac{1 + \dfrac{N_4}{N_6}}{1 - \dfrac{N_4}{N_6}\dfrac{N_1}{N_3}} \tag{2-23}$$

Equation (2-22) can also be combined with Eqs. (2-20) to give

$$n_4 = -\left(\frac{N_1}{N_3}\right) n_6 \tag{2-24}$$

The input shaft is at the left of Fig. 2-24, and the output (arm) shaft is at the right.

An analysis of the internal tooth forces for the two halves of the gear train is also shown in Fig. 2-24. In making this analysis, the following facts were used:

(a) The gear sizes were chosen such that $N_1 = 19$, $N_3 = 157$, $N_4 = 156$, and $N_6 = 20$. From Eq. (2-23), this will produce a speed ratio of $r = 157{:}1$ in which the arm will rotate ccw when the input rotates ccw.

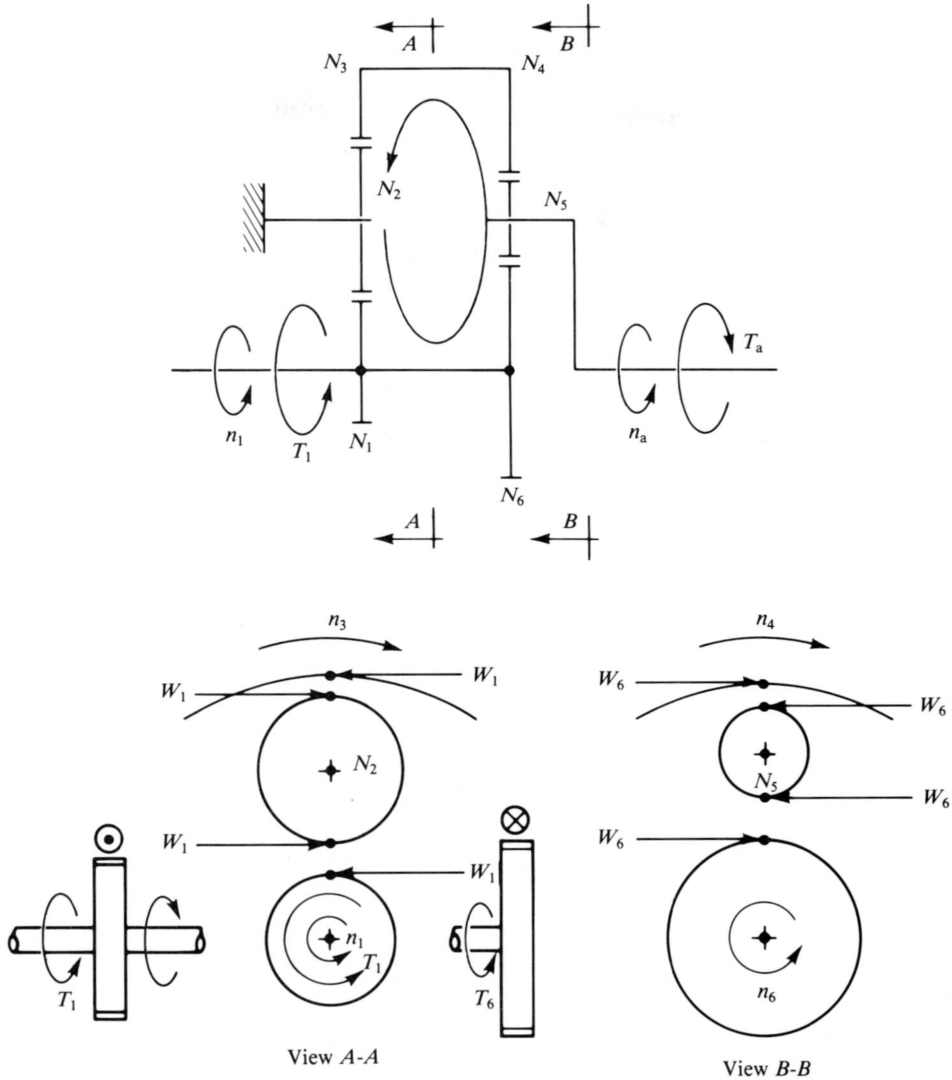

Figure 2-24. Schematic diagram for recirculating power.

(b) The external torque on the arm will be cw because power flows out of the arm terminal. This dictates that the internal tooth forces labeled W_6 be in the directions shown.

(c) The net torque on the compounded ring gears must be zero because the gears simply free-wheel. This dictates that the internal tooth forces labeled W_1 be in directions opposite to W_6.

(d) Free-body diagrams of sun gears N_6 and N_1 dictate that the torque exerted on sun gear N_1 by planet N_2 be ccw (same direction as input torque T_1) and that torque T_6 exerted on sun gear N_1 by sun gear N_6 be cw. This produces the equations

$$T_6 = T_1 + R_1 W_1 \tag{2-25}$$

$$T_6 = R_6 W_6 \tag{2-26}$$

Combining Eqs. (2-25) and (2-26) produces

$$\frac{T_1}{T_6} = 1 - \frac{N_1}{N_6}\frac{N_4}{N_3} \tag{2-27}$$

For the gear sizes chosen, Eq. (2-27) shows that

$$\frac{T_1}{T_6} = 1 - \left(\frac{19}{20}\right)\left(\frac{156}{157}\right)$$

$$\frac{T_6}{T_1} = 17.86 \tag{2-28}$$

Since gears N_1 and N_6 rotate at the same speed, Eq. (2-28) also represents the ratio of power flow through sun gear N_6 to input power P_1!

This gearbox works in the following way. For every 1 hp flowing into the gearbox through terminal 1 there is 16.86 hp flowing into sun gear N_1 from planet N_2 and 17.86 hp flowing out of sun gear N_1 to sun gear N_6. Sun gear N_6 provides 1 hp to the arm (output) and 16.86 hp to planet N_5. We see that there is an internal circulation, ccw, of 16.86 hp through the ring gears and planets. Superimposed on this is the 1 hp which proceeds directly from input to output via the sun gears.

So, although this gearbox produces an appreciable speed reduction of 157:1, it does so at the expense of a large internal horsepower requirement. If this situation is not recognized, failure of the gears may occur.

2-10 TWO-STAGE PLANETARY TRAINS

We noted previously that the maximum practical ratio for a single-stage planetary is around 10:1 (when the ring gear is fixed). Larger ratios can be obtained by simply using single stages in series similar to the way in which large ratios are obtained with worm gears by placing worm gear pairs in series. Typically, this is done with planetaries by compounding the arm of the first stage with the sun gear of the second stage. An illustration is provided in Fig. 2-25.

The epicyclic configuration of Fig. 2-25 is a four-terminal device. With the gears and terminals coded numerically as shown, we can derive the speed formulas as follows.

For the first stage, Eq. (2-14) produces the relationship

$$\frac{n_3 - n_4}{n_1 - n_4} = -\frac{N_1}{N_3} = -N_{13} \tag{2-29}$$

For the second stage, Eq. (2-14) produces

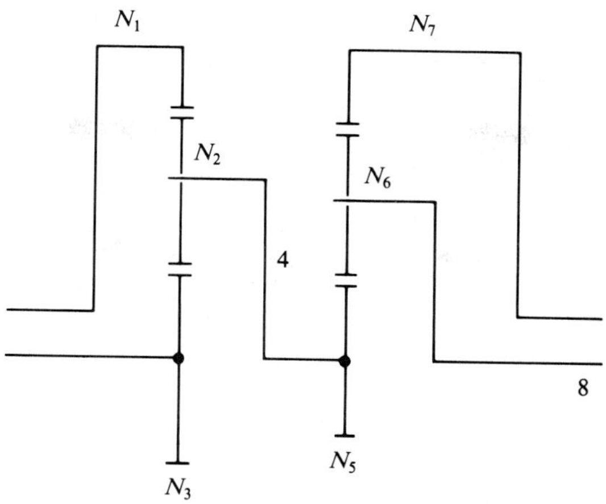

Figure 2-25. Two-stage planetary gear train.

$$\frac{n_5 - n_8}{n_7 - n_8} = -\frac{N_7}{N_5} = -N_{75} \quad \textbf{(2-30)}$$

By eliminating n_4 and n_5, we can combine Eqs. (2-29) and (2-30) to give the four-terminal speed relationship

$$n_8 = \lambda_1 n_1 + \lambda_3 n_3 + \lambda_7 n_7 \quad \textbf{(2-31)}$$

where

$$\lambda_1 = \frac{N_{13}}{\beta} \qquad \lambda_3 = \frac{1}{\beta} \qquad \lambda_7 = \frac{N_{75}(1 + N_{13})}{\beta}$$

$$\beta = (1 + N_{75})(1 + N_{13})$$

In using Eq. (2-31), three terminal speeds may be arbitrarily prescribed; the other terminal speed is then determined by Eq. (2-31).

Since all outputs are concentric, the overall torque and power relationships are

$$T_1 + T_3 + T_7 + T_8 = 0 \quad \textbf{(2-32)}$$

$$T_1 n_1 + T_3 n_3 + T_7 n_7 + T_8 n_8 = 0 \quad \textbf{(2-33)}$$

It can be seen that Eq. (2-32) is equivalent to the three equations

$$T_1 + T_3 + T_4 = 0$$

$$T_5 + T_7 + T_8 = 0$$

$$T_4 + T_5 = 0$$

EXAMPLE 2-8

Suppose we prescribe that the two ring gears of Fig. 2-25 be fixed and assume the following specific numerical data: $N_1 = 120$, $N_2 = 30$, $N_3 = 60$, $N_5 = 48$, $N_6 = 36$,

$N_7 = 120$, $n_1 = 0$, $n_7 = 0$, $n_3 = -150$ rpm (cw), $P_3 = 10$ hp (input). We would like to determine the speed ratio of the device along with the reaction torques T_1 and T_7 required to hold the ring gears at rest.

Solution

Parameters β and λ_3 are determined as follows:

$$\beta = \left(1 + \frac{120}{48}\right)\left(1 + \frac{120}{60}\right) = 10.5$$

$$\lambda_3 = \frac{1}{10.5} = 0.0952$$

$$n_8 = \lambda_1 n_1 + \lambda_7 n_7 + \lambda_3 n_3 = 0.0952(-150) = -14.28 \text{ rpm (cw)}$$

The speed ratio of this device is

$$r = \frac{n_3}{n_8} = \frac{-150}{-14.28} = 10.5$$

Note that this value is also the product of the individual rătios $r = 3$ and $r = 3.5$ of the first and second stages respectively (where their ring gears are fixed).

The input torque T_3 is determined from

$$T_3 = \frac{33{,}000(10)}{2\pi(-150)} = -350 \text{ ft-lb (cw)}$$

The output torque is determined from

$$T_8 = \frac{33{,}000(-10)}{2\pi(-14.28)} = 3675 \text{ ft-lb (ccw)}$$

In a simple single-stage planetary the ratio of ring torque to sun torque is always equal to the parameter α. For the first stage, $\alpha = 2$; therefore,

$$T_1 = 2(-350) = -700 \text{ rpm (cw)}$$

It follows from this that the arm torque on the first stage is

$$T_4 = 700 + 350 = 1050 \text{ ft-lb (ccw)}$$

This arm torque must be balanced by the sun gear torque T_5. Therefore,

$$T_5 = -T_4 = -1050 \text{ ft-lb (cw)}$$

Parameter α for the second stage equals 2.5. Consequently,

$$T_7 = \alpha T_5 = 2.5(-1050) = -2625 \text{ ft-lb (cw)}$$

As a check, the torques must also satisfy Eq. (2-32). We see that

$$T_1 + T_3 + T_7 + T_8 = -700 - 350 - 2625 + 3675 = 0$$

■

2-11 EPICYCLIC DIFFERENTIAL GEARBOXES

Two-stage or compounded planetary gear trains have been conceived in an amazing variety of configurations. One of the more versatile configurations is shown in Fig. 2-26. Note how, in this design, the carrier is a box containing the

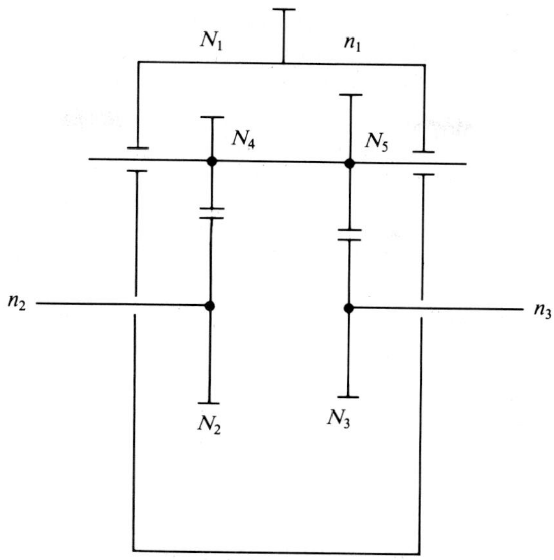

Figure 2-26. Epicyclic differential gear train.

sun gears and planets. This is a three-terminal device in which the carrier and the two sun gears are external terminals.

The speed relationship for this epicyclic is

$$\frac{n_2 - n_1}{n_3 - n_1} = \frac{N_3}{N_5}\frac{N_4}{N_2} = \gamma \tag{2-34}$$

or

$$n_2 = \gamma n_3 + (1 - \gamma)n_1 \tag{2-35}$$

The torque and power relationships are

$$T_1 + T_2 + T_3 = 0 \tag{2-36}$$

$$T_1 n_1 + T_2 n_2 + T_3 n_3 = 0 \tag{2-37}$$

By eliminating n_2 and T_3 from Eqs. (2-35), through (2-37), it is possible to show that

$$\frac{T_3}{T_2} = -\gamma \qquad \frac{T_1}{T_2} = \gamma - 1 \qquad \frac{T_1}{T_3} = \frac{1-\gamma}{\gamma} \tag{2-38}$$

For the gearbox of Fig. 2-26, parameter γ is always positive which means that T_2 and T_3 always have opposite directions. Torque T_1 will be in the same (or opposite) direction as T_3 depending on whether parameter γ is less than (or greater than) unity.

If Eqs. (2-38) are substituted into power equation (2-37), the result will be

$$\underset{(P_1)}{(1-\gamma)n_1} - \underset{(P_2)}{n_2} + \underset{(P_3)}{\gamma n_3} = 0 \tag{2-39}$$

We see that the power equation is the same as the kinematic equation (2-35).

In the examples that follow, we will assume that n_1 is ccw, that $P_1 = T_1 n_1$ is a power input, and that there is only one power output. With these restrictions, the manner in which the epicyclic gearbox of Fig. 2-26 can operate is defined partially in Table 2-9. The direction of torques T_2 and T_3 will be dictated by Eqs. (2-38), and the directions of n_2 and n_3 will depend on whether P_2 and P_3 are inputs or outputs. The vacant entries of Table 2-9 will also depend on the gearbox parameter γ.

TABLE 2-9. Possible Parameters for Figure 2-26.

Parameter	Terminal 1	Terminal 2	Terminal 3
Speed	ccw	—	—
Torque	ccw	—	—
Power	Input (+)	—	—

2-11-1 Rescaling the Range of a Speed Variator

A speed variator is a two-terminal variable-speed transmission of the type we shall discuss in Chapter 3. The input shaft of a speed variator is usually operated at constant speed. The output speed can be varied by adjusting the variator linkage (e.g., by using a knob or handwheel). The speed ratio R of a speed variator is the ratio of the maximum output speed available to the minimum output speed available. Note that this definition differs from the speed ratio of a gearbox which is defined as the input speed divided by the output speed. The speed ratio R of a speed variator is, nominally, 2:1, 4:1, or 6:1.

Case 1. $\gamma < 1$, n_3 *an input terminal*

If we specify $\gamma < 1$, then from Eqs. (2-38) we see that torque T_3 will have to be ccw and torque T_2 will be cw. With n_3 as an input terminal, it follows that n_3 will be ccw, and since terminal n_2 is now the output terminal, its speed n_2 will also have to be ccw. All of the entries of Table 2-9 will have been determined as shown in Table 2-10. They are consistent with Eqs. (2-35) and (2-39).

Suppose we let the input power terminal n_1 be provided from the output of a speed variator while input terminal n_3 is connected to a constant-speed motor, as shown in Fig. 2-27. We will define the minimum and maximum speeds provided by the speed variator as $n_1^{(1)}$ and $n_1^{(2)}$ respectively. The ratio R is then

TABLE 2-10. Parameters for Figure 2-26 ($\gamma < 1$, n_3 as Input).

Parameter	Terminal 1	Terminal 2	Terminal 3
Speed	ccw	ccw	ccw
Torque	ccw	cw ($\gamma < 1$)	ccw ($\gamma < 1$)
Power	Input (+)	Output (−)	Input (+)

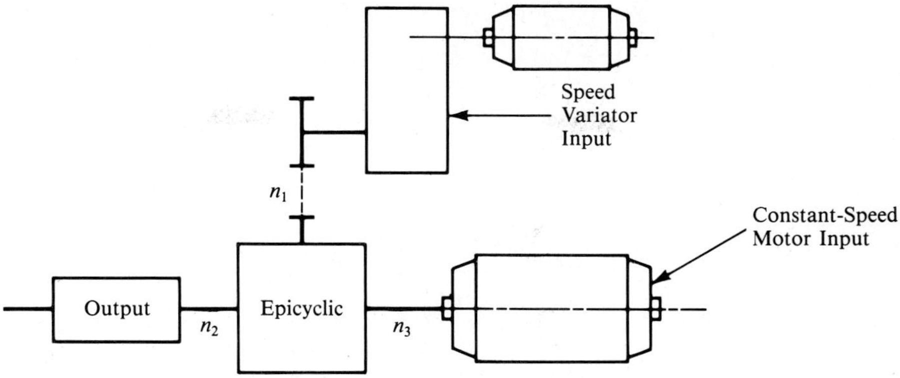

Figure 2-27. Epicyclic differential as a speed range rescaler.

$$R = \frac{n_1^{(2)}}{n_1^{(1)}} \tag{2-40}$$

The gearbox output speeds $n_2^{(1)}$ and $n_2^{(2)}$ corresponding to these two input speeds are given by Eq. (2-35), namely,

$$n_2^{(1)} = \gamma n_3 + (1 - \gamma) n_1^{(1)} \tag{2-41}$$

$$n_2^{(2)} = \gamma n_3 + (1 - \gamma) n_1^{(2)} \tag{2-42}$$

Eliminating n_3 from Eqs. (2-41) and (2-42) gives

$$\frac{n_2^{(2)} - n_2^{(1)}}{n_1^{(2)} - n_1^{(1)}} = 1 - \gamma \tag{2-43}$$

We see that the range $[n_2^{(2)} - n_2^{(1)}]$ of the output can be made very small compared to the input range $[n_1^{(2)} - n_1^{(1)}]$ by choosing γ close to unity. As an illustration, suppose that

$$\gamma = 0.98,\ n_1^{(2)} = 1720\text{ rpm},\ n_1^{(1)} = 430\text{ rpm},\ R = 4:1,\ n_3 = 1800\text{ rpm}$$

Solving Eqs. (2-41) and (2-42), we find that the output speeds are

$$n_2^{(2)} = 1798.4\text{ rpm} \qquad n_2^{(1)} = 1772.6\text{ rpm}$$

or

$$n_2 = 1785.5 \pm 12.9\text{ rpm}$$

The total range in output speed of the gearbox is 25.8 rpm for a range of 1290 rpm in the output of the speed variator. This kind of kinematic leveraging is useful in feedback control systems where speed n_2 is to be accurately controlled. Note that it requires the sun and planet gears to be nearly identical in size (e.g., $N_3/N_5 = N_4/N_2 = 0.99$).

Case 2. $\gamma > 1$, n_2 *an input terminal*

If we specify $\gamma > 1$ and let n_2 be an input terminal, then terminal n_3 must be the output terminal. Proceeding as before (but with $\gamma > 1$), we find that the directional sense of the speeds and torques will be as shown in Table 2-11.

Suppose we let the input terminal n_2 be provided from the output of the speed variator while terminal n_1 is connected to a constant-speed motor, as shown in Fig. 2-28. It is proposed that this configuration be used to control the output shaft n_3 through the value $n_3 = 0$. Now the speed equations become

$$n_2^{(1)} = \gamma n_3^{(1)} + (1 - \gamma)n_1 \tag{2-44}$$

$$n_2^{(2)} = \gamma n_3^{(2)} + (1 - \gamma)n_1 \tag{2-45}$$

Eliminating n_1 from these equations, we find

$$\frac{n_3^{(2)} - n_3^{(1)}}{n_2^{(2)} - n_2^{(1)}} = \frac{1}{\gamma} \tag{2-46}$$

TABLE 2-11. Parameters for Figure 2-26 ($\gamma > 1$, n_2 as Input).

Parameter	Terminal 1	Terminal 2	Terminal 3
Speed	ccw	ccw	ccw
Torque	ccw	ccw ($\gamma > 1$)	cw ($\gamma > 1$)
Power	Input (+)	Input (+)	Output (−)

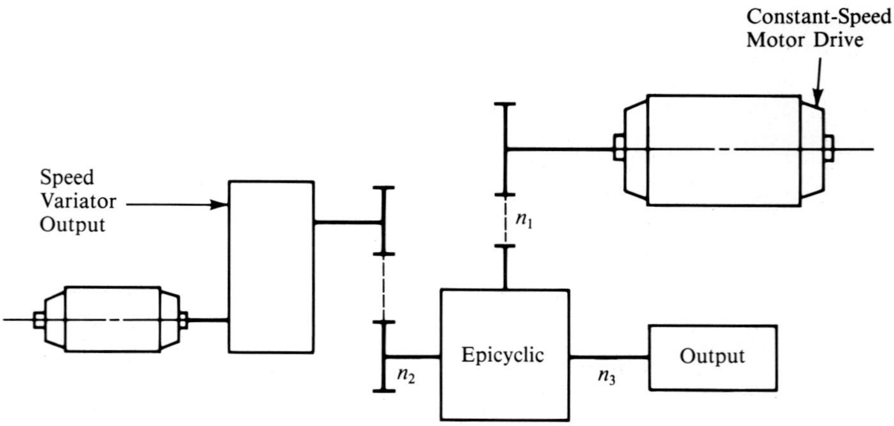

Figure 2-28. Epicyclic differential with a zero-speed output.

As an illustration, suppose we let

$$n_2^{(2)} = 2400 \text{ rpm}, \; n_2^{(1)} = 400 \text{ rpm}, \; n_3^{(2)} = 800 \text{ rpm}, \; n_3^{(1)} = 0$$

Then parameter γ can be computed to be $\gamma = 2.5$. The speed equation becomes

$$n_2 = 2.5n_3 - 1.5n_1 \tag{2-47}$$

This equation shows that when n_3 is equal to zero (and it is an output terminal), speeds n_1 and n_2 must have opposite directions, which is incompatible with the data of Table 2-11. The device simply cannot work in the manner intended when both of the terminals n_1 and n_2 are specified to be input terminals.

An analysis of this same configuration with $\gamma < 1$ will lead to the same conclusion. In fact, the epicyclic gear train of Fig. 2-26 can provide a zero speed at an output terminal only when it is operated as a device having two outputs and only one input. This condition will require placing a load on the output terminal not passing through the zero speed.

The contention we have made is illustrated by Table 2-12, in which both terminals n_2 and n_3 are outputs. Now there is a compatibility between Eq. (2-47) and Table 2-12, and terminal n_3 can be operated through zero speed. To achieve this, however, the directional rotations and correspondences between speeds n_2 and n_3 must be changed to

$$n_2^{(2)} = -2400 \text{ rpm}, \; n_2^{(1)} = -400 \text{ rpm}, \; n_3^{(2)} = 0, \; n_3^{(1)} = 800 \text{ rpm}$$

These changes keep γ positive and also satisfy the requirements of Table 2-12.

Equation (2-47) remains unchanged but now is compatible with Table 2-12. We can solve for speed n_1 by substituting $n_2^{(2)}$ and $n_3^{(2)}$ into Eq. (2-47). This substitution produces $n_1 = 1600$ rpm and Eq. (2-47) becomes

$$n_2 = 2.5n_3 - 2400 \tag{2-48}$$

So we see that in order to obtain the output speed range of zero to 800 rpm (ccw), the speed variator must act as a load. This could be arranged by placing a synchronous load on what is normally the variator input shaft (e.g., a synchronous generator). The power generated at the variator could be fed back to the motor, which drives the carrier terminal ccw at 1600 rpm.

Equation (2-39) shows that when $n_3 = 800$ rpm, the power split between the output terminals n_3 and n_2 is 5:1.

TABLE 2-12. Parameters for Figure 2-26 ($\gamma > 1$, n_2 and n_3 as Outputs).

Parameter	Terminal 1	Terminal 2	Terminal 3
Speed	ccw	cw	ccw
Torque	ccw	ccw ($\gamma > 1$)	cw ($\gamma > 1$)
Power	Input (+)	Output (−)	Output (−)

2-11-2 Correcting the Registration between Rolls

Many machines of concern to the mechanical system designer include rolls (or rollers) that must be synchronized or otherwise kept in register with one another. Control of registration can be obtained by using an epicyclic gearbox such as that shown in Fig. 2-26. The terminals would be arranged and connected as shown in Fig. 2-29.

In this arrangement, the carrier is fixed and the rolls are driven by a single motor and the gearbox as shown. The rolls can be made to rotate at the same speed by having the chain drive, which connects one of the rolls to terminal n_2, compensate for the internal gear ratio γ of the epicyclic.

Registration between the two rolls can be examined by looking at Eq. (2-35) in its angular displacement form:

$$\theta_2 = \gamma\theta_3 + (1 - \gamma)\theta_1 \tag{2-49}$$

This equation can be rewritten in the form

$$\theta_r = \theta_2 - \theta_3 = (\gamma - 1)(\theta_3 - \theta_1) \tag{2-50}$$

where θ_r is the registration between the two rolls. Imagine, for the moment, that

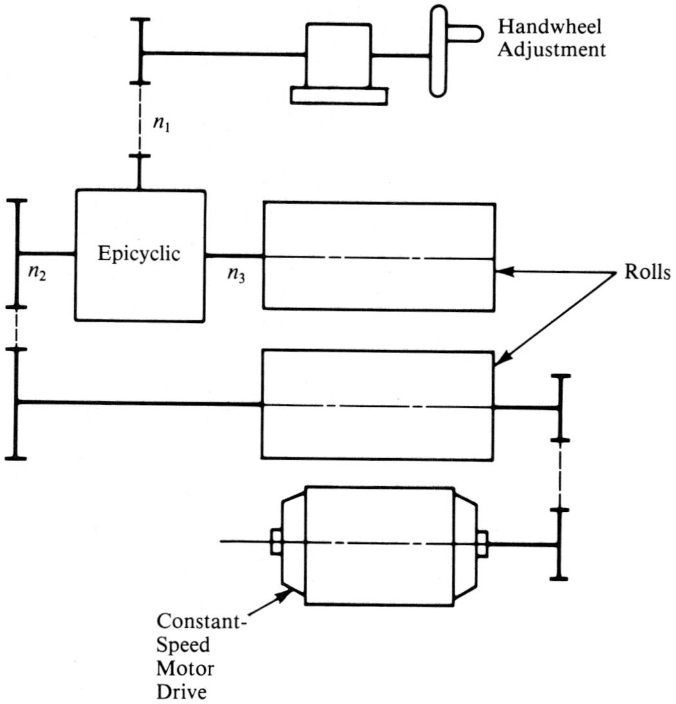

Figure 2-29. Epicyclic differential as a registration adjustor.

θ_3 is fixed. Then the registration θ_r can be changed by changing θ_1, which could be accomplished with the simple handwheel arrangement shown.

Once the proper registration is obtained in the manner indicated above, the carrier is fixed and the two rolls will maintain this registration when they are operated by the drive motor. Any loss of registration because of creep in the power transmission system (possibly from a belt drive) can be continuously corrected by using the handwheel to adjust the carrier angular position.

2-12 EPICYCLIC WHEEL OR DRUM DRIVES

In certain high-torque applications such as winch drives, or mobile earth-moving equipment involving wheel or track drives, epicyclic gearboxes offer significant advantages over conventional parallel-shaft gearboxes. Often these vehicle locomotion drives are powered by hydraulic motors because of their high power to volume ratios. These advantages are lost, however, if the hydraulic motor must be coupled to a conventional gearbox (to provide the high torque required) or is designed to produce high torque directly, in which case it becomes bulky and heavy.

Since epicyclic gearboxes are more compact and offer savings in weight and volume compared to conventional gearboxes, they are often used to provide the high torque required in winch and mobile applications even though they are more expensive. In fact, the compatibility between the epicyclic gearbox and high-speed hydraulic motor has led a few gearbox suppliers to design epicyclic gearboxes specifically for use as winch, wheel, or track drives. One such integrated drive is shown in Fig. 2-30.

The drive unit shown in Fig. 2-30 is a two-stage (or double) planetary gearbox in which the second-stage carrier is manufactured integrally with the flange upon which the wheel, or track, sprocket is to be mounted. The input to

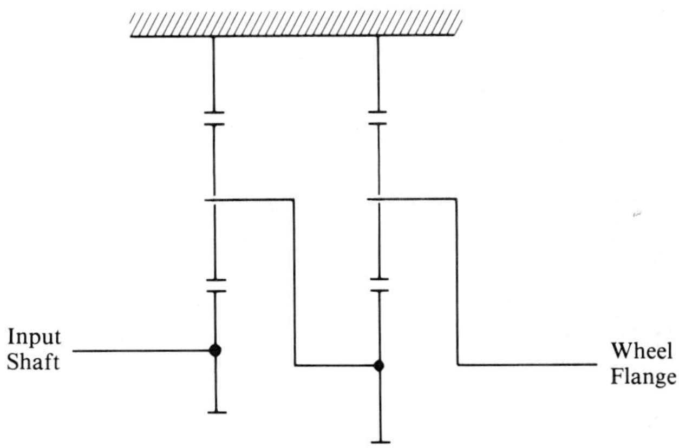

Figure 2-30. Double planetary wheel drive with the arm as the wheel flange.

TABLE 2-13. Typical Track Drive Data. (Courtesy of Funk Division, Cooper Industries.)

TRACK DRIVES

	P25T	P40T	P65T	P120T
Rated Output Torque Lb -Ft	25,000	40,000	65,000	120,000
Rated Output Speed RPM	25	20	13	9
Range of Available Ratios	37:1-79:1* 47:1-78:1**	78:1-106:1* 81:1-165:1**	85:1-115:1* 88:1-179:1**	80:1-194:1
Right Angle Input	Optional	Optional	Optional	Standard
Motor Mounting Pad Configuration	SAE C	SAE D or E	SAE D or E	SAE D or E
No. of Motor Mounting Pads	1	1	1	1
Output Configuration	Sprocket Hub	Sprocket Hub	Sprocket Hub	Sprocket Hub
Weight Pounds	439* 517**	1003* 1143**	1279* 1419**	2044
Side Load Capacity Lbs	90,000	125,000	140,000	350,000
Overall Length Inches	22.94* 25.44**	25.09* 30.46**	28.52* 33.90**	37.25
Mounting Pilot Diameter Inches	13.38	20.50	21.00	23.25
Sprocket Pilot Diameter Inches	11.00	13.75	16.00	19.00
Oil Capacity @ Level Plug) — Gals.	3/4	2 3/8	3 1/4	5 1/4

GEAR SERVICE LIFE VS. OUTPUT TORQUE

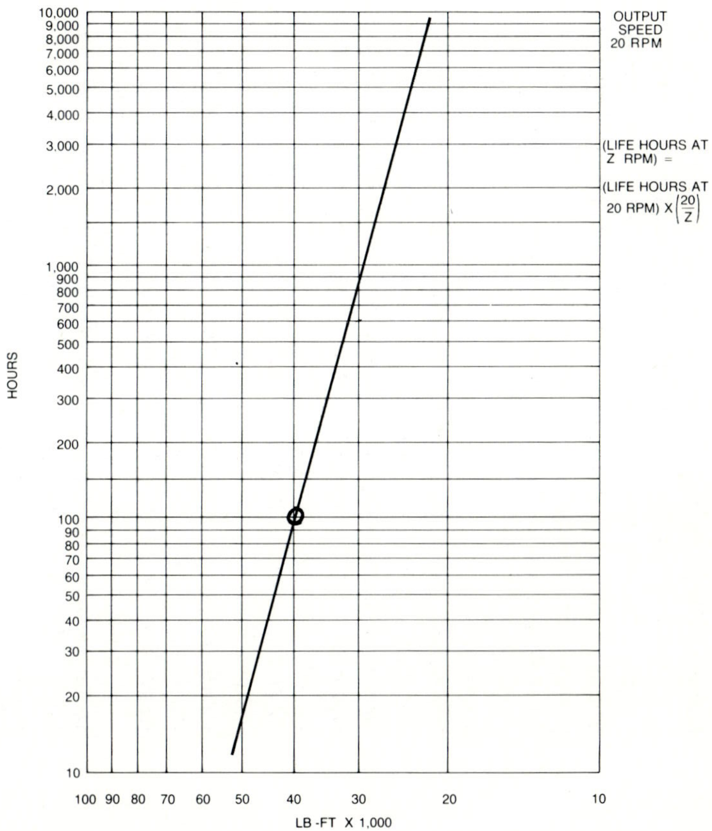

the drive is a hydraulic motor integrated directly into the epicyclic gearbox housing. The schematic diagram for this gear train is also shown in Fig. 2-30.

Typical output torques, output speeds, and gearbox ratios for one manufacturer's track drives are presented in Table 2-13. A typical gear service life versus output torque curve for the 40,000 ft-lb unit is shown beside the table. The curve is based on the surface durability of the gearing.

Alternative epicyclic configurations for wheel drives are shown in Fig. 2-31. The gear train on the left is another two-stage planetary, in which the two ring gears, rather than the second-stage carrier, are connected together and constitute the output flange upon which the wheel is to be mounted. The arrangement on the right is a differential planetary drive in which one ring gear is fixed while the second one constitutes the output wheel flange. All of these devices are two-terminal arrangements where the input terminal will generally be a hydraulic motor.

The concentric arrangement of the components of an epicyclic or planetary gear train makes it possible to insert the whole gearbox inside a winch drum or the drive pulley of a conveyor system. The input can be either an electric or a hydraulic motor operating at high speed. The drum or conveyor pulley rotates at a much lower speed depending on the gear ratio of the epicyclic gearbox inside it.

Finally, wheel and track drives are often examples of split-power epicyclic transmissions, which we will discuss in a later section. The epicyclic shown on the left of Fig. 2-31 is such a transmission. The power provided to the input sun gear is split between the planet carrier and the first ring gear. The portion flowing to the first-stage planet carrier travels through the second sun gear and then to the second ring gear, via the second-stage (fixed) planet, where it is combined with the fraction coming from the first ring gear. In this way, all of the input power eventually reaches the output wheel hub (neglecting losses). Typically, the mechanical efficiencies of these epicyclic wheel drives are around 90 to 95 percent.

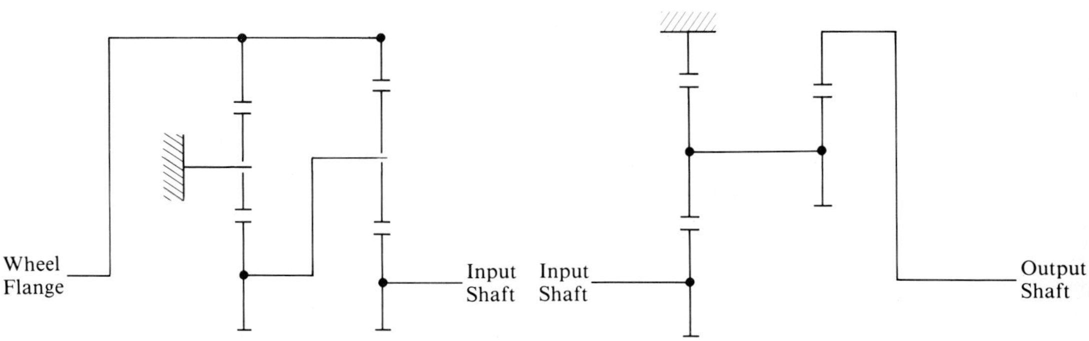

Figure 2-31. Double planetary wheel drive with ring gears as the wheel flange.

2-13 EPICYCLIC BEVEL GEAR DIFFERENTIALS

The approach developed for spur, or helical, epicyclic gear trains is also valid when the epicyclic train contains bevel gears. To illustrate this, consider the four-terminal gear train shown in Fig. 2-32.

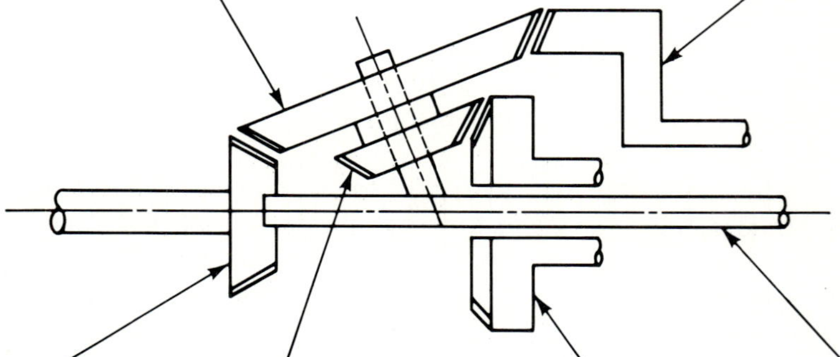

Figure 2-32. Humpage's epicyclic bevel gearing.

In this configuration, known as Humpage's gear train, we have three bevel sun gears and compounded bevel planet gears mounted in a carrier. Our speed formula Eq. (2-14) applies to this configuration as well as to those discussed earlier. If the speed formula is applied twice, the result is

$$\frac{n_1 - n_6}{n_4 - n_6} = -\left(\frac{N_4}{N_2}\right)\left(\frac{N_2}{N_1}\right) = -\frac{N_4}{N_1} \tag{2-51}$$

$$\frac{n_1 - n_6}{n_5 - n_6} = -\left(\frac{N_5}{N_3}\right)\left(\frac{N_2}{N_1}\right) \tag{2-52}$$

Note that there are, potentially, four terminals available as implied by the above two equations (n_1, n_4, n_5, n_6). Only two of the terminals can have speeds arbitrarily assigned because the simultaneous solution of Eqs. (2-51) and (2-52) will determine the other two.

EXAMPLE 2-9

Suppose that sun gear N_4 in Fig. 2-32 is held fixed while terminal 6 (the arm) has a prescribed input speed n_6. The output speeds n_1 and n_5 will then be determined from the equations

$$\frac{n_1}{n_6} - 1 = \frac{N_4}{N_1} \tag{2-53}$$

$$\frac{\dfrac{n_1}{n_6} - 1}{\dfrac{n_5}{n_6} - 1} = -\frac{N_5 N_2}{N_3 N_1} \tag{2-54}$$

Solving these equations, we find that

$$\frac{n_1}{n_6} = 1 + \frac{N_4}{N_1} \tag{2-55}$$

$$\frac{n_5}{n_6} = 1 - \frac{N_4 N_3}{N_5 N_2} \tag{2-56}$$

Assume that $N_1 = 15$, $N_2 = 60$, $N_3 = 12$, $N_4 = 75$, and $N_5 = 24$. Also, take $n_1 = 1200$ rpm (ccw). Then speeds n_6 and n_5 can be determined, from Eqs. (2-55) and (2-56), to be

$$n_6 = 200 \text{ rpm (ccw)} \qquad n_5 = 75 \text{ rpm (ccw)}$$ ■

2-13-1 Vehicle Differential Gear

An epicyclic bevel gear differential of the type found in vehicle drives is shown in Fig. 2-33. In this arrangement, the axle of (say) the back wheels is broken, with each wheel connected to a separate sun gear of the epicyclic train. The two sun gears are called side gears.

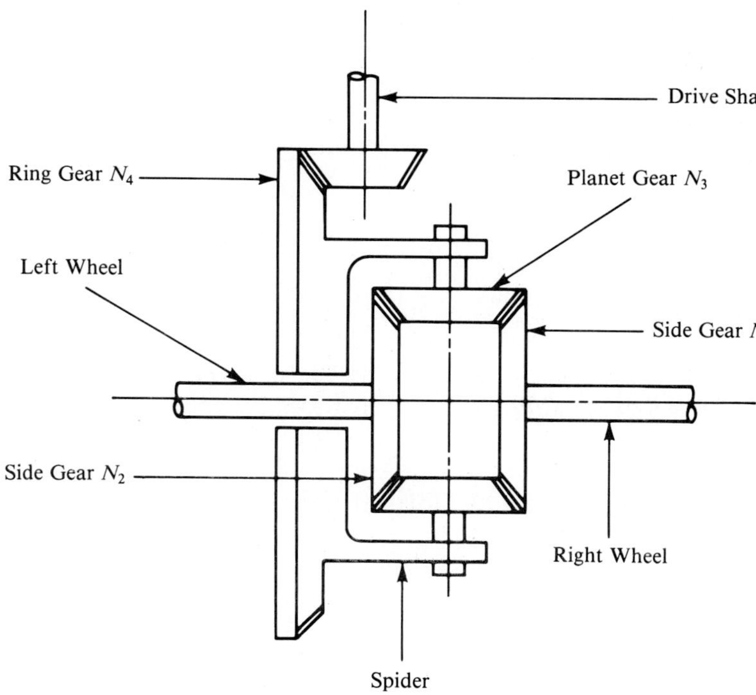

Figure 2-33. Automobile differential gear.

The bevel planet gears are mounted in a carrier (called the spider) which is rigidly connected to a gear N_4 called the ring gear. The ring gear is driven by the drive shaft, coming from the engine, through bevel pinion N_5.

The speed formula for this gear train is

$$\frac{n_1 - n_4}{n_2 - n_4} = -\left(\frac{N_2}{N_3}\right)\left(\frac{N_3}{N_1}\right) = -1 \qquad \textbf{(2-57)}$$

or

$$n_1 + n_2 = 2n_4 \qquad \textbf{(2-58)}$$

The torque and power equations are

$$T_1 + T_2 + T_4 = 0 \qquad \textbf{(2-59)}$$

$$T_1 n_1 + T_2 n_2 + T_4 n_4 = 0 \qquad \textbf{(2-60)}$$

Equations (2-58) through (2-60) can be combined to show that

$$T_1 = T_2 \qquad T_4 = -2T_1 = -2T_2 \qquad \textbf{(2-61)}$$

We see that the torques developed at the back wheels are equal, which explains why an automobile using this simple differential cannot develop traction when one of the wheels is on ice. To overcome this limitation, clutches are incorporated within the transmissions.

When $n_1 = n_2$, Eq. (2-58) will show that the whole device rotates as a rigid body with the planets not rotating. This is the situation when the vehicle is traversing a straight path. However, if the vehicle is traversing a curved path, the back wheels must rotate at slightly different speeds to prevent tire skidding and its accompanying tire wear. In this situation, let Δn_1 and Δn_2 be the slight changes in speed required to accommodate the slight difference in speed caused by the two wheels traversing different radii of turning. Then

$$n_1 + \Delta n_1 + n_2 + \Delta n_2 = 2n_4 \qquad \textbf{(2-62)}$$

or

$$\Delta n_1 = -\Delta n_2 = \Delta n \qquad \textbf{(2-63)}$$

In other words, one wheel speeds up slightly by the differential amount Δn while the other wheel slows dowm by the same amount in order to accommodate the non-skidding turning motion. Torques T_1 and T_2 will always remain equal during the turn.

2-13-2 Non-Bevel Gear Differentials

Many of the epicyclic gear trains discussed earlier can be made to exhibit the same kind of differential behavior as the vehicle bevel gear differential. As an example, consider again the gearbox of Fig. 2-26, whose speed equation (Eq. [2-34]) is

$$\frac{n_2 - n_1}{n_3 - n_1} = \gamma$$

With gears N_2, N_3, N_4, and N_5 selected such that $\gamma = 1/2$ or 2, this epicyclic will operate kinematically just as does the bevel gear differential. For example, if $\gamma = 1/2$, then Eq. (2-34) becomes

$$n_1 + n_3 = 2n_2 \tag{2-64}$$

In this arrangement, sun gear N_3 and carrier become analogous to side gears N_1 and N_2 of the bevel gear differential while sun gear N_2 becomes analogous to the bevel gear spider.

2-14 EPICYCLIC SPEED REDUCERS

Epicyclic gear trains that incorporate compounded planets and two ring gears can provide an unusually compact gear reduction unit in which a broad range of speed reduction ratios can be obtained within the same gearbox housing. One such design is shown in Fig. 2-34. In this design, the input is a sun gear. The first ring gear is fixed while the second ring gear is the output. The planet carrier floats and is not connected to the reducer housing in any way. It is claimed that this floating design facilitates equal loading of the three sets of planets mounted in the carrier.

By applying the planetary speed formula to the two planetary trains of Fig. 2-34, the carrier speed can be eliminated to produce the following expression for the speed ratio of this two-terminal device:

$$r = \frac{n_i}{n_o} = \frac{-\dfrac{N_4}{N_1}(N_3 + N_1)}{\dfrac{N_5}{N_2}(N_3 - N_4)} \tag{2-65}$$

In Eq. (2-65), the difference between the numbers of teeth in the two ring gears appears in the denominator while their product appears in the numerator. High

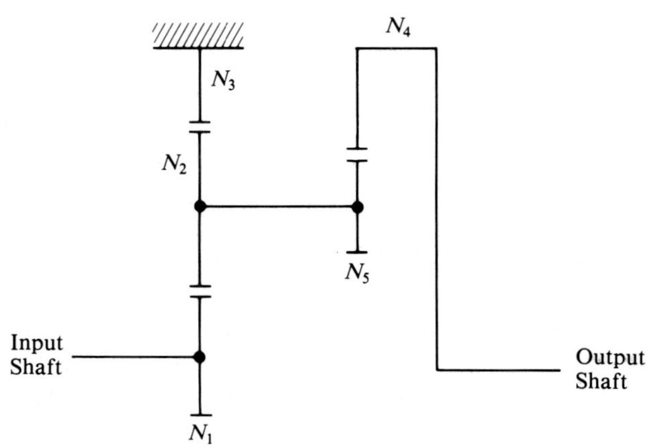

Figure 2-34. Epicyclic speed reducer.

ratios can be achieved by letting the ring gears differ by only a few teeth (or even by only one tooth). Note also that the sign of the speed ratio can be easily changed depending on whether N_3 or N_4 is the larger ring gear. It is apparent that this configuration contains a broad range of speed ratios.

EXAMPLE 2-10

Suppose the numbers of teeth for the gears of Fig. 2-34 are $N_1 = 15$, $N_2 = 43$, $N_3 = 101$, $N_4 = 99$, and $N_5 = 41$. Then the speed ratio of this reducer is

$$r = -\frac{99}{15}\left(\frac{101 + 15}{96.3 - 99}\right) = +283.55$$

If the gears are reversed so that $N_1 = 15$, $N_2 = 41$, $N_3 = 99$, $N_4 = 101$, and $N_5 = 43$, then

$$r = -\frac{101}{15}\left(\frac{99 + 15}{103.83 - 101}\right) = -271.12$$ ■

2-14-1 Application Factors

Planetary and epicyclic gear reduction units can be selected off-the-shelf by using the load class and service factor procedures outlined in Section 2-3. The service factors are the same values as tabulated in Table 2-4 for the helical gear reducer units.

2-15 MATCHING THE COMPONENT CHARACTERISTICS OF MECHANICAL SYSTEMS THAT INCLUDE RATIOING COMPONENTS

In Chapter 1 we discussed the concept of the equilibrium operating point of a mechanical system consisting of a prime mover directly coupled to a load. This was a simple matching problem of prime mover and load characteristics facilitated by the fact that both prime mover and load have the same speed when they are directly coupled.

When a ratioing component, such as a geared speed reducer, is inserted between the prime mover and the load, the ratio r of the speed reducer will be dictated by any of several criteria including specified load speed, power maximization, and torque (or force) requirements. The matching problem now becomes more complex than for a system consisting simply of a load directly coupled to the prime mover.

Suppose we are given a prime mover characteristic

$$T_m = f(\omega_m) \tag{2-66}$$

Let there be a ratioing device whose ratio r is defined by

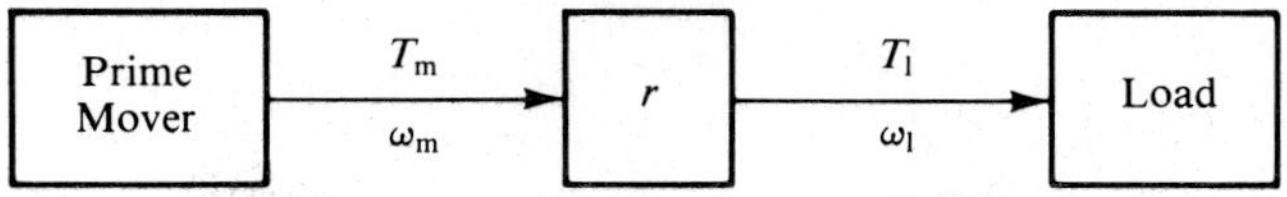

Figure 2-35. Power train with ratioing component.

$$r = \frac{\omega_m}{\omega_\ell} = \frac{T_\ell}{T_m} \tag{2-67}$$

where the subscript "m" denotes the prime mover while the subscript "ℓ" denotes the load (see Fig. 2-35).

By substituting Eqs. (2-67) into Eq. (2-66), we obtain

$$\frac{T_\ell}{r} = f(r\omega_\ell) \tag{2-68}$$

Equation (2-68) is called the prime mover characteristic as seen by the load (or referred to the load). Note that the function f is the prime mover function but the variables are the load variables T_ℓ and ω_ℓ.

Similarly, if we are given a load characteristic curve, we have

$$T_\ell = g(\omega_\ell) \tag{2-69}$$

We can change Eq. (2-69) to a form called the load characteristic as seen by the prime mover:

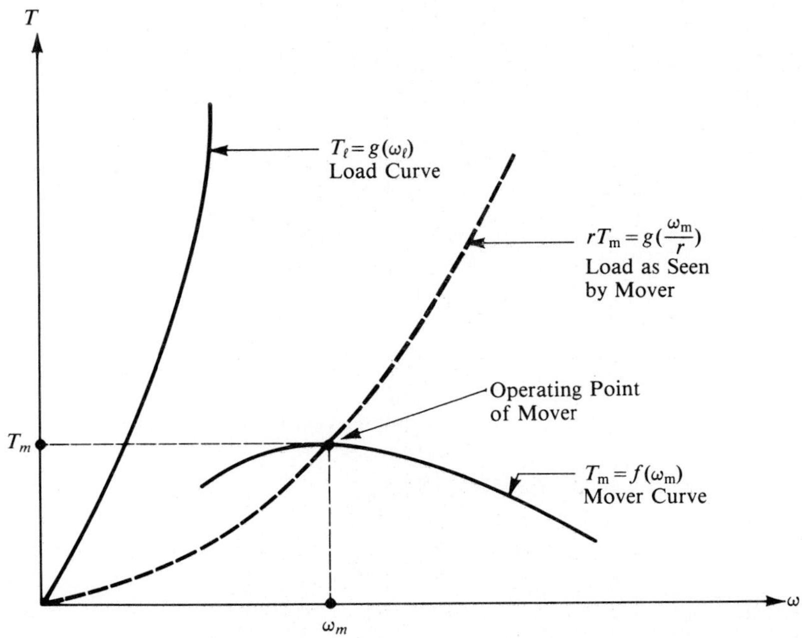

Figure 2-36. Operating point of the prime mover.

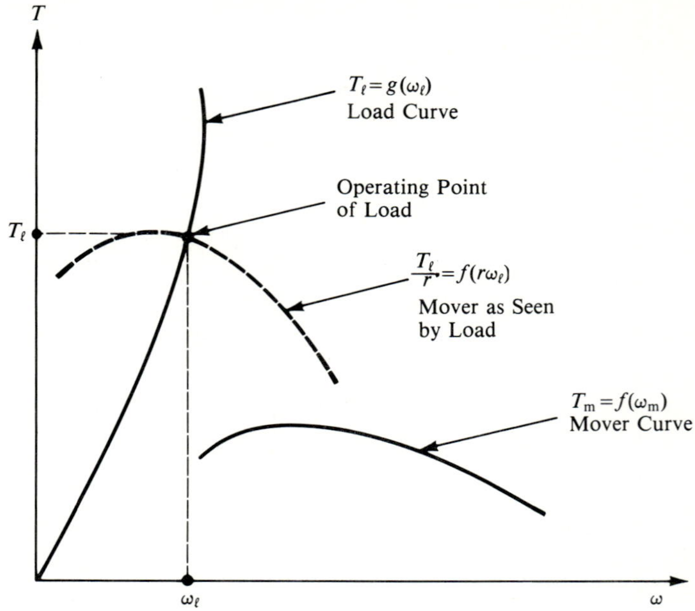

Figure 2-37. Operating point of the load.

$$rT_m = g\left(\frac{\omega_m}{r}\right) \tag{2-70}$$

The function g is the load function but the variables are now T_m and ω_m.

With a ratioing device included in the power transmission between prime mover and load, the operating points of prime mover and load will be two distinct points (when ratio r becomes unity, these operating points coalesce). We can determine these two operating points in the following ways.

The operating point of the prime mover is obtained by the intersection of the prime mover characteristic curve with the load curve as seen by the prime mover. This idea is sketched in Fig. 2-36. Similarly, the operating point of the load is obtained from the intersection of the load curve with the prime mover curve as seen by the load. This idea is sketched in Fig. 2-37. In both of these figures, the actual value of ratio r is implicitly greater than unity. The figures also show how a ratioing element may be required to match a load to a prime mover because the load and prime mover curves (by themselves) do not intersect.

2-15-1 Matching Components for Maximum Power Transmission

Suppose the prime mover and load curves are those shown as the heavy solid lines in Fig. 2-38. We want the system operating point to occur such that the power transmission will be a maximum.

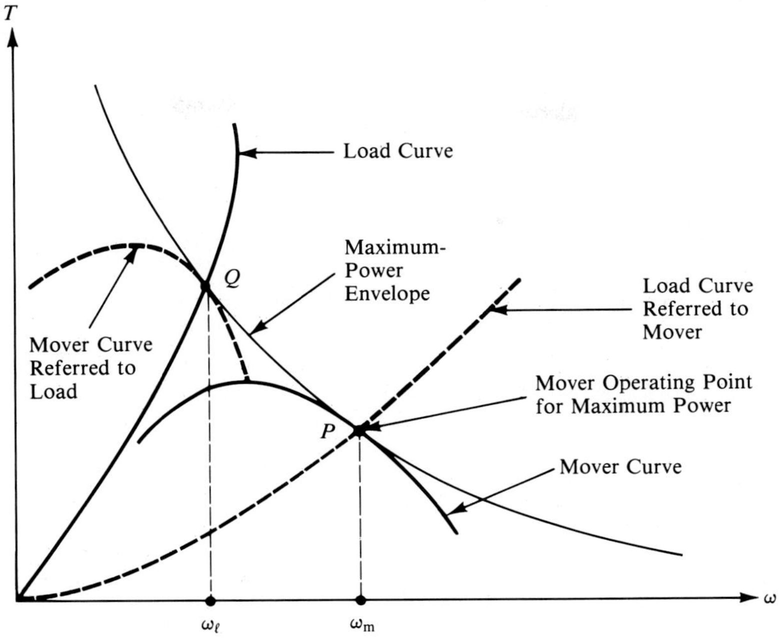

Figure 2-38. Maximizing a system's power output.

The maximum power that the prime mover can supply occurs at the point (P) of the prime mover characteristic curve tangent to the hyperbolic envelope as shown in Fig. 2-38. In other words, the set of parametric curves $T \cdot \omega = \text{constant}$ is a family of rectangular hyperbolas, and the constant will be largest for the hyperbola located farthest from the origin. This is the hyperbola just tangent to the prime mover curve at point P.

Now the ratio r of the ratioing element in the system required to achieve maximum power transmission will be defined such that the load curve, as seen by the prime mover, just passes through point P. The speed corresponding to this point is the speed at which the prime mover must run for maximum power transmission.

The speed of the load for maximum power transmission will occur at point Q, which is defined as the intersection of the load curve and the curve of the mover, as seen by the load, when the ratioing component has the requisite value of r.

EXAMPLE 2-11

Suppose a certain vehicle has a tractive effort F versus speed characteristic V at its driving wheels defined by

$$F = 135 - 0.117(V - 31)^2 \text{ lb}$$

The speed V is in mph. This characteristic can be viewed as the vehicle mover characteristic which already incorporates some ratioing elements characterized by the ratio r_o.

The load characteristic for this vehicle for horizontal travel (as dictated by aerodynamic drag and bearing friction) is given by

$$F = 0.035V^2 \text{ lb}$$

The mover and load characteristics are shown as the heavy solid curves in Fig. 2-39.

(a) At what speed will this vehicle travel with its present power train and load demand?
(b) What is the maximum power this vehicle can develop and at what vehicle speed would this power be realized with this particular power train?
(c) What is the highest speed this vehicle is capable of traveling with its present engine and how must the power train be modified to achieve this speed?
(d) If the ratio of the ratioing components in the power transmission is simply halved in an attempt to double the vehicle speed for a given engine speed, what will actually happen to the vehicle speed?

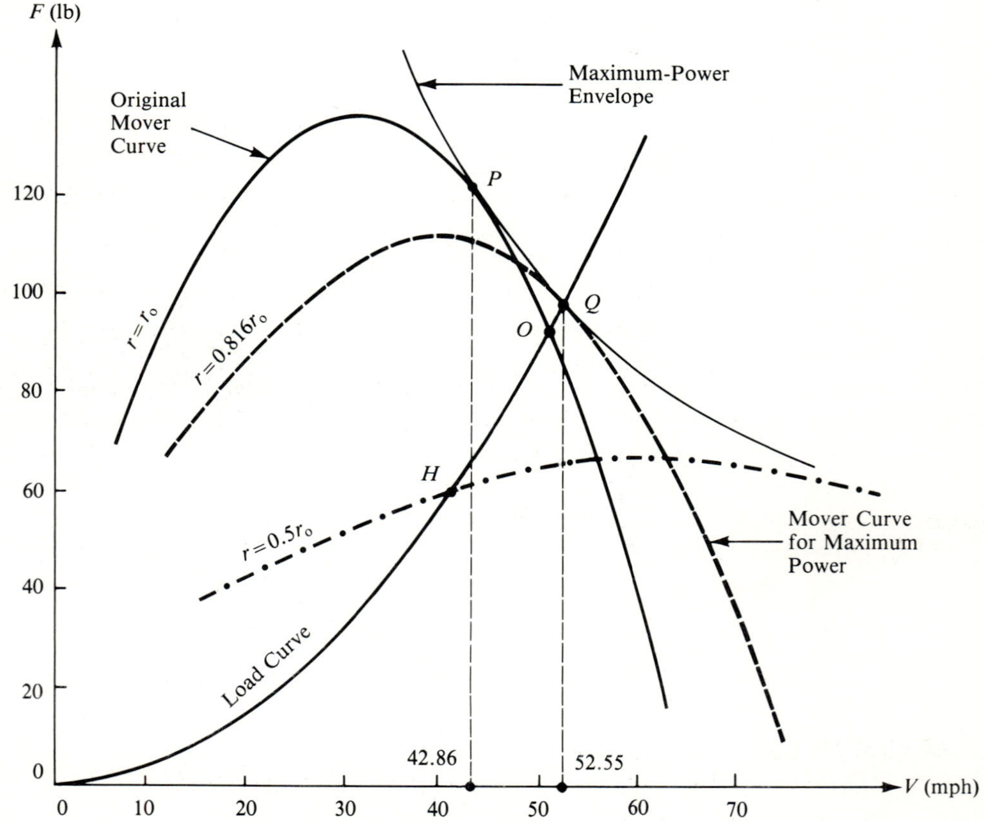

Figure 2-39. Characteristic curves for Example 2-11.

Solution

(a) The speed at which this vehicle will travel with its original power train is obtained by solving mover and load curves simultaneously. In other words, we solve for V in the equation

$$135 - 0.117(V - 31)^2 = 0.035V^2$$

We find that the operating point O occurs at a speed of $V = 50.65$ mph and that the corresponding tractive effort is $F = 89.8$ lb.

(b) The power available to the load is given by

$$P = F \cdot V = [135 - 0.117(V - 31)^2] \cdot V$$

Differentiating this expression with respect to speed and setting the result to zero shows that the maximum power that can be developed is $P =$ 5081 lb-mph and the maximum power would require a vehicle speed of 42.86 mph which could be realized only if the load were increased. This corresponds to point P of Fig. 2-39.

(c) The power extracted by the load is given by

$$P = F \cdot V = [0.035V^2] \cdot V$$

Insofar as the load is concerned, the maximum power available would make possible a vehicle speed of

$$0.035V^3 = 5081$$

or

$$V = 52.55 \text{ mph}$$

This is the highest speed the vehicle is capable of achieving with its current engine. But, achieving it would require that an additional ratio r be incorporated into the power train, where

$$r = \frac{42.86}{52.55} = 0.816$$

Consequently, the overall ratio of the ratioing components must be changed from r_o to $0.816r_o$. This has the effect of changing the original mover curve to the curve shown dashed in Fig. 2-39. The dashed curve intersects the load curve at point Q which is also tangent to the maximum-power envelope. With this change in r, the operating point will move from point O to point Q.

(d) If the ratio r_o is simply halved, the mover curve will become the one shown as the dash-dot line of Fig. 2-39. We see that this curve intersects the load curve at a value of approximately $V = 40$ mph (see point H). Consequently, the operating point moves from point O to point H and our attempt to increase speed is defeated. The speed actually will drop from $V =$ 50.65 mph to about $V = 40$ mph. ■

2-16 SUMMARY

One of the purposes of this book is to expose the variety of ways for connecting a prime mover to a load. When a ratioing component is needed to accomplish the connection, the designer must ensure that the ratioing component can safely transmit the speed and torque and that this is done economically in both dollar and space terms. Geared transmissions are a very old, reliable, positive, and economical means of accomplishing speed and torque changes between the load and its prime mover.

The basic concepts of geared speed reduction can be traced back to the Greeks before the time of Christ. Leonardo da Vinci described the basic concept of bevel gears among his many ingenious engineering artifices. And, with the advent of the steam engine, iron gears with cast teeth began to replace the wooden geared systems of antiquity. For a history of geared transmissions, see Reference 2-9.

In this chapter we have presented a summary of the myriad geared speed reducer configurations that are commercially available. These devices are available in ratings from fractional horsepower to several thousand horsepower and are used in applications ranging from consumer products and industrial machinery to military uses.

2-17 REFERENCES

2-1. *The Speed Reducer Book*, published by Winsmith, Division of UMC Industries, Inc., Springville, NY, 1980.

2-2. *Link-Belt and Stearns Power Transmission Products*, Catalog LB 1300, published by PT Components, Inc., 7545 Rockville Road, Box 802, Indianapolis, IN, 1982.

2-3. *Browning Power Transmission Equipment*, Catalog Number 9, published by Browning Manufacturing Division, Emerson Electric Co., Maysville, KY, 1980.

2-4. *Practice for Enclosed Speed Reducers*, AGMA 420.04, Dec., 1975, published by American Gear Manufacturers Association, 1901 North Fort Myer Drive, Arlington, VA, 1975.

2-5. *Power Transmission Design Handbook*, published by Power Transmission Design, 111 Chester Avenue, Cleveland, OH.

2-6. Peter Lynwander, *Gear Drive Systems*, Marcel-Dekker, New York, 1983.

2-7. R. J. Ferguson, "Short Cuts for Analyzing Planetary Gearing," *Machine Design*, May 26, 1983, pp. 55–58.

2-8. *Mechanical Drives Reference Issue*, *Machine Design*, June 28, 1984.

2-9. Darle W. Dudley, *The Evolution of the Gear Art*, American Gear Manufacturers Association (AGMA), 1969.

PROBLEMS

2-1. From the overhung load factor of 1.5 for a V-belt drive, determine the tension in the slack side of the V-belt as a fraction of the tension in the taut side. (*Answer*: $T_s = 0.20T_t$)

2-2. A hoist system for launching and recovering boats must be able to lift 20,000 lb at a speed of 20 ft/min. The hoist drum has a pitch diameter of 24 inches and is driven from the output shaft of a gear reducer (of the type shown in Table P2-2 on pages

98–99) by a 4.2:1 (reduction) chain drive. The sprocket on the output shaft has a pitch diameter of 15 inches. The reducer is driven at its input by a directly coupled 1150 rpm electric induction motor. This application has a load class of MS and a duty cycle of 10 hours per day. Make a sketch and determine

(a) The size of the reducer (*Answer*: HT-1566)

(b) The overhung load acting on the reducer output shaft (*Answer*: approximately 9524 lb)

(c) The overhung load rating of the chosen reducer (*Answer*: 10,730 lb)

2-3. See Table P2-2. Determine the mechanical efficiency of the HT-1566 reducer having a ratio of 86.5. (*Answer*: 100 percent to within round-off errors)

2-4. Determine the output torque, ratio, and overhung load capacity required to select a speed reducer to meet the following conditions. The driven machine is a car puller cable drum rotating at 50 rpm and requiring 4.5 hp. The service factor is to be 1.5. The prime mover is a 1750 rpm electric motor directly coupled to the reducer input shaft. The reducer output shaft is coupled to the cable drum with an RC-80 roller chain having a 15T (4.81 inch pitch diameter) sprocket on the reducer output shaft and a 35T sprocket on the cable drum shaft. (*Answers*: 3646 in-lb; 15:1; 1516 lb)

2-5. The arrangement of power transmission components shown in Fig. P2-5 is proposed in order to operate a load (24 hours per day) intermittently while the motor runs continuously. The system is started by engaging the clutch and is stopped by simultaneously disengaging the clutch and applying the brake. The motor runs at 1750 rpm and provides 20 hp. When the load is running, it does so at 11 rpm. Assume that the load class is MS. Determine

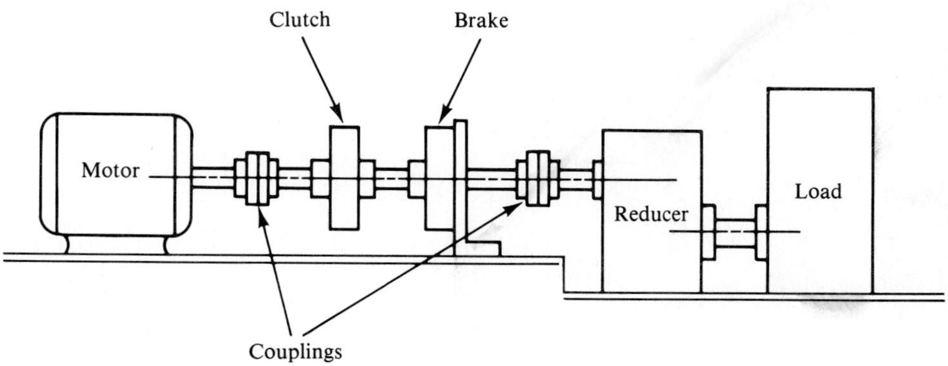

Figure P2-5. Figure for Problem 2-5.

(a) A suitable speed reducer (see Table P2-5 on page 100) for the configuration shown (*Answer*: HT-2250, 159:1)

(b) The mechanical efficiency of the chosen reducer (*Answer*: 98.6 percent)

2-6. The debarker drum in a lumber mill is to be directly coupled to the output of a parallel-shaft reducer (see Fig. P2-6). The input shaft of the reducer is coupled to a 1750 rpm induction motor by a V-belt drive that has a 4.115 inch pitch diameter sheave on the motor and a 10 inch pitch diameter sheave on the reducer input shaft. The debarker drum operates at 28 rpm and, nominally, requires 9500 ft-lb of torque. The system is to operate 3 to 10 hours per day. Select a double-reduction speed reducer from Table P2-6 on page 101.

TABLE P2-2. Tables for Problem 2-2. (Courtesy of PT Components, Inc.)

Rating Tables

Triple reduction/1150 RPM/Type HT cast iron housing or Type HTS welded steel housing

Nominal ratio ▲	Output speed, RPM	Rating description ▫	Speed reducer size, Type HT or HTS													
			1000	1028	1300	1350	1500	1566	1650	1700	1750	2000	2050	2100	2250	2300
31.4	36.6	**Mechanical HP**	**12.8**	**14.7**	**23.4**	**27.0**	**35.7**	**43.0**	**50.2**	**55.4**	**60.7**	**74.5**	**82.6**	**95.5**	**107**	**117**
		Output torque	22070	25300	40040	46085	61685	73925	85860	94810	103785	128660	142670	164865	184450	202175
34.7	33.1	**Mechanical HP**	**11.7**	**13.4**	**21.2**	**24.4**	**32.4**	**39.0**	**45.6**	**50.4**	**55.2**	**68.0**	**75.4**	**87.5**	**97.5**	**107**
		Output torque	22185	25440	40290	46375	62110	74430	86465	95470	104515	129605	143720	166760	185800	203650
38.4	29.9	**Mechanical HP**	**10.4**	**12.0**	**19.2**	**22.5**	**29.3**	**35.2**	**41.4**	**45.7**	**50.0**	**61.0**	**67.6**	**78.1**	**87.5**	**96.0**
		Output torque	22325	25595	40540	47505	62525	74930	87065	96135	105235	130685	144915	167460	187505	205525
42.5	27.0	**Mechanical HP**	**9.60**	**11.0**	**17.4**	**20.1**	**26.6**	**32.0**	**37.7**	**41.6**	**45.6**	**56.0**	**62.1**	**72.0**	**80.4**	**88.0**
		Output torque	22425	25710	40770	46925	62900	75385	87610	96735	105895	131495	145815	168500	188735	206870
47.1	24.4	**Mechanical HP**	**8.80**	**10.0**	**16.0**	**18.7**	**24.4**	**29.3**	**34.4**	**38.0**	**41.6**	**51.2**	**56.8**	**65.6**	**73.5**	**80.6**
		Output torque	22525	25830	40970	47930	63235	75785	88130	97310	106525	132310	146720	169540	189970	208220
52.1	22.1	**Mechanical HP**	**8.00**	**9.13**	**14.5**	**16.7**	**22.2**	**26.6**	**31.2**	**34.5**	**37.8**	**46.6**	**51.7**	**60.0**	**67.0**	**73.4**
		Output torque	22630	25945	41180	47405	63600	76215	88645	97875	107145	133136	147630	170595	191215	209590
57.7	19.9	**Mechanical HP**	**7.20**	**8.30**	**13.0**	**15.0**	**20.0**	**24.0**	**28.3**	**31.2**	**34.2**	**42.3**	**47.0**	**54.2**	**61.0**	**66.6**
		Output torque	22735	26065	41400	47655	63960	76650	89160	98445	107765	133960	148545	171650	192460	210955
70.6	16.3	**Mechanical HP**	**6.00**	**6.80**	**11.0**	**12.5**	**16.6**	**20.0**	**23.3**	**25.8**	**28.2**	**35.0**	**38.7**	**45.0**	**50.2**	**55.0**
		Output torque	22925	26285	41780	48510	64580	77390	90090	99475	108885	135480	150230	174640	194770	213485
86.5	13.3	**Mechanical HP**	**5.00**	**5.60**	**9.00**	**10.2**	**13.3**	**16.0**	**19.2**	**21.2**	**23.3**	**28.1**	**31.2**	**36.0**	**37.0**	**40.4**
		Output torque	23095	26480	42120	48480	63480	76075	90945	100420	109930	133365	147890	170895	174840	191640
95.7	12.0	**Mechanical HP**	**4.50**	**5.13**	**8.10**	**9.37**	**12.4**	**15.0**	**17.5**	**19.3**	**21.1**	**25.6**	**28.4**	**33.0**	**33.6**	**37.0**
		Output torque	23165	26560	42275	48805	65425	78410	91350	100865	110420	134250	148870	172025	176320	193260
106	10.8	**Mechanical HP**	**4.01**	**4.60**	**7.33**	**8.44**	**11.3**	**13.5**	**16.0**	**17.5**	**19.2**	**23.4**	**26.0**	**30.0**	**30.6**	**33.5**
		Output torque	23250	26660	42435	48845	65690	78720	91730	101290	110880	135275	150000	173345	177730	194805

TABLE P2-2 (continued)

Overhung Load Ratings

Triple reduction/Type HT cast iron housing

Nominal ratio	Overhung load ratings of low speed shafts, pounds ▲										
	Speed reducer size										
	HT 1000 HT 1028	HT 1300 HT 1350	HT 1500 HT 1566	HT 1650 HT 1700 HT 1750	HT 2000 HT 2050 HT 2100	HT 2250 HT 2300 HT 2350	HT 2450 HT 2500	HT 2900 HT 2950 HT 3000	HT 3400 HT 3450 HT 3500	HT 4100 HT 4150 HT 4200	HT 4900 HT 4950 HT 5000
1750, 1150, 870, 580 RPM											
All nominal ratios	5700	8380	10730	13730	17380	21060	26650	35190	45650	47290	74750

TABLE P2-5. Table for Problem 2-5. (Courtesy of PT Components, Inc.)

Rating Tables

Triple reduction/1750 RPM/Type HT cast iron housing or Type HTS welded steel housing

Nominal ratio ▲	Output speed, RPM	Rating description □	Speed reducer size, Type HT or HTS													
			1000	1028	1300	1350	1500	1566	1650	1700	1750	2000	2050	2100	2250	2300
117	15.0	**Mechanical HP**	**5.50**	**6.30**	**9.00**	**10.3**	**13.1**	**16.0**	**21.3**	**23.5**	**25.8**	**31.5**	**35.0**	**40.4**	**41.3**	**45.3**
		Output torque	22990	26360	37590	43265	55220	66180	88955	98220	107520	131810	146160	168895	173460	190125
130	13.5	**Mechanical HP**	**5.00**	**5.70**	**9.00**	**10.3**	**12.2**	**15.0**	**18.0**	**19.7**	**21.6**	**26.4**	**29.3**	**34.0**	**37.0**	**40.5**
		Output torque	23080	26465	41330	47575	56510	67720	82140	90695	99285	121615	134855	155835	170315	186680
144	12.2	**Mechanical HP**	**4.00**	**4.50**	**7.02**	**8.08**	**11.7**	**14.0**	**16.0**	**17.7**	**19.4**	**22.8**	**25.3**	**29.2**	**35.7**	**39.1**
		Output torque	19995	22925	35910	41335	59660	71500	82350	90930	99540	116760	129475	149615	180850	198230
159	11.0	**Mechanical HP**	**3.71**	**4.25**	**6.84**	**7.87**	**10.4**	**12.4**	**14.6**	**16.2**	**17.7**	**21.7**	**24.1**	**28.0**	**30.4**	**33.4**
		Output torque	20990	24070	38710	44555	58800	70470	83310	91990	100700	123325	136755	158025	171750	188250
176	9.9	**Mechanical HP**	**3.00**	**3.40**	**5.65**	**6.50**	**8.30**	**10.0**	**11.8**	**13.0**	**14.2**	**18.0**	**20.0**	**23.0**	**26.1**	**28.7**
		Output torque	18650	21385	35600	40975	52420	62820	74570	82340	90140	112540	124795	144205	163790	179530
195	9.0	**Mechanical HP**	**2.70**	**3.01**	**4.82**	**5.55**	**7.50**	**9.00**	**10.7**	**11.8**	**13.0**	**15.7**	**17.4**	**20.1**	**22.7**	**25.0**
		Output torque	18755	21505	33780	38880	52480	62895	75100	82925	90780	109815	121770	140715	158635	173880
215	8.1	**Mechanical HP**	**2.43**	**2.80**	**4.11**	**4.73**	**6.40**	**7.63**	**9.80**	**10.8**	**11.8**	**13.5**	**15.0**	**17.3**	**21.3**	**23.3**
		Output torque	18850	21615	31820	36625	49400	59205	75570	83440	91345	103865	115175	133095	165170	181040

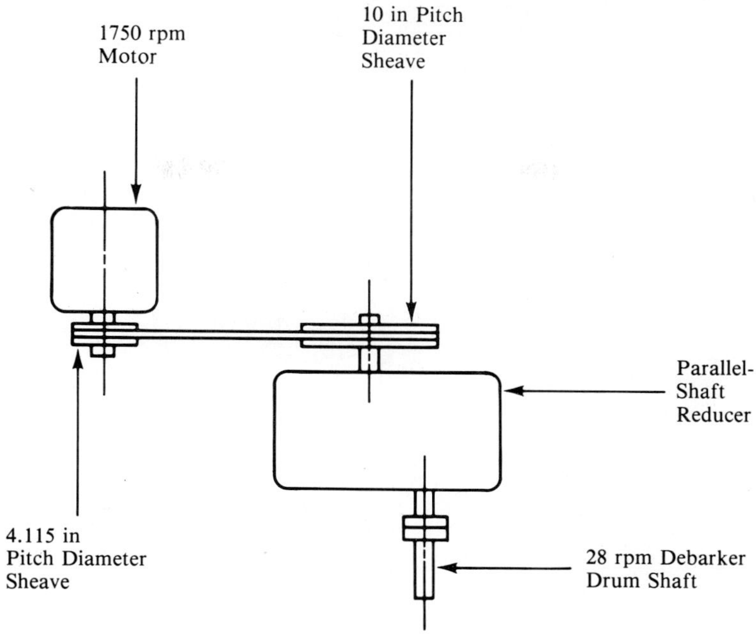

Figure P2-6. Figure for Problem 2-6.

TABLE P2-6. Table for Problem 2.6. (Courtesy of Browning Mfg. Div., Emerson Electric Co.)

720 RPM Driver — Double Reduction

Nominal Ratio	Exact Ratio	Nominal Output RPM	Input HP		Output Torque In.-Lbs.	Maximum■ Overhung Load-Lbs.	
			Mechanical	Thermal		Output Shaft	Input Shaft
5010 Reducer							
9.30	9.427	77	132	—	108922	16314	1173
11.39	11.488	63	106	—	106593	16314	1420
13.95	14.037	51	87.5	—	107508	16314	1435
17.09	17.053	42	75.0	—	111531	16314	1501
20.93	21.199	34	60.1	—	111532	16314	1564
25.63	25.674	28	50.1	—	112603	16314	1625
31.39	30.797	22	40.5	—	109194	16314	1731
38.44	38.372	18.7	31.6	—	106137	16314	2193
5011 Reducer							
9.30	9.194	77	190	—	153353	19720	1005
11.39	11.204	63	151	—	148096	19720	1836
13.95	13.690	51	126	—	150996	19720	1985
17.09	16.632	42	101	—	147039	19720	2448
20.93	20.676	34	81.0	—	146708	19720	2827
25.63	25.040	28	67.6	—	148327	19720	2975
31.39	31.597	22	54.5	—	150985	19720	3461
38.44	38.266	18.7	45.0	—	150729	19720	3461
5012 Reducer							
9.30	9.239	77	262	—	211898	20696	1897
11.39	11.527	63	203	—	204733	22944	2417
13.95	13.963	51	165	—	202685	24930	2693
17.09	16.231	42	143	—	203171	25653	2514
20.93	20.538	34	113	—	206748	25653	2828
25.63	25.667	28	91.2	—	204899	25653	3063
31.39	31.815	22	75.1	—	209149	25653	3387
38.44	38.739	18.7	60.1	—	203788	25653	3725

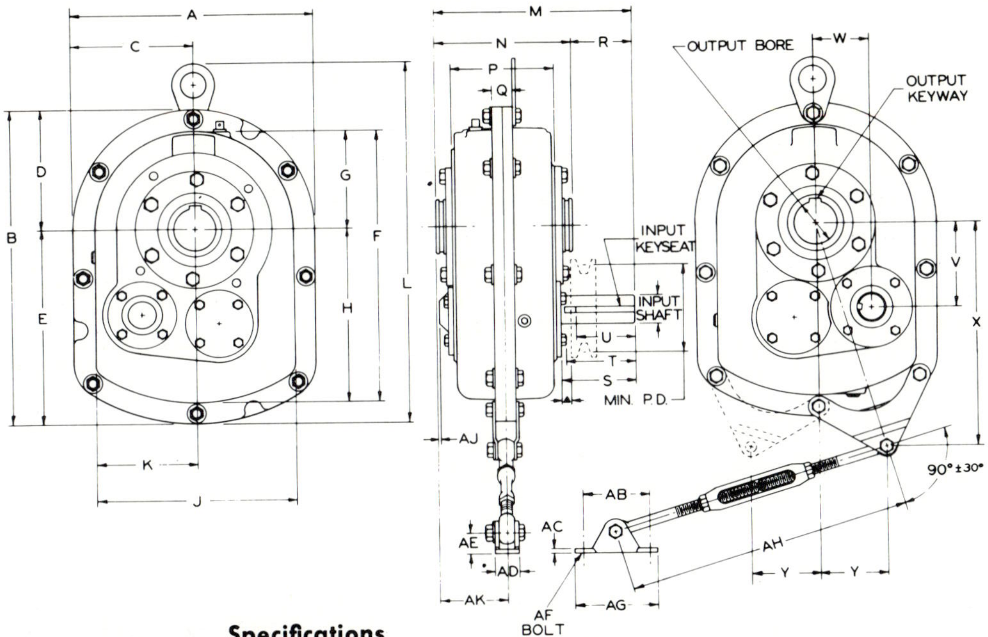

Specifications

Table No. 1

Part No. ★	Dimensions																
	A	B	C	D	E	F	G	H	J	K	L	M	N	P	Q	R	S
104SM	8 1/8"	10 1/8"	4 1/16"	4 1/16"	6 1/16"	8 1/2"	3 1/4"	5 1/4"	6 1/2"	3 1/4"	12"	7 17/32"	5 15/16"	4 1/16"	3/4"	1 19/32"	2 3/16"
107SM	9 1/16	11 23/32	4 17/32	4 17/32	7 3/16	10 3/32	3 23/32	6 3/8	7 7/16	3 23/32	13 19/32	8 5/16	6 5/16	4 1/4	3/4	2	2 23/32
115SM	11 1/8	14 3/8	5 9/16	5 9/16	8 13/16	12 1/16	4 13/32	7 21/32	8 13/16	4 13/32	17 1/4	9 7/32	6 13/16	4 5/8	1	2 13/32	2 15/16
203SM	12 7/8	16 3/4	6 7/16	6 7/16	10 5/16	14 1/4	5 3/16	9 1/16	10 3/8	5 3/16	19 3/8	10 13/16	7 11/16	5 3/8	1 1/4	3 1/8	3 15/16
207SM	14	18 7/16	7	7	11 7/16	15 15/16	5 3/4	10 3/16	11 1/2	5 3/4	21 1/16	11 11/16	8 1/4	5 7/8	1 1/4	3 7/16	4 1/8
215SM	16 7/8	22 1/4	8 7/16	8 7/16	13 13/16	19 1/4	6 15/16	12 5/16	13 7/8	6 15/16	24 13/16	12 15/16	9 9/16	6 13/16	1 1/2	3 3/8	4 1/4
307SM	17 3/4	23 5/16	8 7/8	8 7/8	14 7/16	20 1/16	7 1/4	12 13/16	14 1/2	7 1/4	25 3/4	15 7/8	11 5/16	8 5/16	1 1/2	4 9/16	5 1/8
315SM	17 3/4	23 5/16	8 7/8	8 7/8	14 7/16	20 1/16	7 1/4	12 13/16	14 1/2	7 1/4	25 3/4	16 3/4	12 3/8	9 1/8	1 1/2	4 3/8	5 3/16
407SM	22	29 1/8	11	11	18 1/8	25 5/8	9 1/4	16 3/8	18 1/2	9 1/4	31 7/16	19 1/16	13 9/16	10 13/64	1 1/2	5 1/2	6 11/16
415SM	23 1/2	31	11 3/4	11 3/4	19 1/4	27 1/2	10	17 1/2	20	10	33 5/16	21 1/2	14 5/16	10 15/16	1 1/2	7 3/16	8 1/4
507SM	27 1/4	35 1/4	13 5/8	13 5/8	21 5/8	31 3/4	11 7/8	19 7/8	23 3/4	11 7/8	37 9/16	22 21/32	15 1/8	11 11/16	1 1/2	7 17/32	8 3/8

Exact Ratios — Stock Reducers

Table No. 2

Part No. ★	RATIO SYMBOL			
	Single Reduction	Double Reduction		
	05	09	15	25
104SM	4.867	8.760	14.532	24.840
107SM	5.059	8.830	14.689 ●	24.188 ●
115SM	5.125	8.786	14.573 ●	25.563 ●
203SM	5.063	8.685	14.900 ●	24.652 ●
207SM	5.056	8.611	14.615 ●	24.995 ●
215SM	4.889	8.602	14.899 ●	24.857 ●
307SM	4.947	8.460	14.771 ●	24.737 ●
315SM	4.733	8.635	14.564 ●	25.755 ●
407SM	—	—	15.012	24.997
415SM	—	—	14.926	24.990
507SM	—	—	14.800	24.777
608SM	—	—	15.275	25.556

●THESE SIZES CARRIED IN STOCK FOR VERTICAL OPERATION ALSO. ADD "V" AFTER "SM" IN PART NO., FOR EXAMPLE 107SMV15.

Figure P2-7. Figure and table for Problem 2-7. (Courtesy of Browning Mfg. Div., Emerson Electric Co.)

2-7. A double-reduction, 25:1, model 207SM shaft-mounted reducer is to be applied as indicated in Fig. P2-7. The reducer is mounted on a screw conveyor shaft rotating at 60 rpm to which 10 hp are transmitted. The prime mover is an induction motor running at 1750 rpm under the load. Assume that the V-belt sheave mounted on the reducer input shaft has a pitch diameter of 6.3 inches. The reducer is held fixed by the torque arm which incorporates an overload release switch. Use a service factor of 1.0.

(a) Determine the overhung load on the reducer input shaft. Assume the overhung load factor is 1.5. (*Answer*: approximately 200 lb)

(b) Determine the motor sheave diameter. (*Answer*: 5.4 inches)

(c) Assume that the reducer input shaft rotates cw and that the output rotates cw when the reducer is viewed from the front. What load should be preset into the overload release if the overload release is to be tripped by a 25 percent overload? (*Answer*: 980 lb)

2-8. The numbers of teeth on the gears of the epicyclic gear train sketched in Fig. P2-8 are $N_1 = 95$, $N_2 = 24$, $N_3 = 17$, and $N_4 = 13$. Gear N_4 is driven at 100 rpm (cw) and torque T_4 is 1000 in-lb (cw). Determine

(a) The angular velocity of the arm if the ring gear N_1 is held fixed (*Answer*: $n_5 = 15.85$ rpm, ccw)

(b) The horsepower developed at the arm (*Answer*: 1.59 hp output)

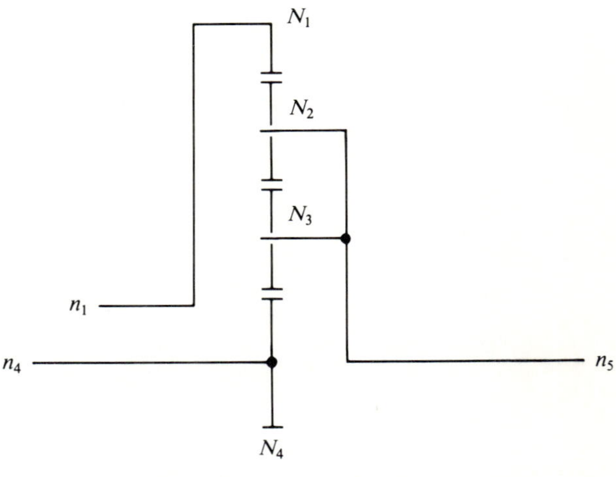

Figure P2-8. Figure for Problem 2-8.

2-9. A three-terminal epicyclic train is shown in Fig. P2-9.

(a) Determine the speed relationship for the terminals.

(b) Suppose $N_1 = N_5 = 36$, $N_2 = N_3 = N_4 = 32$, $N_6 = 64$, $n_1 = 400$ rpm (ccw), and $n_6 = 0$. What is the angular velocity of the arm? (*Answer*: 155 rpm, ccw)

(c) If the input torque is $T_1 = 1000$ ft-lb (ccw), what will be the arm torque T_7 for the conditions of part (b)? (*Answer*: 2580 ft-lb, cw)

2-10. Show that for a single-stage planetary gear train the ratio of ring gear torque to sun gear torque is always equal to the parameter $\alpha = N_r/N_s$. (Hint: Make a sketch of the

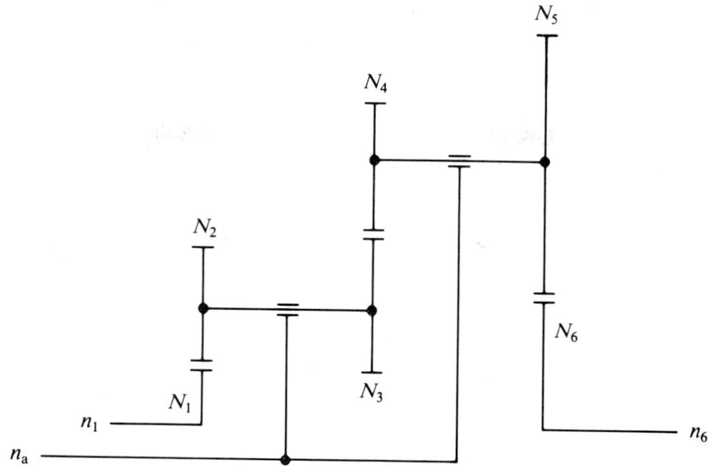

Figure P2-9. Figure for Problem 2-9.

tooth forces on the various gears and relate these forces to the external torques.) Also show that the ratio of arm torque to sun gear torque is $-(1 + \alpha)$.

2-11. For a single-stage planetary gear train, show that the sun gear and ring gear rotational motions must be in opposite directions unless both are inputs or both are outputs. Also show that when the sun gear and ring gear rotate in the same direction, the arm must also rotate in the same direction.

2-12. Show that for a single-stage planetary when sun gear and ring gear have opposing rotational motions, the arm may be rotating either cw or ccw depending on the magnitudes of sun and ring gear rotational speeds.

2-13. Verify that the directions given for rotational speed and torque in Table 2-7 are correct.

2-14. Show that when the angular velocity of a planet, relative to its carrier, is zero, its absolute angular velocity will be the same as that of the carrier.

2-15. Derive Eq. (2-12) from the relationship

$$\omega_s = k_r \omega_r + k_a \omega_a \qquad \textbf{(P2-15)}$$

where k_r and k_a are constants. Equation (P2-15) states that the relationship among the angular velocities for a single-stage planetary gear train is linear. (Hint: Let velocities ω_r and ω_a be zero, in turn, and analyze the gear train.)

2-16. Suppose a single-stage planetary operates with equal sun and ring gear speeds (same sense also). What will be the speed of the arm? Interpret this result.

2-17. In Example 2-5, suppose that instead of being fixed the ring gear rotates cw at a speed of 133.33 rpm. If 100 hp are input to the sun gear, which rotates at 3600 rpm (ccw), determine if the power passing through the ring gear terminal is an input or output power. What is its magnitude? (*Answer*: 11.1 hp output)

2-18. A split-power transmission is sketched in Fig. P2-18. Show that the speed relationship for this transmission is

$$\frac{n_1}{n_o} = 1 + \frac{N_3}{N_1} + \left(\frac{N_3}{N_1}\right)\left(\frac{N_6}{N_4}\right)$$

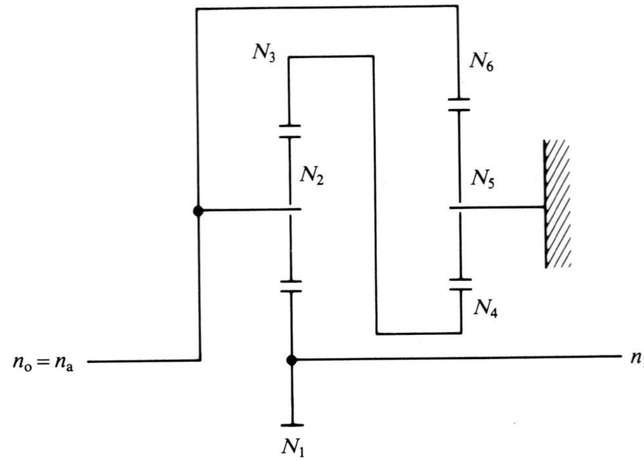

Figure P2-18. Figure for Problem 2-18.

2-19. A split-power transmission is sketched in Fig. P2-19. Show that the speed relationship for this gear train is

$$\omega_1 = \left[\frac{\alpha_1 \alpha_2}{1 + \alpha_1 + \alpha_2}\right]\omega_3 + \left[\frac{(1+\alpha_1)(1+\alpha_2)}{1 + \alpha_1 + \alpha_2}\right]\omega_7$$

where

$$\alpha_1 = \frac{N_3}{N_1} \qquad \alpha_2 = \frac{N_4}{N_6}$$

Notice that in this transmission the connected ring gear N_4 and the arm of planet N_2 simply free-wheel.

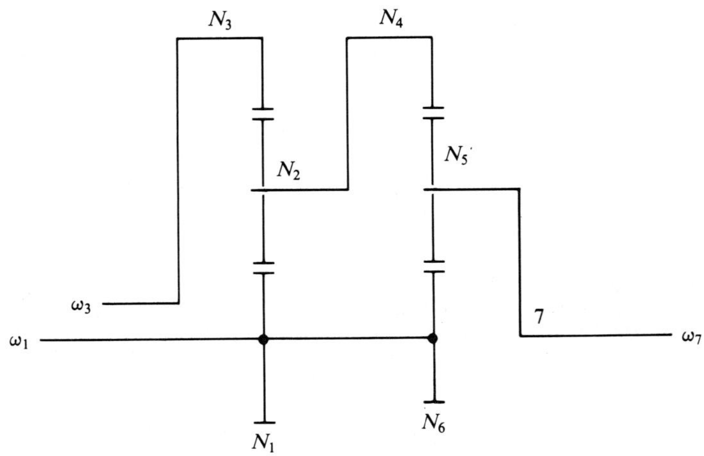

Figure P2-19. Figure for Problem 2-19.

2-20. In the epicyclic gearbox shown in Fig. 2-26, let $N_2 = 36$, $N_3 = 24$, $N_4 = 24$, and $N_5 = 36$. Sun gear N_3 is to be an output terminal and is to be varied in the interval

$$n_3^{(1)} = 0 \leq n_3 \leq 1500 \text{ rpm (ccw)} = n_3^{(2)}$$

Determine the required speed range for terminal n_2 if the carrier rotates at a constant speed of $n_1 = 1200$ rpm (ccw). (*Answer*: $n_2^{(1)} = 666.66$ rpm, ccw; $n_2^{(2)} = 1333.33$ rpm, ccw) Is terminal n_1 an input or an output? What about terminal n_2?

2-21. The epicyclic gear train shown in Fig. P2-21 has been widely used in power transmissions. Assume the device to be 100 percent efficient and show that the ratio of output speed to input speed is

$$\frac{n_6}{n_1} = \frac{1 - \dfrac{N_3 N_4}{N_1 N_6}}{1 + \dfrac{N_3}{N_1}}$$

Additionally, show that the fraction (of the total power transmitted) transmitted by sun gear N_4 is

$$\frac{T_4 n_4}{T_6 n_6} = \frac{\dfrac{N_4}{N_6}\left(1 + \dfrac{N_3}{N_1}\right)}{1 - \dfrac{N_3 N_4}{N_1 N_6}}$$

If $N_1 = N_6$, $N_4 = N_3$, and $N_1 = 2N_3$, how much power is transmitted by ring gear N_1?

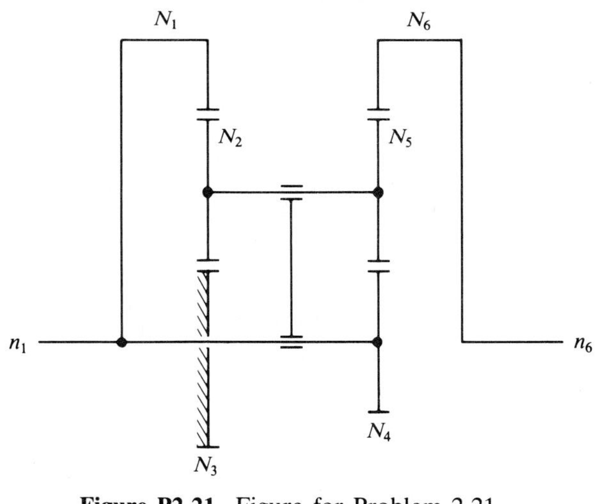

Figure P2-21. Figure for Problem 2-21.

2-22. Consider a simple single-stage planetary in which the numbers of teeth on the ring, planet, and sun gears are as follows: $N_r = 100$, $N_p = 32$, $N_s = 36$. The sun gear rotates at 400 rad/sec (ccw) and the ring gear is fixed. The power transmitted is 14.4 kW.

(a) Determine the angular speed of the arm.

(b) Determine the angular speed of the planet relative to the planet center.

(c) Determine the external torque exerted on the ring gear.

(d) Determine the net torque exerted on the arm.

(e) Draw free-body diagrams of the arm, planet, and sun gears and show the internal forces and external torques acting on them.

2-23. Assume that $\gamma < 1$ and show that it is impossible for the epicyclic differential of Fig. 2-26 to provide an output power in a range including zero unless there is only one input terminal and two output terminals.

2-24. Derive Eq. (2-65). Also show that it is always true that

$$N_3 - N_4 = N_2 - N_5$$

If the ring gears differ by only one tooth, show that the speed ratio is

$$r = -\frac{N_4}{N_1}(N_4 + N_1 + 1)$$

2-25. Two gears, *A* and *B*, are connected to the side gears of two bevel gear differentials as shown in Fig. P2-25. The terminals of the two differentials are to be connected together by gearing in such a way that gears *A* and *B* can be actuated by only two handwheels, *X* and *Y*. The system is to work in the following way:

(a) With handwheel *X* fixed, rotating handwheel *Y* cw (or ccw) will cause both gears *A* and *B* to rotate ccw (or cw).

(b) With handwheel *Y* fixed, rotating handwheel *X* cw (or ccw) will cause gear *A* to rotate cw (or ccw) while gear *B* will rotate ccw (or cw).

Design the gearing required to connect the differential terminals to achieve these requirements.

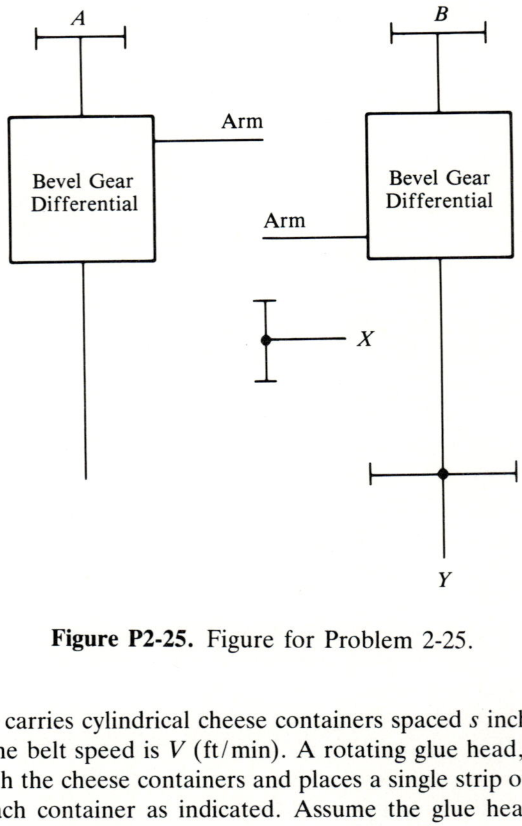

Figure P2-25. Figure for Problem 2-25.

2-26. A belt conveyor carries cylindrical cheese containers spaced *s* inches apart as shown in Fig. P2-26. The belt speed is *V* (ft/min). A rotating glue head, of diameter d_2, is synchronized with the cheese containers and places a single strip of glue in the center of the top of each container as indicated. Assume the glue head makes only one rotation per glue application.

This system is to be mechanized using a single motor which drives one side gear of a simple bevel gear differential. The differential is to be applied in such a way

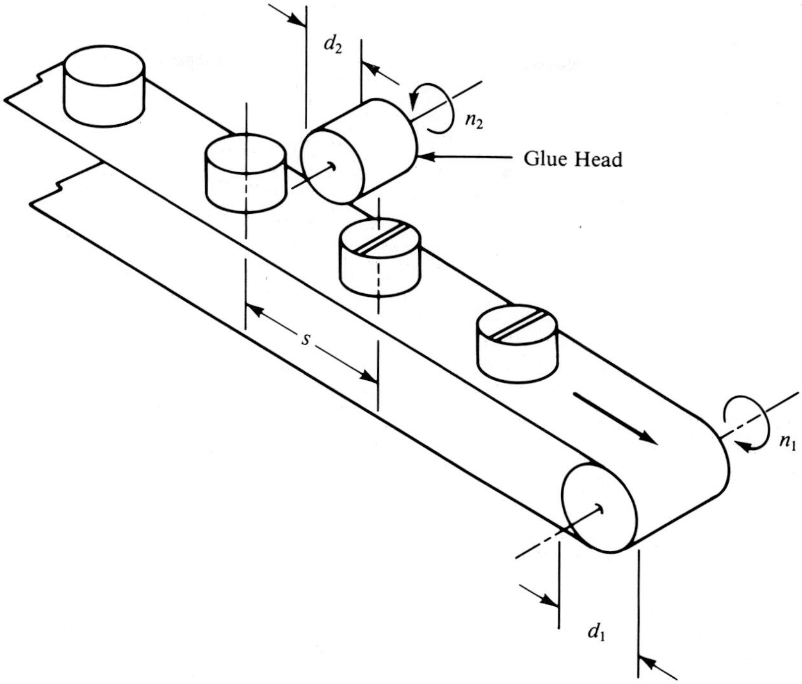

Figure P2-26. Figure for Problem 2-26.

that registration between the glue head and the cheese containers can be adjusted (by turning a handwheel) to insure that the glue applications will be in the center of the cheese containers. Show how a differential and motor can be applied to this system to achieve the above requirements. Also determine the diameter ratio d_1/d_2.

2-27. Determine the torque required to restrain a 100 hp, spur gear, parallel-shaft gearbox from overturning if it consists of a single set of gears. The gearbox ratio is 10:1 and the input speed is 1750 rpm. Make a sketch and show all torques acting on the gearbox.

2-28. Repeat Problem 2-27 for a gearbox with a double-reduction set of parallel spur gears. All other parameters remain the same.

2-29. A prime mover has the torque-speed characteristic

$$T_{\mathrm{m}} = 140 - 5.2N_{\mathrm{m}}^2 \text{ N-m}$$

where the speed is in rev/sec. It is to drive a load at maximum power whose characteristic is

$$T_\ell = 1.73N_{\mathrm{l}}^2 \text{ N-m}$$

where N_ℓ is in rev/sec.

(a) Determine the required gear ratio r of the ratioing device.

(b) Determine the power transmitted (in watts).

(c) Derive an expression for the load curve as seen by the prime mover.

2-30. The kinematic relationship for the speeds of a three-terminal gearbox is a linear one, namely,

$$n_1 + \alpha n_2 + \beta n_3 = 0$$

Show that the constants α and β are related in the form

$$\alpha = -e$$

$$\beta = e - 1$$

where e is the train value of the transmission when terminal n_3 is held fixed (in proceeding from shaft n_2 through shaft n_1). The value e is taken positive when n_1 and n_2 have the same rotational senses.

In other words, the speed relationship is

$$n_1 - en_2 + (e - 1)n_3 = 0$$

The three coefficients of a three-terminal epicyclic are always related in this fashion (i.e., by a single parameter). (Hint: Look at the case of rigid body rotation.)

2-31. Show that for a simple planetary consisting of a sun, ring, and carrier the value e is equal to

$$e = -\frac{N_r}{N_s}$$

(Hint: Let n_3 be the carrier.)

2-32. Show that the power equation (2-15) can be put in the form

$$T_1 n_1' + T_2 n_2' = 0$$

where n_1' is the speed of terminal 1 relative to the arm and n_2' is the speed of terminal 2 relative to the arm.

This equation shows that, relative to the arm, there must be a power inflow balanced by a power outflow. Check Example 2-7 to verify if this relative power equation is satisfied in that example.

2-33. Show that the equation of Problem 2-30 can be written as

$$\frac{n_1 - n_3}{n_2 - n_3} = e$$

2-34. When $e = -1$, what kind of epicyclic does Problem 2-33 represent?

2-35. Show that for any three-terminal gearbox (epicyclic) the torque relationships are

$$\frac{T_2}{T_1} = -e, \quad \frac{T_3}{T_1} = e - 1, \quad \frac{T_3}{T_2} = \frac{1-e}{e}$$

where e is defined as in Problem 2-30.

3

Mechanical Variable-Speed Transmissions and Clutches

At the end of Chapter 2, we observed that the ratio r of the ratioing components in a power transmission will be dictated, in part, by the load characteristic(s). Consequently, if the load characteristic (or prime mover characteristic) is variable (or changeable), it will be expedient to be able to change ratio r accordingly. And since these changes in loading may be continuous rather than discrete, we would like our ratioing components to be capable of continuous variation (or adjustment) in ratio r.

Geared transmissions are available in which discrete changes can be made in the gear ratio r by sliding various gears into (or out of) contact by moving them along their splined axes. This is how the common four-speed or five-speed manual transmission of an automobile works. Of course, a clutch is also needed to disengage the engine (prime mover) from the driving wheels while the transmission gear ratio is being changed. Otherwise we would be shifting under power which could damage the gearing. Geared transmissions of this kind are discussed in engineering textbooks on kinematics and, therefore, will not be discussed further here. Insofar as application and service factors are concerned, these transmissions are treated in the same way as the gearboxes discussed earlier.

3-1 VARIABLE-PITCH-DIAMETER PULLEY DRIVES

Ideally, our ratioing component would be capable of providing a continuously variable (often called infinitely variable) speed ratio r. Perhaps the simplest and most common transmission of this type is the belt or chain drive which utilizes a variable-pitch-diameter (or split) pulley. One such pulley is shown in Fig. 3-1.

One of the pulley faces shown in Fig. 3-1 is fixed to the hub while the other

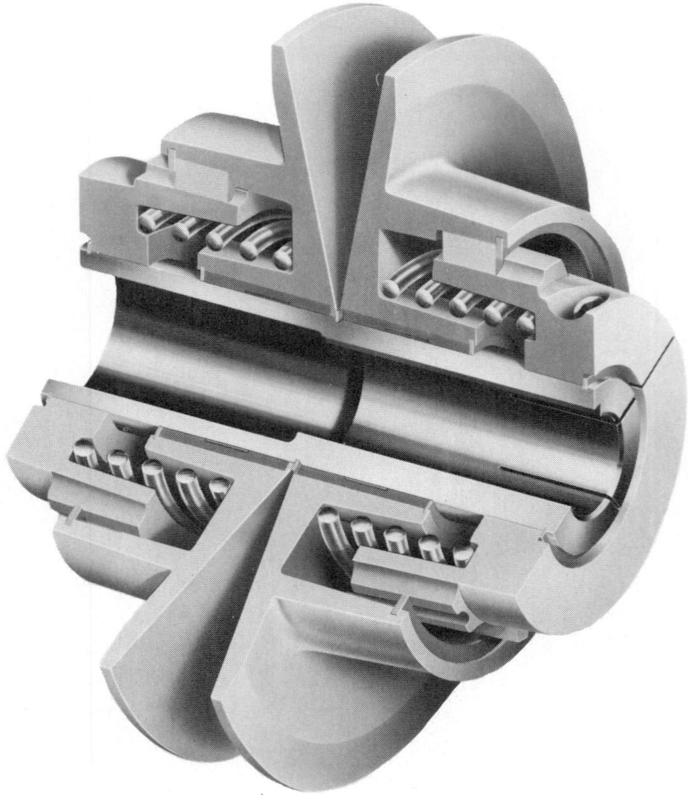

Figure 3-1. Variable-pitch-diameter pulley. (Courtesy of T. B. Wood's Sons Company.)

one is spring-loaded and movable axially along the hub. The depth at which the belt rides in the pulley depends on the distance between the pulley halves. This, in turn, depends on the center distance between this variable-pitch pulley and the other pulley (which has a fixed pitch diameter), to which the belt is connected.

Figure 3-2 shows a variable-pitch pulley transmission in which a fixed-pitch(diameter) pulley is located at the left and a variable-pitch(diameter) pulley is located at the right. As shown, the variable-pitch pulley is mounted on a motor shaft while the fixed-pitch pulley is mounted at the input shaft of a centrifugal blower.

The motor of Fig. 3-2 is mounted on a base capable of shifting the motor in such a way as to control the center distance between the fixed and variable pulleys. In Fig. 3-2, the motor is in its rightmost position and the belt is riding low (minimum pitch diameter) in the variable-pitch pulley. As the motor is cranked back toward the left using a lead screw on a handwheel, the center distance will be decreased and the belt will be caused to ride higher in the variable-pitch pulley as the two halves of the pulley come together. This action increases the pitch diameter of the driving pulley, and ratio r of the belt drive will decrease. A line diagram of this mechanism is shown in Fig. 3-3.

Figure 3-2. Variable-pitch pulley transmission. (Courtesy of T. B. Wood's Sons Company.)

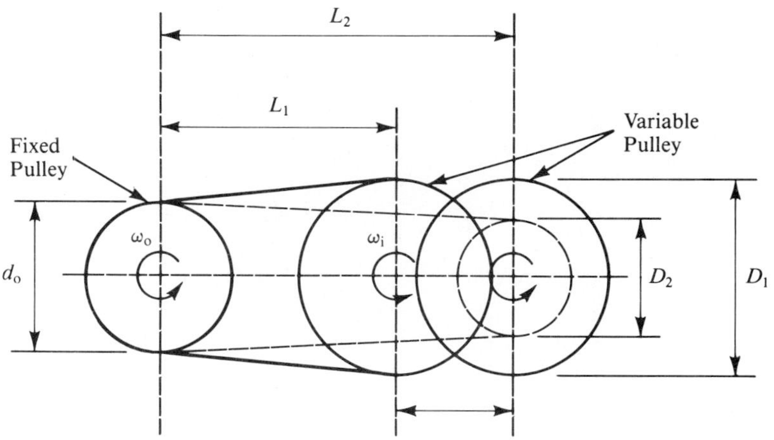

$$r_1 = \frac{\omega_i}{\omega_o} = \frac{d_o}{D_1} \qquad r_2 = \frac{\omega_i}{\omega_o} = \frac{d_o}{D_2}$$

Figure 3-3. Variable-pitch pulley transmission.

3-1-1 The PIV Drive (Transmission)

One disadvantage of the belt drive is that it is susceptible to creep which destroys the registration between the two pulleys. All friction (non-positive) drives are subject to this action.

Another version of the split pulley idea which is a positive drive is shown in Fig. 3-4. Here we have both pulleys as variable-pitch devices while the center distance between them is held fixed. The worm and worm-wheel mechanism shown is connected to a linkage which causes one of these split pulleys to have its faces closed at the same time the other pulley has its faces opened. In this way we obtain more kinematic leveraging than for a single variable-pitch pulley alone while maintaining the center distance fixed.

Additionally, the device shown in Fig. 3-4 is a positive drive rather than a simple belt frictional drive like that shown in Fig. 3-2. A uniquely designed toothed chain is used with cogged pulley faces, which prevents the creep associated with a belt drive. This is called a positive, infinitely variable (PIV) drive because there is no slip or creep attendant to its use.

EXAMPLE 3-1

Determine the speed ratios provided by a PIV transmission such as that sketched in Fig. 3-4 as the variable-pitch sheaves assume their limiting, anti-symmetrical configurations.

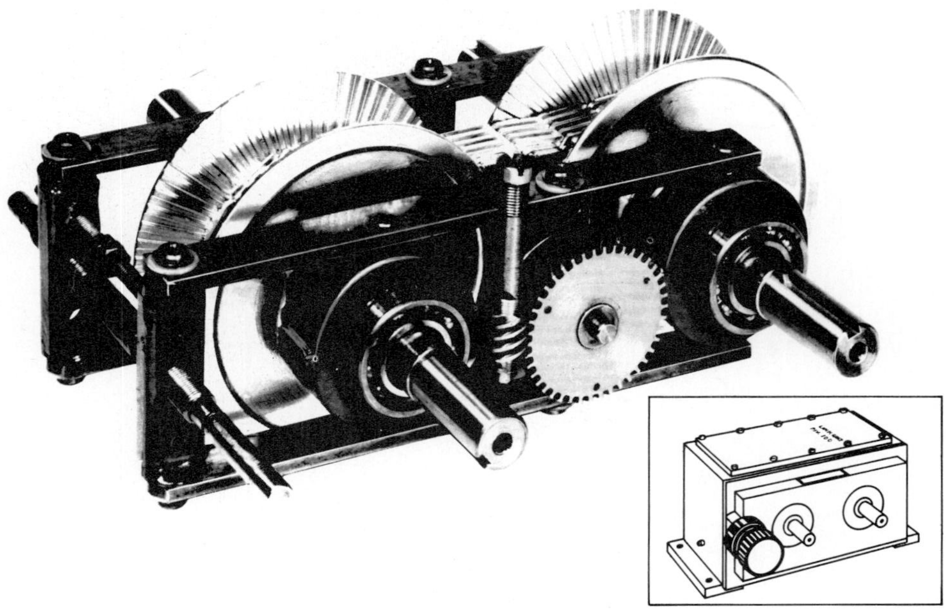

Figure 3-4. PIV variable-speed transmission. (Courtesy of PT Components, Inc.)

Solution

Consider the two limiting configurations sketched in Fig. 3-5. Symbol d denotes the minimum diameter these sheaves can assume while D denotes their maximum diameter. We let α be the ratio D/d.

Assume that the left sheave (pulley) is the driver and that the right sheave is driven. Then, for a given input speed n_i (rpm), we will have the extreme output speeds n_o as

$$\frac{n_{o(2)}}{n_i} = \alpha \qquad \frac{n_{o(1)}}{n_i} = \frac{1}{\alpha} \tag{3-1}$$

We let r be the ratio of the input speed n_i to the output speed n_o. Then

$$r_1 = \alpha \qquad r_2 = \frac{1}{\alpha} \tag{3-2}$$

The ratio R of the PIV (also called a speed variator) is defined as the maximum output speed divided by the minimum output speed (for a fixed input speed). Therefore,

$$R = \frac{n_{o(2)}}{n_{o(1)}} = \frac{\alpha}{1/\alpha} = \alpha^2 = \frac{r_1}{r_2} \tag{3-3}$$

Thus, we find that

$$R = \alpha^2 \quad \text{or} \quad \alpha = \sqrt{R} \tag{3-4}$$

Speed variators are specified by their ratio R rather than by ratio r (as in the case of a speed reducer).

As an illustration, suppose we have a 4:1 speed variator of the PIV type and that it is driven at an input speed of $n_i = 860$ rpm. Then

$$\alpha = \sqrt{4} = 2 = R$$

and

$$n_{o(2)} = 1720 \text{ rpm} \qquad n_{o(1)} = 430 \text{ rpm}$$

■

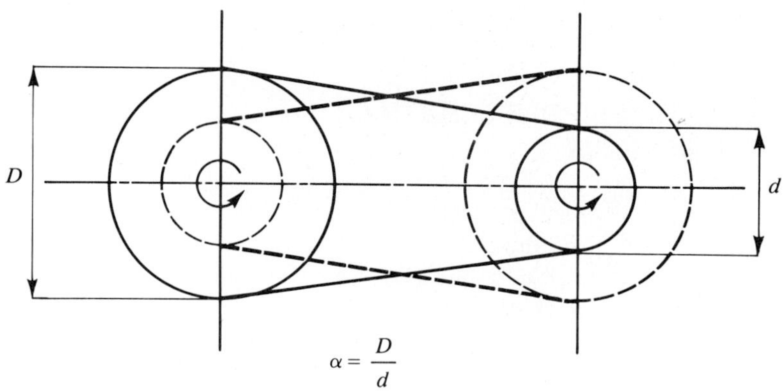

Figure 3-5. Figure for Example 3-1 (limiting PIV configurations).

3-1-2 Speed Regulation of Variable-Pitch-Diameter Drives

Suppose a variable-pitch belt transmission is operating with a fixed sheave on the driving motor and a variable-pitch sheave on the driven member. The adjustment mechanism for the variable-pitch sheave is held fixed. The intention is that for a given input speed (fixed), the output speed will remain constant because, presumably, ratio R of the transmission is fixed.

If the load on this drive changes, or fluctuates, what will actually happen is that there will be accompanying output speed changes, or fluctuations, despite the fact that ratio R, nominally, is being held fixed. The fluctuations occur as the torque increases because the tension in the belt (or chain) increases and the side thrust of the sheave pulley halves of the fixed pulley will make the belt slightly narrower and the belt will descend slightly lower in the fixed motor sheave. The result will be a slight slackening on the spring-loaded driven sheave so the two halves will close slightly and make the belt ride slightly higher in the variable-pitch sheave. The small effect on the two pulleys causes ratio R to change slightly even though this was not intended.

We will define the speed regulation of a variable-pitch drive to be the percent change in output speed from the no-load to the full-load (rated load) condition while the input speed remains fixed. As an illustration, if a variable-pitch drive has a speed regulation of 2 percent at maximum output speed and the nominal maximum and minimum output speeds at no-load are 1200 rpm and 800 rpm respectively, then the change in output speed from the no-load to the full-load condition will be

$$\Delta n_{\text{max}} = 1200(0.02) = 24 \text{ rpm}$$

The actual maximum output speed under full-load will then be 1176 rpm rather than 1200 rpm.

Since the speed regulation might be different at minimum output speed (because of centrifugal belt forces), it must be specified for both maximum and minimum output speeds.

3-2 VARIABLE-SPEED DIFFERENTIAL TRANSMISSIONS

In some mechanical systems, extremely accurate control is required of certain speed variables. Examples abound in printing presses, metering pumps, alternator drives, textile machines, film-processing equipment, and many other applications involving the controlled stretch of a web being processed in a roll system (paper, film, foil, wire, fiber, yarn, etc.).

One way of achieving this accurate control, which is capable of fine tuning, is to compress the speed range of the output of a conventional speed variator. We have already seen how this kinematic leveraging can be obtained with a differential (see Chapter 2, Section 2-11-1). An additional benefit of combining a differential with a speed variator is that the combination can be designed in such a way as to bypass a major part of the power being transmitted around the speed variator, which is the component likely to cause fluctuations in speed under large power changes.

Special transmissions are available, off-the-shelf, that provide a compression of the speed variator speed range, used in the transmission, by a factor of 1000 or more. They are called differential-draw transmissions. Figure 3-6 is a cut-away photograph of a differential-draw transmission that combines a PIV speed variator (shown at the left) with a bevel gear differential and associated spur gearing. These units are available in sizes up to 45 hp and with control sensitivities that provide a 1 rpm change in output speed for one complete turn of the control handwheel. Even greater sensitivity can be obtained by incorporating a Vernier control.

Most variable-speed drives such as a PIV will experience changes in output speed in response to changes in load. The narrow output speed range of the differential-draw drive minimizes the effect of these load variations on output speed. This is a critical requirement in machines requiring synchronization of one or more parts (e.g., the rolls of a web processing machine). A typical speed regulation plot is shown in Fig. 3-7.

The abscissa of Fig. 3-7 is the output speed variation expressed as a percentage of the average output speed, and the ordinate is the percent change in output speed in response to a change in loading from no-load to full-load.

Figure 3-6. Differential-draw transmission. (Courtesy of Fairchild Industrial Products Company.)

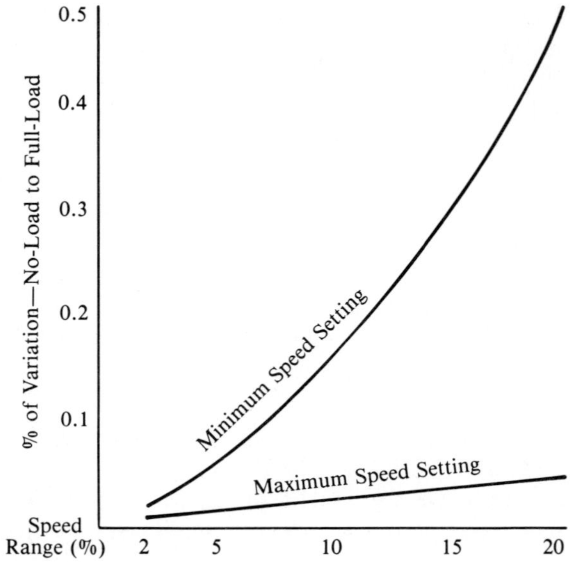

Speed Range (%)	Max. Speed Setting (%)	Min. Speed Setting (%)
2	0.005	0.025
5	0.01	0.07
10	0.02	0.17
20	0.05	0.5

Figure 3-7. Speed regulation of a differential-draw transmission. (Courtesy of Fairchild Industrial Products Company.)

EXAMPLE 3-2

A 9 hp differential-draw transmission has an input speed of 900 rpm. The maximum output speed is 1772 rpm while the minimum output speed is 1720 rpm as the control screw experiences 13.2 turns. What is the variation in output speed from no-load to full-load when the speed variator is set for maximum output speed and what is the speed variation when the variator is set for minimum output speed?

Solution

The speed range of this transmission is

$$SR = \frac{1772 - 1720}{1746} = 0.03 = 3 \text{ percent}$$

From Fig. 3-7, we see that the variation in output speed will be only about 0.007 percent when the control screw is set for maximum output speed. Consequently, we can expect that the change in output speed will be only about

$$0.00007(1772) = 0.124 \text{ rpm}$$

as the load goes from no-load to 9 hp load. Similarly, when the control screw is set for minimum output speed, we can expect a change in output speed of only around 0.04 percent. Therefore, the change will be only about

$$0.0004(1720) = 0.688 \text{ rpm}$$

as the load goes from no-load to 9 hp load. ■

Variable-speed differential transmissions can also be designed to produce a wide speed range of the output including the case in which the minimum output speed is zero. These transmissions are used where it is desirable to bring a machine shaft to rest without turning off the prime mover. A typical wide-speed-range device will provide, say, a maximum output speed of 518 rpm and a minimum output speed of zero when the input speed is 690 rpm (see Problem 3-2). The horsepower capacity of this particular device is 5 hp. This type of transmission is extremely useful when it is necessary to "inch" the speed in the vicinity of zero.

3-3 TRACTION DRIVES

We shall define a traction drive to be a ratioing component consisting of two or more smooth conjugate surfaces in rolling contact while separated by a very thin, highly compressed lubricant film. The rolling surfaces are generally balls, cones, cylinders, discs, or toroids and are usually made from hardened steel. The lubricant, or traction fluid, can be oil, silicone, or some other synthetic fluid.

Note that this is a somewhat narrow use of the word *traction* which in ordinary usage could also mean the traction between a tire and a dry road or the kind of traction occurring in a belt drive. The unique aspect of a traction drive is that the compressed lubricant between the rollers does not act as a lubricant in the ordinary sense, which we would expect to cause slip (and not traction). Under the extreme pressures exerted on this fluid film (of the order of 150,000 psi) by the normal forces exerted on the surfaces, the lubricant acts as a plastic solid which in conjunction with the elastic deflection of the mating surfaces will permit traction to occur. This is called elastohydrodynamic traction (EHD).

The traction between mating surfaces in a traction drive is not perfect in the sense of there being a total absence of slip (or creep). Slip prevents the kind of synchronous angular velocity transmission we obtain with a toothed drive, but since the slip is normally small (of the order of 1 percent), we call it creep rather than slip.

So the disadvantage of a traction drive is that a small creep will exist between mating surfaces similar to that in a V-belt drive. Shock loading can also be a problem in that it may cause a gross slip to occur at the traction contact which could be injurious both to the traction drive and to the machinery it is driving.

Advantages of the traction drive are that it is very smooth in operation and relatively noiseless compared to toothed (geared) drives. The traction drive is also generally designed as a variable-speed ratioing device.

Traction drives find application in a wide range of machines where compactness, ruggedness, speed control, and smooth operation are required. The principal competitor of the traction drive is the electronically controlled electric motor (see Chapter 11). But traction drives are used in places where electronically controlled motor drives cannot go.

Many different traction drive configurations suitable for variable-speed transmissions have been conceived by ingenious engineers and inventors. These have been surveyed in several publications (see References 3-1 through 3-3). It should be kept in mind that any geared drive configuration can be made into a traction drive configuration simply by replacing the gears with smooth wheels. Consequently, we find the epicyclic scheme a particularly attractive one in traction drives.

3-3-1 Traction Drive Types

The major types of traction drives are

- **(a)** Ball-disk
- **(b)** Ball variator
- **(c)** Roller-disk
- **(d)** Ring-roller
- **(e)** Ring-cone
- **(f)** Toroidal
- **(g)** Planetary

Since these types are discussed in great detail in References 3-1 through 3-3, we will discuss only three of the more interesting and useful ones here. Table 3-1 lists the characteristics of some of the more common types.

The Ball Variator. The ball variator is the most widely used traction drive. It was introduced in the late 1940s by Jean Kopp in Switzerland. Since that time, several U.S. companies have been licensed to produce it. More than a million units have been produced.

TABLE 3-1. Typical Traction Drive Characteristics.

Traction Drive Type	Typical Power Rating (hp)	Speed Ratio, r	Mechanical Efficiency (%)	Drift under Load (%)
Ball-disc	0.1 to 5	100:1	75 to 85	0.1 to 1
Ball variator	0.5 to 16	9:1 to 12:1	75 to 94	0.3 to 4
Ring-roller	0.1 to 5	9:1 to 100:1	85 to 90	0.1
Ring-cone	0.1 to 20	12:1 to 100:1	80 to 85	0.1 to 1
Toroidal	0.3 to 5.5	7:1	80	3
Planetary	25	6:1	75 to 85	0.5

Figure 3-8. Cut-away of a ball variator. (Courtesy of Winsmith, Peerless-Winsmith, Inc.)

A cut-away photo and schematic diagram of the ball variator are presented in Figs. 3-8 and 3-9. The device consists of an input shaft connected rigidly to a disk which is in tractive contact with several balls located around the disk periphery. Each ball rotates on an axis whose tilt is controllable by a suitable linkage actuated by a handwheel. The balls rotate but do not orbit. The output shaft is connected to a similar disk in tractive contact with the balls.

As shown in Fig. 3-9, variable output speed can be obtained for a fixed input speed by controlling the tilt of the ball axes. When the ball axes are parallel to the input and output shaft axes, the ratio of the device is 1:1. As the ball axes are

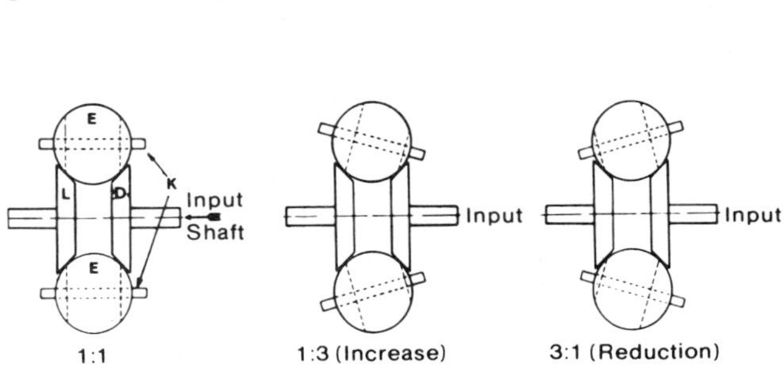

Figure 3-9. Principle of speed variation. (Courtesy of Winsmith, Peerless-Winsmith, Inc.)

tilted in one direction, a 1:3 speed increase at the output can be obtained, whereas if the axes are tilted in the other direction, a decrease of up to 3:1 is obtained.

In addition to the rather complicated mechanical linkages required to tilt the ball axes, means must also be provided for producing the normal forces at the traction interface between balls and disks. Although the mechanisms are quite complex to manufacture and assemble, the unit is compact and accurate. It has been widely used in metering pumps, printing presses, and automated production lines.

Several variations on the theme of the ball variator are available. In one design, the ball is replaced by a tapered roller (see Reference 3-1).

The Ring-Roller Variator. One of the oldest traction drives still in use in the United States is the ring-roller drive developed in 1935 by The Graham Transmissions Company. This ingenious variable-speed drive uses epicyclic kinematic principles to obtain forward, zero, and reverse speeds of the output shaft for fixed input speed by traversing a ring axially along a tapered cone roller (some authors classify this drive as a ring-cone drive).

A cut-away picture of the Graham ring-roller traction drive is shown in Fig. 3-10. The device consists of compounded planets, one of which is geared while the other is a traction roller (cone). The planet gear is in contact with a ring gear

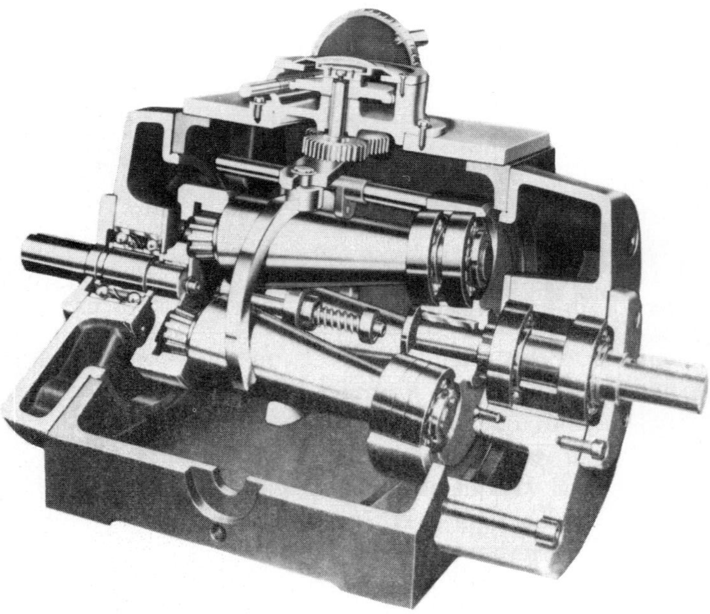

Figure 3-10. Graham ring-roller traction drive. (Courtesy of the Graham Company.)

which is the output of the transmission. The planet carrier is the input to the transmission and the smooth roller (cone) is in contact with a traction ring which is fixed with respect to rotation but shiftable axially along the roller axis. As the traction ring is shifted, the diameter of the traction roller planet changes.

A schematic representation of the ring-roller drive is shown in Fig. 3-11. Ring R_1 and planet P_1 are the traction drive components while ring R_2 and planet P_2 are gears. Ring R_1 can be shifted axially but does not rotate.

If we let R_2, P_2, P_1, and R_1 represent numbers of teeth (or, effectively, numbers of teeth), our epicyclic speed formula shows that

$$\frac{n_2 - n_c}{\overset{0}{\not{n}_1} - n_c} = +\left(\frac{R_1}{P_1}\right)\left(\frac{P_2}{R_2}\right) \tag{3-5}$$

where n_c is the carrier (input) angular speed (rpm) and n_2 is the output speed (rpm) of ring gear R_2. Solving for n_2, we find that the speed ratio r of the device is given by

$$\frac{1}{r} = \frac{n_2}{n_c} = 1 - \left(\frac{R_1}{P_1}\right)\left(\frac{P_2}{R_2}\right) \tag{3-6}$$

Note that the output can be made zero by requiring that

$$P_1 = \left(\frac{R_1}{R_2}\right)P_2 \tag{3-7}$$

Suppose this value is chosen to represent the diameter of the midpoint of the cone roller planet P_1. Then, as the ring is shifted to the right of this midpoint position, the ratio n_2/n_c will be negative while for positions to the left of this midpoint the ratio will be positive.

We see that this device has the capability of completely reversing the direction of rotation of the output (n_2) simply by shifting the position of ring R_1. Maximum forward speed is obtained with the ring shifted all the way to the left while maximum reverse speed occurs with the ring shifted all the way to the right. Note how the axis of the cone roller is inclined so that the surface of the roller is parallel to the external shaft axes, permitting contact with the ring at all axial positions.

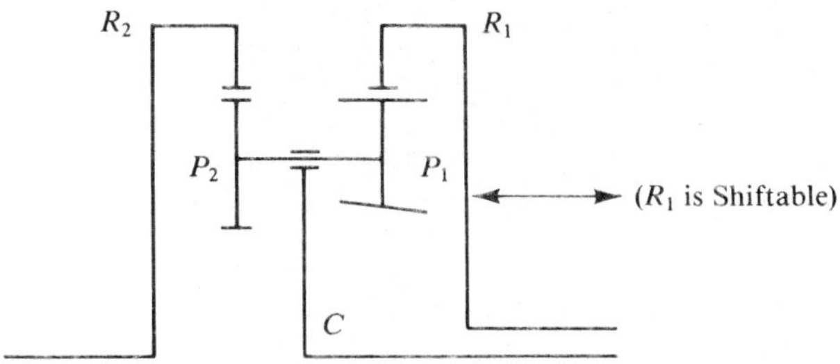

Figure 3-11. Schematic diagram of a ring-roller drive.

Although the Graham traction drive has many advantages kinematically, it is available only in sizes up to 5 hp.

The Ring-Cone Variator. Another traction drive speed variator that uses epicyclic kinematic principles is the ring-cone variator shown in Fig. 3-12. This device is similar to Humpage's epicyclic bevel gear train discussed in Chapter 2. We have compounded planet rollers, one of which is conical. The conical planet is in contact with an internal ring-roller which does not rotate but which can be shifted along the conical surface to vary the speed at which the planets orbit. Note that there is no shaft connected to the carrier. The input and output shafts are connected to roller discs which are in tractive contact with the compounded traction planets.

The speed relationship for this variable-speed transmission is (see Problem 3-6)

$$\frac{n_o}{n_i} = \frac{D_i}{D_o}\left(\frac{d_2 D_o - d_3 D_r}{d_1 D_r}\right) \tag{3-8}$$

The various dimensions are shown in Fig. 3-13.

Ring-cone drives are available in sizes up to around 20 hp and, as Eq. (3-8) shows, the ring can be shifted to produce zero output speed. These traction drives have been used in machine tools, extruders, mixers, centrifuges, pumps, packaging machines, and printing presses.

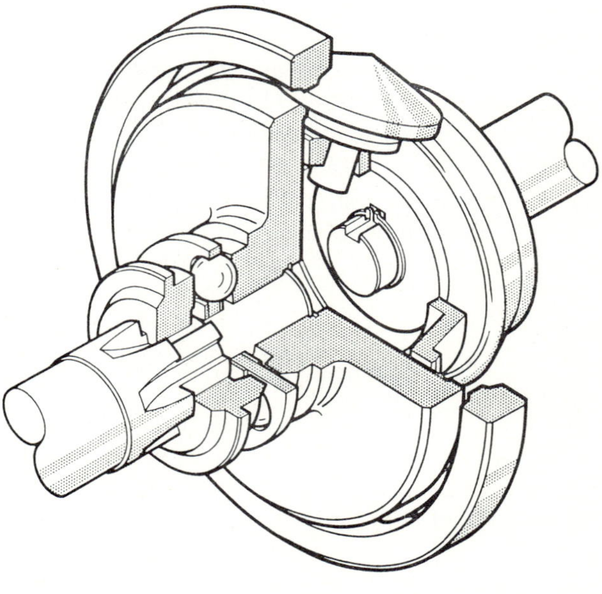

Figure 3-12. Graham ring-cone traction drive. (Courtesy of the Graham Company.)

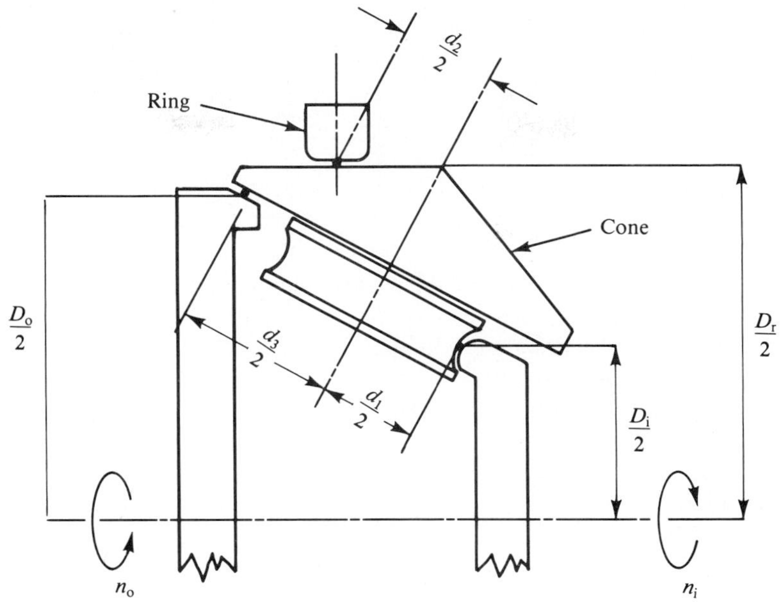

Figure 3-13. Schematic diagram of a ring-cone drive.

3-4 FRICTIONAL COUPLINGS

Almost all mechanical systems will require some means for connecting and disconnecting loads from prime movers. In some systems the prime mover will be incapable of developing an appreciable torque at low speeds. Consequently, the prime mover is started and brought up to speed while disconnected from its load. After being started, the prime mover can develop enough torque so that it can be connected to the load. A classic example is the automobile whose prime mover (the engine) must be capable of being connected and disconnected from the driving wheels (which provide the load through traction with the road). If the automobile has a manual transmission, a frictional clutch provides the required coupling.

Mechanical systems will also generally possess inertia effects resulting from masses in motion. The effects can be very large, and at some point in the operation of the system the inertias will need to be brought safely to rest. This can be done by introducing a frictional coupling between the moving masses and a reference which is at rest. In this case, the frictional coupling is called a brake.

The difference between a clutch and a brake is not functional but, rather, only semantic. A coupling device used to bring two moving systems to a common speed is called a clutch. The very same device will be called a brake if only one of the systems is moving and the device is used to bring the moving system to rest.

In later chapters, we will discuss mechanical system couplings in which coupling traction is provided by fluid or electromagnetic interactions. These

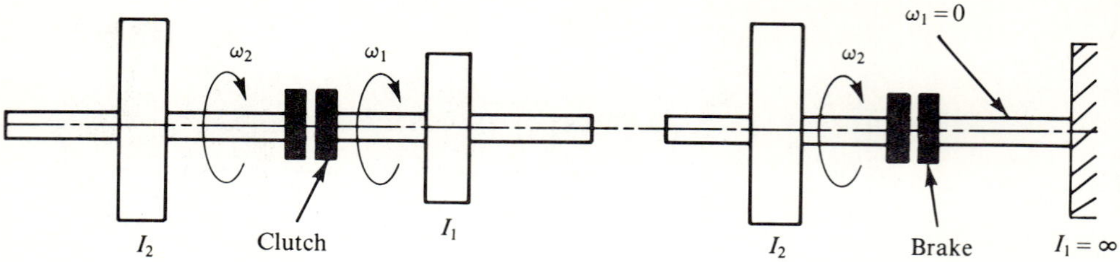

Figure 3-14. Clutch and brake configurations.

couplings may also be called clutches or brakes even though they may produce a slip between the two engaged systems after the clutch (or brake) has fulfilled its engagement function. For the couplings to be discussed here, the engagement will be mechanical through the action of mechanical friction between solid mating surfaces. When these devices have fulfilled their engagement function, there will be no slip between the mating surfaces; the surfaces will be locked together by mechanical friction.

Schematic diagrams of clutch and brake configurations are shown in Fig. 3-14. When clutches and brakes are fulfilling the function for which they were designed, engagement will occur, say, in order to bring two rotating shafts to a common speed. Thus, the process of engagement necessarily introduces transient behavior in the mechanical system of which the clutch is a mechanical component. Therefore, an engineering study of frictional couplings should include an analysis of clutch system dynamics as well as the rather detailed force and torque analysis of the various clutch configurations themselves. The force versus torque relationships can be found in any good book on machine design, so our focus here is on engagement dynamics and on the kinds of clutch configurations available for incorporation in mechanical systems.

3-4-1 Engagement Dynamics

In designing mechanical systems that incorporate clutches, we may be interested in several transient performance parameters including (a) the energy dissipated in the clutch during clutching or braking, (b) the total relative displacement occurring between the clutch halves during clutching, and (c) the time required to bring a load up to its steady operating speed. This last parameter will consist generally of two consecutive time intervals: the clutching time δt required to synchronize the prime mover and load speeds, plus the time Δt to accelerate the load after the clutch is finally engaged.

Determining the acceleration time to bring a load up to speed is a rather complex problem when a clutch is involved because both prime mover and load torques may be changing during the starting process. To understand this process, refer to Fig. 3-15.

In Fig. 3-15, we assume that the input shaft is rotating at speed ω_1 while the output rotates at speed ω_2. Both speeds are cw when viewed from the right, with $\omega_1 \geq \omega_2$.

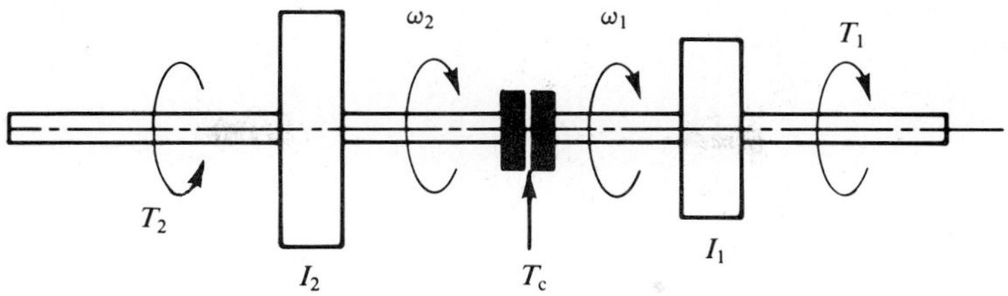

Figure 3-15. Engagement of two rotating masses.

The torque developed in the clutch is designated as T_c. For the system shown, clutch torque T_c on the right clutch face is ccw while clutch torque T_c exerted on the left clutch face is cw. The system has input torque T_1 (cw), and the output torque is T_2 (ccw). Power flows from the input, through the clutch, to the output.

Free-body diagrams of each half of the system will show that

$$T_c - T_2 = I_2 \frac{d\omega_2}{dt} \tag{3-9}$$

$$T_1 - T_c = I_1 \frac{d\omega_1}{dt} \tag{3-10}$$

where I_1 and I_2 are the moments of inertia of the input and output sides of the system. If we add Eqs. (3-9) and (3-10), we obtain

$$T_1 - T_2 = I_1 \frac{d\omega_1}{dt} + I_2 \frac{d\omega_2}{dt} \tag{3-11}$$

Now we make several assumptions and provide several definitions. Time t_i is defined as the time at which the clutch just begins to engage. We designate $\omega_1(t_i) = \omega_1$ and $\omega_2(t_i) = \omega_2$ as the initial speeds at time t_i (these can be considered to be constants). Time t_e is defined as the time at which engagement is completed, that is, $\omega_1(t_e) = \omega_2(t_e)$. Note that after engagement is completed, both $\omega_1(t)$ and $\omega_2(t)$ may still be changing with time (even though they are equal to each other) because, as Eq. (3-11) shows, torque T_1 may exceed torque T_2.

After the clutch has fully engaged, speeds $\omega_1(t)$ and $\omega_2(t)$ are equal and Eq. (3-11) becomes

$$T_1 - T_2 = (I_1 + I_2) \frac{d\omega}{dt} \qquad (t \geq t_e) \tag{3-12}$$

where

$$\omega = \omega(t) = \omega_1(t) = \omega_2(t) \tag{3-13}$$

We define time t_f as the time when input and output torques T_1 and T_2 are equal, so at this time (and beyond)

$$\omega = \omega(t) = \omega_f = \text{constant} \tag{3-14}$$

Time t_f is the time at which the steady-state operating point has been reached. A sketch of the system behavior is shown in Fig. 3-16.

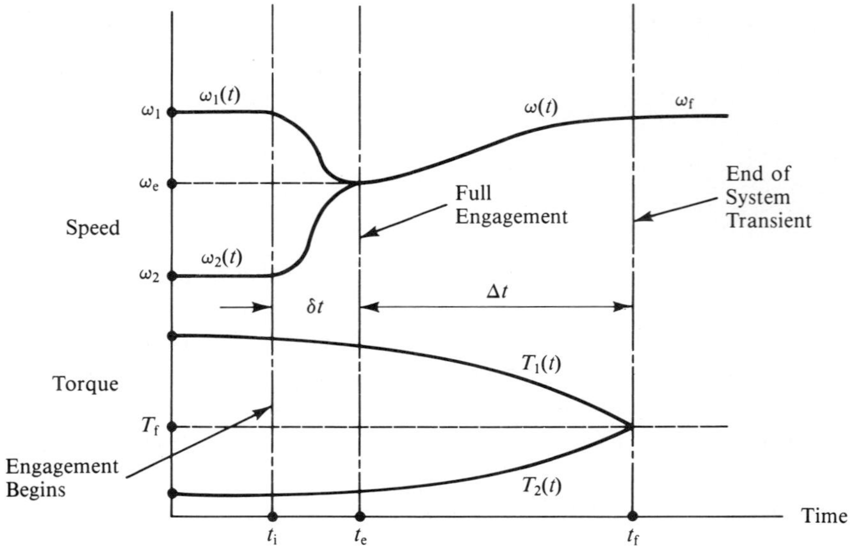

Figure 3-16. Typical system characteristics.

In many systems, the clutching time δt will be relatively small compared to the total time $\delta t + \Delta t$ required to engage and accelerate the system to its steady operating point (t_f, ω_f, T_f). Therefore, we might profitably assume that torques T_1 and T_2 remain constant in the interval δt. If we assume, additionally, that the clutch torque T_c is constant during this short clutching time, then Eqs. (3-9) and (3-10) can be written as

$$d\omega_2 = \left(\frac{T_c - T_2}{I_2}\right) dt \tag{3-15}$$

$$d\omega_1 = \left(\frac{T_1 - T_c}{I_1}\right) dt \tag{3-16}$$

Actually, the clutch torque will rise rapidly from zero (fully disengaged) to the value T_c. When the clutch is fully engaged, the clutch torque will drop suddenly from T_c to the shaft torque being transmitted.

Integrating Eqs. (3-15) and (3-16) produces

$$\omega_2(t) = \omega_2 + \left(\frac{T_c - T_2}{I_2}\right)(t - t_i) \tag{3-17}$$

$$\omega_1(t) = \omega_1 + \left(\frac{T_1 - T_c}{I_1}\right)(t - t_i) \tag{3-18}$$

When $t = t_e$, the angular velocities $\omega_1(t)$ and $\omega_2(t)$ are both equal to ω_e. Therefore, Eqs. (3-17) and (3-18) can be equated to produce

$$\delta t = \frac{I_1 I_2(\omega_1 - \omega_2)}{T_c(I_1 + I_2) - (I_1 T_2 + I_2 T_1)} = t_e - t_i \tag{3-19}$$

If the external torques T_1 and T_2 are absent, then Eq. (3-19) reduces to

$$\delta t = \frac{I_1 I_2(\omega_1 - \omega_2)}{T_c(I_1 + I_2)} \tag{3-20}$$

In this situation, the output shaft is accelerated at the expense of deceleration of the input shaft.

Equation (3-20) can be written in slightly different form as

$$\delta t = \frac{\omega_1 - \omega_2}{T_c\left(\dfrac{1}{I_1} + \dfrac{1}{I_2}\right)} \tag{3-21}$$

This formula shows that the clutching time is proportional to the initial speed difference, inversely proportional to clutch torque, and proportional to the inertias involved.

The angular displacement of the two masses I_1 and I_2 during the clutching time can be determined by integrating Eqs. (3-17) and (3-18). This will show that

$$\theta_2(t_e) = \omega_2 \delta t + \left(\frac{T_c - T_2}{I_2}\right)\frac{\delta t^2}{2} \tag{3-22}$$

$$\theta_1(t_e) = \omega_1 \delta t + \left(\frac{T_1 - T_c}{I_1}\right)\frac{\delta t^2}{2} \tag{3-23}$$

The relative displacement $\delta\theta = \theta_1 - \theta_2$ is then

$$\delta\theta = (\omega_1 - \omega_2)\delta t - \frac{\delta t^2}{2}\left(\frac{T_c(I_1 + I_2) - (I_1 T_2 + I_2 T_1)}{I_1 I_2}\right) \tag{3-24}$$

The work done at the clutching interface during time δt is

$$W = T_c \delta\theta \tag{3-25}$$

or

$$W = T_c(\omega_1 - \omega_2)\delta t - \frac{T_c^2 \delta t^2}{2}\left(\frac{I_1 + I_2}{I_1 I_2} - \frac{I_1 T_2 + I_2 T_1}{T_c I_1 I_2}\right) \tag{3-26}$$

If we combine Eqs. (3-19) and (3-26), we get

$$W = \frac{T_c(\omega_1 - \omega_2)^2}{2\left(\dfrac{T_c - T_1}{I_1} + \dfrac{T_c - T_2}{I_2}\right)} \tag{3-27}$$

When the external torques T_1 and T_2 are absent, Eq. (3-27) becomes

$$W = \frac{I_1 I_2(\omega_1 - \omega_2)^2}{2(I_1 + I_2)} \tag{3-28}$$

Note that this last equation has the form $W = \frac{1}{2} I\omega^2$. It represents the dissipation of energy in the clutch in order for engagement to occur by the mechanism of friction.

Dissipation of mechanical energy within the clutch during clutching causes the temperature of the clutch parts to rise, with a potential for overheating. One of the critical design aspects of clutches and brakes is the heat transfer design and choice of materials such that temperature increases are accommodated safely. If high surface temperatures are created continuously in the braking surfaces, a phenomenon called fade will result. Fade is a significant decrease in the coefficient of friction with attendant loss of clutching (or braking) power. For more information on this important subject, see Reference 3-4.

3-4-2 Referred (Equivalent) Torques and Inertias

Suppose the mechanical system includes several inertias and a ratioing device in addition to a clutch. Then the simple configuration of Fig. 3-15 will not apply and a configuration like that in Fig. 3-17 may be appropriate. Here the ratioing component is a simple chain drive and there is an additional inertia I_2' between the input inertia I_1 and output inertia I_3'.

Now Eqs. (3-9) and (3-10) become

$$T_c - T_2' = I_2' \frac{d\omega_2}{dt} \tag{3-29}$$

$$T_1 - T_c = I_1 \frac{d\omega_1}{dt} \tag{3-30}$$

We can also write the additional equation

$$T_3' - T_3 = I_3' \frac{d\omega_3}{dt} \tag{3-31}$$

The output torque for the system of Fig. 3-17 is T_3. Torques T_2' and T_3' are internal to the system; they are related by the equation

$$T_2' = \left(\frac{\omega_3}{\omega_2}\right) T_3' \tag{3-32}$$

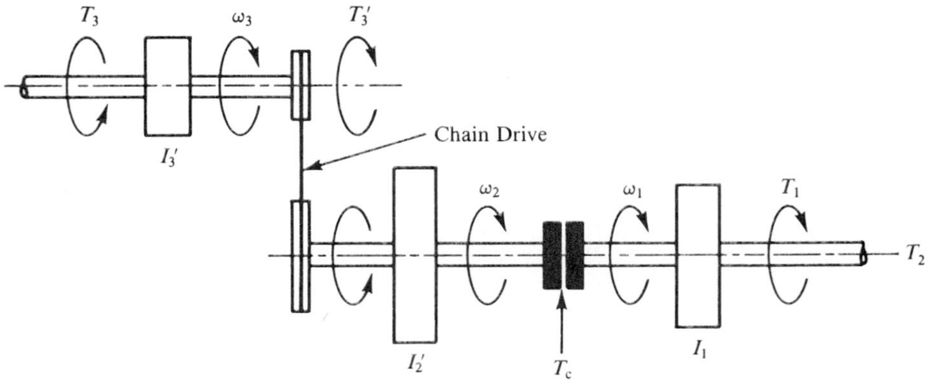

Figure 3-17. Several rotating masses on several shafts.

The relationship between speeds ω_2 and ω_3 is dictated by the ratio r of the ratioing component, in the following manner:

$$\omega_2 = r\omega_3 \tag{3-33}$$

If Eqs. (3-29) through (3-33) are combined, the result is

$$T_c - \left(\frac{1}{r}\right)T_3 = \left(I_2' + \left(\frac{1}{r}\right)^2 I_3'\right)\frac{d\omega_2}{dt} \tag{3-34}$$

Comparing Eq. (3-34) to Eq. (3-9), it becomes apparent that all of the results of our preceding analyses (based on Fig. 3-15) will remain valid provided we define

$$T_2 = T_{r(2)} = \left(\frac{1}{r}\right)T_3 \tag{3-35}$$

$$I_2 = I_{r(2)} = I_2' + \left(\frac{1}{r}\right)^2 I_3' \tag{3-36}$$

Then Eq. (3-34) can be written

$$T_c - T_{r(2)} = I_{r(2)}\frac{d\omega_2}{dt} \tag{3-37}$$

Equation (3-37) has the same form as Eq. (3-9). Torque $T_{r(2)}$ is called the output torque (load) referred to shaft 2, and $I_{r(2)}$ is called the inertia referred to shaft 2. Note that the inertia of the load, referred to shaft 2, would be just the second term on the right side of Eq. (3-36). We see that the load, reflected across the ratioing component, involves ratio r linearly, whereas the inertia, reflected across the ratioing component, involves ratio r quadratically.

A free-body diagram of the input side of Fig. 3-17 produces the same result as for Fig. 3-15, namely,

$$T_1 - T_c = I_1\frac{d\omega_1}{dt} \tag{3-38}$$

Combining Eqs. (3-37) and (3-38) gives

$$T_1 - T_{r(2)} = I_1\frac{d\omega_1}{dt} + I_{r(2)}\frac{d\omega_2}{dt} \tag{3-39}$$

This is the same form as Eq. (3-11).

Suppose we define clutch slip s as

$$s = \frac{\omega_1 - \omega_2}{\omega_1} \tag{3-40}$$

Then it follows that

$$\omega_2 = (1 - s)\omega_1 \tag{3-41}$$

and

$$\frac{d\omega_2}{dt} = (1 - s)\frac{d\omega_1}{dt} - \omega_1\frac{ds}{dt} \tag{3-42}$$

If Eq. (3-42) is substituted into Eq. (3-39), the result is

$$T_1 - T_{r(2)} = [I_{r(2)}(1-s) + I_1]\frac{d\omega_1}{dt} - I_{r(2)}\omega_1\frac{ds}{dt} \tag{3-43}$$

Now, we have already derived a relationship, Eq. (3-19), for the clutching time δt corresponding to the condition $s > 0$. The remaining acceleration time Δt required to bring the load to its steady-state operating speed, after the clutch has become fully engaged, is determined from Eq. (3-43) with $s = 0$. That is,

$$T_1 - T_{r(1)} = I_{r(1)}\frac{d\omega_1}{dt} \tag{3-44}$$

where

$$T_{r(1)} = T_{r(2)} \qquad I_{r(1)} = I_{r(2)} + I_1 \tag{3-45}$$

Note that since both shafts rotate at the same speed when $s = 0$, the output torque referred to shaft 2 is the same as that referred to shaft 1, and the load inertia, now referred to shaft 1, is simply the sum of the value of shaft 1 plus the value referred to shaft 2.

In Eq. (3-44), torque $T_{r(1)}$ is often called the load torque as seen by the input shaft (shaft 1), and inertia $I_{r(1)}$ is called the load inertia as seen by the input shaft. Torque T_1 is the input torque and ω_1 is the input speed. A sketch of the system equivalent to that of Fig. 3-17 is shown in Fig. 3-18 for the case $s = 0$.

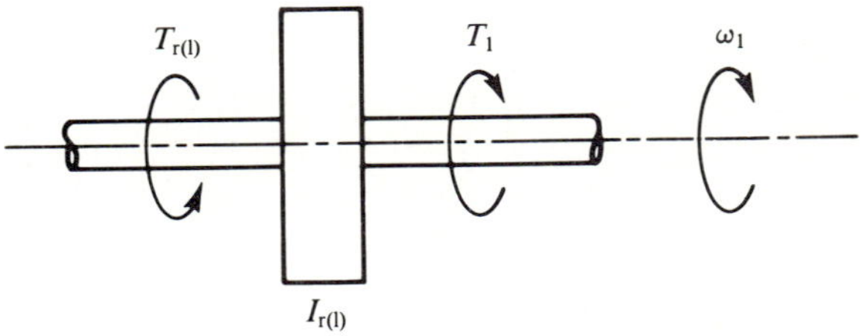

Figure 3-18. Equivalent system ($S = 0$).

EXAMPLE 3-3

A winch system is configured as shown in Fig. 3-19. The winch drum rotates at 9 rpm and its wire rope grooves have a pitch diameter of 12 inches. The drum must be capable of winching, and holding, a 21,000 lb load (i.e., tension in the wire rope).

The reducer is a triple-reduction, parallel-shaft gearbox containing helical gearsets. Between the induction motor and the gearbox are a disk clutch, a brake, and a one-way pawl clutch. The pawl clutch enables the load to be stopped and held. The disk clutch makes it possible to start the load (full-load) from the stopped position. The brake is necessary in order to control the load, during payout, once the pawl has been released (when the disk clutch is disengaged).

The steady operating speed of the motor (under load) is 1750 rpm. The estimated rotational moments of inertia of the various components are as follows:

(a) Reducer referred to reducer input shaft = 0.326 lb-ft^2
(b) Couplings = 0.20 lb-ft^2 each

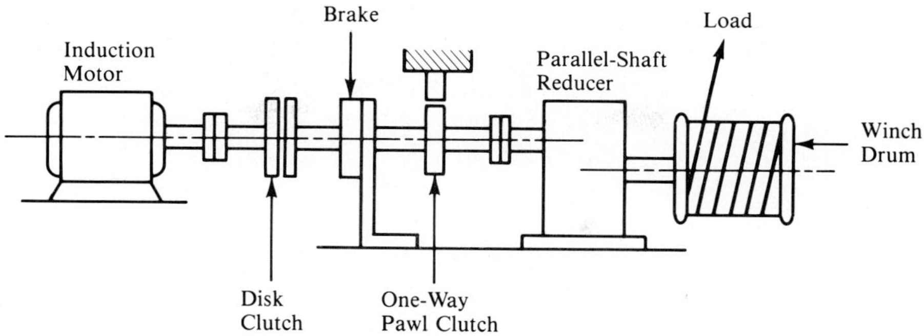

Figure 3-19. Winch system.

(c) Pawl clutch = 1.0 lb-ft^2
(d) Brake = 0.5 lb-ft^2
(e) Disk clutch, motor half = 12 lb-ft^2
(f) Disk clutch, reducer half = 0.5 lb-ft^2
(g) Motor rotor = 6 lb-ft^2

The radius of gyration of the winch drum is 5 inches and the mass of the winch drum is 300 lb.

Assume that during the disk clutch engagement process the clutching torque is five times the torque transmitted by the clutch during steady-state operation and that the average motor torque is 70 ft-lb during this same period of time.

(a) Determine the moment of inertia of the drum referred to the output shaft of the clutch.
(b) Determine the load torque referred to the output shaft of the clutch.
(c) Assume the motor speed just prior to engagement of the clutch is 1800 rpm and determine the minimum speed the motor will experience during the engagement process.
(d) How much heat will be generated during the engagement process?
(e) Suppose the load is stopped and held by the pawl clutch with the disk clutch disengaged. When the pawl clutch is released, it is desired to pay out the load at a constant speed of 18 rpm. What will be the rate of heat generation in the brake?
(f) What is the maximum deceleration rate of the load when the paying out process is stopped if the braking torque is not to exceed the load torque by more than a factor of 5?

Solution

(a) The moment of inertia of the winch drum is

$$I_d = 300\left(\frac{5}{12}\right)^2 = 52 \text{ lb-ft}^2$$

The reducer input speed is 1750 rpm. Therefore, the ratio of the reducer is

$$r = \frac{1750}{9} = 194.44$$

The drum inertia referred to the reducer input shaft (and hence the clutch output shaft) is then

$$I_{rd} = \left(\frac{1}{r}\right)^2 I_d = \frac{52}{(194.44)^2} = 0.0014 \text{ lb-ft}^2$$

We see that this inertia is negligible when referred to the reducer input shaft. However, it does not include any potential inertia of the load attached to the wire rope.

The total inertia of the components attached to the clutch output shaft is

$$I_2 = 0.0014 + 0.326 + 0.20 + 1.0 + 0.5 + 0.5$$
$$= 2.53 \text{ lb-ft}^2 = 0.0786 \text{ slug-ft}^2$$

The total inertia of the components attached to the clutch input shaft is

$$I_1 = 6 + 0.20 + 12 = 18.2 \text{ lb-ft}^2 = 0.565 \text{ slug-ft}^2$$

(b) The load torque is

$$T_l = 21{,}000\left(\frac{6}{12}\right) = 10{,}500 \text{ ft-lb}$$

Referred to the reducer input shaft (or clutch output shaft), this is

$$T_{rl} = \left(\frac{1}{r}\right) T_l = \frac{10{,}500}{194.44} = 54 \text{ ft-lb}$$

(c) The prime mover (motor) torque is 70 ft-lb during clutching; therefore, from Eq. (3-19), the clutching time δt is

$$\delta t = \frac{0.565(0.0786)\left(\frac{1800(2\pi)}{60} - 0\right)}{5(54)(0.565 + 0.0786) - 0.565(54) - 0.0786(70)} = 0.061 \text{ second}$$

Then the motor speed at the end of clutching will be given by Eq. (3-18):

$$\omega_1 = \frac{1800(2\pi)}{60} + \left(\frac{70 - 5(54)}{0.565}\right)(0.061) = 167 \text{ rad/sec} = 1594 \text{ rpm}$$

(d) The heat generated is given by Eq. (3-27):

$$W = \frac{5(54)\left(\frac{1800(2\pi)}{60} - 1\right)^2}{2\left(\frac{5(54) - 70}{0.565} + \frac{5(54) - 54}{0.0786}\right)} = 1546 \text{ ft-lb}$$

(e) The rpm of the brake will be

$$\omega_b = 194.44(18) = 3500 \text{ rpm} = 366.52 \text{ rad/sec}$$

and the torque exerted by the brake must be

$$T_b = 54 \text{ ft-lb}$$

Thus, the brake work will be

$$W_b = 54(366.52) = 19{,}792 \text{ ft-lb/sec}$$

(f) The deceleration rate is determined from Eq. (3-37):

$$5(54) - 54 = 0.0786 \frac{d\omega_2}{dt}$$

$$\frac{d\omega_2}{dt} = 2755 \text{ rad/sec}^2$$

Then the drum deceleration is

$$\frac{d\omega_d}{dt} = \frac{2755}{194.44} = 14.17 \text{ rad/sec}^2$$

The linear deceleration of the load will then be

$$\frac{d(\text{load})}{dt} = 14.17\left(\frac{6}{12}\right) = 7.085 \text{ ft/sec}^2$$

■

3-5 CLUTCH TYPES

Many different types of clutches and brakes have been developed to fulfill the various functions required in application. In addition to the wide variety of clutch configurations, there are several methods of clutch actuation. Table 3-2 lists some of the more common types. References 3-5 and 3-6 can be consulted for detailed information.

The function of a clutch is to permit the smooth and gradual connection of two rotating members in order to transmit power. A brake is a clutch with one of the members simply at rest. Clutches may be designed based on fluid, electromagnetic, or mechanical (or dry friction) traction principles. Here we are concerned only with the latter type; fluid and electromagnetic clutches will be discussed in later chapters.

The basic engineering design of dry friction clutches involves simple geometrical, statical, material, thermal, and wear considerations, which are discussed in any number of good machine design textbooks (see References 3-4 and 3-7). In general, the objective is to develop a useful relationship for the torque capability of the clutch, T_c, in terms of the actuating force, F_a, supplied to it. Of course, this relationship involves geometrical factors and principles of statics as well as the coefficient of friction, f. Additionally, learning the detailed technology of specific clutch devices means understanding the appropriateness of the engineering assumptions that must inevitably be made.

It is not our purpose here to become involved in the details just mentioned, because they are thoroughly covered in the references cited. We merely want to consider some of the many devices available and their methods of actuation. Simply keep in mind that for all of these devices there is a well-known functional relationship of the type

$$T_c = F(F_a, f, \text{geometry}) \tag{3-46}$$

TABLE 3-2. Clutch Configurations and Actuation Methods.

Clutch types

1. Friction
   - Disk or other axial types
   - Band and drum
   - Cone types
   - Drum with internal shoes
   - Drum with external shoes
2. Over-running
   - Ratchet and pawl
   - Wrap spring
   - Sprag
   - Roller-ramp
3. Others
   - Magnetic particle
   - Jaw types
   - Detent
   - Overload

Actuation methods

1. Mechanical linkage
2. Hydraulic
3. Pneumatic
4. Electromagnetic
5. Centrifugal
6. Mechanical spring

3-5-1 Frictional Engagement Clutches

Disk Clutch. The most common type of frictional clutch is the disk type. The basic configuration is shown in Fig. 3-20.

If it is assumed that the pressure created at the clutch interface (by the clamping force F_a) is uniform over the clutching area, it is easy to show that the clutching torque T_c is given by

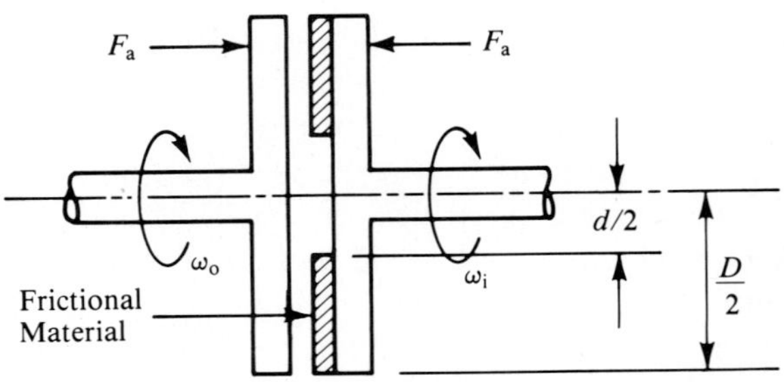

Figure 3-20. Basic disk clutch configuration.

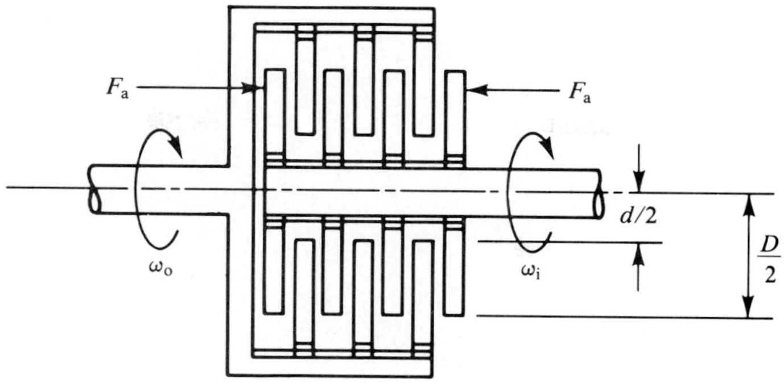

Figure 3-21. Multiple-disk clutch configuration.

$$T_c = \frac{f \cdot F_a}{3} \left(\frac{D^3 - d^3}{D^2 - d^2} \right) \tag{3-47}$$

where f is the coefficient of friction of the friction material.

More braking torque can be created for a given axial clamping force simply by increasing the number of frictional disk interfaces. The scheme in Fig. 3-21 shows four disks internally splined to the input shaft and three disks externally splined to a barrel connected to the output shaft. When the clutch is disengaged, the disks are free to shift along their splines slightly (axially) in order to separate themselves and prevent torque from being transmitted. For clutch engagement, the disks are clamped together by an axial force F_a which can be supplied mechanically, pneumatically, hydraulically, or electromagnetically. The movable disks shift axially along their splines until they physically touch and are clamped together. The clutching torque now becomes

$$T_c = \frac{f \cdot F_a}{3} \left(\frac{D^3 - d^3}{D^2 - d^2} \right) N \tag{3-48}$$

where N is the number of frictional contact interfaces created.

Multiple-disk clutches can be used as brakes simply by fixing one set of the disks to the chassis of the machine upon which the brake is mounted. These disk brakes are often designed in such a way that the clamping force F_a is provided by powerful mechanical springs. This device is called a spring-set brake. To disengage a spring-set brake, the spring force F_a must be opposed by a mechanical, hydraulic, or pneumatic de-actuating force. Such a brake is often called a fail-safe brake because it will automatically be set should the de-actuating force be disrupted accidentally. An example is shown in Fig. 3-22.

Cone Clutch. Another clutch configuration in which engagement occurs by axial translation is the cone (and cup) clutch sketched in Fig. 3-23. Analysis of the statics and geometry of this device will show that if the clutching pressure is assumed uniform, the relationship between clutch torque T_c and actuating force F_a is

$$T_c = \frac{f \cdot F_a}{3 \sin \alpha} \left(\frac{D^3 - d^3}{D^2 - d^2} \right) \tag{3-49}$$

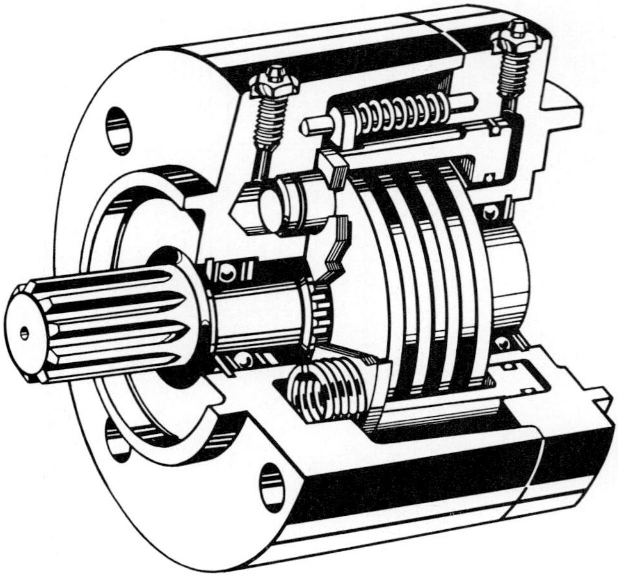

Figure 3-22. Fail-safe brake. (Courtesy of Auto Specialties Mfg. Co.)

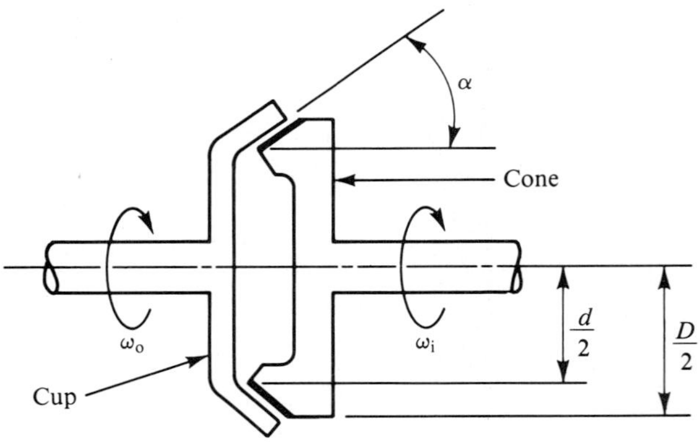

Figure 3-23. Cone clutch.

Cone clutches must be designed such that the cone angle is confined to the range 10 to 15 degrees. If the angle is made too small (8 degrees or less), the clutch may grab or require a large force to disengage it.

Internal Shoe and Drum Clutch (Centrifugal). The clutch or brake mechanism that uses a drum and internally expanding shoes is well known as a result of its application for many years in automotive braking. An illustration of this clutching device is shown in Fig. 3-24 where the internal shoes are actuated (expanded) by centrifugal action.

Figure 3-24. Internal shoe and drum brake. (Courtesy of the Hilliard Corporation.)

Centrifugal clutches of the type shown in Fig. 3-24 are used in applications requiring one or more of the following:

(a) Soft-start
(b) Overload protection
(c) Automatic or delayed engagement at a predetermined speed

The soft-start qualities of the centrifugal clutch help allow standard NEMA Design B motors and smaller motors to be used because motor size will not be dictated by extreme starting load conditions. Soft start also permits the use of smaller service factors in selecting gear reducers, belt, and chain drives.

The delayed engagement feature makes these clutches suitable in conveyor applications (or similar mechanical systems) in order to retard speed should the conveyor try to run away because of overload. In this application, the drum would be fixed, so the device would function as a brake.

Under overload conditions for an extended period of time, many prime movers will be damaged unless protected by an unloading device. Protection can be accomplished by using a centrifugal clutch as a coupling.

When used as a brake, the drum and internal shoe mechanism can be actuated mechanically, hydraulically, or pneumatically.

External Shoe and Drum Brake. The drum and shoe type clutch or brake device can be arranged such that the shoes are external to the drum and must be contracted to engage shoes and drum. One such design is shown in Fig. 3-25. The brake shown is a spring-set, solenoid-released component often found in crane and mill applications where service requirements are severe. For many years, these brakes used asbestos brake linings, but they are now available using non-asbestos materials.

Figure 3-25. External shoe and drum brake. (Courtesy of the Square D Company.)

The torque versus actuating force relationships for drum and shoe brakes (or clutches) are implicit rather than explicit because of the complicated geometry. These relationships can be found in any good machine design book.

3-5-2 Miscellaneous Clutches

In addition to the conventional dry friction clutches discussed above, a large number of ingenious couplings have been devised for special machine functions. Many of these couplings are designed so that slip will occur for one direction of rotation of the input while engagement will occur for rotation in the opposite direction. Perhaps the simplest of these clutches is the ratchet and pawl type, shown in Fig. 3-26. The mechanism shown has two pawls so that the permitted rotation can be either cw or ccw.

An over-running clutch transmits torque in one direction only. It releases when input rotation is reversed or when the output overspeeds the input in the drive direction. The ratchet and pawl mechanism can be used as an over-running clutch. Another type of over-running clutch, called a roller-ramp clutch, is shown in Fig. 3-27.

Assume that the inner part of the clutch in Fig. 3-27 is rotating cw. Then the rollers will be rolled up the ramps machined into the inner part so they will lock (by a wedging action) the inner and outer parts together as in a traction drive. If

Figure 3-26. Ratchet and pawl clutch. (Courtesy of the Lowell Corporation.)

the outer part overspeeds in the cw direction, the rollers will be released (rolling down the ramps) so the output can free-wheel independently of the inner portion (which is the input shaft). The same condition will occur if the inner part is rotated ccw.

A variety of clutches use this over-running lock and release type of action. One of the most versatile examples is the sprag clutch in which the rollers of Fig. 3-27 are replaced by specially designed dogs called sprags (see Reference 3-6).

Figure 3-27. Roller-ramp clutch. (Courtesy of Zurn/ Marland Corporation.)

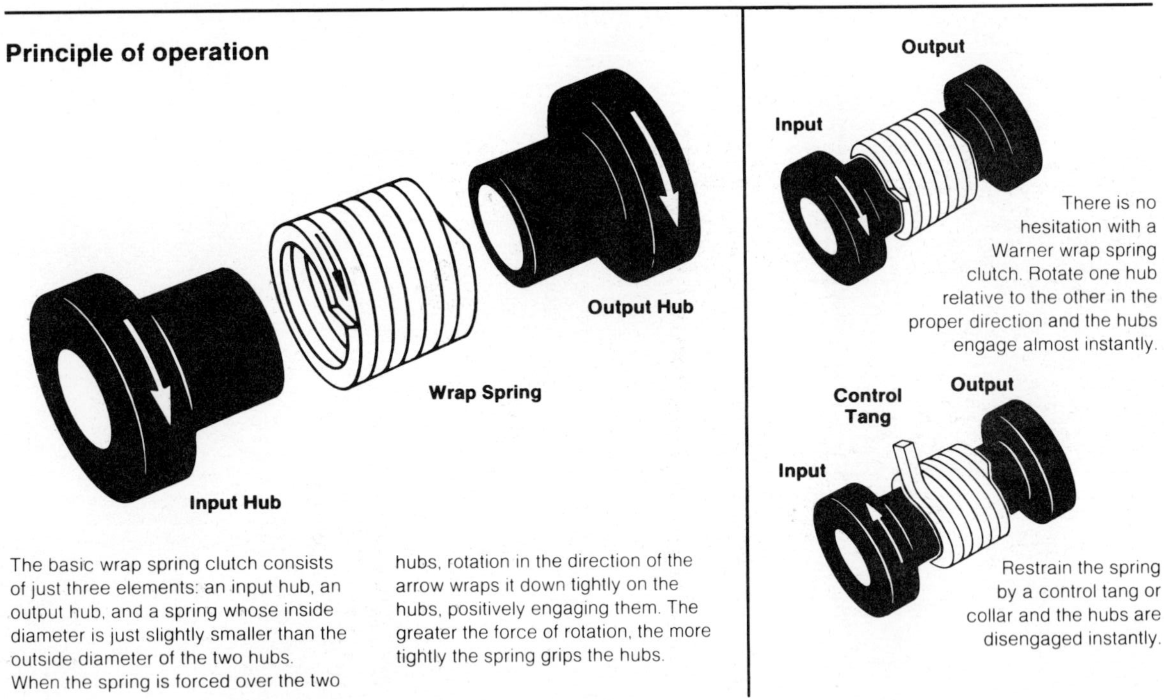

Figure 3-28. Wrap spring clutch. (Courtesy of Warner Electric Brake and Clutch Company, South Beloit, Illinois.)

The roller-ramp clutch principle is also used in single-revolution clutches. These clutches have application in many machines requiring control of intermittent motion from a constantly rotating input shaft. When the clutch is actuated, positive engagement will occur for precisely one revolution, after which time the clutch automatically disengages. This principle can be used to obtain both a fractional and an integral number of revolutions on demand.

Another useful clutch design in which an input shaft can be engaged for a controlled number of revolutions (while providing over-running free-wheeling in the other direction) is the wrap spring design shown in Fig. 3-28. The basic wrap spring clutch consists of three elements: an input hub, an output hub, and a spring with an inside diameter just slightly smaller than the outside diameter of the two hubs. When the spring is forced over the two hubs, rotation in the direction of the arrow wraps the spring tightly on the hubs, positively engaging them. The greater the torque, the more tightly the spring grips the hubs.

The clutch is released by a tang on the spring which must be externally restrained. This release causes the gripping action of the spring to disappear so the hubs are disengaged. This type of actuation can be incorporated into more complex clutch-brake configurations to achieve more complex engagement and braking applications.

3-6 SUMMARY

The variable-speed and clutch devices discussed in this chapter are but a few of the dazzling array of devices available to the mechanical systems designer. The neophyte designer should begin to develop a library of these components by obtaining catalogs from manufacturers. The kinematic and dynamic capabilities of these mechanisms can be determined only from engineering data contained in these catalogs.

Catalog data can also teach the designer many things not covered in a formal engineering education. As an example, an excellent treatise on one-way clutches is available from an industrial source (see Reference 3-8).

3-7 REFERENCES

3-1. F. W. Heilich III and E. E. Shube, *Traction Drives*, Marcel-Dekker, New York, 1983.

3-2. S. H. Loewenthal, "A Historical Perspective of Traction Drives and Related Technology," reprint from *Advanced Power Transmission Technology*, Report CP-2210, NASA, Cleveland, OH, 1981.

3-3. F. Najlepszy, "Traction Drives Roll Up Impressive Gains," *Machine Design*, October 24, 1985.

3-4. V. M. Faires, *Design of Machine Elements*, 4th ed., Macmillan, New York, 1965.

3-5. W. C. Orthwein, *Clutches and Brakes*, Marcel-Dekker, New York, 1986.

3-6. *Clutches and Brakes*, Formsprag Division of Dana Corporation, series of reprints from *1973–74 Power Transmission and Bearing Handbook*, Industrial Publishing Co., Pittway Corporation, 1972.

3-7. A. H. Burr, *Mechanical Analysis and Design*, Elsevier, New York, 1981.

3-8. *One-Way Clutch Design Guide Types, Selection, Applications*, available from Spring/Brummer Division of Borg-Warner Corporation, Bellwood, IL.

PROBLEMS

3-1. The 9 hp differential-draw transmission sketched in Fig. P3-1 consists of a 4:1 PIV speed variator (whose input shaft is run at 900 rpm) and 10:1 reduction spur gearing between the output shaft of the speed variator and one of the side gears of the bevel gear differential. Note that the input shaft of the PIV is also directly connected to the spider (carrier) of the bevel gear differential and that the output of the entire transmission is the other side gear. Determine

(a) The maximum and minimum output speeds of the PIV output shaft

(b) The maximum and minimum speeds of the side gear of the bevel gear differential to which the spur gearing is connected

(c) The maximum and minimum speeds of the side gear of the bevel gear differential to which the output shaft of the differential-draw transmission is connected

(d) The percent variation in speed of the output shaft of the PIV

(e) The percent variation in speed of the output of the entire transmission

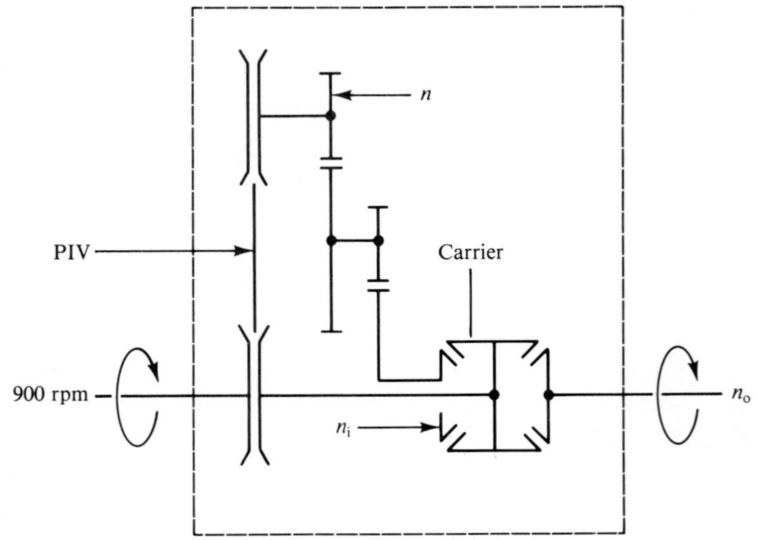

Figure P3-1. Figure for Problem 3-1.

3-2. A 5 hp wide-speed-range, variable-speed differential transmission is sketched in Fig. P3-2. Its characteristics are shown in the accompanying table. Assume that the ratio of the PIV in this transmission is $R = 4:1$. For the conditions indicated for drive size 4 of the table, determine the numerical value of the gear ratio $N_{12} = N_1/N_2$.

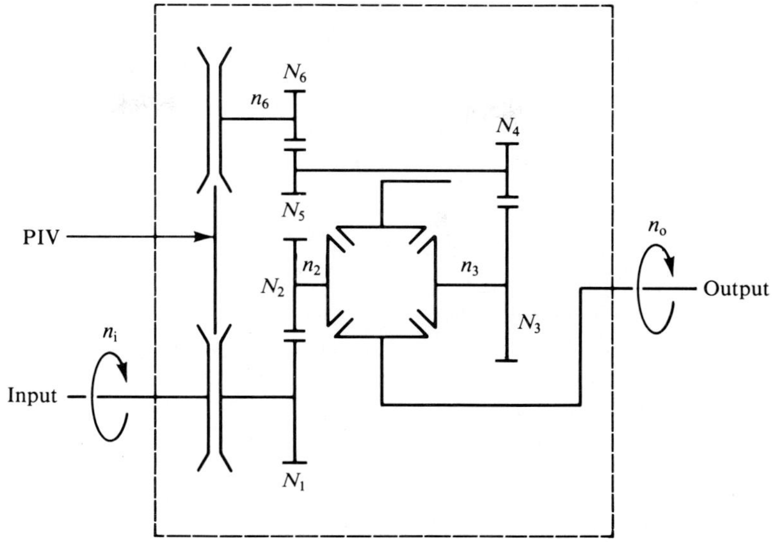

Wide Speed Range, High Speeds

Drive Size	Maximum Constant Speed Shaft (rpm)*	Output Horsepower, Torque, and Speed: At Maximum Speed Setting hp	Torque (lb-in)	rpm	At Minimum Speed Setting hp	Torque (lb-in)	rpm	Constant Torque Capacity (lb-in)
½	690	1	155	403	0	315	0	155
1	860	2	210	605	0	415	0	210
2	860	3.38	330	645	0	660	0	330
3	860	5	490	645	0	975	0	490
4	690	6.75	820	518	0	1645	0	820

*Lower input speed results in proportionately reduced output speed and horsepower capacity.

Figure P3-2. Figure and table for Problem 3-2.

3-3. Explain why a PIV speed variator might produce a fluctuation in output speed for fixed input speed as the power being transmitted goes from no-load to full-load.

3-4. The power train sketched in Fig. P3-4 includes a prime mover, a ratioing component (gearbox), a load, and a clutch. The clutch makes possible the disengagement and engagement of the prime mover from the load. Discuss the pros and cons of locating the clutch between the prime mover and the gearbox or between the gearbox and the load. Discuss in terms of speed, torque, power, and the ratio r of the ratioing component.

3-5. Design a ring-roller traction drive that will provide a speed range of the output from -400 rpm to $+400$ rpm (through 0 rpm). Specify the following and make a sketch of the geometric configuration:

(a) Diameter of the roller ring

(b) Diameter and taper of the cone roller

(c) Length of the cone roller

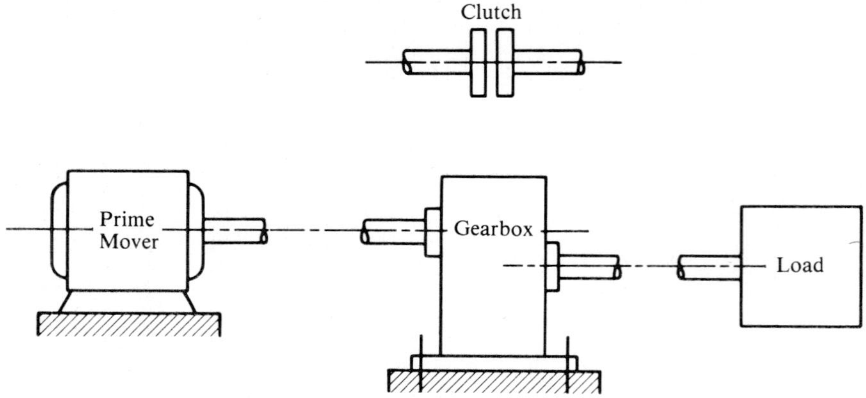

Figure P3-4. Figure for Problem 3-4.

(d) Number of teeth and diametral pitch of the planet gear
(e) Number of teeth and diametral pitch of the ring gear

The design is to be kinematic only, unless you also want to investigate power transmission capability. Use an input speed of 1150 rpm.

3-6. Determine the speed relationship between input and output speed of the ring-cone speed variator sketched in Fig. 3-13. Use the epicyclic speed formula of Eq. (2-14), with numbers of gear teeth replaced by diameters at contact points in the ring-cone traction drive. (Hint: This device is similar to Humpage's epicyclic bevel gear train. But since there is no shaft attached to the carrier, eliminate the carrier speed from the speed formula.)

3-7. A simple planetary gear train is operated with the carrier (arm) as the input and the sun gear as the output. Assume the ring gear to be fixed. Determine the moment of inertia of the gearbox referred to the input shaft. Use the following symbols:

M_p = the mass of each planet gear

I_p = the moment of inertia of a planet gear about its axis

n = the number of planets

V_p = the linear velocity of the planet center

ω_p = the angular velocity of a planet about its axis

I_s = the moment of inertia of the sun gear

ω_s = the angular velocity of the sun gear

I_c = the moment of inertia of the carrier

ω_c = the angular velocity of the carrier

3-8. A frictional clutch is used to drive a load through a gearbox whose speed ratio is r. The moment of inertia of the load is I_ℓ and the load torque is T_ℓ. Assume that the inertia of the gears in the gearbox is I_g when referred to the gearbox input shaft. Determine the equivalent torque T_e and inertia I_e of this system as seen by the clutch (see Fig. P3-8).

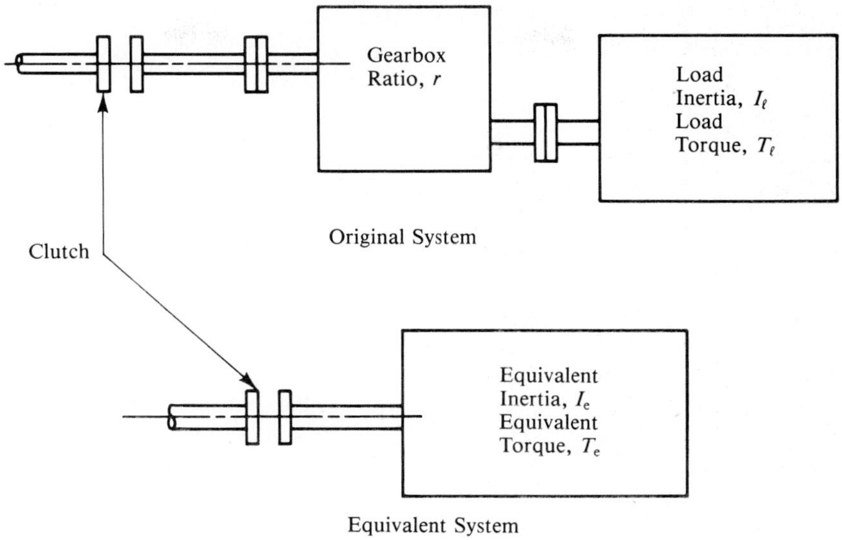

Figure P3-8. Figure for Problem 3-8.

3-9. A disk clutch is to be used as an overload protection device for a conveyor system drive. The conveyor is driven by a 1200 rpm, 7.5 hp induction motor by means of a gearbox having a ratio r of 160. The disk clutch is set for an axial force F_a of 275 lb by means of springs, and the brake lining has radial dimensions $d = 2.5$ inches, $D = 3.75$ inches. The clutch is used as a coupling on the motor shaft between motor and gearbox. If the coefficient of friction for impending slip is $f = 0.4$, at what percentage of rated motor torque will this clutch slip? Assume there are $N = 4$ disk interface surfaces.

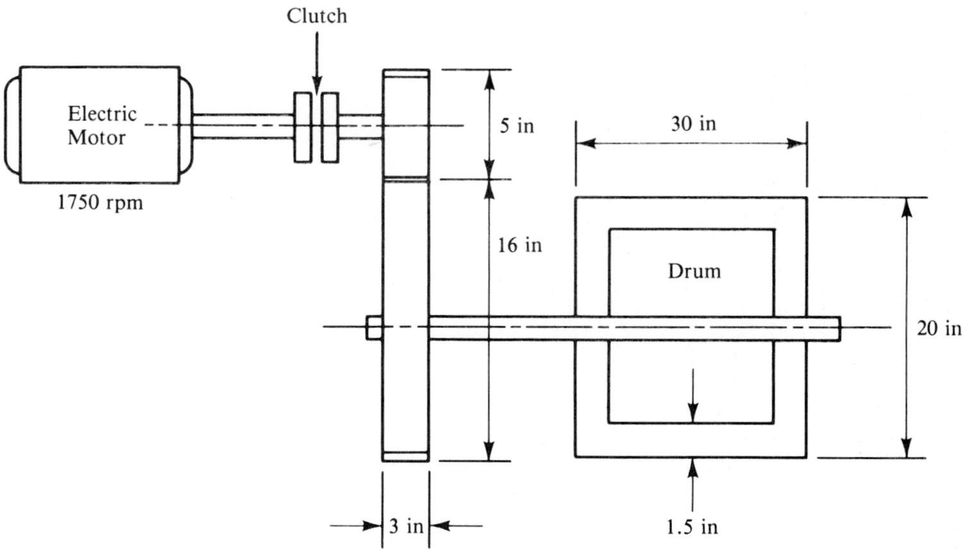

Figure P3-11. Figure for Problem 3-11.

3-10. Using band brake theory, show that the torque capacity of a wrap spring clutch (when engaged) is

$$T = \frac{pbD^2}{4}(e^{2\pi Nf} - 1)$$

where N is the number of wraps (coils) of the spring, f is the coefficient of friction, D is the spring diameter (mean), b is the width of a single wrap, and p is the pressure between spring and hub at assembly.

3-11. Figure P3-11 shows a drum that is driven by a 1750 rpm induction motor through a clutch and gear pair coupling. Assume that the materials are steel and that the drum is accelerated from rest to full-load speed in 1 second. What clutching torque will be required to do this? You can neglect the rotary inertia of the clutch.

3-12. A winch drum is lowering a 1000 lb load at a speed of 60 ft/min. If the rotary inertia of the drum is 100 lb-ft^2, what will be the braking torque (exerted on the drum shaft) required to bring the load to rest uniformly in 0.50 second?

Part B

HYDRAULIC COMPONENTS

In the previous three chapters, we focused on strictly mechanical power transmission components. The design of such components and their incorporation into mechanical systems have traditionally been within the province of the mechanical engineer. Significant design work with mechanical systems, however, will inevitably involve electrical, hydraulic, pneumatic, thermal, and control components in addition to mechanical components. Therefore, the mechanical engineer who is successful in advancing into a supervisory design, or management, position will be the one who deliberately chooses to broaden his or her technical knowledge to include electrical, hydraulic, thermal, pneumatic, and control components.

To illustrate this point, consider a typical machine for which a chief engineer, or manager of engineering, would have overall project responsibility (see Fig. B-1). The machine shown is a plastic injection molding machine. It involves a conveyor and hopper for supplying and holding granular plastic material from which plastic parts are to be made. From the hopper, the granular material is fed into a cylinder where it is softened by electrical resistance heaters or steam heaters. The dies, within which the plastic part will be formed, are closed and clamped using hydraulic cylinders. After the dies are closed, the plastic material is forced into the die cavity with tremendous force by a hydraulically actuated plunger. The plastic is cured with heaters while in the die cavity. After curing, the dies are hydraulically retracted and the part is extracted.

Not only must the engineer design (or select) all of these components but he or she must control the sequence of events that are to occur. And this must be done in such a way as to mitigate all hazards so that the machine operator will not contact them. Additionally, the design engineer must select the hydraulic pumps, valves, cylinders, and motors. And, finally, the machine will incorporate various electrical switches, sensors, and control relays for controlling the position of cylinders, ram, and safety closures.

In the next four chapters, we shall discuss the fluid power components and transmissions incorporated into mechanical systems. We will begin by discussing a common fluid power circuit which can be used to obtain the alternating extension and retraction of two hydraulic cylinders. We will then continue through both hydrostatic and hydrodynamic drives and end the discussion with hydroviscous clutches.

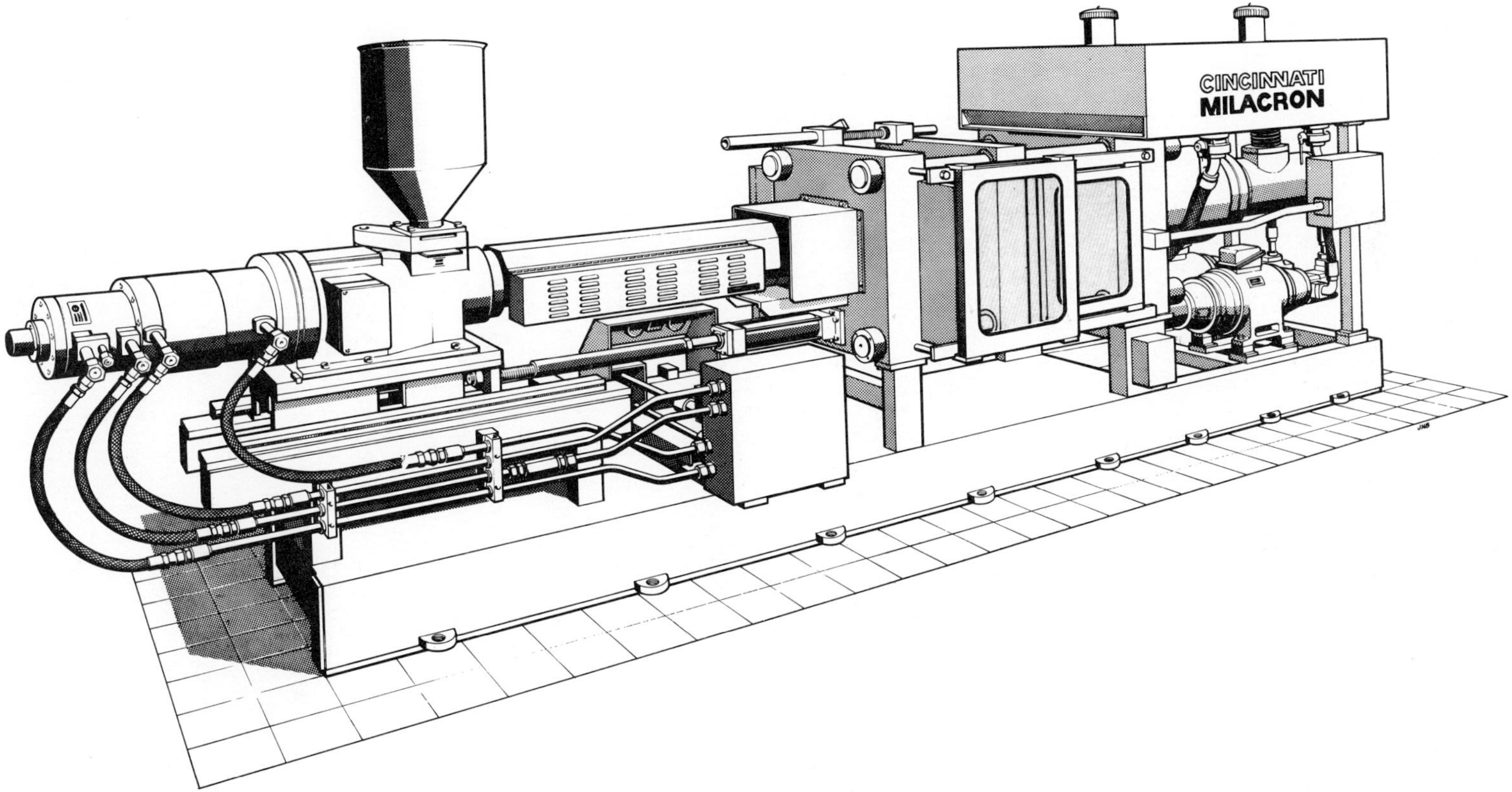

Figure B-1. Injection molding machine. (Courtesy of Cincinnati Milacron.)

4

Fluid Power Circuit Components

Our objective in this chapter is to present some simple fluid power circuits in order to show what kinds of components are involved and to develop an intuitive feel for how they behave.

Figure 4-1 shows a fluid power circuit diagram, a sequence diagram, and an electrical ladder circuit diagram for causing two hydraulic cylinders to be actuated alternately. Qualitatively, the system operates in the following way. The cycle is initiated by means of a START pushbutton (IPB). This action causes a valve to admit oil to the rodless (blank) end of cylinder 1 which forces the cylinder's piston to the right. Eventually, a cam mounted on the rightward-moving piston of cylinder 1 contacts the limit switch labeled 1LS. Actuation of 1LS causes two events to be initiated simultaneously. The valve for cylinder 1 is reversed, which causes the cylinder's piston to reverse direction and move to the left. At the same time, the valve for cylinder 2 is caused to admit oil to its blank-end, forcing the cylinder's piston to the left. Eventually, the cam on the leftward-moving piston of cylinder 2 contacts limit switch 2LS. Actuation of 2LS causes a reversal of direction of the piston of cylinder 2, back toward the right. In the meantime, the piston of cylinder 1 will stop its leftward movement when it reaches the left end of the cylinder. Later, the piston of cylinder 2 will also reach its extreme rightward position, completing the cycle. When both pistons are stalled against their cylinder heads, the output of the pump, which has been powering the pistons, will be vented through a relief valve back to the reservoir (tank).

We will return to this system later and describe in greater detail how it is controlled. But, first, it will be helpful to understand what some of the unique features of such hydraulic systems are, since they account for the widespread use of fluid power in certain types of mechanical systems (e.g., in mobile earth-moving equipment).

Some advantages of using hydraulic (or fluid) power in mechanical systems are:

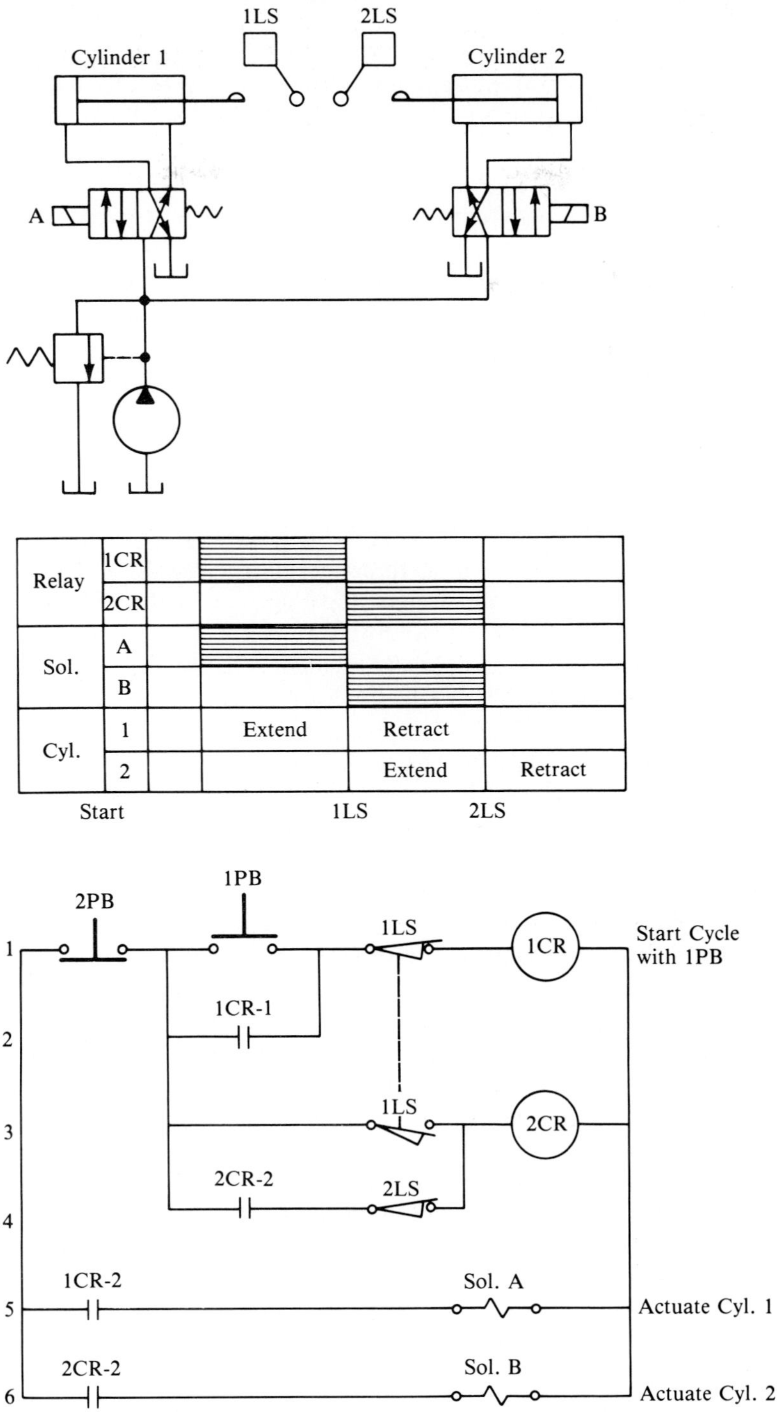

Figure 4-1. Two hydraulic cylinders acting sequentially.

(a) *High torque density.* The torque developed by a hydraulic actuator is proportional to pressure differences, of which values of 5000 psi and higher can be readily sustained. In electrical actuators, torque development is limited by magnetic saturation of the actuator materials.

(b) *High torque to inertia ratio.* This is a consequence of the same factors outlined in item (a). The high torque to inertia ratio means that hydraulic actuators have a high speed of response with high acceleration and deceleration and speed reversal. In the controls sense, high gain and bandwidth can be obtained by using hydraulic actuators in servo-control systems.

(c) *High power density.* Again, this is a consequence of the factors outlined in item (a), enhanced by the good heat convection provided by the oil flow. Horsepower to weight ratios of up to 2.5 to 3 hp/lb can be obtained. These ratios are critical in mobile and airborne power applications.

(d) *High stiffness.* The torque-speed characteristic is steep, which means that relatively constant speed can be maintained independent of load. Additional control advantages accrue to the high stiffness of hydraulic actuators used by closed-loop systems.

(e) *Convenience in connecting and controlling remotely located fluid power components.* Hydraulic lines are more flexible than the links, chains, and levers required in linking mechanical components. Of course, this flexibility does not exceed that of electrical components.

The major disadvantages of using fluid power components are:

(a) *Contamination of the hydraulic fluid.* It is difficult to prevent dirt and crud from clogging precision clearances designed into fluid actuators and valves. Extensive filtering systems must be used to insure cleanliness of the oil. This is, perhaps, the most serious disadvantage of using fluid power.

(b) *Leaks.* Both internal and external leakages are inevitable. Special provisions must be made to accommodate such leakages. Line ruptures and bursts can also create the hazard of high-velocity oil jets (and most likely the oil will be hot!). The oil might also be flammable!

(c) *Controllability.* Hydraulic systems are not as easy to control as electrical systems. And if hydraulic components are integrated with electrical controls, special interfaces are required. Also, electrical power sources are readily available and dependable in very large supply whereas the hydraulic system made independent of electrical prime movers will necessarily be limited by the engine prime movers available.

In summary, hydraulic systems are attractive in applications requiring high torque amplification or the packing of high power into a small space. Where lower power densities are acceptable, either mechanical, electrical, or electromechanical actuators are usually preferred.

4-1 PUMPS

The necessity of having a fluid flowing in, or as part of, a mechanical system arises frequently. Simply consider air conditioning systems, pipelines, water supply systems, boilers, heat exchangers, injection molding machines, hydraulic presses, machine tools, industrial machinery, automobiles, airplanes, forklifts, backhoe diggers, bulldozers, cranes, and so on. In some of these systems the fluid flow is an end in itself (e.g., in a pipeline), while in others it may be only an adjunct to a more fundamental purpose (e.g., the control of a machine tool). Nevertheless, all of these systems fall within the design province of the mechanical engineer.

In systems where the fluid flow itself represents the major system function, we usually find non-positive-displacement pumps (NPDPs, also called dynamic pumps). Often, these pumps will be delivering rather large volumetric flows, at relatively low heads, of relatively low viscosity fluids. An example is the centrifugal pump used to supply feed water to a boiler. It is not this kind of pump, or system, that is of interest to us, per se, even though the dynamic relationships for such pumps are pertinent to fluid couplings and torque converters, which are of interest.

In systems where the fluid flow is being utilized as part of a control or torque (or force) amplification scheme, we almost always find positive-displacement pumps (PDPs) being used. An example is the gear pump used to supply oil to the piston driving a linkage of a backhoe digger. In any event, we see that pumps are hydromechanical components about which the mechanical systems designer must be knowledgeable.

4-1-1 Non-Positive-Displacement Pumps (NPDP)

A good example of a NPDP device is the centrifugal type illustrated in Fig. 4-2. For centrifugal pumps of the kind shown, there is a substantial clearance between the periphery of the impeller and the casing (volute). As the pressure rises, against which this type of pump works, there will be an increased back-leakage between the impeller and casing such that the flow delivered will drop off. The existence of this substantial back-leakage precludes there being a definite relationship, independent of head, between the fluid delivered and the angular displacement of the impeller. As a matter of fact, this type of pump can have its exit completely blocked off, so that there will be no flow delivered at all, without destroying or stalling the pump. With the exit blocked off, the impeller will simply churn away without delivering any output flow; all of the input power will be dissipated into the heating of the fluid. You can easily show this with a hair dryer which is simply a small pump of the NPDP type.

Don't try this with a positive-displacement pump (PDP)! The pump will stall (and possibly damage its prime mover), or the pump casing will break (or even possibly explode). A PDP is one in which, ideally, the amount of fluid delivered by the pump, per revolution of the impeller shaft, is independent of the pressure head against which the pump works; its volumetric flow depends only on pump

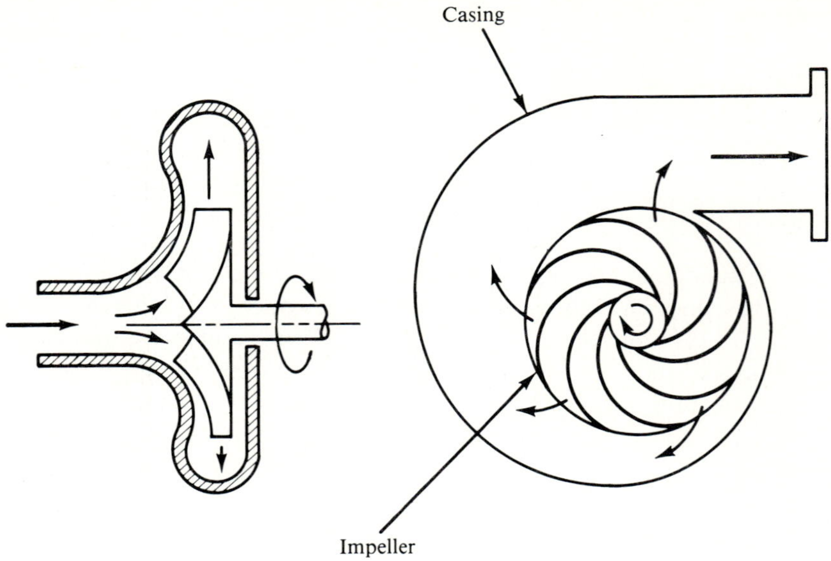

Figure 4-2. Schematic diagram of a centrifugal pump.

rpm. A close-fitting piston in a cylinder is an example of such a pump. Since there can be virtually no back-leakage, the fluid must be delivered if the piston moves (of course, there may in reality be a small back-leakage past the piston rings so that even a piston in a cylinder will not be a perfect PDP). Systems incorporating PDPs will always include relief valves to vent the fluid being pumped if a flow blockage develops somewhere in the system.

Figure 4-3 shows the basic difference between the operating characteristics of NPDP and PDP types. Note the drooping characteristic of the NPDP which makes the developed head drop off with flow rate. The flow rate of the PDP is independent of the developed head (when internal leakage is neglected). The

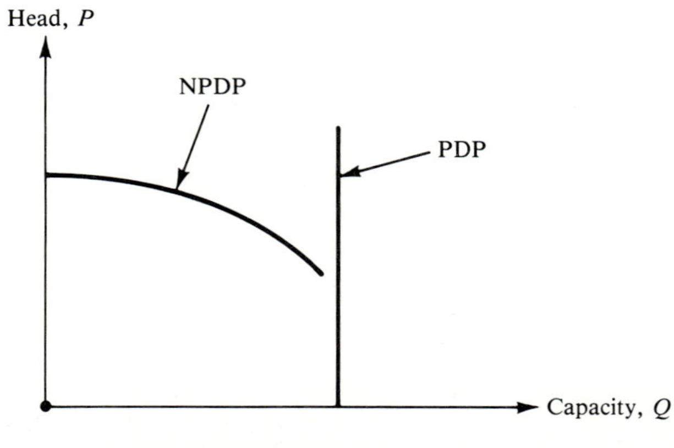

Figure 4-3. Comparison of pump characteristics.

centrifugal pump (NPDP) does not depend, for its operation, upon a close fit between the impeller and casing as does the PDP. The principle upon which it operates is dynamic rather than just hydrostatic.

We are interested in pumps as components in mechanical systems and in those characteristics of pumps that will permit us to integrate a pump into a system in a compatible (if not optimum) way. For a centrifugal pump, the most important characteristic will be its head (P) versus capacity (Q) output characteristic curve. Some of the essential information about this characteristic can be obtained from elementary principles of fluid mechanics. All that is really required is a kinematic analysis of impeller and fluid velocities along with the angular momentum principle of fluid mechanics. These principles will be developed in terms of the nomenclature shown in Fig. 4-4 for an idealized NPDP impeller configuration.

In Fig. 4-4, a single blade of an impeller is shown rotating at angular velocity ω. The blade velocity of the outer tip of the backward-curved blade is $u_2 = \omega r_2$, while at the inner tip the blade velocity is $u_1 = \omega r_1$.

Quantities w_1 and w_2 are the fluid velocities, relative to the blade, at entrance and exit of the impeller. As the diagram shows, quantities V_1 and V_2 are the absolute fluid velocities at inlet and exit as compounded vectorially from the blade and relative velocities. With the blade angles β_1 and β_2 defined as shown in the figure, we have

$$V_{t(2)} = u_2 - V_{n(2)} \cot \beta_2 \tag{4-1}$$

$$V_{t(1)} = u_1 - V_{n(1)} \cot \beta_1 \tag{4-2}$$

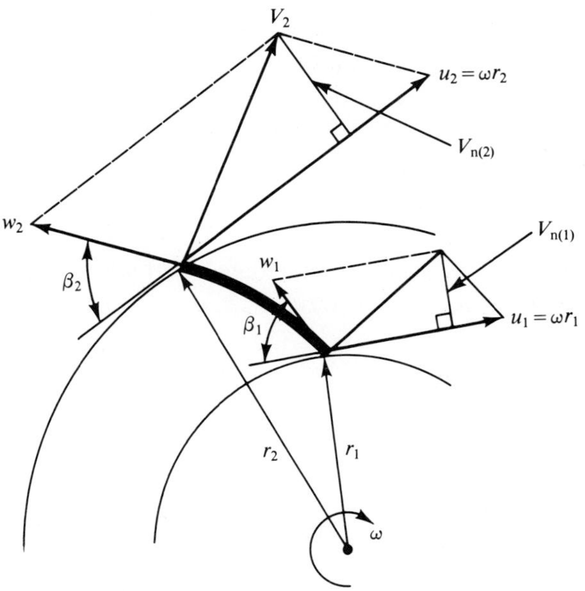

Figure 4-4. Velocity diagram of an idealized impeller blade.

where quantities V_t and V_n are components of the absolute fluid velocity V along the directions tangent and normal to the circles of radius r centered about the axis of the pump impeller.

Quantities V_t and V_n are the important velocities because the angular momentum and volumetric flow are expressed in terms of them. The angular momentum theorem states that the torque T that must be applied to this fluid, in order to achieve the flow configuration shown, is

$$T = \rho Q(r_2 V_{t(2)} - r_1 V_{t(1)}) \tag{4-3}$$

while the volumetric flow rate Q depends on the normal components as follows:

$$Q = 2\pi r_1 b_1 V_{n(1)} = 2\pi r_2 b_2 V_{n(2)} \tag{4-4}$$

Dimension b represents the blade width, and Eqs. (4-4) express the steady continuity equation for an incompressible fluid.

The hydraulic power developed (often called the water horsepower) is

$$\text{HP}_w = \omega T = \rho Q(u_2 V_{t(2)} - u_1 V_{t(1)}) \tag{4-5}$$

The power can also be expressed in terms of the pump total head P, fluid specific weight γ, and volumetric flow rate Q as

$$\text{HP}_w = \gamma Q P \tag{4-6}$$

where

$$P = \frac{p_2}{\gamma} + \frac{V_2^2}{2g} + y_2 - \left(\frac{p_1}{\gamma} + \frac{V_1^2}{2g} + y_1\right) \tag{4-7}$$

In the above analysis the frictional effects have been neglected.

We can estimate the effect of the exit blade angle β_2 by assuming that $\beta_1 = 90°$, which is tantamount to saying that the inlet angular momentum is negligible (or that straightener vanes have been used at the pump inlet). Then, by combining Eqs. (4-1) and (4-3) through (4-6), and neglecting $V_{t(1)}$, we find that

$$P = \frac{u_2^2}{g} - \left(\frac{u_2 \cot \beta_2}{2\pi r_2 b_2 g}\right) Q \tag{4-8}$$

Admittedly, Eq. (4-8) has been developed neglecting frictional effects, but it nevertheless shows some important effects of the blade geometry as shown in Fig. 4-5.

The effects shown will not persist for large flow rates because we know the head must ultimately fall off for large flow rates in all cases. This behavior is so because the pump efficiency, which we assumed to be $\eta = 100$ percent, begins to drop off rapidly at the higher flow rates.

The ultimate droop in the characteristic curve for the forward-curved blade at large values of Q (dashed line in Fig. 4-5) means that there will be a possibility of having two distinct flow rates corresponding to the same head. This is a potential source of instability in the operation of a system using such a pump. Consequently, most centrifugal pumps are designed to have backward-curved impeller blades or to operate on the drooping portion of the curve for a

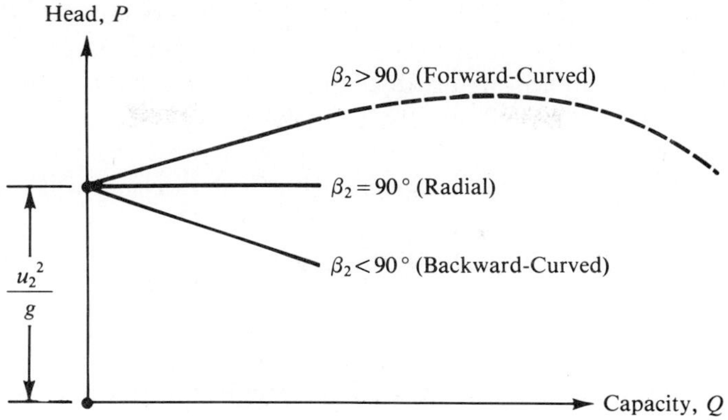

Figure 4-5. Theoretical effect of the blade exit angle.

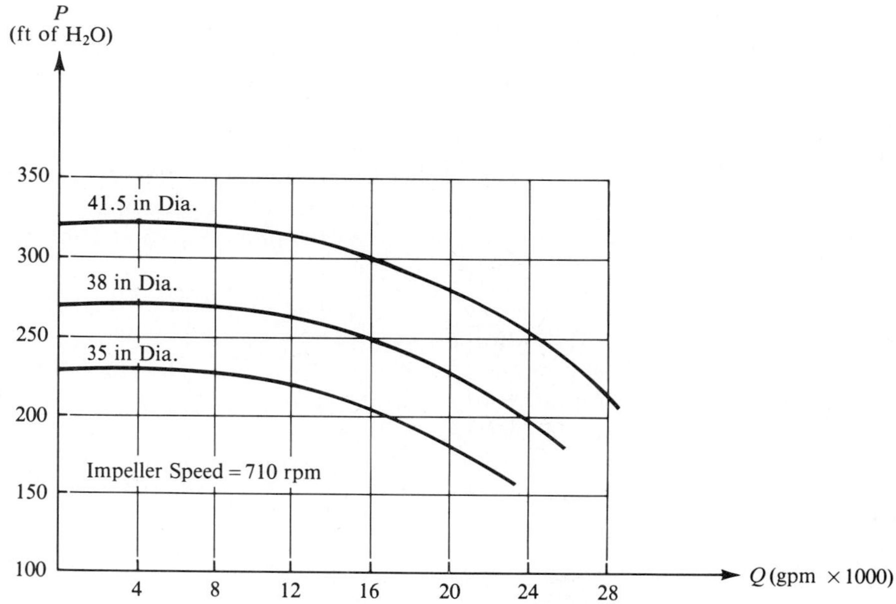

Figure 4-6. Performance curves for a centrifugal pump.

forward-curved impeller. Some actual quantitative curves for a backward-curved impeller are presented in Fig. 4-6.

4-1-2 Matching a Pump to a System Characteristic

Many fluid mechanical systems consist of extensive piping or duct circuits containing valves, straight pipe, and various fittings (elbows, tees, etc.). One of the system designer's tasks is to select a pump (or blower if the fluid is a gas) that will provide a specified flow rate in the piping system. A good example is the air conditioning system in a building in which cooling is provided by ducting cool, dehumidified air through a duct network. The cooling requirements will dictate the volumetric flow (cfm) that must be provided. For this cfm, there will be a definite pressure drop in the duct system. In selecting a blower to propel the fluid in the duct system, the designer must choose one that will simultaneously provide the required pressure drop and flow rate. This is the fundamental problem of matching a converter characteristic to a load characteristic, which we discussed in Chapter 1.

To illustrate the concepts required, consider the pumping system sketched in Fig. 4-7. By applying Bernoulli's equation to the piping system shown, the total head P that must be provided by the pump to produce flow in the system can be shown to be

$$P = H + h_f = P(Q) \tag{4-9}$$

where term h_f represents the frictional dissipation in the piping system, which is a function of the flow rate Q. The term H represents elevation head. Equation (4-9) is called the piping system characteristic (or load characteristic) because the head P, which the pump must provide, is a function of the flow Q in the system.

Note that even if the volumetric flow rate Q is zero in this open system, the pump will still have to provide the head H just to overcome the change in elevation of the fluid as it is pumped from the lower to the upper reservoir. If this were a closed piping system circuit, then H would be zero and the head P would have to overcome only the frictional losses in the system.

With a knowledge of the actual piping configuration (pipe lengths and diameters, elbow and fitting losses, valve losses, etc.), it is possible to actually calculate Eq. (4-9) and display it as shown in Fig. 4-8. The parabolic nature of the load curve is representative of turbulent flow in the piping system. For laminar flow (which will occur rarely), the characteristic curve will be linear. In any event, the characteristic curve of the load can be determined independently of the pump

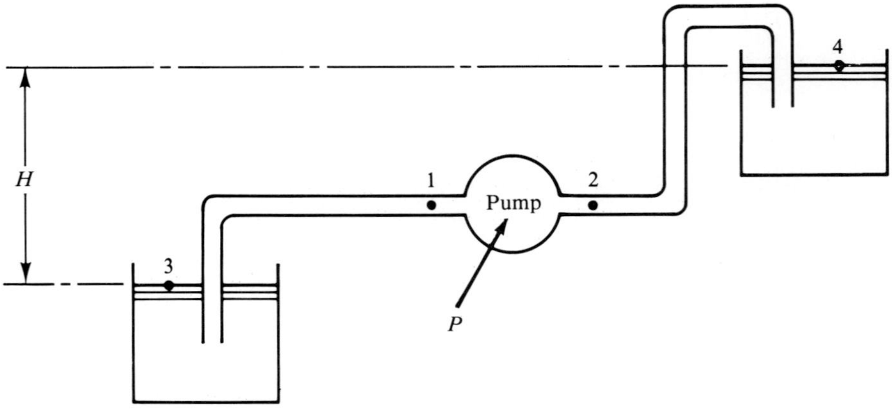

Figure 4-7. Open pumping system.

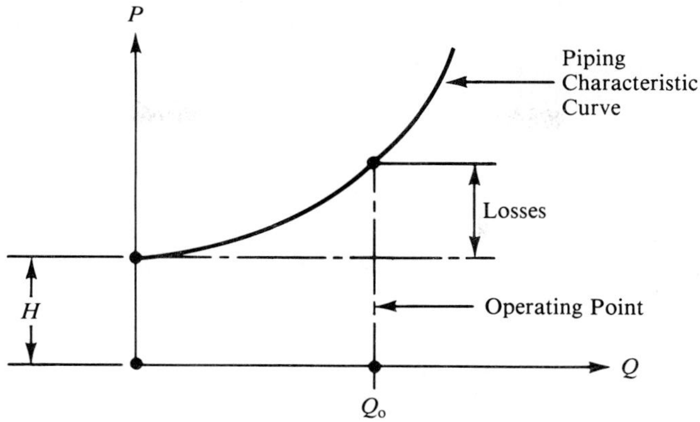

Figure 4-8. Piping characteristic (load) curve.

connected to it (by either calculation or experiment). The value $Q = Q_o$ at which this system will operate in the steady state will depend on the pump installed in it.

The output characteristic curve of P versus Q for the pump (or blower) can similarly be determined independently of the piping system either by experiment or by theoretical calculation. A typical converter characteristic for a pump having backward-curved blading is shown in Fig. 4-9. In contrast to the ascending piping load characteristic, the pump (or blower) characteristic is a drooping one.

When the pump of Fig. 4-9 is installed in the piping system of Fig. 4-8, the operating point at which pump and system will be in equilibrium (steady state) is defined by the intersection of the pump characteristic and system characteristic curves (see Fig. 4-10). The dashed curves of Fig. 4-10 represent the load curves that would be obtained by incorporating a throttling valve (or other throttling device) into the piping system. As the throttling valve is closed down, the flow in the system will diminish as the pressure drop increases (in proceeding from point $0 \rightarrow 1 \rightarrow 2$).

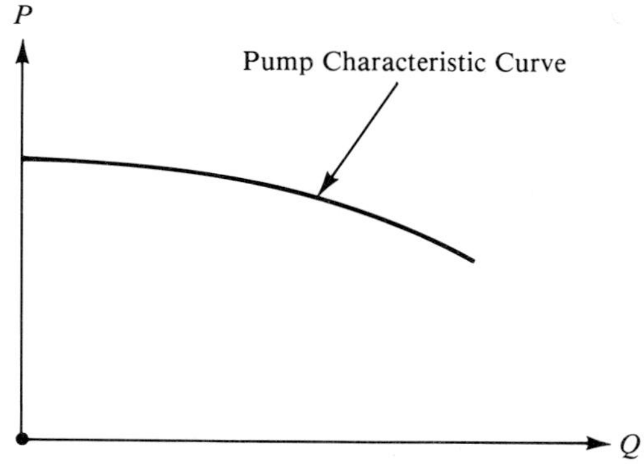

Figure 4-9. Characteristic curve of a centrifugal device.

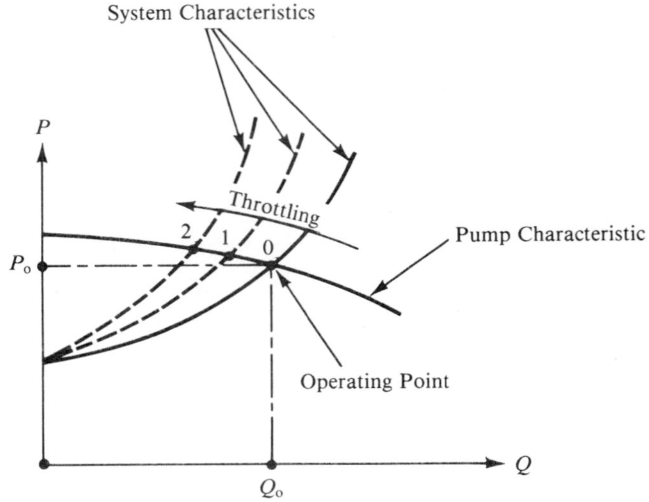

Figure 4-10. Matching pump and system characteristics.

EXAMPLE 4-1

The output characteristic of a certain centrifugal pump is

$$P = (0.45N - 2.75) + 0.14Q - 0.065Q^2 \tag{1}$$

where P is the pump head in kN/m^2, N is the pump impeller speed in rev/sec, and Q is the volumetric flow rate in thousands of liters/min. The pump is used to meet a piping system load requirement expressed as

$$P = \frac{Q^2}{14.4} \tag{2}$$

(a) If the speed of the pump is $N = 20$ rev/sec, what will be the steady-state operating point of this system?

(b) Suppose the volumetric flow rate is to be lowered to 6000 liters/min by lowering the pump speed. At what speed will the pump have to operate and what power will be consumed if we assume the pump to be 100 percent efficient?

(c) Suppose the volumetric flow rate is to be lowered to 6000 liters/min by inserting a throttling resistance into the piping system while the pump speed is maintained constant at 20 rev/sec. What power will now be consumed if we assume the pump to be 100 percent efficient?

Solution

(a) Equating characteristic curves (1) and (2) for a speed of $N = 20$ rev/sec and solving for Q produces

$$Q = 7.35 \times 10^3 \text{ liters/min}$$

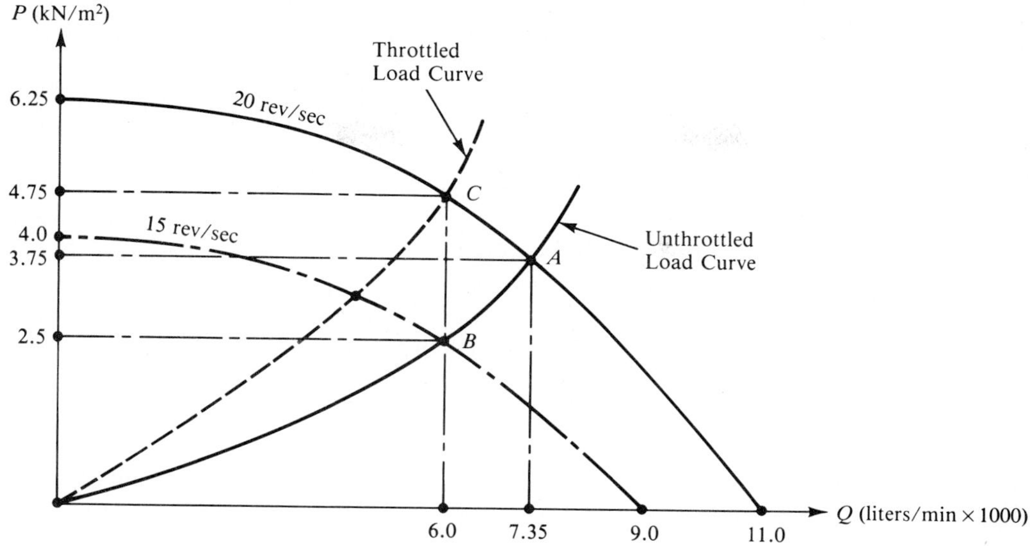

Figure 4-11. Characteristic curves for Example 4-1.

The corresponding head is

$$P = 3.75 \text{ kN/m}^2$$

This operating point, point A, is sketched in Fig. 4-11.

(b) Again equating characteristic curves (1) and (2) and substituting the value $Q = 6$, we obtain the pump speed

$$N = 15 \text{ rev/sec}$$

The corresponding steady-state operating head can be calculated to be (point B in Fig. 4-11)

$$P = 2.5 \text{ kN/m}^2 = 0.255 \text{ m of water}$$

Assuming the fluid being pumped is water ($\gamma = 9800 \text{ N/m}^3$), the power consumed is then

$$\text{Power} = \gamma PQ = (9800)(0.255)\left(\frac{6}{60}\right) = 250 \text{ W}$$

(c) If the flow is to be reduced by throttling the 20 rev/sec pump, the operating point will be point C in Fig. 4-11. The operating head will be determined from Eq. (1) to be

$$P = 4.75 \text{ kN/m}^2 = 0.4847 \text{ m of water}$$

The power consumed is then

$$\text{Power} = (9800)(0.4847)\left(\frac{6}{60}\right) = 475 \text{ W}$$

We see that some 225 W will be dissipated if the throttling scheme is used rather than changing the pump speed. ■

4-1-3 Positive-Displacement Pumps (PDP)

We will define a fluid power system to be a hydraulic system used to transmit, amplify, and control motion, force or torque, and power (usually as an adjunct to the operation of a machine). Often, the fluid power system is simply an alternative to a mechanical or electrical system that could perform the same function. But the fluid power system may be preferred because of its flexibility, ruggedness, power density, or portability.

At the heart of the fluid power system is a positive-displacement pump. A PDP is a pump in which, ideally, the volumetric flow delivered by the pump, per revolution of the pump shaft, is independent of the pressure head against which the pump works. Ideally, then, the volumetric flow rate of a PDP will depend only on pump rpm. This dependence on rpm will require that there be no internal back-leakage.

A close-fitting piston in a cylinder is an example of a PDP. If the fit between piston and cylinder is perfect, there will be no chance for back-leakage; then the piston displacement (calculated from dimensions only) is related directly to the amount of fluid discharged. This is in direct contrast to the NPDP, in which there is a large clearance between impeller and casing with resultant back-leakage.

The idealized head versus discharge curve for a PDP is sketched in Fig. 4-12(a). We see that the characteristic curve is simply a vertical line on the *P*-*Q* plot. This means that the flow in a hydraulic system powered by such a pump is not dictated by a matching of the pump characteristic to the system characteristic in the usual sense. Instead, the flow is dictated entirely by the pump rpm and not by the head against which the pump works.

The head at which the ideal PDP must operate is determined entirely by the system (load) characteristic at the flow rate Q dictated by the pump rpm. If the external load is small, then the head developed by the PDP will be small, and vice versa. In particular, the PDP cannot work against an infinite resistance as can a NPDP device. If the flow becomes blocked in the PDP circuit, the pressure will

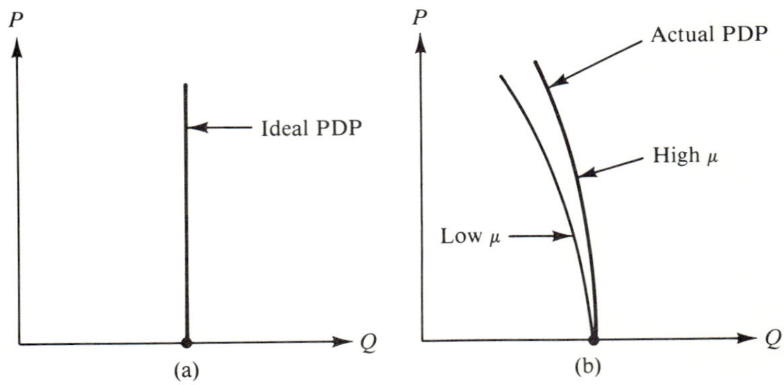

Figure 4-12. PDP characteristic curves.

rise drastically and the pump will crack, or blow up, unless it is protected by a relief valve. Keep in mind that a PDP will tend to develop whatever head is necessary to satisfy a load placed upon it.

Of course, there will always be some back-leakage, albeit small, in any actual PDP. The amount of leakage will depend on the head against which the pump works; for higher heads, the internal back-leakage will be higher. This higher back-leakage causes the pump characteristic curve to bend slightly to the left as shown in Fig. 4-12(b). The extent of the leakage also depends on the viscosity of the fluid being pumped, as is shown.

We reiterate that the pressure head developed by a PDP is dictated almost solely by the load against which the pump works. If the pump outlet is blocked, the outlet pressure will rise to an extremely high value sufficient to either stall the pump motor or break the pump. Safety measures must always be taken to mitigate this hazard. There are basically two ways of doing this: by using a relief valve or by using a pressure-compensated pump (both of which will be discussed in detail later).

EXAMPLE 4-2

At a shaft speed of 1800 rpm, a PDP has a theoretical displacement of 0.003 gal/rev. The internal back-leakage of this pump is 5 percent of the theoretical displacement when the pump is working against a head of 3000 psi. Assume the leakage is proportional to the pump head and determine

(a) The characteristic pump curve with P in psi and Q in gpm
(b) The head at which this pump will operate with a load whose characteristic is

$$P = 68Q^2 \qquad (P \sim \text{psi},\ Q \sim \text{gpm})$$

Solution

The theoretical volumetric discharge with no back-leakage is assumed to occur at zero pump head. This value is

$$Q = 1800(0.003) = 5.4 \text{ gpm}$$

Therefore, the back-leakage at 3000 psi is

$$Q = 0.05(5.4) = 0.27 \text{ gpm}$$

Since the pump characteristic is linear, it is given by

$$Q = 5.4 - \left(\frac{0.27}{3000}\right)P$$

or

$$P = 60{,}000 - 11{,}111.11Q$$

The operating point is obtained by equating load and pump characteristics. Thus, solving for P, we find that $P = 1862$ psi. The resulting discharge is $Q = 5.232$ gpm. These various values are sketched in Fig. 4-13. ■

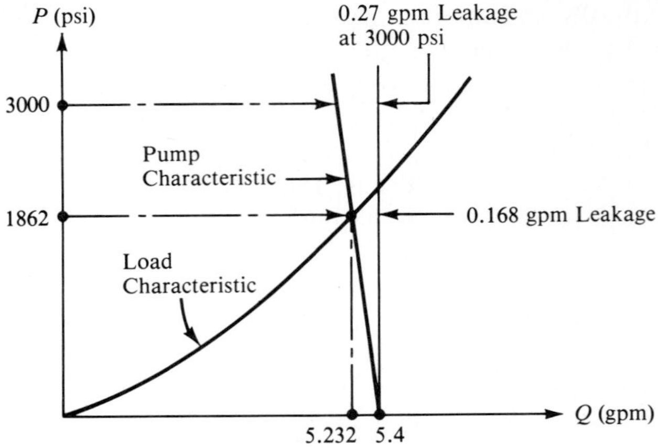

Figure 4-13. Characteristics for Example 4-2.

4-1-4 Types of PDPs

Positive-displacement pumps can be classified into four major types: (a) gear pumps, (b) lobe pumps, (c) vane pumps, and (d) piston pumps. Additionally, some of these pumps are available in both fixed-displacement and variable-displacement versions. In a fixed-displacement pump, the discharge can be changed only by changing the pump shaft rpm, while in the variable-displacement type the discharge is dependent on pump geometry, which can be changed during operation, as well as upon shaft rpm.

Gear Pumps. A typical gear pump configuration is shown in Fig. 4-14. Oil enters the pump at the inlet and is trapped between the rotating gear teeth and pump housing, where it is then carried around the periphery of the rotating gears to the pump outlet, at which point it is discharged. This is called an external gear pump because the oil is moved by trapping it between the gears and external housing.

The gear pump is of the fixed-displacement type. Its output can be changed only by changing the speed of the rotating gears, and this is generally not done in most fluid power circuits. The outlet pressure that such a pump can develop ranges from several hundred psi up to about 3000 psi. Capacities range from a fraction of a gpm up to around 150 gpm.

The volumetric flow rate of a gear pump can be estimated by assuming that the spaces between gear teeth and the outside housing have the same volume as the gear teeth. Then the flow rate can be shown to be

$$Q = \frac{\pi(D_o^2 - D_i^2)nF}{4(231)} \tag{4-10}$$

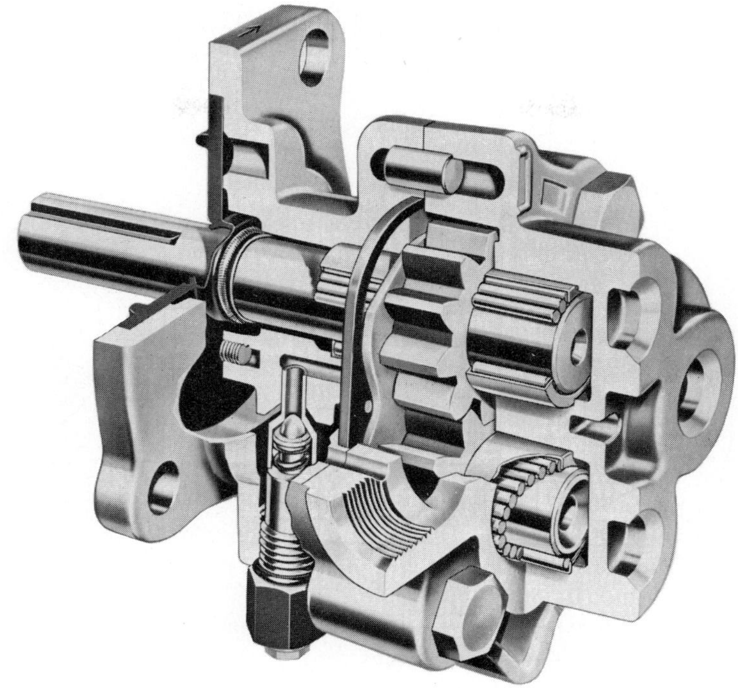

Figure 4-14. External gear pump. (Courtesy of Webster Electric Company, Inc.)

where Q = the volumetric flow rate, in gpm
D_o = the gear outside diameter, in inches
D_i = the gear root diameter, in inches
F = the gear face width, in inches
n = the gear rpm

The actual flow rate will be somewhat less than the theoretical value suggested by Eq. (4-10) because of the internal back-leakage between the gears and housing. This lost flow rate is called slip or slippage.

In another version of the gear pump, fluid is moved from inlet to outlet by trapping it between a small externally toothed gear fitted eccentrically within a larger, internally toothed gear. In this device, both gears rotate in the same direction. This gear pump is called an internal gear pump.

Lobe Pumps. A lobe pump operates on the same principle as the gear pump, but the lobe profiles are not involute gear teeth as in the case of a gear pump. This permits designs having higher displacement than with gear pumps. Two lobe pump versions are shown in Fig. 4-15.

In the Gerotor pump, which operates on the same principle as an internal

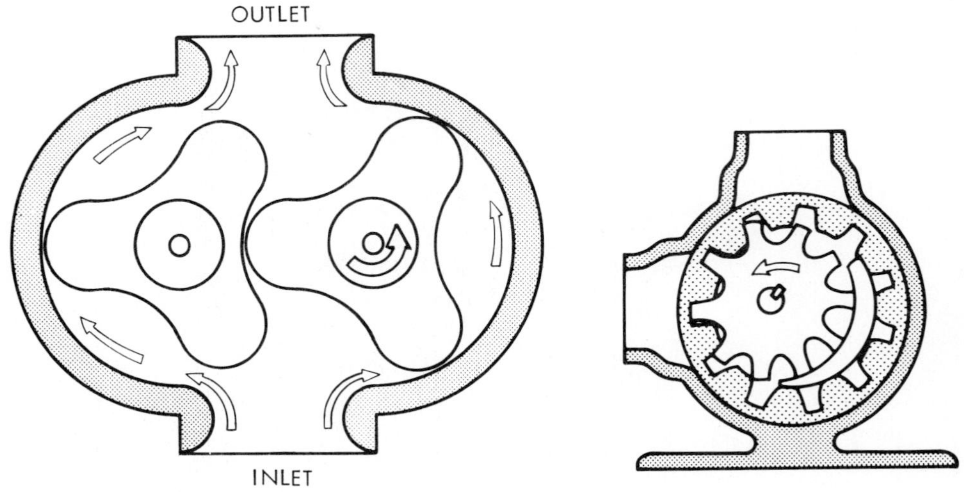

Figure 4-15. External and internal lobe pumps. (Courtesy of Vickers, Inc.)

gear pump, the inner rotor is driven and carries the outer rotor around with it. There is one less lobe on the inner rotor than on the outer rotor, creating a space within which fluid can be trapped and moved from inlet to outlet. The advantage of this design is that the relative sliding velocities are low, which results in quiet operation and long life. The Gerotor pump is often used in applications where a small, compact pump is needed.

Vane Pumps. Gear and lobe pumps are inherently constant-flow-rate devices. The only way that flow rate can be made variable with these pumps is to change the rpm of the driving motor shaft. Vane pumps have the advantage that they can be made both relatively cheaply and in variable-displacement versions. A vane pump designed with provision for variable output flow rate is shown in Fig. 4-16.

The schematic drawing in Fig. 4-16(a) illustrates the principle of operation. The rotor barrel has vanes inserted in slots. As the rotor turns, the vanes are forced by centrifugal force to contact the hardened ring of the housing. As the vanes proceed from the point of shortest distance between rotor and housing, the fluid chambers between successive vanes and the housing increase in size. Oil flows into the chambers through a peripheral suction port. After the chambers have been fully expanded, the suction port is no longer accessible and the chambers between vanes begin to contract, forcing oil out through the exit port.

Note that there is an eccentricity ε between the center of the rotor barrel and the external ring. It is clear from the figure that if this eccentricity could be varied, then the device could be a variable-displacement pump even for constant rotor rpm. For example, if eccentricity ε is zero, then both rotor and housing are concentric and the chambers between vanes will not change in volume as the rotor turns. No fluid can be delivered by this pump. Maximum displacement will,

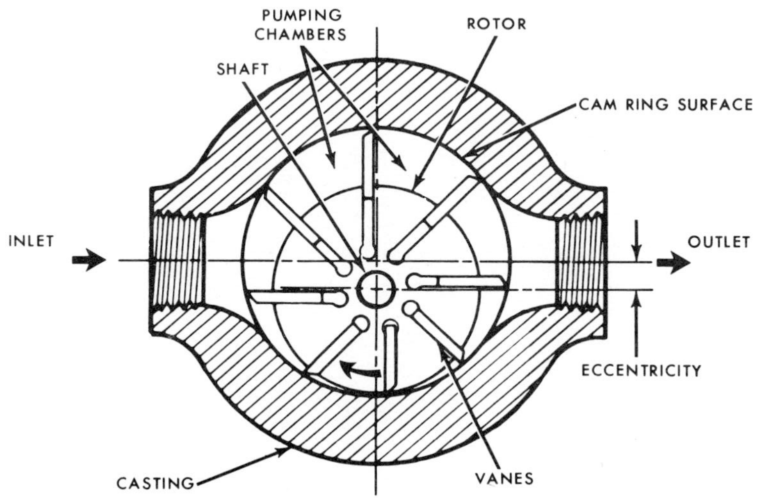

(a)

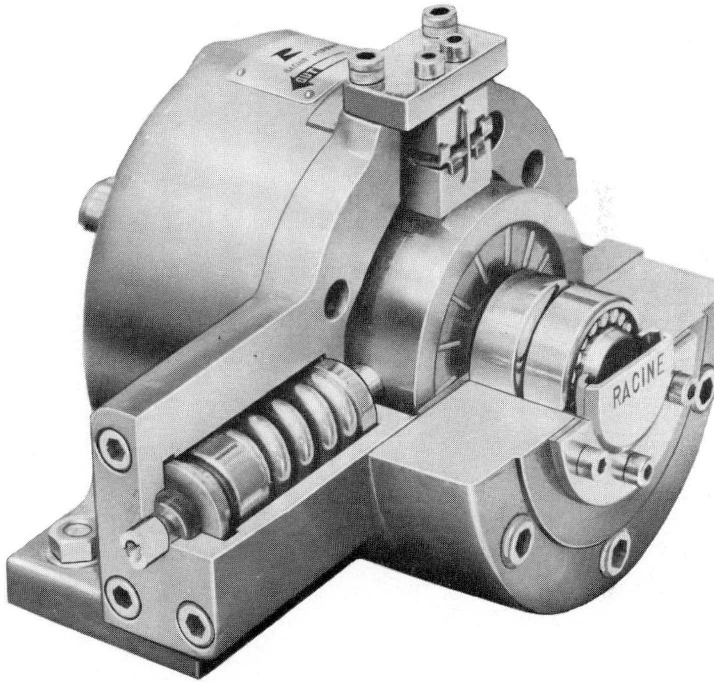

(b)

Figure 4-16. Variable-displacement vane pumps. ([a] Courtesy of Vickers, Inc; [b] courtesy of Racine Hydraulics, Dana Corp.)

correspondingly, occur with maximum eccentricity ε. So, if eccentricity ε can be made controllable while the pump operates, the pump will have a controllable output flow rate.

One disadvantage of the vane pump is that it is inherently an unbalanced device insofar as forces acting on the rotor barrel are concerned because the high outlet pressures acting on the projected area of rotor and vanes is not balanced by the lower inlet pressure. In fixed-displacement versions, this problem can be overcome by special designs of the outer ring geometry (make it elliptical) and by having multiple inlet and exit ports.

The vane pump shown in Fig. 4-16(b) is of the variable-displacement type. In this design, the cam ring can be shifted by piloting oil to a piston acting on one side of the ring. The piston acts against a spring-loaded pin on the other side of the ring. The operating eccentricity of the pump then depends on the balance between the force caused by spring deflection and the hydraulic force on the opposing side of the cam ring.

Vane pumps are made in a large variety of types and by a large number of manufacturers. They are available with heads of around 1000 psi in the unbalanced versions and heads of up to 3000 psi in the balanced versions.

Radial-Piston Pumps. Piston pumps are not only the oldest form of hydraulic pump but, today, they provide the highest degree of sophistication found in hydraulic pump designs. Piston pumps can operate at high working pressures with only small back-leakage (high volumetric efficiency). They can also be designed and manufactured as variable-displacement devices. For these reasons, they are the most expensive hydraulic pumps.

We will be concerned with two types of piston pumps: the radial-piston pump and the axial-piston pump. A radial-piston pump is shown in Fig. 4-17. In this design, the rotor assembly consists of a cylinder barrel which contains radially oriented pistons and a reaction ring against which the radial pistons are caused to seat by centrifugal force. The cylinder barrel that contains the pistons rotates about a stationary shaft, called the pintle, which is ported for suction and discharge. Again, notice that the centerline of the cylinder barrel is eccentric with respect to the centerline of the reaction ring and its housing (similar to the vane pump). If provisions are made in the design so that the retaining ring can be shifted such as to change the eccentricity ε, then the stroke of the pistons (as they reciprocate radially) will be controllable, and so will the discharge. This can be accomplished in a manner similar to the arrangement with a vane pump. The retaining ring is acted upon by a spring-loaded pin on one side. This spring is balanced against a piston on the opposite side that is controllable in a variety of ways (handwheel, hydraulic piston, etc.).

As the pump pistons move radially outward, the suction port in the pintle will be open, and the discharge port closed, to these pistons. Fluid flows into the cylinders. Then, as the pistons move into the quadrants where they will be caused to move inward, the discharge port will be open to them and the fluid is discharged. Note that the thrust shoes on the outer ends of the radial pistons do

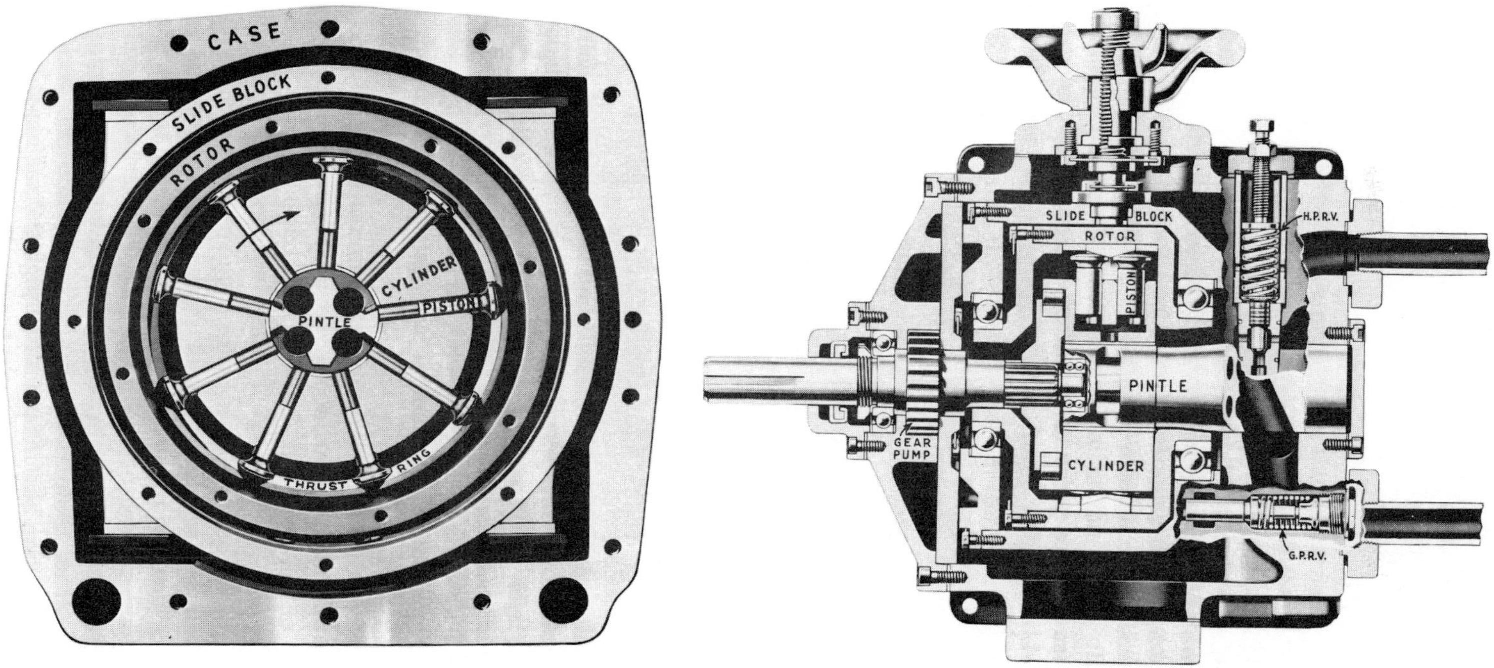

Figure 4-17. Radial-piston, variable-stroke pump.
(Courtesy of The Oilgear Company.)

not slide against the reaction ring. The reaction ring is actually rotating at the same rpm as the cylinder barrel.

Finally, note that servo-control of a radial-piston pump of variable displacement is a definite possibility (as is also true of the vane pump). This servo-control would be accomplished by controlling the eccentricity of the reaction ring mechanically, hydraulically, or electrically by means of a feedback control. Of course, manual control would also be possible simply by adjusting the reaction ring using a handwheel.

Axial-Piston Pumps. In an axial-piston pump, the axes of the pistons are parallel to the axis of rotation of the cylinder barrel that holds them. Here, we will discuss only two of the several kinds of axial-piston pumps available. These are the swash-plate pump and the bent-axis pump.

A swash-plate pump is shown in Fig. 4-18. In this device, the cylinder barrel

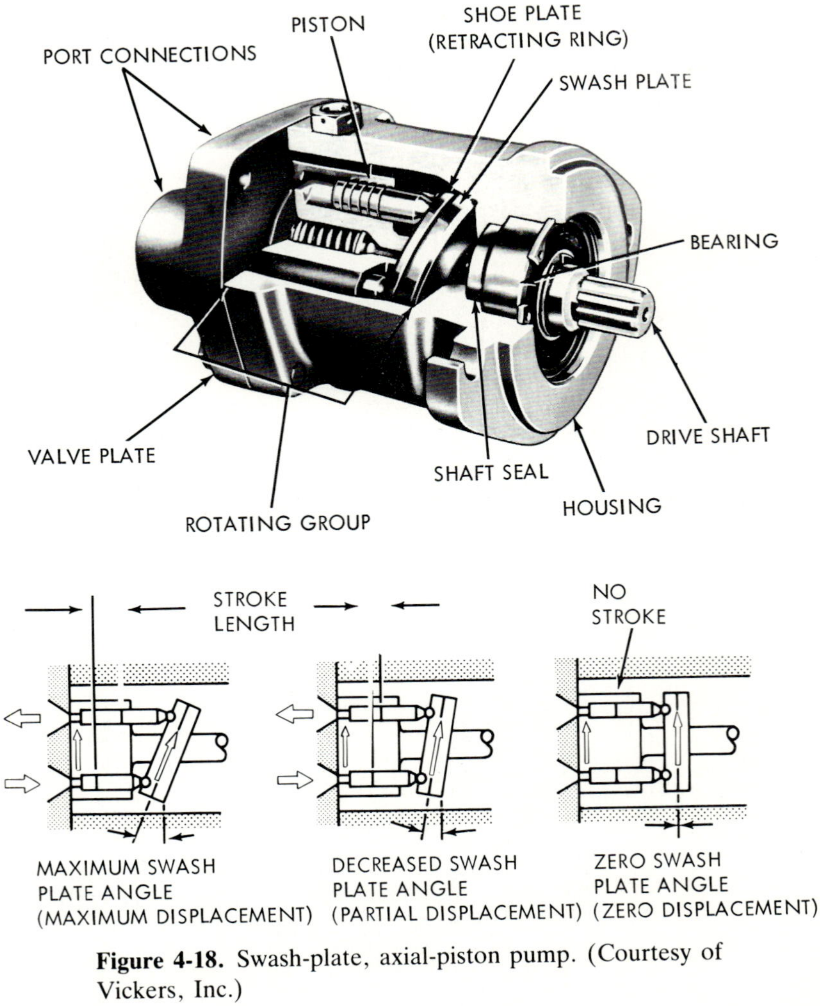

Figure 4-18. Swash-plate, axial-piston pump. (Courtesy of Vickers, Inc.)

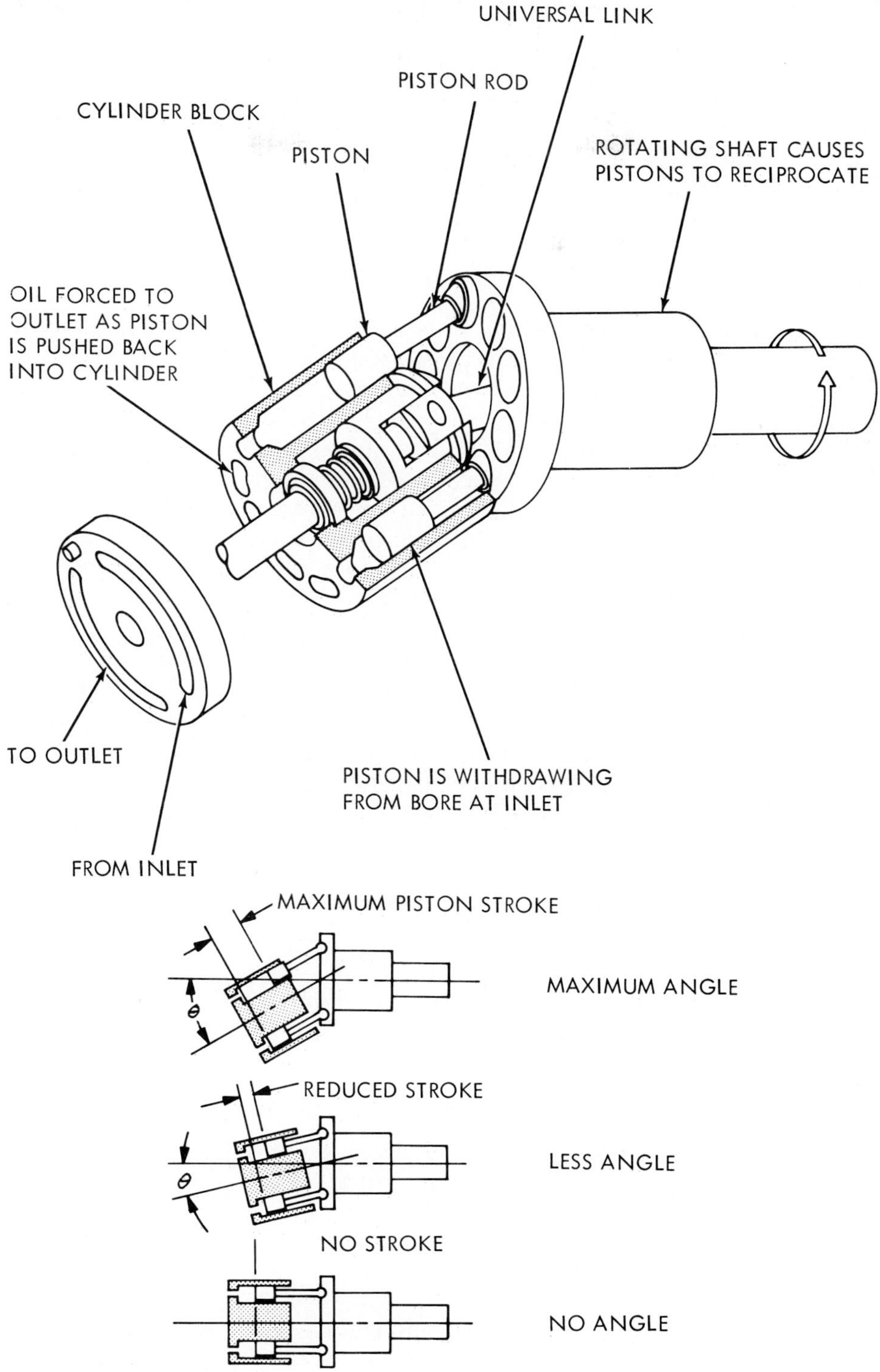

Figure 4-19. Bent-axis, axial-piston pump. (Courtesy of Vickers, Inc.)

containing the pistons is turned by the pump drive shaft. The pistons are caused to reciprocate axially by a shoe plate to which the pistons are connected by ball joints. The shoe plate bears against a swash plate situated at an angle to the axis of the cylinder barrel. Thus, during a half-revolution of the cylinder barrel, the pistons are permitted to extend by the obtuse angle of the swash plate while drawing in oil through the intake ports. During the subsequent half-revolution of the cylinder barrel, where the pistons are in contact with the acute angle of the swash plate, they move in the opposite direction, discharging oil through the exit ports.

The swash-plate pump can be made into a variable-displacement pump by having the angle of the swash plate controllable. When the plane of the swash plate becomes perpendicular to the axis of the cylinder barrel, the pistons will cease to reciprocate and the delivery will go to zero. Maximum delivery will occur when the plane of the swash plate is situated at its maximum possible angle with respect to the cylinder barrel axis. As shown in the inset of Fig. 4-18, the angle of the swash plate can be controlled by mounting it in a yoke that pivots in accordance with the balance of moments produced by the reciprocating pistons bearing against it and by a control piston that actuates the yoke. The control piston is called the servo-piston because the pump can be servo-controlled in this fashion.

A bent-axis pump is shown in Fig. 4-19. The bent-axis pump works on the same general principle as the swash-plate pump but its pistons are caused to reciprocate axially by driving the cylinder barrel through a universal joint and connecting shaft situated at an angle to the rotating driving flange into which the pistons are seated by ball-jointed connecting rods. In this design, the sliding friction, existing between shoe plate and swash plate of the swash-plate pump, is eliminated. By controlling the angle between the drive shaft and the cylinder barrel, the bent-axis pump can be made to be a variable-displacement device. When the axes are parallel, the pistons cannot reciprocate and the discharge is zero. The angle of the bent axes can be controlled either manually or in the servo-control mode.

4-1-5 Characteristics of PDPs

A positive-displacement pump is a power converter which we can represent as a black box as shown in Fig. 4-20. The ideal displacement (without internal back-leakage) of the PDP is given by

$$Q_i = N \cdot q \text{ (cfm)} \tag{4-11}$$

where q is the theoretical pump displacement per revolution and N is the pump rotational speed in rpm.

In the more versatile PDPs, the theoretical pump displacement q is a variable that can be controlled through a variable geometry parameter e. In a vane or radial-piston pump, parameter e is the eccentricity ε between rotor and stator. In an axial-piston pump, parameter e is related to the angle α between two

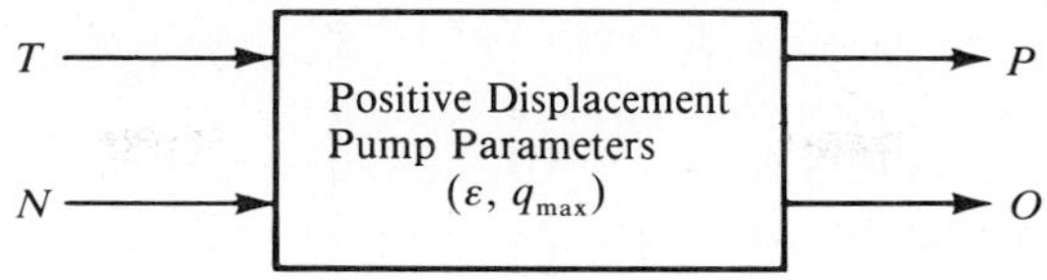

Figure 4-20. Pump as a black box.

axes. In the bent-axis pump, α is the bent-axis angle, while in the swash-plate pump, α is the angle between swash plate and rotor barrel.

We will define e_{max} to be the maximum value that the geometry parameter can assume. The stroke S of the pump is then defined to be

$$S = \frac{e}{e_{max}} = \frac{\varepsilon}{\varepsilon_{max}} = \frac{\alpha}{\alpha_{max}} \tag{4-12}$$

When $S = 1$, the pump is said to be fully stroked and it delivers its maximum volumetric discharge. When $S = 0$, the rotor is centered in the vane or radial-piston pump or the angle of the axial-piston pump is zero. In either case, the discharge will be zero.

We will define q_{max} to be the theoretical displacement of the pump at full stroke (i.e., the maximum theoretical pump displacement per revolution). Then

$$q = S \cdot q_{max} \tag{4-13}$$

Substituting Eq. (4-13) into Eq. (4-11), we obtain

$$Q_i = N \cdot (S \cdot q_{max}) \tag{4-14}$$

The volumetric efficiency η_v of the PDP is defined as the ratio of the actual discharge Q (with back-leakage) to the ideal discharge Q_i. Therefore,

$$\eta_v = \frac{Q}{Q_i} = \frac{Q}{N \cdot S \cdot q_{max}} \tag{4-15}$$

The actual torque T required to produce a given pump head P will be greater than the ideal torque T_i that could produce the head P in the absence of frictional effects in the pump. Therefore, the torque efficiency of the PDP is defined as

$$\eta_T = \frac{T_i}{T} \tag{4-16}$$

In the absence of both frictional and back-leakage losses, the idealized input power $T_i\omega$ to produce the head P will be equal to the idealized output power PQ_i. Therefore,

$$2\pi T_i N = T_i \omega = PQ_i \tag{4-17}$$

Note that N is pump speed in rpm while ω is pump speed in rad/min.

Substituting Eq. (4-14) into Eq. (4-17), we find the expression for the idealized torque to be

$$T_i = \frac{P(S \cdot q_{max})}{2\pi} \tag{4-18}$$

The overall efficiency η_o of the pump is defined as the actual output power PQ divided by the actual input power $T\omega = 2\pi TN$. That is,

$$\eta_o = \frac{PQ}{T\omega} = \frac{PQ}{2\pi TN} \tag{4-19}$$

If we substitute Eqs. (4-15), (4-16), and (4-17) into Eq. (4-19), we find that

$$\eta_o = \eta_v \eta_T \tag{4-20}$$

In other words, the overall pump efficiency is simply the product of the volumetric and torque efficiencies.

In Reference 4-1, a detailed kinematic analysis is made in which it is shown that the maximum theoretical displacement, q_{max}, of a vane or radial-piston pump is related to the maximum eccentricity, ε_{max}, the number of vanes (or pistons), n, and the area A associated with each vane (or piston). In other words, it is shown that

$$q_{max} = 2An\varepsilon_{max} \tag{4-21}$$

Then, for a vane or radial-piston pump, we find that the ideal discharge and ideal torque are given (in slightly different form) by Eqs. (4-14) and (4-18):

$$Q_i = \omega\left(\frac{An\varepsilon}{\pi}\right) \qquad T_i = P\left(\frac{An\varepsilon}{\pi}\right) \tag{4-22}$$

The factor in parentheses is a pump configuration parameter.

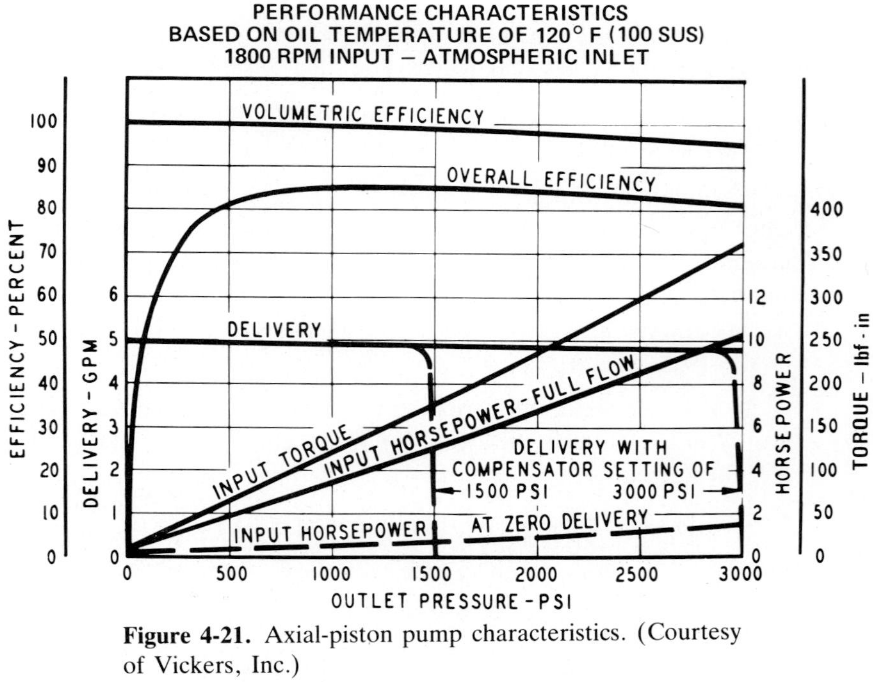

Figure 4-21. Axial-piston pump characteristics. (Courtesy of Vickers, Inc.)

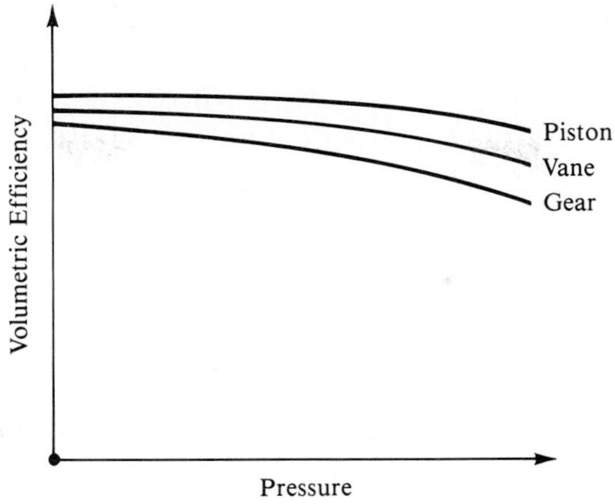

Figure 4-22. Typical volumetric efficiency curves.

A typical set of PDP characteristic curves is presented in Fig. 4-21 for an axial-piston pump. Note that the orientation of the coordinate axes is different here than that discussed earlier (Fig. 4-4) in that the P axis is now the abscissa while the Q axis is the ordinate.

Typical volumetric and overall efficiencies of gear, vane, and piston pumps are shown in Fig. 4-22. We see that the piston pump is the highest quality pump with the gear pump being the lowest quality. The vane pump falls in between. Of course, the cost of the pumps is in the reverse order of their quality.

EXAMPLE 4-3

Assume that the pump whose characteristics are shown in Fig. 4-21 is operated at a constant speed of 1800 rpm and is fully stroked ($S = 1$). Determine its displacement per revolution (q_{max}). Also determine the torque efficiency, internal leakage, and output horsepower at a working head of 3000 psi.

Solution

Assuming there is no internal back-leakage at zero head, the discharge will be 5 gpm at 1800 rpm. Therefore (recall that 1 cubic foot = 7.48 gallons),

$$q_{max} = \frac{5(1728)}{7.48(1800)} = 0.64 \text{ in}^3/\text{rev}$$

The volumetric and overall efficiencies are seen to be approximately $\eta_v = 0.95$ and $\eta_o = 0.81$ at $P = 3000$ psi. Therefore,

$$\eta_T = \frac{0.81}{0.95} = 0.853 \text{ at } P = 3000 \text{ psi}$$

With a volumetric efficiency of 0.95, the internal leakage is 5 percent at $P = 3000$ psi. Therefore, the internal leakage is

$$\delta q = 0.05(0.64) = 0.032 \text{ in}^3/\text{rev}$$
$$= 0.05(5) = 0.25 \text{ gpm}$$

The input torque at $P = 3000$ psi is approximately $T = 360$ lb-in. Therefore, the input horsepower is

$$\mathrm{HP_i} = \frac{2\pi(1800)(360)/12}{33{,}000} = 10.28 \text{ hp}$$

This checks the value shown in Fig. 4-21. The output horsepower is then given by

$$\mathrm{HP_o} = \eta_o \mathrm{HP_i} = 0.81(10.28) = 8.32 \text{ hp}$$

We can check this against $P \cdot Q$. Then

$$\mathrm{HP_o} = \frac{3000(144)(5 \times 0.95)}{33{,}000(7.48)} = 8.31 \text{ hp}$$

■

4-1-6 The Pressure-Compensated Pump

The implementation of a sequentially controlled fluid power system involves actuators such as hydraulic cylinders (or motors) interfaced to PDPs by means of directional control valves. As we shall see later when we discuss hydraulic circuits, some directional control valves will totally block the output of the PDP, so unless some action is taken to protect the pump, it will be subjected to dangerously high pressures. One way in which to mitigate this hazard is to provide a pressure relief valve that will limit the pressure the pump can experience and that will divert the pump's output back to the pump source of oil. However, the use of a relief valve generally is accompanied by a dissipation of pump power which, in many instances, is undesirable.

An alternative method to the use of a relief valve in protecting a pump is to use a pressure-compensated pump. There are also additional benefits accruing to the use of a pressure-compensated pump which go beyond the mere function of protection. These benefits, as usual, come at the expense of increased costs because a pressure-compensated pump must necessarily be of the variable-displacement type.

Pressure compensation is, in effect, a method of converting a pump having a PDP characteristic into one having a characteristic similar to that of the NPDP. What we do is design the PDP so that its delivery Q will be responsive to the discharge pressure against which it works. For the radial-piston or unbalanced vane pump, this is done by making the ring position responsive to outlet pressure as shown schematically in Fig. 4-23 (also see Fig. 4-16[b]).

In the radial device of Fig. 4-23, the axis of rotation of the rotor is fixed relative to the pump housing. The ring against which the vanes (of a vane pump) or the pistons (of a radial-piston pump) bear can be shifted depending on the balance between opposing forces provided by a spring-loaded piston (on the right) and a hydraulically actuated piston (on the left). The hydraulically actuated piston is subjected to the pump outlet pressure. Therefore, as the outlet pressure rises, the ring will begin to shift to the right against the force of the mechanical spring. In this fashion, the eccentricity ε of the pump will be decreased, and so will the pump discharge (in accordance with Eq. [4-22]). At the same time, the outlet pressure that the pump can provide will begin to be limited.

Pressure compensation is achieved with the variable-displacement axial-

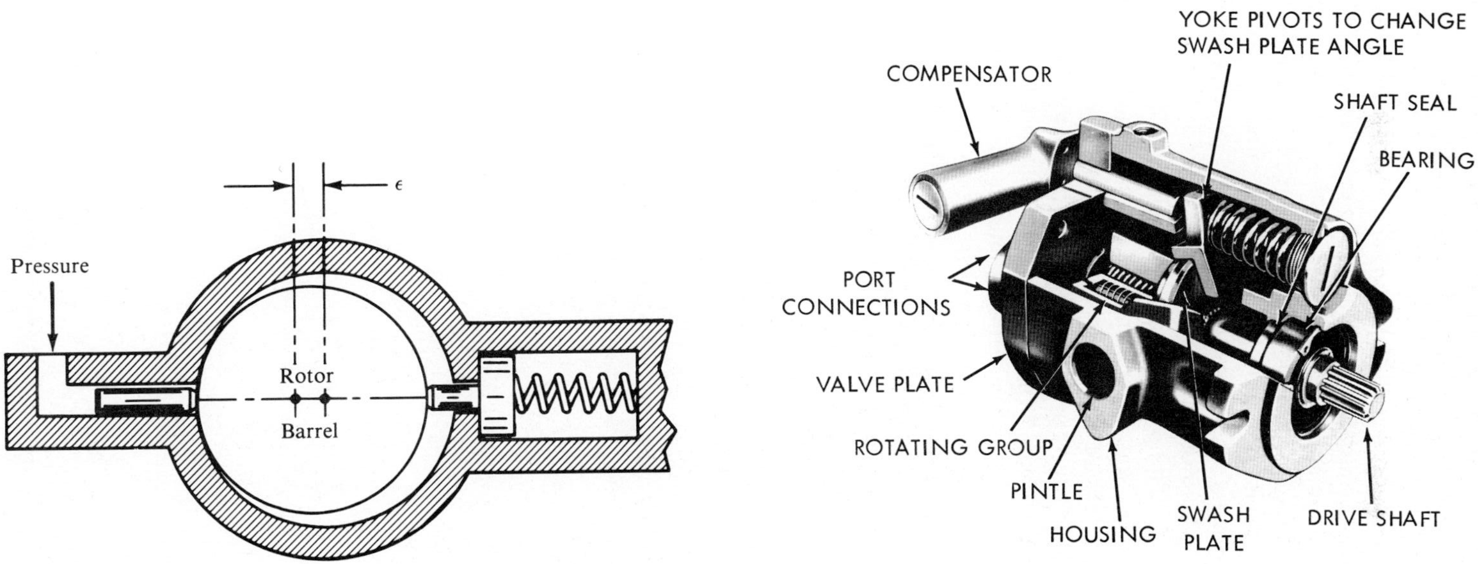

Figure 4-23. Pressure compensation operation. (Courtesy of Vickers, Inc.)

piston pump by controlling the angle between the axes of the bent-axis pump or between the swash plate and cylinder barrel in the swash-plate pump (see Fig. 4-23). The internal hydraulic servo-cylinder actuates the yoke containing the swash plate in opposition to the hydraulic forces developed by the reciprocating pistons as they discharge oil into the pump exit ports. In the design shown in Fig. 4-23, increased pump outlet pressure is communicated to the servo-piston as reduced pressure (by a spool valve arrangement), and angle α tends to be diminished, reducing the pump discharge as outlet pressure rises.

In both types of pumps, an eccentricity parameter (ε or α) tends to be diminished as pump outlet pressure increases and this, in turn, tends to diminish pump output, Q. The precise manner in which this action occurs depends on the spring constant of the mechanical spring in the arrangement of Fig. 4-23. The resultant characteristic pump curve will be as shown in Fig. 4-24.

In Fig. 4-24, the cutoff pressure is defined to be the pump outlet pressure that will cause stroke S of the pump to just begin to be decreased. For any pressure below this cutoff value, the pump acts as a fixed-displacement pump and delivers full flow rate. As the pressure begins to rise above the cutoff value, eventually eccentricity ε will go to zero, which cuts off the flow entirely. The pressure at this point is called the dead-head pressure. At dead-head pressure, the rotor is centered within the ring and simply rotates without the radial pistons (or vanes) reciprocating. There will be no flow and a small dissipation of power due to bearing oil friction.

The slope of the pump curve between cutoff and dead-head pressure depends on the spring constant of the spring used to balance the ring against pump exit pressure.

When a pressure relief valve is installed for fail-safe protection of a pressure-compensated pump, its setting should be some 250 to 300 psi higher than the pump dead-head pressure in order to avoid overlapping actuation of pump and relief valve.

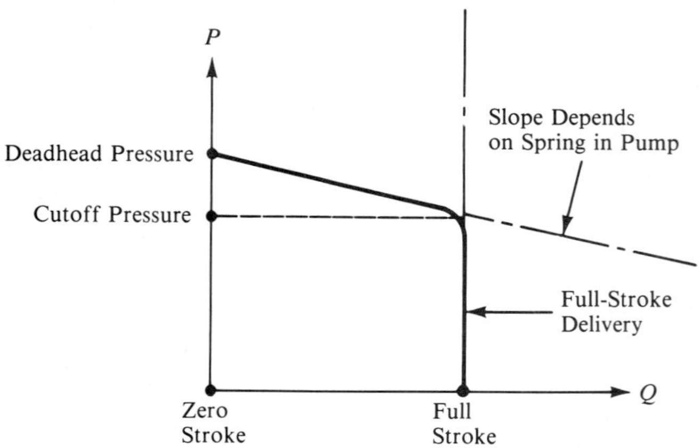

Figure 4-24. Pressure compensation characteristic.

EXAMPLE 4-4

To move the reaction ring of a radial-piston pump, the actuating force F_a (see Fig. 4-23) must overcome the sum of the spring force F_s plus the hydrostatic forces of the rotating radial pistons developed in reaction to the discharge against the pump outlet head. It has been found empirically that if the actuating force F_a is taken to be in the interval $0.5pA$ to $0.75pA$ (where A is the area of a radial piston and p is the discharge pressure), it will be more than sufficient to actuate the ring against the radial piston hydrostatic forces plus stiction and inertia.

Now suppose a spring is to be selected for a radial-piston pump such that it will be pressure compensated with a dead-head pressure of 3000 psi and a cutoff pressure of 2400 psi. Discharge at full stroke is to be 5 gpm; the eccentricity is to be 1/2 inch; and the diameter of a radial pumping piston is to be 1.2 inches. Determine

(a) The diameter of the actuating piston
(b) The spring constant of the mechanical spring
(c) The pre-compression of the spring for full discharge

Solution

We assume the actuating force to be $F_a = 0.75pA$. Then

$$F_a = 0.75(3000)\left(\frac{\pi(1)^2}{4}\right) = 1767 \text{ lb}$$

Remembering that when the pump is at dead-head pressure of 3000 psi the ring is centered, the eccentricity is zero, and the discharge is zero, the actuating force will be working against only the spring force (because no reaction force is exerted by the radial pistons), then we know the fully compressed spring force will be $F_s = 1767$ lb. In this position, the actuating piston will experience the dead-head (pump outlet) pressure and be in balance with the spring force. Therefore, the actuating piston area will have to be

$$A_a = \frac{1767}{3000} = 0.589 \text{ square inch}$$

The actuating piston diameter corresponding to this area is $D_a = 0.866$ inch.

When the actuating piston experiences the incipient cutoff pressure of 2400 psi, it will provide an actuating force of $F_a = 0.589(2400) = 1414$ lb. If we neglect the reaction force of the radial pistons at cutoff pressure position, this actuating force must again be balanced by the spring. Thus, as the spring expands by the amount of the eccentricity (1/2 inch), its force diminishes from 1767 lb to 1414 lb. The spring constant is then

$$k = \frac{1767 - 1414}{0.5} = 706 \text{ lb/in}$$

The total compression of the spring when it is fully compressed is then

$$l = \frac{1767}{706} = 2.5 \text{ inches}$$

This means that the spring is pre-compressed by some 2 inches when it is in the cutoff position (or when the pump is fully stroked). ■

4-1-7 Choosing an Appropriate PDP

We have seen that positive-displacement pumps are available in essentially four basic configurations (gear, lobe, vane, and piston) and in two types (fixed-displacement and variable-displacement). The marketplace also makes some of these pumps available as double pumps (a high- and a low-capacity pump in one unit), and most variable-displacement pumps can be operated in the pressure-compensated mode.

Keep in mind that, as a rule of thumb, gear pumps are the least expensive and piston pumps are the most expensive, with vane pumps somewhere in between. These relative costs are closely tied to the relative volumetric efficiencies; the piston pump has the highest and the gear pump the lowest volumetric efficiency with vane pumps again in the middle.

The three most important factors dictating the choice of a particular PDP are pump pressure rating, displacement (i.e., volumetric discharge capacity), and power. The pressure rating of the pump will be determined by the force or torque requirements of the actuator(s) that the pump drives. Displacement is dictated by the speed at which a load is to be moved by an actuator. The power requirement is determined by the maximum product of pressure and displacement that the pump will experience in driving its actuator(s).

One of the most important factors to be considered in choosing a PDP is the volumetric efficiency. Since discharge will drop off with pressure, so will performance. It will be necessary to choose the pump that can deliver the requisite discharge at the highest pressure against which the pump works. It also is true that overall efficiency, and hence operating cost, is proportional to volumetric efficiency. And since volumetric efficiency is higher for the more expensive pumps, pump selection will involve an assessment of capital costs versus operating costs.

When the question of capital costs versus operating costs is going to be a very important one in selecting a pump, the pressure-compensated mode of operating a variable-displacement pump should be considered. It is an alternative to the fixed-displacement pump (vented by a pressure relief valve) which may involve considerable dissipation of power. Power dissipation can be mitigated using the more expensive pressure-compensated pump.

Duty rating and pump life are also important considerations in choosing a PDP. These two factors are actually related by the fatigue mechanism of failure, and generally the pump manufacturer must be consulted when intermittent operation or unusual hydraulic fluids are being contemplated. One manufacturer defines intermittent duty as 10 percent of the overall time that application of pressure does not exceed 6 seconds. More generally, duty is classified as light (25 percent), medium (50 percent), and heavy (75 percent), with the parenthetical percentages indicating the fraction of maximum capacity the pump will experience.

One of the most important factors determining pump maintenance and life requirements is the filtration system used in the hydraulic circuit. It is imperative that an excellent filtration system be used.

And, finally, keep in mind that the pump manufacturer will be anxious to assist you in choosing the appropriate pump.

4-1-8 Pump Symbols

Hydraulic circuits are sketched schematically in a manner similar to that in which electrical circuits are drawn. This means using simple symbols for the various components. The symbols for PDPs are shown in Fig. 4-25.

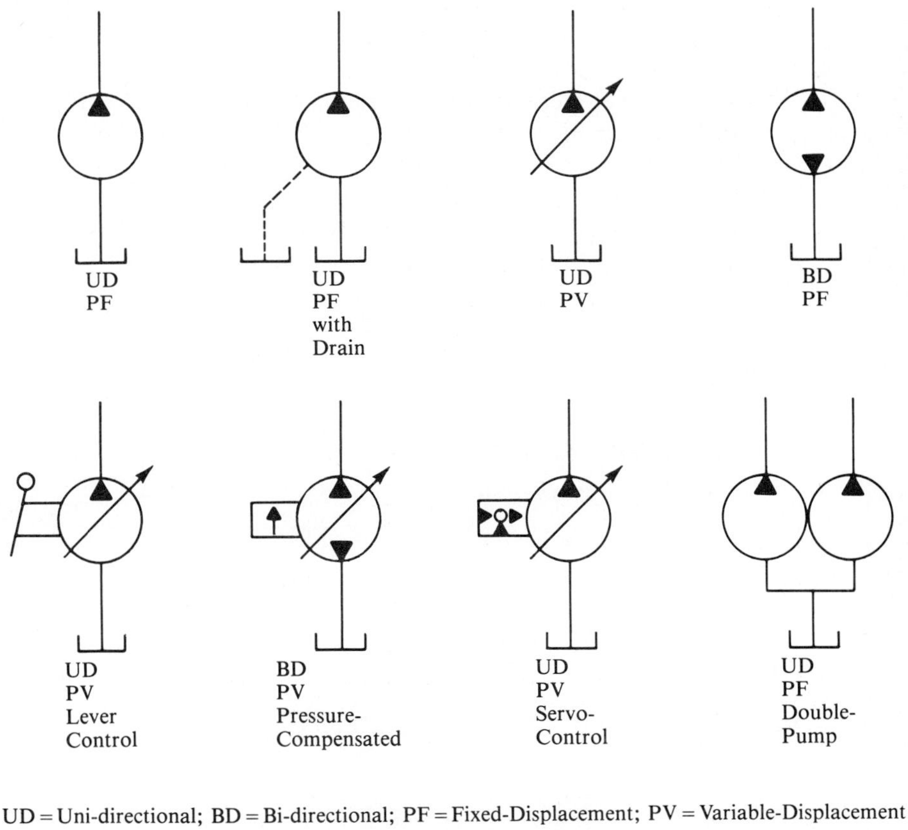

Figure 4-25. Symbols for positive-displacement pumps.

4-2 ACTUATORS

Although we discussed pumps first, the design of a fluid power system actually begins with an actuator because it is the device that actually causes our mechanical system to be energized. The choice of a hydraulic actuator over a mechanical or electrical actuator is made in the context of the advantages and disadvantages of fluid power components listed at the beginning of this chapter.

Actuators can be linear or rotary. For our purposes, a linear actuator is simply a hydraulic cylinder. A rotary actuator may be of limited angular displacement or continuous in angular displacement. When the rotary displacement is continuous, we call the actuator a hydraulic motor. Hydraulic motor designs generally have the same type of internal configurations as hydraulic pumps even though, functionally, they are black box opposites.

4-2-1 Hydraulic Cylinders

A hydraulic cylinder is a fluid power actuator that converts supply pressure and pump volumetric discharge into force and linear velocity. It consists of a piston and piston rod(s) located within a ported cylinder as shown in Fig. 4-26.

To control the operation of a hydraulic cylinder valves must be placed in the supply and exhaust lines that connect the cylinder to the pump and reservoir (tank). The valve shown in Fig. 4-26 is called a two-position, four-way, directional control valve. It is shown in its rightmost position where it connects the pump discharge to the cylinder blank-end (supply) port and the reservoir to the rod-end (exhaust) port. The piston is being driven to the right for the valve position

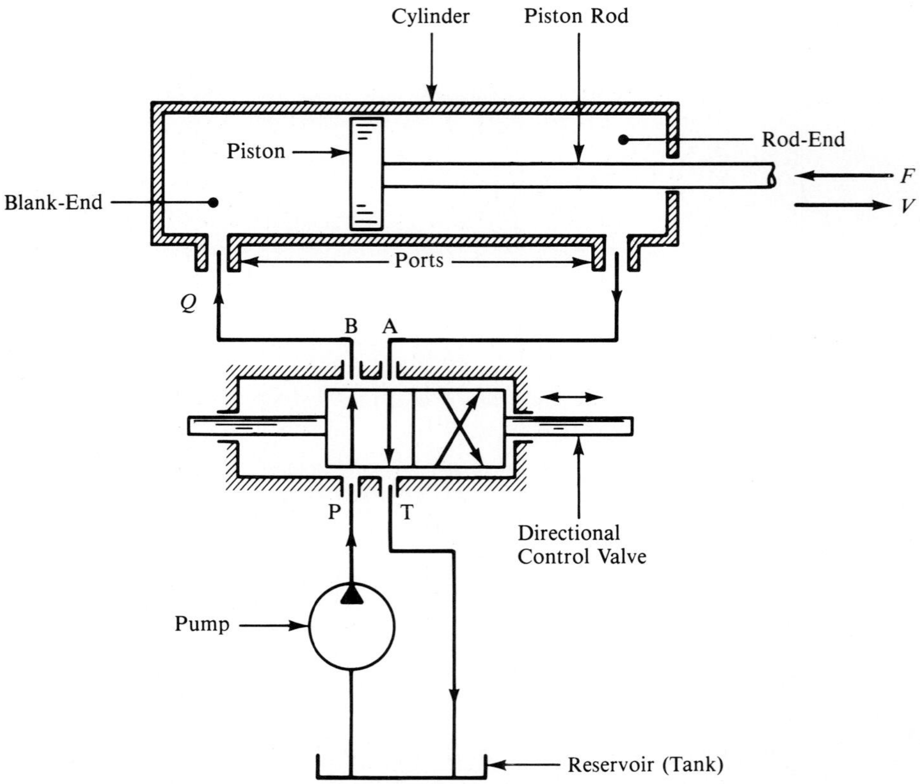

Figure 4-26. Hydraulic cylinder nomenclature.

shown. If the directional control valve is shifted to its leftmost position, then the pump discharge will connect to the rod-end port and the tank will be connected to the blank-end port. In this valve position, the piston will be caused to move to the left.

Hydraulic cylinders are marketed in a wide variety of capacities and types. Force capabilities range from a few ounces to thousands of tons. Strokes range from inches to many feet, and diameters range from an inch to several feet. Cylinder types include (a) single- and double-acting, (b) single- and double-rod, (c) telescoping, (d) cushioned and non-cushioned, and (d) mountings (foot, flange, clevis, trunnion, and rabbit types).

A single-acting cylinder can be powered in only one direction. The return stroke is usually caused by gravity. This kind of cylinder is often used as the ram of a large hydraulically actuated press or lift. A double-acting cylinder can be powered in both directions and generally requires a directional control valve. The double-acting cylinder is the most common application.

In a cushioned cylinder, the cylinder inlet and exit ports are partially blocked off as the piston nears the ends of its stroke. The resulting metering effect of the fluid as it exits through the exhaust port acts as a braking force to give a controlled deceleration to a fast-moving piston which might otherwise bang into the cylinder heads. One version of a cushioned cylinder is shown in Fig. 4-27.

Cylinder Statics and Kinematics. Most hydraulic cylinder calculations are made under the assumption that the piston moves at constant speed. This assumption reduces what would otherwise be a dynamics problem to a statics problem. For example, consider the situation sketched in Fig. 4-28.

If we assume the piston moves toward the right at constant speed against an external load F directed toward the left, a force balance will show that the pressure in the blank-end is

$$p_b = \frac{F}{A} + \left(\frac{A - a}{A}\right) p_r \tag{4-23}$$

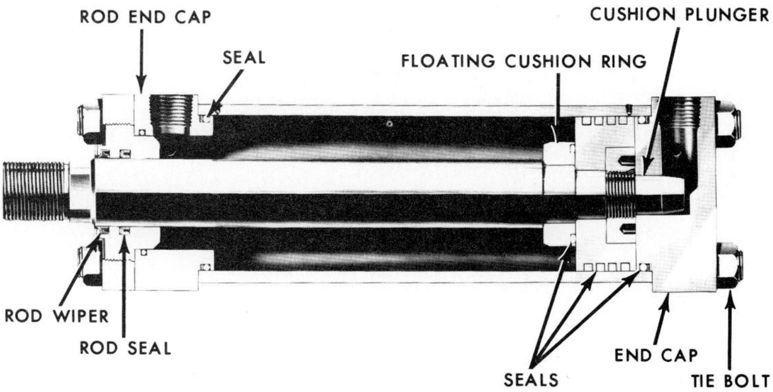

Figure 4-27. Cushioned hydraulic cylinder. (Courtesy of Vickers, Inc.)

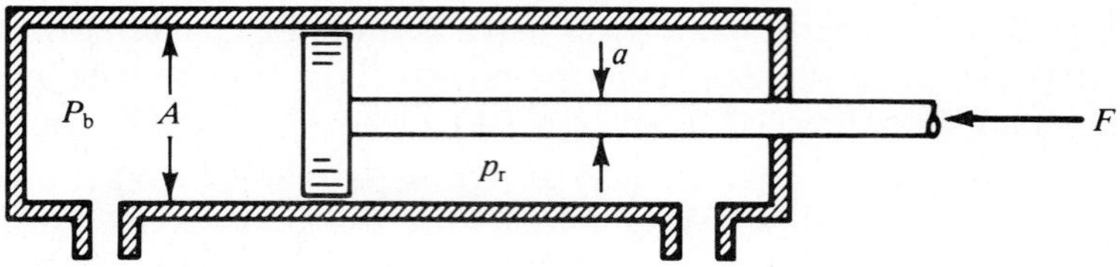

Figure 4-28. Cylinder statics.

where p_r is the pressure in the rod-end, A is the area of the piston, and a is the area of the piston rod. Often, pressure p_r will be negligible compared to F/A, so p_b is just F/A. On the other hand, note that if F is zero, then the rod-end pressure p_r must exceed the blank-end pressure p_b for constant piston velocity to occur.

If the piston moves toward the left at constant speed against an external load F directed toward the right, a force balance will then show that the pressure in the rod-end is

$$p_r = \frac{F}{A-a} + \left(\frac{A}{A-a}\right)p_b \tag{4-24}$$

Here, p_b will be the exhaust pressure, which may be negligible compared to F/A, in which case p_r will be just $F/(A-a)$. We see that it takes a larger rod-end pressure, compared to a blank-end pressure, to drive a given load F. In analyzing hydraulic cylinder actuators, it is always important to draw a free-body diagram of the piston in order to determine pressures because, frequently, the exhaust pressure (whether it be in the blank-end or the rod-end) will not be negligible.

The speed of the piston of a hydraulic cylinder is determined by equating cylinder displacement rate to pump discharge. For a given pump discharge (cylinder inflow) of Q cfm, the piston speed in fpm is given by

$$V\ (\text{fpm}) = \frac{Q\ (\text{cfm})}{A\ (\text{ft}^2)} \tag{4-25}$$

For a cylinder having a rod on only one side of the piston, we see that the piston will have a different speed, for a given pump flow Q, depending on whether flow is into the blank-end or the rod-end. Note that

$$V = \frac{Q}{A}\ (\text{blank-end}) \qquad V = \frac{Q}{A-a}\ (\text{rod-end}) \tag{4-26}$$

So, when fluid is ported into the rod-end, the piston moves faster than when the fluid is ported into the blank-end.

In terms of more common quantities used for flow and area, the piston speed equation is

$$V\ (\text{fpm}) = 19.25\left(\frac{Q\ (\text{gpm})}{A\ (\text{in}^2)}\right) \tag{4-27}$$

The Regenerative Circuit. If the area of the rod-end of a single-rod hydraulic cylinder is made one-half of the blank-end area, the same speed (for a given Q) can be obtained in both directions of piston motion by directing the discharge from the rod-end back to the inlet port of the blank-end when the piston moves toward the rod-end. This is called a regenerative circuit (see Fig. 4-29).

The cylinder shown is a single-rod, double-acting cylinder actuated with a two-position, three-way, directional control valve. In the left figure, both the pump output and the return flow from the rod-end of the cylinder are being ported into the blank-end as the piston moves to the right. Thus, the pump supply is augmented to compensate for the increased inflow volume needed to move the piston to the right at the same speed that a lesser flow will be able to move the piston to the left. In the figure at the right, only the pump supply is ported into the rod-end as the piston moves leftward.

An analysis of the volumetric flows involved in Fig. 4-29 will show that the relationship between leftward piston velocity V_ℓ and rightward piston velocity V_r is

$$\frac{V_\ell}{V_r} = \frac{\dfrac{a}{A}}{1 - \dfrac{a}{A}} \qquad \textbf{(4-28)}$$

When the area of the rod is one-half the piston area, the ratio $a/A = \frac{1}{2}$, and then $V_\ell = V_r$.

The other valve shown in the lower right of each figure is a pressure relief valve. Its function is to open when the piston has reached either end of the cylinder, in order to bypass the pump output around the cylinder back to the tank. When the piston reaches either end of the cylinder, the pump outlet pressure will tend to rise, which will actuate the pressure relief valve.

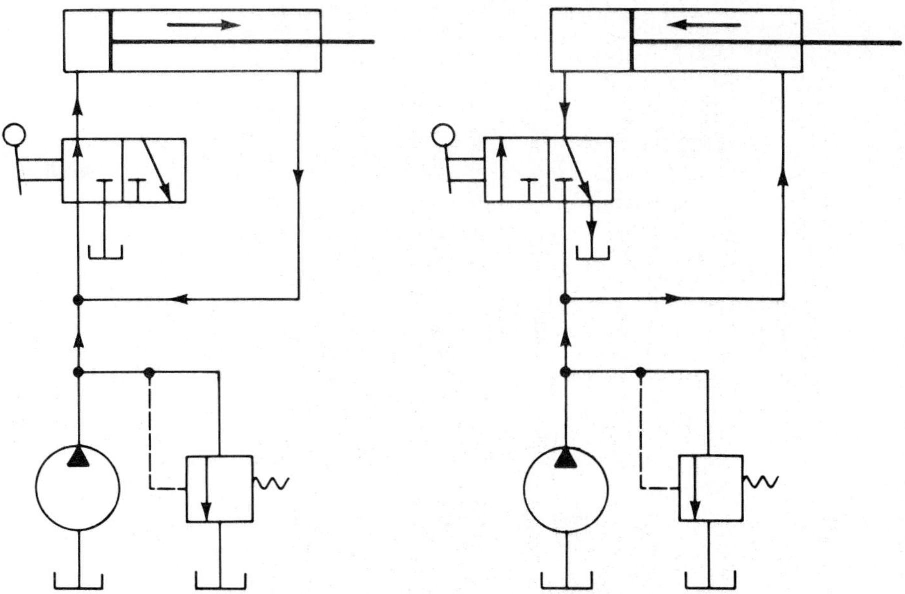

Figure 4-29. Regenerative circuit.

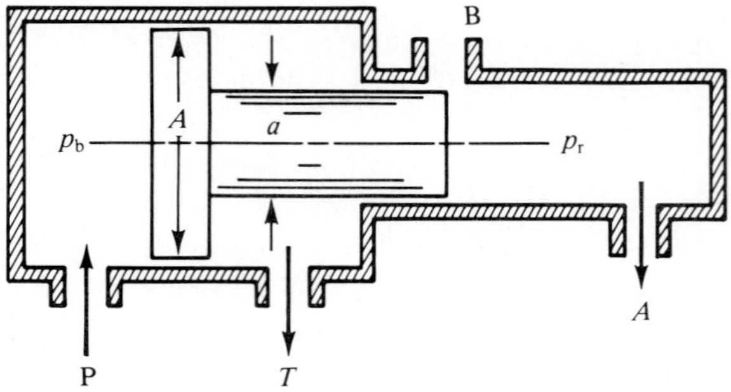

Figure 4-30. Hydraulic intensifier.

Hydraulic Intensifiers. A hydraulic intensifier is a hydraulic cylinder configuration whereby pressure can be multiplied at the expense of displacement. The function of the intensifier is to provide working pressures that exceed those the pump can safely supply directly. A single-acting intensifier is sketched in Fig. 4-30.

The intensifier has a large piston of area A integral with a smaller piston of area a. Oil is ported from the pump port P into the blank-end of the large piston. As the piston moves to the right, port B is covered and fluid is discharged at high pressure p_r through port A, which is connected to the actuator being used. Ports T and B permit refill on the retraction stroke. The ratio of pressures involved is simply inverse to the intensifier area ratio. That is,

$$\frac{p_r}{p_b} = \frac{A}{a} \tag{4-29}$$

Choosing Hydraulic Cylinders. The following factors must be considered in choosing and specifying hydraulic cylinders:

(a) *Pressure*. Working pressures must be specified because cylinders are pressure rated.
(b) *Stroke*. The required stroke will be dictated by the load requirements. There is a relationship between stroke and rod diameter dictated by the necessity to avoid rod buckling.
(c) *Piston velocity*. Cylinders are rated with respect to maximum piston velocity.
(d) *Cushioning*. If high cylinder speeds and loads are involved, serious consideration should be given to selecting a cushioned cylinder.
(e) *Loading and duty cycle*. Various types of cylinders are available depending on the specific nature of the loading. Also, cylinder life is an important specification dictated by duty cycle.
(f) *Mounting*. Since cylinders are available in a large variety of mounting configurations, the mounting must be specified.

Symbols for Hydraulic Cylinders. Symbols used in hydraulic circuits for representing various kinds of cylinders are shown in Fig. 4-31.

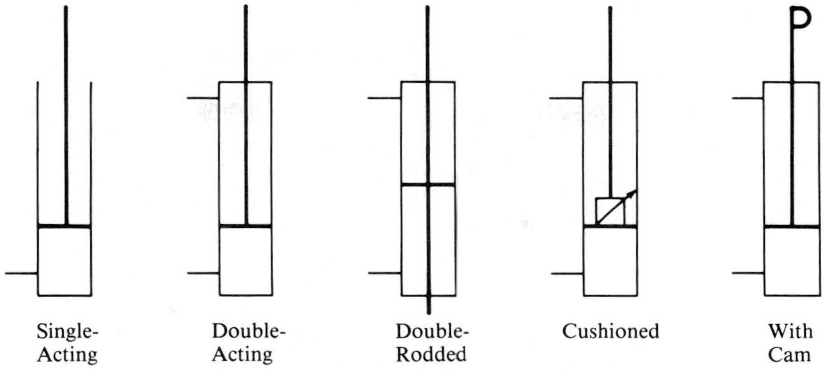

Figure 4-31. Symbols for hydraulic cylinder actuators.

4-2-2 Limited-Rotation Actuators

A limited-rotation actuator (also called a rotary actuator) is a fluid power actuator that converts pump supply pressure and discharge into torque and limited rotational displacement. This type of device might be used, for example, to actuate the wrist of a hydraulically powered mechanical robot.

Rotary actuators are available in gear, vane, and piston configurations. Hydraulic cylinders can also be converted to rotary actuators by tying the piston to a rack that drives a rotating pinion. Output torques are available in the range of about 10^2 to 10^6 in-lb, and rotational displacement depends on the design (from a fraction of a radian to several radians).

Two types of vane rotary actuators are shown in Fig. 4-32. In the single-vane design, rotation is usually limited to around 300 degrees, while in the double-vane design, the rotation is limited to around 150 degrees. As in the case of a hydraulic cylinder, control of a rotary actuator generally requires interfacing the actuator ports with a directional control valve.

The torque produced by a rotary actuator is given by

$$T = Ar\,\Delta p \tag{4-30}$$

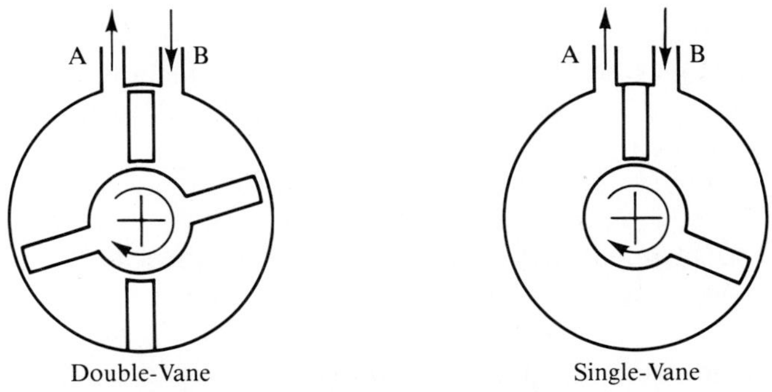

Figure 4-32. Two types of vane rotary actuators.

Figure 4-33. Symbol for the rotary actuator.

where the factor Ar is a geometry parameter available from the manufacturer and Δp is the pressure drop across the vane. Similarly, the angular speed ω is given by

$$\omega = \frac{Q}{Ar} \tag{4-31}$$

The criteria used in selecting rotary actuators are the same as those delineated earlier for hydraulic cylinders. The circuit symbol used to indicate a rotary actuator is shown in Fig. 4-33.

4-2-3 Positive-Displacement Hydraulic Motors (PDM)

A hydraulic motor is a rotary actuator having continuous angular displacement. Virtually all of the positive-displacement pumps discussed earlier can be designed to operate as hydraulic motors. These include the gear, lobe, vane, and piston versions.

Gear motors can be of the external or internal type, similar to gear pumps. Torque is developed by the supply pressure acting on the gear teeth surfaces. As was true with the gear pump, a gear motor is inherently a fixed-displacement device, with speed dictated essentially by volumetric flow through the motor. Gear motors are available in speeds up to around 5000 rpm and pressures of around 2500 psi.

Vane motors are also similar to their pump counterparts. They are more versatile than the gear pump in being available in both the fixed- and variable-displacement types. They are also available as high-torque, low-speed devices, which is an important consideration in some applications. As an example, vane motors are available that will operate in the speed range of 10 to 100 rpm and provide an output torque of some 50,000 in-lb.

Piston-driven hydraulic motors are the most sophisticated and expensive of the hydraulic motor types. They are often chosen when a variable-displacement rotary actuator is required. These motors are available in axial-piston and radial-piston types. Piston motors offer the highest volumetric and overall efficiencies and a wide range of operating speeds (from zero to 5000 rpm) and pressures (from zero to 5000 psi).

4-2-4 Characteristics of Positive-Displacement Motors

All of the equations developed for positive-displacement pumps in Section 4-1-5 also apply to positive-displacement hydraulic motors. In particular, the following pairs of equations apply:

$$Q_i = N \cdot S \cdot q_{max} \qquad T_i = \frac{P \cdot S \cdot q_{max}}{2\pi} \tag{4-32}$$

$$Q_i = \omega\left(\frac{An\varepsilon}{\pi}\right) \qquad T_i = P\left(\frac{An\varepsilon}{\pi}\right) \tag{4-33}$$

Recall that N is motor speed in rpm, q_{max} is motor displacement per revolution, P is motor supply pressure, ω is motor speed in rad/min, S is motor stroke, and ε is motor eccentricity. In a variable-displacement motor, the stroke S (and hence eccentricity ε) can be controlled between limits of zero and the maximum value $S = 1$.

Equations (4-32) and (4-33) show some interesting features of hydraulic motor behavior. For example, if the volumetric flow rate Q_i is held constant, the

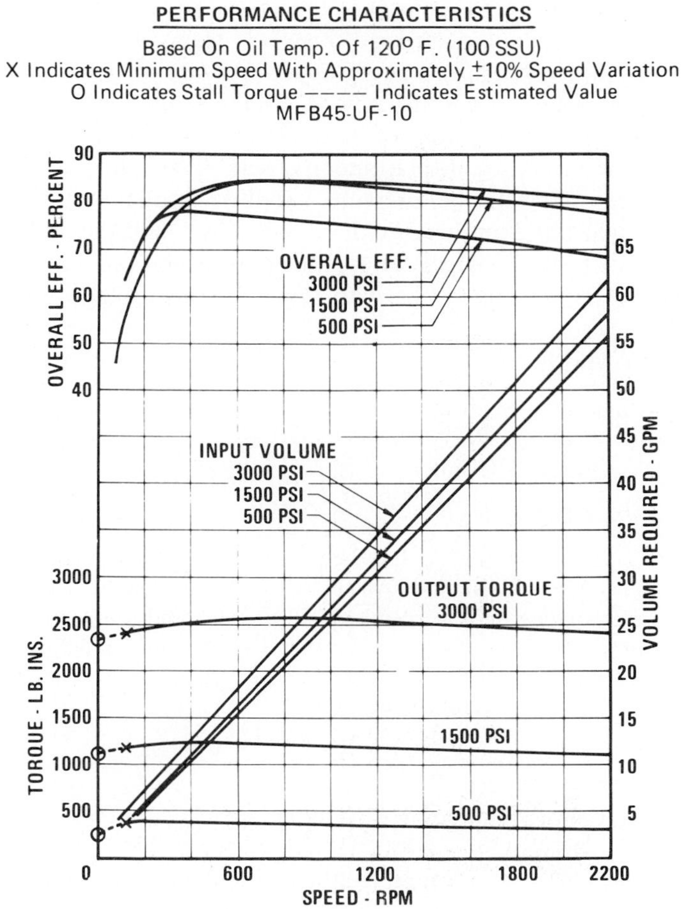

Figure 4-34. Axial-piston motor characteristics. (Courtesy of Vickers, Inc.)

motor speed N (or ω) will increase as stroke S (or eccentricity ε) is diminished. If the eccentricity is permitted to go to zero (which centers the rotor barrel, of the motor, in its reaction ring), the speed theoretically becomes infinite. This tendency must be avoided.

The output characteristic of a hydraulic motor is its torque-speed curve. Equations (4-32) and (4-33) show that if displacement is fixed and inlet pressure is held constant, motor torque T_i will be theoretically constant. Thus, the ideal hydraulic motor characteristic is a horizontal line on the T-N plot.

Characteristic curves for an actual hydraulic piston motor are presented in Fig. 4-34. We see that the motor torque is not precisely constant and independent of speed as the theoretical equations state. The reason is that frictional effects and internal leakage of the motor will vary slightly with speed. Since frictional effects increase with speed, we would expect the output torque to droop (as shown) at the higher speeds.

As was the case with PDPs, we expect the highest performance from piston motors and the lowest from gear motors, with vane motors in between.

The input volume curves of Fig. 4-34 show that a greater input volume Q_i is required for a given speed N as supply pressure is increased. This is the result of decreasing volumetric efficiency η_v (increasing back-leakage) as pressure increases. Keep in mind that torque and pressure are essentially proportional in a hydraulic motor when displacement is held fixed. Thus, as the load torque on the motor rises, so will the pressure supplied at the motor inlet. The increasing pressure will cause an increase in back-leakage, and the motor speed will diminish slightly due to this leakage. Thus, the actual speed regulation of a hydraulic motor is not perfect as it would be with a perfect (no leakage) motor.

EXAMPLE 4-5

Assume that the axial-piston motor of Fig. 4-34 is supplied with oil at the constant rate of 40 gpm. What is its speed regulation as its load torque varies between 320 in-lb and 2500 in-lb?

Solution

From Fig. 4-34, we see that the inlet pressures corresponding to the two load torque values are 500 psi and 3000 psi. The input volume curves then show that the corresponding motor speeds will be 1600 and 1400 rpm respectively. Since the average speed in this pressure range is 1500 rpm, we define the speed regulation as

$$\text{Speed regulation} = \frac{1600 - 1400}{1500} = 13.33 \text{ percent} \qquad \blacksquare$$

4-2-5 Compensator Control of Positive-Displacement Motors

With a variable-displacement motor having controllable eccentricity, it is possible to servo-control the motor. Suppose, for example, that the load torque on the motor increases. Then, as the torque equation of Eqs. (4-33) shows, either the supply pressure P or the eccentricity will have to be increased. The tendency for the pressure to increase under these circumstances can be used to actuate a change in eccentricity, thus maintaining the supply pressure constant. Of course, this action will cause the speed of the motor to decrease, as the equation for Q_i shows. Since this may not be desirable, a more appropriate measure might be to servo-control the pump which supplies fluid to the motor. We will discuss servo-control in more detail later.

4-2-6 Choosing an Appropriate Positive-Displacement Motor

The same kinds of considerations must be made in choosing a hydraulic motor as were discussed for pumps in Section 4-1-7. Gear motors, vane motors, and piston motors are similar to their pump counterparts insofar as volumetric, torque, and overall efficiencies are concerned and in relationship to their relative capital costs. Actually, the process of choosing a hydraulic motor can seldom be carried out independently of the process of pump selection since the motor requires a pump to drive it. As we shall see later, the combination of a pump and motor is called a hydrostatic transmission, and it is the overall performance of the transmission that overrides the individual performance of pump and motor. Since there will inevitably be back-leakage in both pump and motor, and frictional losses in both, the output of the transmission depends on more than the motor performance alone.

Note in Fig. 4-34 that the starting torque (or stall torque) is less than the torque at operating speeds. Consequently, if a load is to be started under a full-load torque, a motor must be chosen that is somewhat oversized compared to one that can provide the load torque at operating speed.

An additional complicating feature affecting the choice of a hydraulic motor as a component in a mechanical power transmission is the effect of all of the rotating inertias of other components in the power train (e.g., clutches, gearboxes, and load). Thus, if a system must be brought up to operating speed rapidly, or if it must rapidly respond to changing load conditions, the choice of components will be different than for a system not having these requirements.

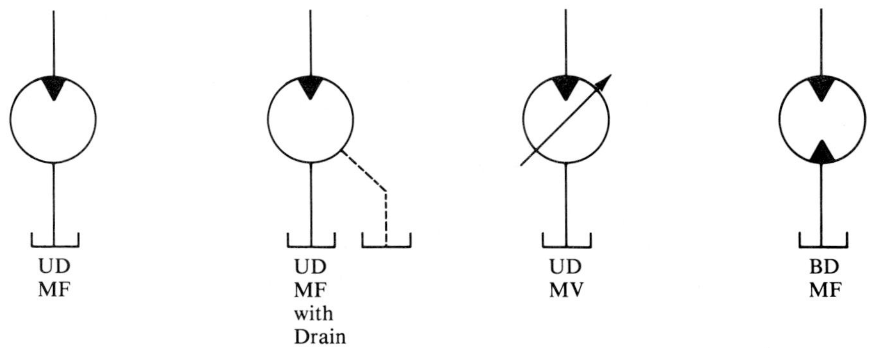

UD = Uni-directional; BD = Bi-directional; MF = Fixed-Displacement; MV = Variable-Displacement.

Figure 4-35. Symbols for hydraulic motors.

4-2-7 Hydraulic Motor Symbols

Symbols for hydraulic motors are the same as for pumps with the exception that the triangles (arrowheads) point inside to show that the motor is receiving flow and power rather than delivering it (see Fig. 4-35).

4-3 VALVES

We have already sketched some hydraulic circuits that include valves without having described how the valves actually work. Now we want to discuss valves in some detail. We are interested primarily in pressure control valves, flow control valves, and directional control valves.

4-3-1 Pressure Control Valves

Pressure control valves have two primary functions in hydraulic circuits: to protect the circuit against dangerously excessive pressures and to control both the amount and the direction of hydraulic fluid discharge. We noted earlier that it is generally necessary to incorporate a pressure relief valve into a hydraulic circuit in order to limit pressure and to protect circuit components (particularly the pump). Relief valves are almost always located at the pump discharge, and when they open they will pass fluid back to the reservoir, thus bypassing the fluid power circuit.

Relief Valves. The two basic kinds of pressure relief valves are the direct-acting (or simple) relief valve and the piloted relief valve. A sketch of a direct-acting relief valve is provided in Fig. 4-36 along with its circuit symbol as defined by the American National Standards Institute (ANSI).

In the direct-acting pressure relief valve, the circuit pressure acts directly against a valve poppet seated in a valve seat and directly opposed by a spring. The spring force can be adjusted by compressing the spring by means of an adjusting screw.

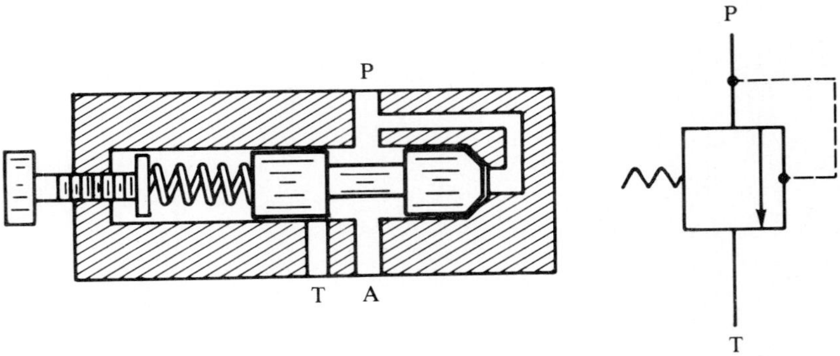

Figure 4-36. Simple relief valve.

The cracking pressure is defined as the circuit hydraulic pressure (at the pump exit) that will just overcome the resisting spring force and unseat the valve poppet. At cracking pressure, the valve opening will be small and only a small amount of fluid will be passed through the valve back to the reservoir. As the circuit pressure increases, the valve opening will become larger because the spring will be compressed more. The full-flow (or rated) pressure is the circuit pressure at which substantially all of the flow coming from the pump will be bypassed through the relief valve. The difference between cracking pressure and full-flow pressure is called the relief valve pressure override (or bandwidth).

Direct-acting relief valves can be caused to chatter and fluctuate under certain conditions, and in some cases the pressure override might be objectionable because of the power losses associated with it. Remember that when fluid is bypassed back to the tank through a relief valve, all of the power represented by the product of relief valve pressure and flow rate is dissipated in the form of heat. It is therefore extremely desirable to minimize pressure override unless the relief valve is being used as a form of flow control.

Some of the objectionable features of the direct relief valve can be eliminated by using a piloted relief valve, a schematic diagram of which is presented in Fig. 4-37. This device has a small pilot valve (lower left) for cracking the valve plus a larger valve spool (upper right) which is controlled by the pilot valve. The upper right-hand spool (or piston) has an orifice drilled through it so that the high-pressure port can communicate with the small pilot valve.

For pressures below the cracking pressure (dictated by the adjustable spring of the pilot valve), the pilot valve remains closed and so does the main valve spool because pressures are balanced on it and it is held closed by its own light spring. When the circuit pressure rises above the pilot valve setting, the pilot valve opens, exhausting the fluid between the two pistons. The pilot valve opening unbalances

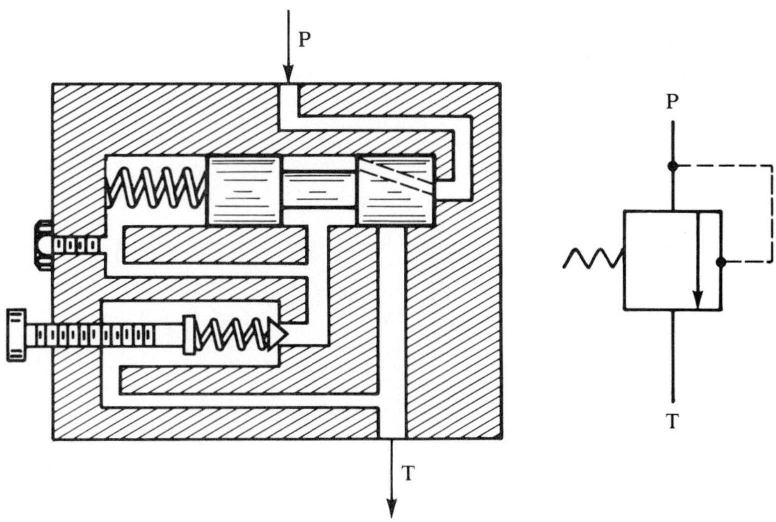

Figure 4-37. Pilot-operated relief valve.

the main valve, and the main valve is actuated to the left, opening the inlet port directly to the exhaust port (and, hence, to the tank). The pressure override required to accomplish this is minimal (of order of magnitude 20 psi), and complete bypassing of the pump outlet to the tank is obtained without the necessity to compress a heavy spring.

Note that this valve contains a vent connection which allows an external control to be used to actuate the valve instead of the pilot valve. The vent connection makes it possible to use the same valve body for an unloading valve of the type discussed in the next section.

Unloading Valves. An unloading valve is a pressure control type of valve that operates in a manner somewhat similar to a piloted relief valve. One type of unloading valve is sketched in Fig. 4-38.

In contrast to the piloted relief valve, which responds to a pilot pressure in the circuit just upstream of the relief valve, an unloading valve responds to a pilot pressure coming from a remote source. Also, in contrast to a relief valve, an unloading valve does not necessarily require high pressure at the pump outlet for actuation. Consequently, when a pump is unloaded using an unloading valve, the horsepower loss through the valve can be minimized by designing the valve for low pressure drop when it is open. Of course, to do all of this will require that the pilot pressure be maintained from the remote source.

Note that the chamber containing the spring is ported to the discharge side of the valve which will normally be connected to the tank (reservoir). Porting is necessary so that the small leakage past the spool into this chamber due to inlet pressure will not build up in the spring chamber (and hence prevent the valve from opening). This valve is said to be internally drained because the path from the spring chamber to the discharge side of the valve lies inside the valve itself. As is shown in the symbol for the unloading valve, the internal drain is indicated as a dotted line while the pilot line is dashed. Remember that there is no fluid flow in a pilot line. Although the pilot line is filled with fluid, it is only a hydraulic transmitter of pressure.

Figure 4-38. Pressure-unloading valve.

EXAMPLE 4-6

When the actuator of the fluid power circuit of Fig. 4-39 operates, it requires a pump outlet pressure of 1500 psi and a pump discharge of 5 gpm. When the actuator is to be stopped, the three-position, four-way, directional control valve is centered (as shown), completely blocking the pump discharge from entering the actuator circuit. Consequently (since the pump will generally not be turned off), the pump discharge must be diverted back to the tank through the relief valve or through the unloading valve depending on which valve is open.

The relief valve setting is some 500 psi above the working pressure of the actuator so it will open at 2000 psi. If the unloading valve remains closed, the pump discharge will have to flow through the relief valve at 5 gpm through a pressure drop of 2000 psi. The power dissipated will therefore be

$$\text{HP} = Q \cdot p = \frac{\left(\frac{5}{7.48}\right)(2000)(144)}{(60)(550)} = 5.83 \text{ hp}$$

The power dissipation can be reduced substantially by diverting the pump outlet flow to the tank through the unloading valve rather than through the relief valve. Suppose the circuit is designed so that when the directional control valve is centered, the pilot line of the unloading valve becomes pressurized. This will cause the unloading valve to open, and the pump discharge will now flow back to the tank through only the pressure drop across the unloading valve, which might be of order of magnitude 20 psi. Then the horsepower loss will be diminished to only a hundredth of that associated with flow through the relief valve (0.0583 hp). ■

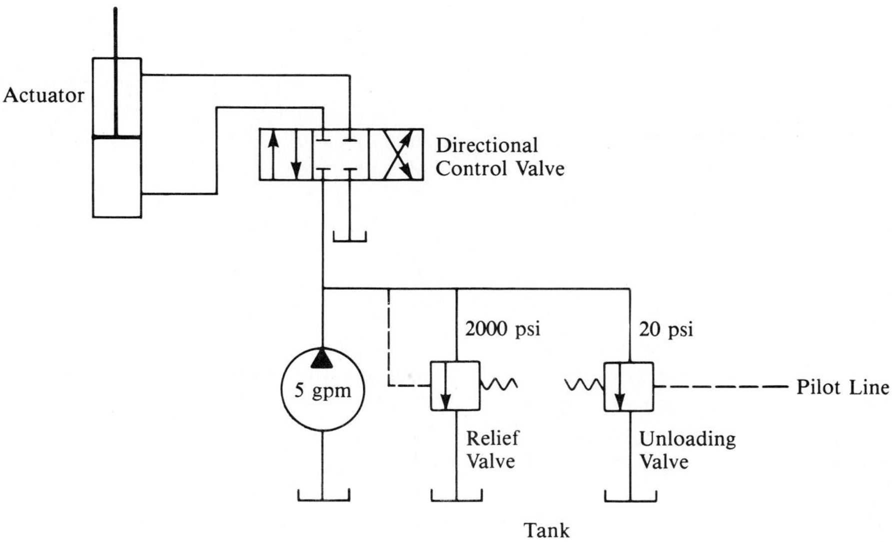

Figure 4-39. Hydraulic circuit for Example 4-6.

Sequence Valves. A sequence valve is simply a valve of the pressure relief type in which the vented fluid, instead of returning to the reservoir (or tank), is ported to some other portion of the hydraulic circuit (called the secondary circuit). A primary circuit pressure is used to actuate a sequence valve, and after the valve sequences, this primary pressure is maintained at the valve secondary port. An example is sketched in Fig. 4-40.

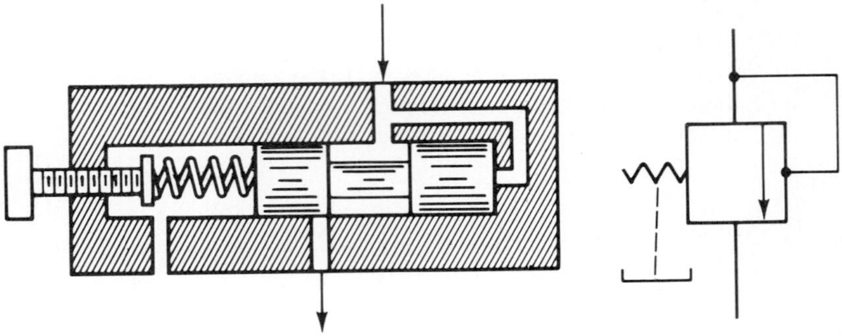

Figure 4-40. Sequence valve.

Sequence valves are used to direct circuit flow in a sequential manner. A typical application is to control the sequence in which two actuators are to operate, as discussed in Example 4-7. Note that the spring chamber of the sequence valve must be drained to prevent the build-up of pressure caused by leakage past the spool from pressure at the primary port. In a sequence valve, this must be an external drain back to the reservoir.

EXAMPLE 4-7

A typical sequence valve application is shown in Fig. 4-41. Cylinder A is the primary actuator and moves first when the directional control valve D is shifted to the right. Note that the setting of the sequence valve S must be high enough for the valve to remain closed while actuator A is working against its load. If, for example, the load on actuator A is 10,000 lb and the cylinder is a hydraulic cylinder whose piston area is 5 square inches, then the pump pressure required to move this load will be 2000 psi and the sequence valve will have to be set at, say, 2200 psi in order that it remain closed while actuator A is moving its load.

When actuator A gets to the end of its stroke, it will stall and the pump outlet pressure will increase beyond 2000 psi. When the pump outlet pressure reaches 2200 psi (practically instantaneously), the sequence valve S will open and actuator B will become operative. Of course, the system must be designed such that the pressure on actuator B is sufficient to keep sequence valve S open but not high enough to actuate pressure relief valve R. This requires that the setting of R exceed that of S (by, say, some 250 psi) and that the load on actuator B plus the pressure drop across S amount to at least a pump outlet pressure of 2200 psi.

The check valve is in the circuit so that the cylinders can be retracted by moving directional control valve D to its leftmost position. Note that in this particular circuit, there is no control of the order in which the two cylinders A and B

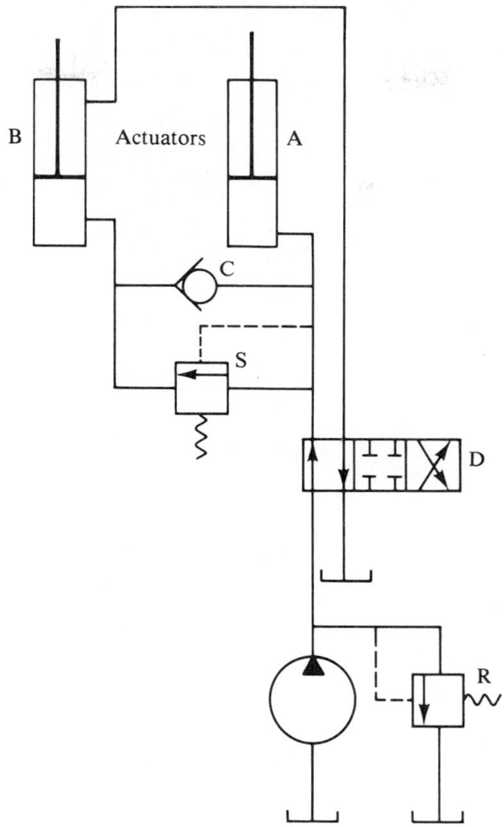

Figure 4-41. Sequence valve application.

will be retracted. But the working strokes of cylinders A and B are in a sequence in which A completes its stroke first and then B will complete its stroke. This circuit would be used in a machine where cylinder A actuates a clamp and cylinder B operates on work held in the clamp. ■

Counterbalance Valves. A counterbalance valve permits free flow in one direction and a restricted flow in the opposite direction. An example is shown schematically in Fig. 4-42.

The counterbalance valve is commonly used in conjunction with a single-acting cylinder used to elevate and lower a load. When the pump is off, we do not want the load to fall under gravity; the counterbalance valve prevents this by maintaining a back-pressure against the load. When the pump is on, the load can be elevated (flow is through the check valve) or powered down (flow is through the spool portion of the valve) depending on the position of the directional control valve.

A typical hydraulic circuit that incorporates a counterbalance valve is shown

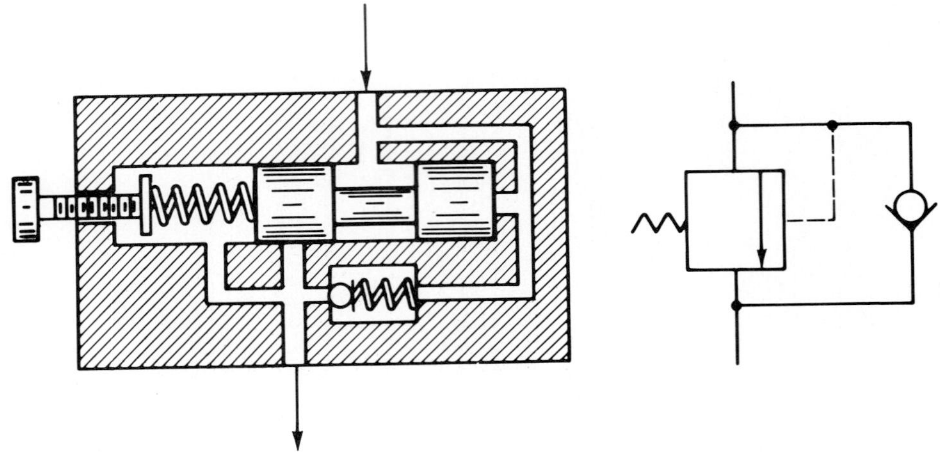

Figure 4-42. Counterbalance valve.

in Fig. 4-43. During descent, the flow in the circuit is cw (left figure) and the load W is counterbalanced by the resistance in the valve produced by the (adjustable) spring acting on the spool. During ascent, the flow in the circuit is ccw through the check valve, which offers negligible resistance to flow.

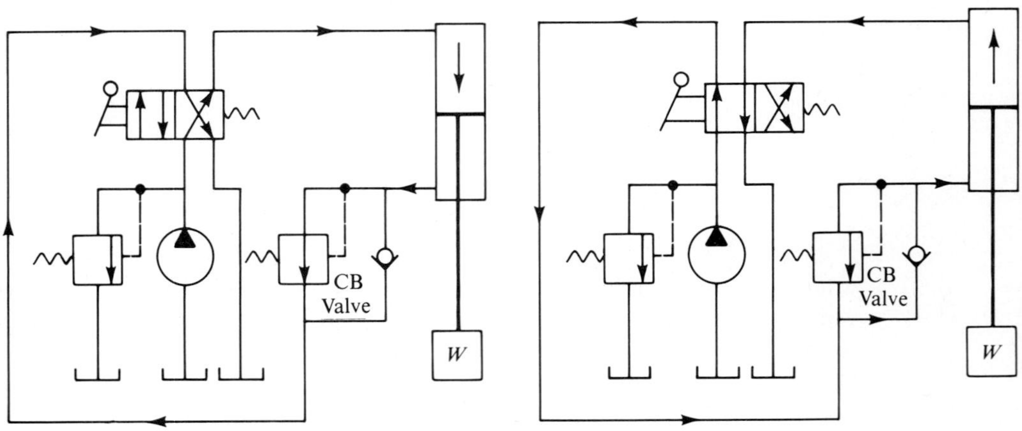

Figure 4-43. Counterbalance valve application.

Pressure-Reducing Valves. All of the valves discussed up to this point are said to be normally closed valves. They all open when a prescribed pressure (usually dictated by a spring) is applied via a pilot line. In contrast to this type of normally closed valve is the pressure-reducing valve, which is normally open and whose function is to sense and control pressure downstream from the valve. A schematic diagram of this type of valve is presented in Fig. 4-44.

The position of the spool of this pressure-reducing valve depends only on the spring and the downstream pressure. If the downstream pressure tends to

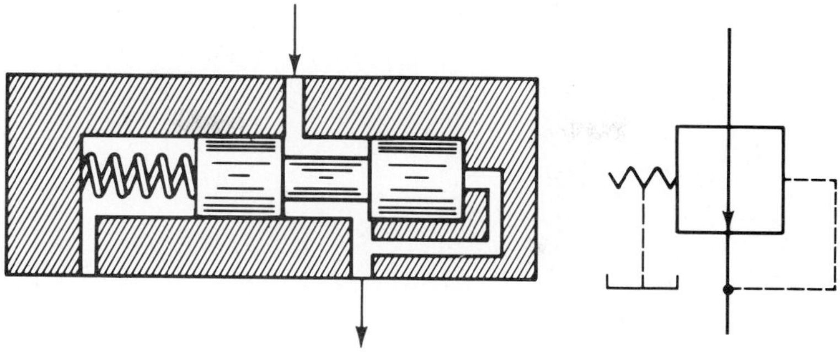

Figure 4-44. Pressure-reducing valve.

increase, the spool will move slightly to the left, throttling the exit port and thus increasing the resistance of the valve. For a fixed inlet pressure, throttling will cut down the flow through the valve while the pressure at the exit will be essentially maintained. If the pressure at the exit decreases, the spool moves to the right, permitting an increased flow but maintaining the exit pressure because exit pressure is always dictated by the balance between spring and exit pressure. Note that this valve must be externally drained.

4-3-2 Flow Control Valves

In many of the hydraulic circuits employed by the mechanical designer, it will be necessary to control the speed of an actuator. We have already seen that the relationship between the speed V of a hydraulic piston and the volumetric flow rate Q is

$$V\ (\text{fpm}) = 19.25\left(\frac{Q\ (\text{gpm})}{A\ (\text{in}^2)}\right) \tag{4-27}$$

Similarly, an equation can be developed between the rpm of a hydraulic motor and the volumetric flow rate. If the displacement per revolution of the motor is q, then

$$n\ (\text{rpm}) = 231\left(\frac{Q\ (\text{gpm})}{q\ (\text{in}^3/\text{rev})}\right) \tag{4-34}$$

Although this speed control could be achieved by controlling the pump rpm, it is far more common to have the pump operate essentially at constant rpm. Therefore, speed must generally be controlled by diverting part of the pump output around the actuator (i.e., to partially bypass the actuator). The simplest way to do this is to introduce a resistance in the line to (or from) the actuator such that part of the pump output is diverted to the tank over the relief valve. In effect, we have a circuit with two branches in parallel, with the actuator in one of the branches. Then the relative flow through the two branches depends on the relative resistances of the branches. The simple electrical analog is sketched in Fig. 4-45.

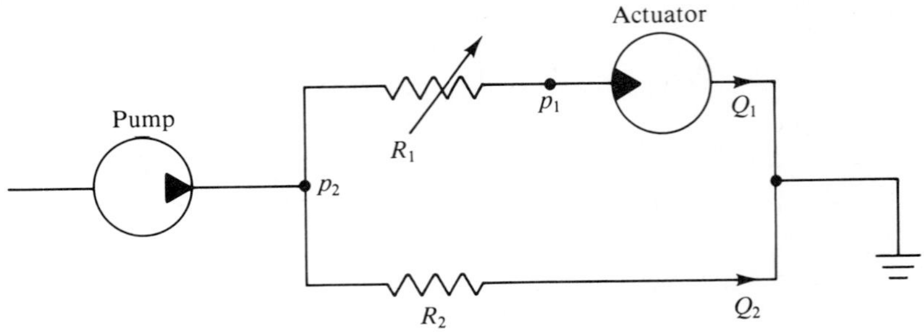

Figure 4-45. Electrical analog of flow control.

Resistance R_2 represents the relief valve while resistance R_1 represents the flow control device (generally, a variable resistance). If a simple variable-resistance orifice is placed in the actuator branch to provide resistance R_1, it will not be possible to maintain a set flow rate Q_1 independent of the load pressure p_1. Since p_1 might possibly increase, the pressure drop (and, hence, flow rate) across resistance R_1 will decrease because the pump outlet pressure p_2 is being maintained essentially constant by the pressure relief valve R_2. Decreased flow has the effect of making the actuator speed dependent on the actuator working pressure, which is undesirable.

Restrictor Type Pressure-Compensated Flow Control Valve. The problem with using a simple orifice for flow control can be surmounted by using a pressure-compensated flow control valve. An example is sketched in Fig. 4-46.

The valve shown is, in effect, two resistances in series, with one of the resistances larger than the other. The high resistance is provided by the balanced piston (at the top) which acts in a manner somewhat similar to a pressure-reducing valve. This is called the compensator spool. The low resistance (at the

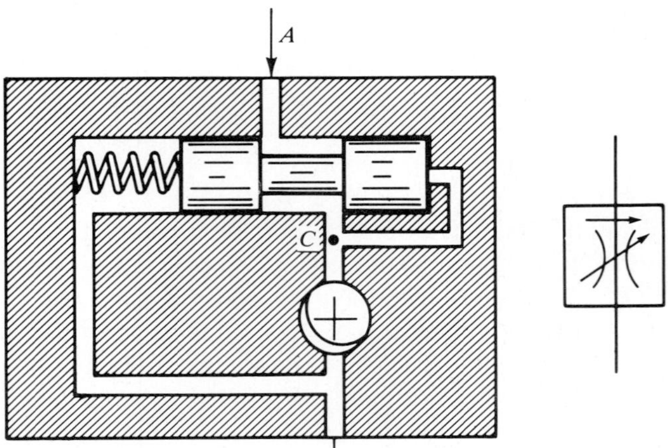

Figure 4-46. Pressure-compensated flow control valve.

bottom) is an adjustable orifice. The compensator spool has a spring which is set for, say, 20 psi. The spring controls the pressure drop across the adjustable orifice to be 20 psi regardless of fluctuations in the exit pressure of the valve (as at point B).

If the exit pressure rises, the pressure at the orifice inlet (point C) will also rise to a value 20 psi above the exit pressure (point B). Since the pressure drop across the orifice is maintained at 20 psi, the flow through the orifice will also be maintained constant. The rise in exit pressure at point B will cause the balanced piston to open slightly to accommodate to the required fixed flow rate. This slight opening is necessary since, with an increase of valve exit pressure, the pressure drop across the valve will have decreased because the inlet pressure (point A) is being maintained constant by the relief valve in the circuit. Thus, we have a variable-resistance valve which, for a given valve setting, can maintain flow independent of valve exit pressure (which is the actuator working pressure).

Bypass Type Pressure-Compensated Flow Control Valve. There is another version of the pressure-compensated flow control valve which, in effect, incorporates the system pressure relief valve within the flow control valve. Thus, the portion of the pump discharge not needed to maintain actuator speed is bypassed back to the tank directly through the flow control valve rather than by means of a separate relief valve as is required when the restrictor type of pressure-compensated pump is used. This valve has the advantage that the pump need operate at only some 20 psi above the valve exit pressure (working load pressure), with subsequent power savings over the arrangement whereby a separate relief valve is used. One disadvantage of this type of valve is that it cannot be used in an actuator exhaust line because of the danger of actuator runaway. We will discuss this danger later.

Proportional Valves. In the superficial context in which we discussed directional control valves previously, they were valves of the on-off type; the valve was either fully on or fully off. A proportional valve is a valve of the directional control type in which the position of the valve can be continuously controlled rather than simply being ON or OFF. Control of the valve position means that the resistance of the valve is being controlled and, hence, so will be the flow through the valve. Thus, in a proportional valve, the functions of directional control and flow control are incorporated into the same valve.

The next section covers directional control valves in more detail.

4-3-3 Directional Control Valves

Directional control valves are usually located between the pump and actuator(s) of a hydraulic circuit. The primary function of a directional control valve is to control the direction of flow to an actuator (i.e., to determine which actuator port will be the inlet port). In Fig. 4-43, for example, the two-position, four-way, directional control valve dictates that the direction of flow be into the blank-end port of the cylinder in the left figure and into the rod-end port in the right figure. In other words, it is the directional control valve that is used to cause the

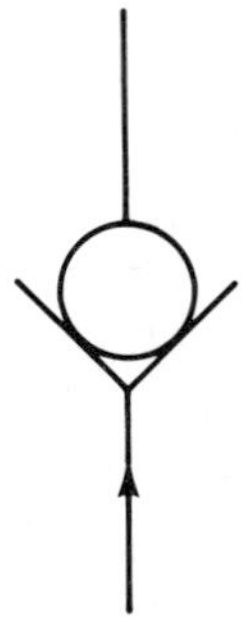

Figure 4-47. Check valve.

hydraulic cylinder to extend, retract, and stop. In a hydraulic motor, the directional control valve will be what is used to determine the direction of rotation.

The simplest form of a directional control valve is a manually operated shut-off valve. Also, a check valve can be considered to be a one-way directional control valve. The symbol for a check valve is shown in Fig. 4-47. The check valve is the hydraulic analog of the electrical diode. It permits flow in one direction but not in the other.

The only form of directional control valve of interest to us in this book is the spool valve type, shown in Fig. 4-48. The spool is the horizontal shaft which is sectioned into cylinders having two different diameters. As the spool is shifted, various ports are opened, or closed, depending on whether or not a section of the shaft having reduced diameter is placed in communication with a port.

These spool type directional control valves are made in a tremendous variety of types (see References 4-2, 4-5, and 4-6). The spool might be shifted manually, pneumatically, hydraulically, mechanically, or electrically. One important method of actuating the spool is by means of electrical solenoids, which makes possible the coupling of hydraulic and electrical control circuits. The possibility of such coupling provides a powerful design tool for the mechanical systems designer.

Directional control valves are classified in terms of number of positions,

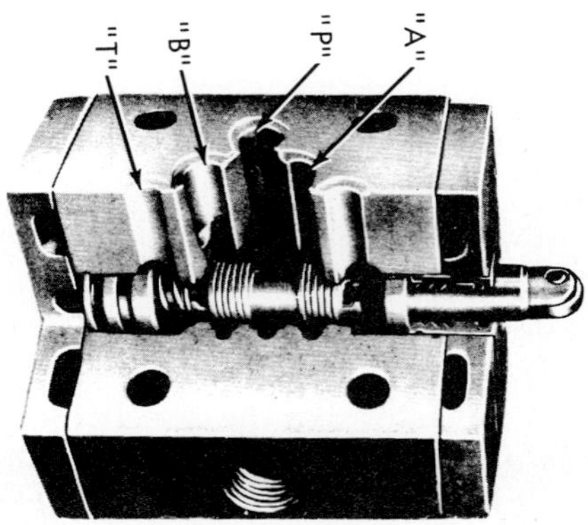

Figure 4-48. Mechanically actuated spool valve. (Courtesy of Vickers, Inc.)

number of ways, and method of actuation. Additionally, these valves can be used in either a discrete (on-off) or a proportional mode. In this section, we will consider directional control valves of the discrete type only. For example, a three-position valve must be discretely in one of three positions (right, left, or centered) where it is either totally ON or totally OFF. As we discussed earlier, a proportional valve can be made to occupy an infinite number of continuously distributed intermediate positions lying somewhere between (but including) ON and OFF.

Two-Position, Two-Way Valve. A very simple spool valve is shown in Fig. 4-49. It is nothing more than a simple on-off control valve. The valve is a normally closed device because the spool is held to the right by the spring. It is manually actuated by pushing the plunger button to the left, which places inlet and outlet ports in communication. This is the ON position. When the force on the plunger is removed, the spool shifts to the right and the valve is OFF. The two positions are the ON and OFF positions. This is called a two-way valve because it has only two ports.

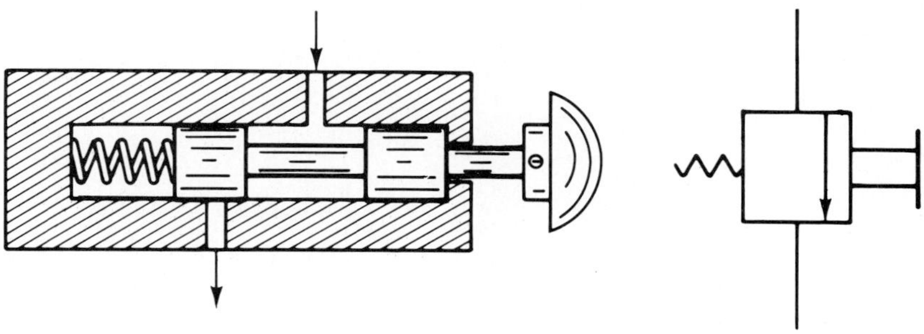

Figure 4-49. Two-way, two-position, manual valve.

Two-Position, Three-Way Valve. A spring-offset, two-position, three-way, cam-operated spool valve is sketched in Fig. 4-50. The two-position, three-way valve is always ON; it simply provides two different possible flow paths. This is

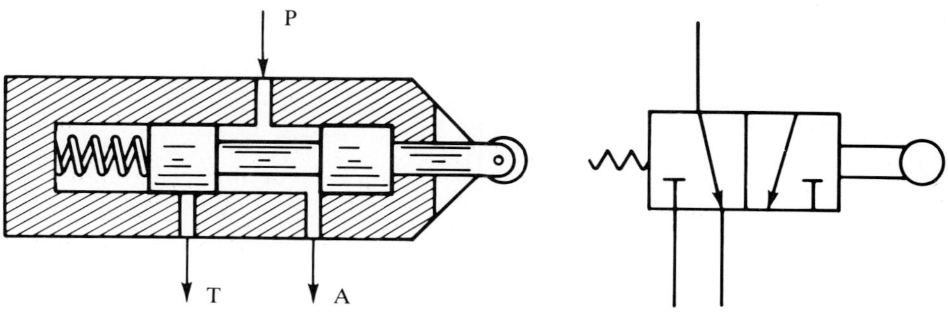

Figure 4-50. Three-way, two-position, cam-actuated valve.

the kind of valve we discussed in connection with Fig. 4-29. The symbols P, T, and A simply indicate the number of ways. The symbols mean pump, tank, and actuator respectively. The valve shown in Fig. 4-50 would be actuated by a cam (which, perhaps, is attached to a moving member of the machine being controlled by this valve). This valve makes possible the coupling between the mechanical and hydraulic circuits of the machine. Of course, this valve could also be actuated by electrical solenoid, by hydraulic pilot line, or manually.

Three-Position, Four-Way Valve. One of the most useful and most commonly used spool valves is the three-position, four-way valve. A typical valve of this kind is shown in Fig. 4-51. The valve shown is called a spring-centered, pilot-actuated device. When the hydraulic pilot pressures on the ends of the spool are equal, the valve is centered by the springs and all four ports are blocked. Hence, this is called a blocked-center valve.

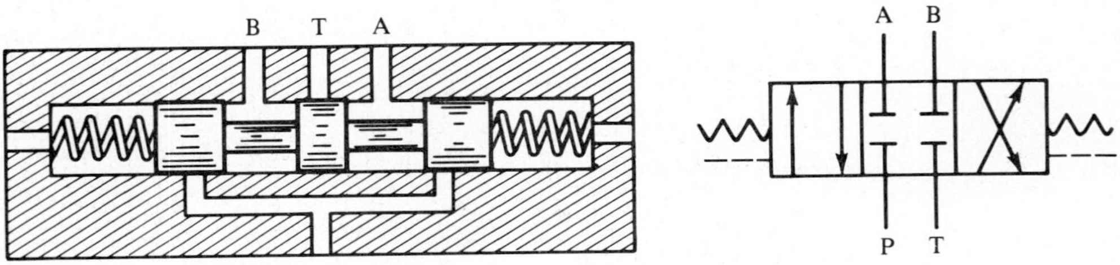

Figure 4-51. Four-way, three-position, pilot-actuated valve.

If the pilot pressures are unequal so that the spool is shifted to the left, then the pump port P is placed in communication with actuator port A while the tank port T is placed in communication with actuator port B. In this situation, port A would be the inlet port of the actuator and port B would be the actuator exhaust port. With the valve shifted to the right, the opposite circumstances prevail.

Again, this valve could be actuated by electrical solenoids, by mechanical linkages, or manually just as well as by hydraulic pilot lines. And various center position configurations are available, as indicated in Fig. 4-52.

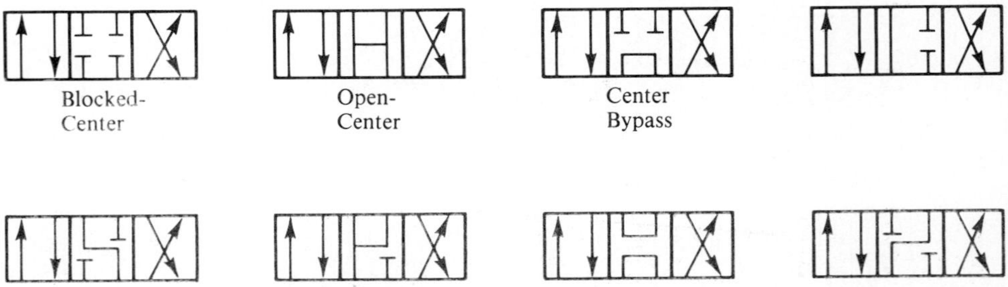

Figure 4-52. Various center conditions for four-way valves.

Suppliers of three-position, four-way valves will often provide a variety of spools that can be used interchangeably in a given valve body. The spools may all be the same in their shifted positions but have different center conditions as indicated in Fig. 4-52. The open-center and center bypass would be used instead of the blocked-center configuration when it is desired to return the pump output directly to the tank through a small pressure drop (because the pressure drop across the control valve center is small) rather than via a pressure relief valve (as would be necessary with a blocked-center valve). Note that with a blocked-center valve, the pump is available to service actuators other than the one being blocked. This service is not possible with the bypass or open-center types.

The other center positions shown in Fig. 4-52 will be used for special circuit conditions which are beyond the scope of this book.

Methods of Directional Control Valve Actuation. We have already suggested that directional control valves may be actuated and controlled in a variety of ways. Figure 4-53 represents various methods (and the appropriate symbols) of actuating valves.

The detent valve, shown in Fig. 4-54, is of particular interest. Note how the spring-loaded ball holds the spool in place until the actuator has sufficient force to overcome the spring force holding the ball in its groove. The valve shown is pilot actuated.

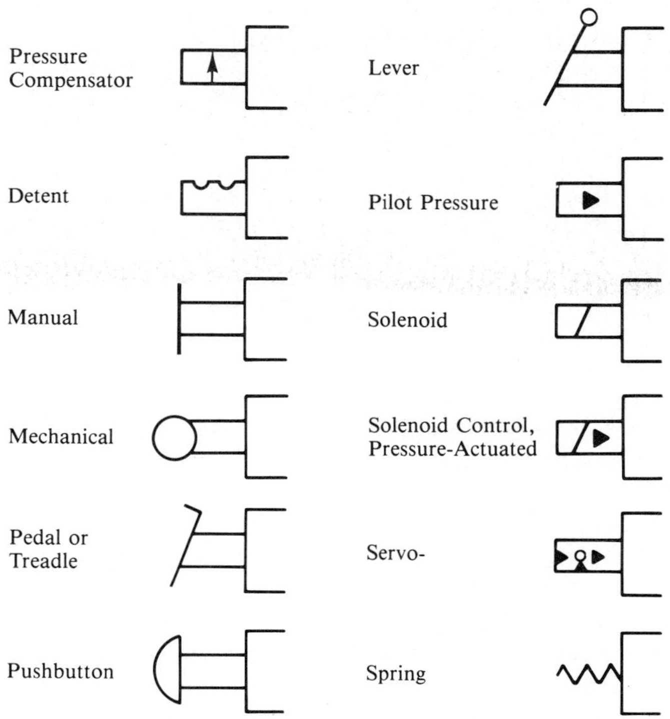

Figure 4-53. Methods of valve actuation.

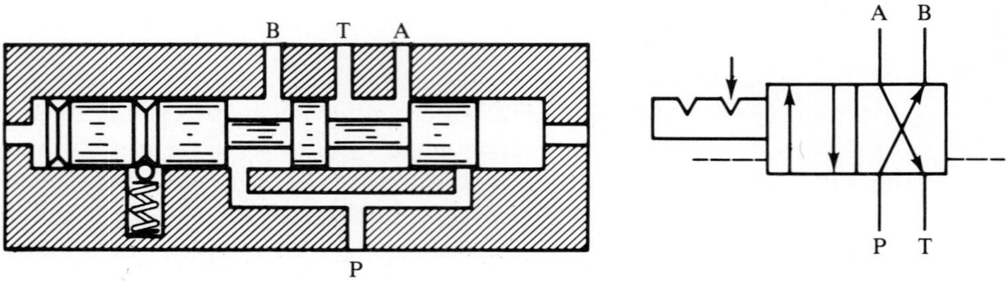

Figure 4-54. Detent valve.

4-4 ACCUMULATORS

A fluid power accumulator is a valved chamber that can store a measurable volume of oil under pressure. In many fluid power circuits, the accumulator serves the same function as does a flywheel in a mechanical power system. In other words, an accumulator permits the installation of a smaller pump than the fluid power circuit requirements would dictate in the absence of an accumulator. The smaller pump can be charging the accumulator during the part of the system cycle where actuators are not requiring any power. Then when the system calls for power, it obtains it from the accumulator in quantities in excess of what the pump would be able to supply directly. The accumulator is simply an energy storage device analogous to a flywheel (or a capacitor in an electrical system).

Perhaps the simplest accumulator concept is a weighted piston in a large vertical cylinder. The pressure is determined by the bore of the cylinder and the weight situated upon the piston. The accumulator would be loaded by pumping oil into the cylinder and raising the weight. Discharge of the accumulator would be accomplished simply by opening a valve and letting the piston descend under its weight, thus discharging fluid into the fluid power circuit. This type of accumulator produces a constant pressure discharge because the weight is constant.

In the piston/cylinder type of accumulator, the weight could be replaced by a spring, resulting in a spring-loaded accumulator. In both cases, this type of accumulator suffers from potential oil leakage past the piston. This disadvantage can be overcome by using a gas-charged accumulator which is simply a chamber having a pressurized gas above the oil-gas interface. These various accumulator concepts are sketched in Fig. 4-55.

The gaseously charged accumulator design in Fig. 4-55 presents a potential hazard in that gas may get into the hydraulic system if the liquid level in the accumulator falls too low. This hazard can be avoided by placing the gas in a bladder or by having a separating diaphragm (see Fig. 4-56).

The accumulator can be used as an auxiliary power source to supplement a pump in a circuit where the cycle time consists of a peak load time intermittently

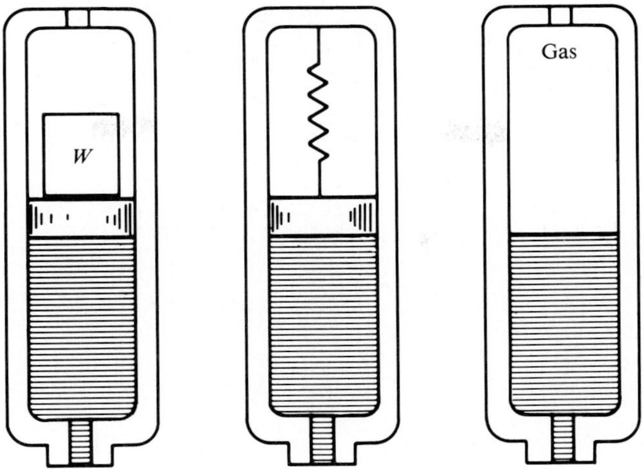

Figure 4-55. Accumulator designs.

with dwell time. See Fig. 4-57. This circuit has a closed-center directional control valve plus a check valve that permits the accumulator to be charged while the actuator is passsive (dwell period). The pressure level in the accumulator is controlled by the pilot-actuated unloading valve. Note that the pump is pressure compensated so a pressure relief valve is not needed in the circuit.

When the directional control valve is moved to an ON position, the full pressure of the accumulator is made available to the actuator. When the pressure drops to the pre-set level, the unloading valve closes and the pump supplies fluid either to the accumulator or to the actuator depending on the actuator requirements. If the active cycle of the actuator is short, the accumulator may supply all

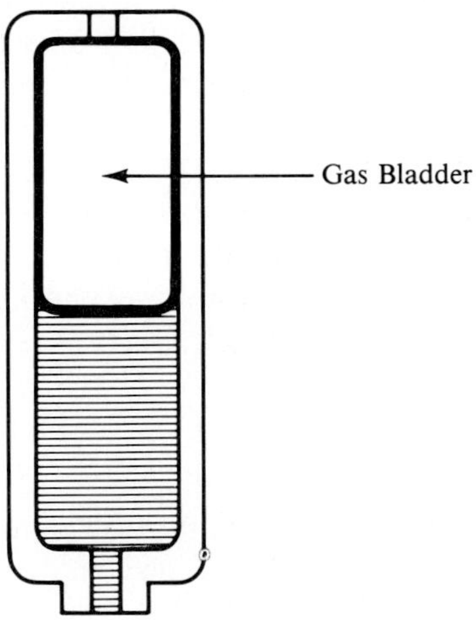

Figure 4-56. Bladder type accumulator.

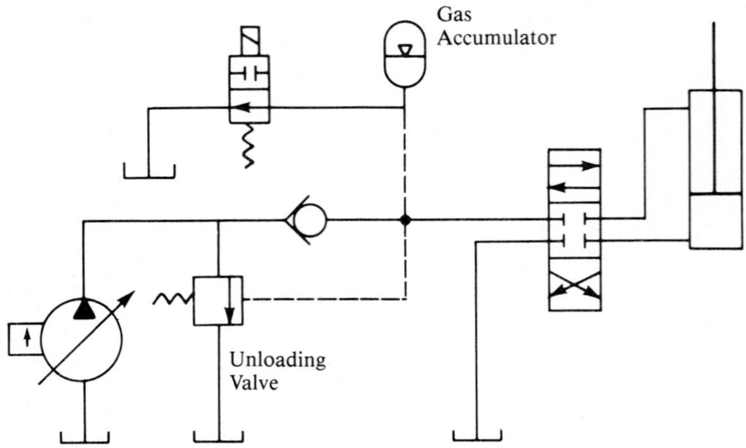

Figure 4-57. Closed-center circuit with accumulator.

of the fluid required. The solenoid-actuated, two-position valve is present in order that the accumulator can be unloaded when the circuit is shut down.

In most fluid power circuits, the pump outlet pressure is dictated by the load(s) on the actuator(s). This can be shown from a free-body analysis of the actuator elements (e.g., piston) and the fact that the pump outlet is directly connected to the actuator inlet. But in a circuit that uses an accumulator as an auxiliary power source, we see that the pump outlet pressure is not dictated by the actuator force requirements.

Accumulators are used in fluid power circuits having the following functional requirements:

(a) Auxiliary power source as discussed for Fig. 4-57
(b) Pump pressure surge suppression or actuator shock load attenuation
(c) Emergency power requirements to mitigate loss of pump or power incidents

For gaseously charged accumulators, the state conditions can be calculated by assuming that gas contraction and expansion occur isothermally, polytropically, or isentropically. If compression or expansion occurs slowly, the isothermal assumption is appropriate, and if these processes occur rapidly, the isentropic assumption applies.

EXAMPLE 4-8

The rapid actuation requirements for the control surfaces of an airplane dictate the use of a gas accumulator. For design purposes, the accumulator is to be sized assuming that the maximum demand of the control system (for oil) will be 12 in^3/sec. The pump capacity of the circuit is fixed at 2 in^3/sec.

Force requirements for the control surface actuators dictate that the accumulator pressure be confined to the interval between 2000 psi and 3000 psi. The maximum time demand on the accumulator will be 10 seconds.

Assume that while the plane is on the ground at 80°F, the accumulator is empty of oil and has a nitrogen gas pressure of 1500 psi. Determine the accumulator size (cubic inches) assuming that the accumulator is charged isothermally while on the ground. Assume that the discharge process occurs isentropically from an initial gas temperature of 80°F. What is the nitrogen gas temperature after the accumulator has been depleted by the airplane control requirements? Also, how long does it take to charge the accumulator while the plane is on the ground?

Solution

Consider the accumulator states sketched in Fig. 4-58. Letting V_n be the gaseous volume, we see that

$$V_{n(3)} = V_{n(2)} + 12(10)$$

Now

$$p_3 V_{n(3)}^{1.4} = p_2 V_{n(2)}^{1.4}$$

Therefore,

$$2000(V_{n(2)} + 120)^{1.4} = 3000(V_{n(2)})^{1.4}$$

Solving, we find that

$$V_{n(2)} = 360 \text{ cubic inches} \qquad V_{n(3)} = 480 \text{ cubic inches}$$

For the isothermal compression, we have

$$p_1 V_{n(1)} = p_2 V_{n(2)}$$

where

$$V_{n(1)} = V_a$$

Here, V_a represents the accumulator volume. Therefore,

$$1500 V_a = 3000(360)$$
$$V_a = 720 \text{ cubic inches}$$

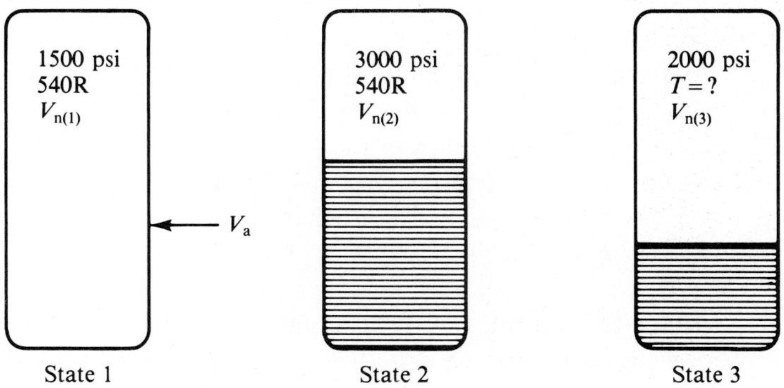

Figure 4-58. Accumulator states for Example 4-8.

For an isentropic expansion, we have the relationship

$$\frac{T_3}{T_2} = \left(\frac{V_2}{V_3}\right)^{0.4}$$

$$T_3 = 540\left(\frac{360}{480}\right)^{0.4}$$

$$= 481\text{R} = 21°\text{F}$$

The time required to charge the accumulator while the plane is on the ground is

$$t = \frac{360}{2} = 180 \text{ seconds} = 3 \text{ minutes}$$

■

4-5 SOME USEFUL HYDRAULIC CIRCUITS

In addition to the several hydraulic circuits discussed earlier in conjunction with particular hydraulic components, such as valves, there are literally hundreds of well-known hydraulic circuits for accomplishing a range of common mechanical system functions. Many of these circuits are discussed in the excellent treatise written by Henke (see Reference 4-2). There are circuits for pressure and speed control; for sequencing actuators; for synchronizing actuators; for rapid advance, slow feed; for actuator deceleration or braking; for press operation; for boom actuation; and for many other functions.

Here we will focus on a few circuits that are either common in use or that cogently illustrate the kind of analyses needed in synthesizing a hydraulic circuit. We will begin by discussing three well-known speed control circuits used for controlling the speed of an actuator.

4-5-1 Speed Control Circuits

Implicit in the following discussion of speed control is the choice of speed control by means of diverting some portion of the constant pump discharge over the system relief valve. Speed control by means of variable pump discharge is the alternative not discussed here.

Speed control of an actuator is accomplished by controlling the flow through the actuator, which was discussed in Section 4-3-2 in terms of the relative resistances in parallel circuit branches. However, the location of the flow controlling resistance was not discussed. This resistance may be located upstream of the actuator (meter-in), downstream of the actuator (meter-out), or in parallel with the actuator (bleed-off).

Meter-In Circuit. A typical meter-in circuit is sketched in Fig. 4-59. A meter-in circuit can be designed in such a way as to provide very accurate speed control of an actuator. Consequently, it is often used in machine tool hydraulic circuits. However, the meter-in circuit is appropriate only for actuator loads that

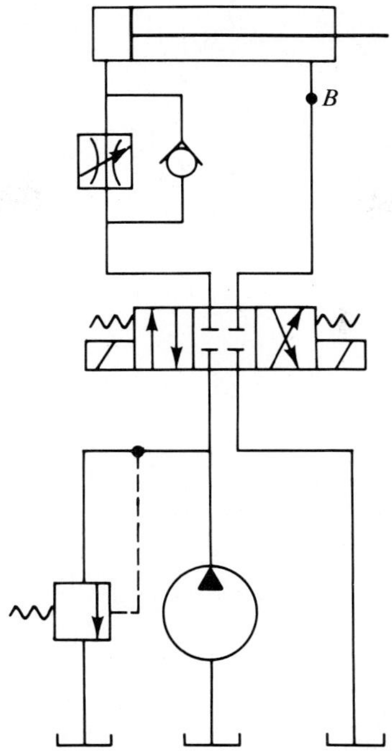

Figure 4-59. Meter-in circuit.

are continuously resistive in nature. If there is a possibility that the load may become over-running at any point in the machine cycle, the meter-in circuit cannot provide the requisite braking of such an inertia load. An example is the drill, being pushed by a hydraulic cylinder, that suddenly breaks through the material being drilled and becomes an over-running load.

A check valve is often installed in parallel with the flow control valve in order that the retraction stroke of the actuator, as shown in Fig. 4-59, is unimpeded (not metered). Without the check valve, the circuit would become a metering-out circuit of the kind discussed below.

Meter-Out Circuit. In a meter-out circuit, the flow control and check valves of Fig. 4-59 are placed in the exhaust line (point B) rather than in the cylinder supply line (as shown). With this type of circuit, there is a resistance between the actuator and the tank to which the actuator exhausts. The resistance provides a braking effect against a sudden over-running load.

A disadvantage of the meter-out circuit is that the rod-end pressure may exceed the pump supply pressure if a single-rod cylinder is used in which the power stroke proceeds toward the rodded side. This potential pressure intensification in the rod-end can produce a bursting hazard in the rod-end.

EXAMPLE 4-9

Determine the minimum resistance of a flow control valve necessary in order that speed control of a hydraulic cylinder can be effected in a metering-out circuit. Let

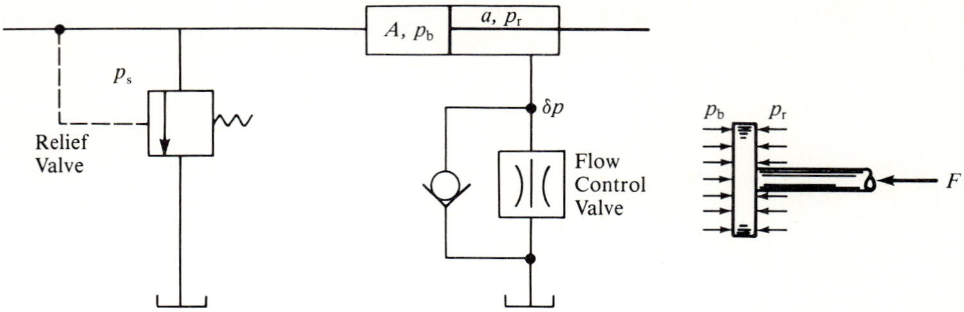

Figure 4-60. Hydraulic circuit for Example 4-9.

the load on the piston be F, the piston area be A, the rod area be a, the relief valve setting be p_s, and the pressure drop across the flow control valve be δp.

A sketch of the circumstances is presented in Fig. 4-60. A free-body diagram of the piston shows that the relationship between pressure in the rod-end and pressure in the blank-end is

$$p_r = \left(\frac{A}{A-a}\right)p_b - \frac{F}{A-a}$$

But the pressure p_r in the rod-end is just the pressure drop δp across the flow control valve (these are gage pressures). Therefore,

$$\delta p = \left(\frac{A}{A-a}\right)p_b - \frac{F}{A-a}$$

Now the meter-out circuit can function as a speed control circuit only if the relief valve is open, which means that p_b must be high enough to actuate the relief valve (i.e., $p_b = p_s$). Therefore, we require δp to be

$$\delta p > \left(\frac{A}{A-a}\right)p_s - \frac{F}{A-a}$$

As a numerical illustration, suppose $A = 15$ square inches, $a = 10$ square inches, $F = 1000$ lb, and $p_s = 2000$ psi. Then

$$\delta p > \frac{15}{5}(2000) - \frac{1000}{5} > 5800 \text{ psi}$$

This is the minimum resistance that must be provided by the flow control valve for the relief valve to be open. If δp is only 3000 psi, then $p_b = 1066$ psi and the relief valve remains closed (and there is no flow control).

This illustration also shows the pressure intensification produced in the rod-end when the meter-out circuit is used with a small load F. When the relief valve is open, the blank-end pressure is only 2000 psi whereas the rod-end pressure is 5800 psi. ■

Bleed-Off Control. In a bleed-off flow control method, the flow-controlling resistance is placed in parallel with the actuator (rather than in series as with meter-in and meter-out circuits). Then the pump discharge not needed by the actuator is returned to the tank through the flow control valve rather than through the relief valve. The advantage of this method of flow control is that the pump operates only at a pressure dictated by the load and not by the relief valve (as is true in meter-in and meter-out circuits).

A disadvantage of bleed-off flow control is that speed control cannot be effected over a broad range because of the possibility that for some conditions the relief valve might be open (if the load becomes high enough). Also, this method does not protect against over-running loads.

Proportional Spool Valve Speed Control. In the simple meter-in, meter-out, and bleed-off circuits, speed control is obtained by placing a pressure-compensated flow control valve in series or parallel with the directional control valve. In these circuits, the directional control valve is simply of the on-off type of spool valve actuated by solenoids, pilot pressure, or similar means. The flow control valve restricts the flow to the actuator to some fraction of the pump output with the remaining portion being bypassed over the relief valve (or over the flow control valve in the bleed-off arrangement).

Another mode of flow control is to combine both flow and directional control within a single valve. This can be achieved by changing the mode of operation of the directional spool valve from simple on-off to proportional control. In other words, instead of displacing the spool to its extreme positions, spool location is made infinitely variable. With the spool valve only slightly open, the spool valve will act as a restrictor as well as a directional valve. As the valve is opened continuously, its restrictive properties will continuously diminish, so the

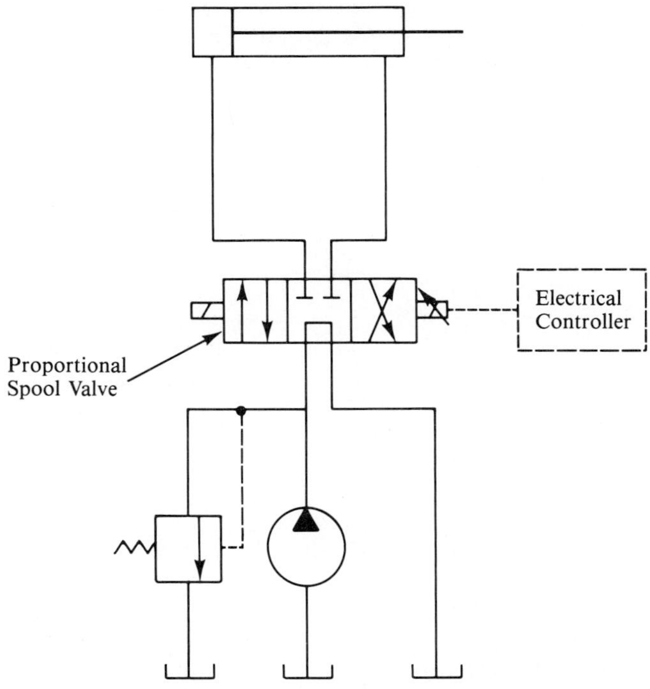

Figure 4-61. Meter-in circuit using a proportional valve.

flow through it will continuously increase up to the maximum value when fully open.

To make the spool position infinitely variable, some means must be incorporated within the valve to make its actuation force infinitely variable (or to provide a feedback control of the spool position). One common way of achieving this is with an electrical torque motor responsive to current in its armature. The valve can be designed in such a way as to make spool position proportional to armature current.

A simple circuit incorporating this kind of valve is shown in Fig. 4-61. The circuit can be made to function as a meter-in or a meter-out circuit depending on how the valve is designed and actuated. The electrical controller that actuates the spool valve in this particular circuit is under the independent control of the operator and is not tied to any other point of the hydraulic circuit (or of the machine of which this circuit is a part). This control scheme is called open-loop control. As we shall see later, with a slight modification the control scheme shown can be converted from open-loop to closed-loop control (servo-control).

Rapid-Advance, Slow-Feed Circuits. When hydraulic circuits are used to power machine tools, the productivity requirements dictate the use of a rapid approach of the tool to the work just prior to the slower feed (when machining is actually done). Various techniques are available for achieving this type of speed control. One version, which uses the metering-out principle, is sketched in Fig. 4-62.

During rapid advance (initiated by valve E), there is no load on the cylinder and the relief valve A remains closed. The cylinder exhaust flow is through the low-resistance, two-position, two-way valve B. The slow feed is initiated when the cam C actuates valve B and closes it. The exhaust flow is now through the

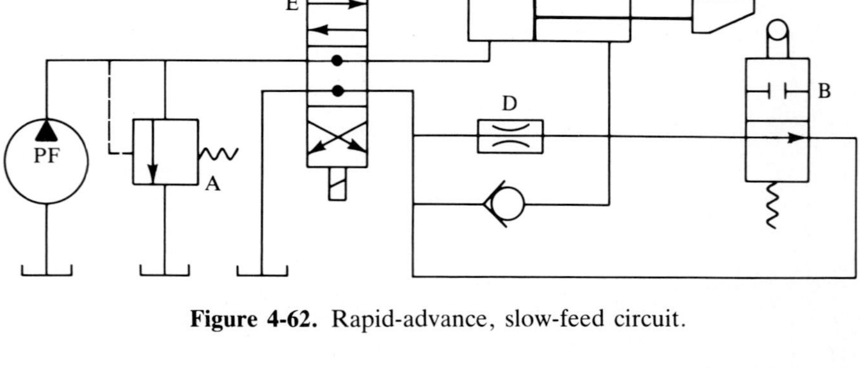

Figure 4-62. Rapid-advance, slow-feed circuit.

metering-out valve D which has resistance sufficient to open relief valve A. Relief valve A bypasses part of the pump output to the tank through the relief valve, and the speed of the piston is reduced (valve D is adjustable in order to obtain the desired feed rate).

The retraction stroke is initiated by directional control valve E being shifted to the left. Flow now proceeds through the low-resistance check valve F, and the return stroke is even more rapid than the rapid-advance stroke because the rod-end area is smaller than the blank-end area. Various metering-out valves are available that are capable of providing both coarse and fine feed rates of the actuator.

A circuit and actuator that will provide a high-speed extension and a low-speed retraction are sketched in Fig. 4-63. During the rapid advance to the right, there is no load on the double piston and the sequence valve A remains closed. Only the smaller piston is used, and the double piston advances rapidly to the right. When the double piston contacts the load, sequence valve A opens and then the high-force, low-speed feed is accomplished. When the double piston reaches the end of the feed stroke, the directional control valve is shifted upward and we have low-speed retraction to the left.

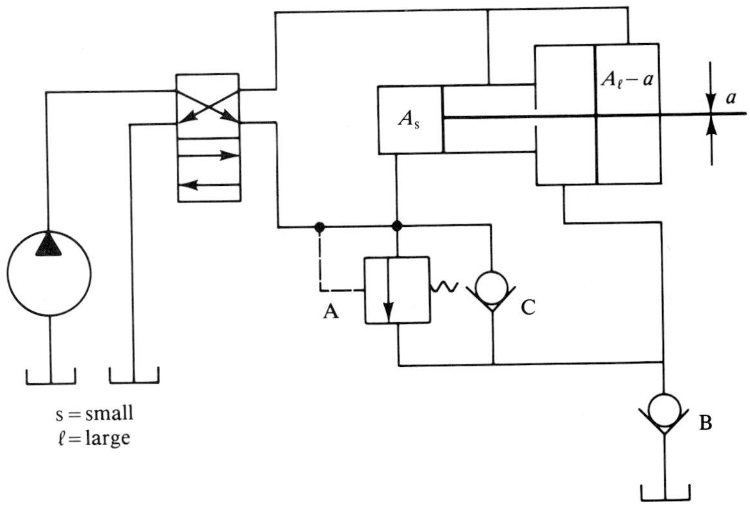

Figure 4-63. High-speed, low-speed circuit.

EXAMPLE 4-10

The circuit sketched in Fig. 4-64 is a rapid advance and feed circuit (to the right) with rapid return (to the left). The system utilizes a double pump, with each pump having a 5 gpm capacity. The hydraulic cylinder has a 10 square inch piston area, a 5 square inch rod area, and a stroke of 4 feet. During the first 3 feet of motion to the right, the speed (rapid advance) of the piston is 19.25 fpm against a load of 5000 lb (both pumps are in the supply circuit). Then, during the last foot of motion to the

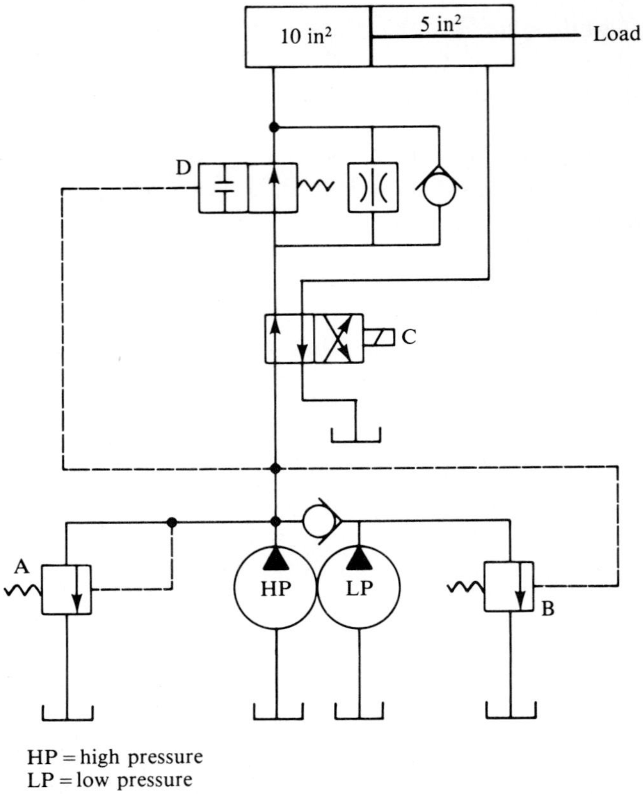

Figure 4-64. Rapid-advance, slow-feed circuit.

right, the speed is slowed to the feed of 5 fpm against a load of 10,000 lb (with only the high-pressure (HP) pump in the circuit). Rapid return is accomplished by shifting directional control valve C to the left. Valve D is pilot actuated. Determine

(a) The discharge (gpm) over relief valve A during the feed stroke
(b) The speed of the rapid return
(c) The spring settings of valves B and D
(d) The spring setting of relief valve A

Solution

During the feed stroke, the relationship between piston feed speed and discharge into the cylinder is

$$V\text{ (fpm)} = \frac{19.25Q\text{ (gpm)}}{A\text{ (in}^2\text{)}}$$

Therefore, the discharge into the cylinder during feed is

$$Q = \frac{(5)(10)}{19.25} = 2.6\text{ gpm}$$

Since only the 5 gpm HP pump is in the circuit at this time, the bypass over the relief valve is

$$Q_A = 5.0 - 2.6 = 2.4 \text{ gpm}$$

During the rapid return, both pumps will be in the circuit (because flow is through the low pressure-drop check valve), and the speed of the return is

$$V = \frac{19.25(10)}{5} = 38.5 \text{ fpm}$$

During rapid advance, both pumps are in the circuit and valve D is open. The pump outlet pressure is therefore

$$p = \frac{5000}{10} = 500 \text{ psi}$$

During the feed stroke, only the HP pump is in the circuit, and the metering-in valve is in the circuit. The pressure on the downstream side of the metering valve is

$$p = \frac{10{,}000}{10} = 1000 \text{ psi}$$

There will be some pressure drop across the metering valve which will have to be determined from the metering valve's characteristics. We assume the pressure drop to be 150 psi. Then the pump outlet pressure will be 1000 + 150 = 1150 psi during the feed stroke.

With the conditions as determined above, the unloading valve B and on-off valve D must remain closed above 500 psi (but below 1150 psi), so we set them at 700 psi. The relief valve will have to be open during the feed stroke, so we set it somewhat below 1150 psi, say, at 1000 psi.

Finally, note that there is no protection against an over-running load with this circuit design. ■

4-5-2 Analysis of Hydraulic Press Circuits

Hydraulic circuits are used in a wide variety of mechanical systems designed to amplify force, many in the metal-working field. In this section, we will consider two hydraulic circuits in order to expose the type of analysis required to select components and to set them.

Hydraulic Forging Press. A hydraulic press circuit is shown in Fig. 4-65 in which a piston (plus ram) weighing 800 lb is to be used to hydraulically forge metal parts between two platens. Before forging, the upper platen must be traversed downward through a distance of 8 inches in 1.04 seconds. During the actual forging process, the forging force will be 50,800 lb as the upper platen moves an additional 1 inch. The area of the blank-end of the cylinder is 20 square inches, and the rod-end piston area is 10 square inches.

The circuit is a regenerative one, through the sequence valve A, during the 8 inch downward approach stroke. After the ram contacts the forging, the increased pump outlet pressure is piloted to the bypass valve B and the bypass valve B opens against its spring setting to permit the pressure to be relieved in the rod-end

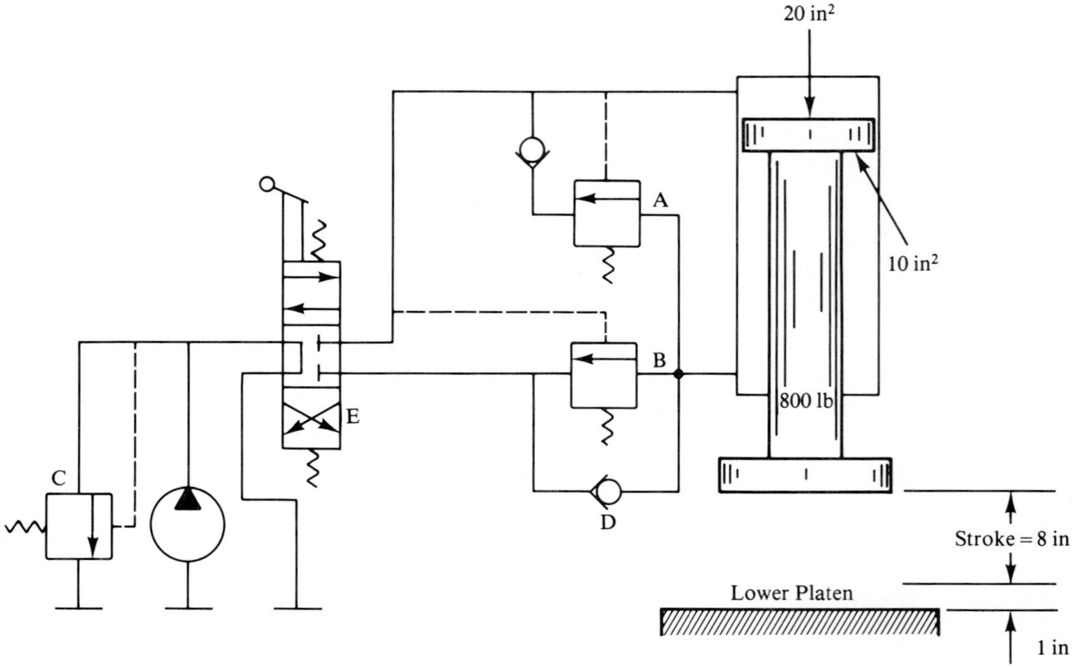

Figure 4-65. Hydraulic forging press.

of the cylinder so that the full power of the pump outlet pressure can do the work of forging.

After forging is completed, the four-way directional control valve E is positioned to raise the piston and ram. This positioning causes the pump outlet to be ported into the rod-end of the cylinder through the check valve D.

Assume that sequence valve A acts as a counterbalance valve and will open when the pilot pressure reaches 320 psi during the 8 inch downstroke. Assume that the pressure drop across bypass valve B is negligible when valve B is open and that all piping pressure drops are negligible. We wish to determine the following:

- **(a)** Pressure drop across counterbalance valve A during the 8 inch approach downstroke
- **(b)** Pressure in the cylinder rod-end during the 8 inch approach downstroke
- **(c)** Pump capacity in gpm
- **(d)** Spring setting of bypass valve B
- **(e)** Spring setting of relief valve C
- **(f)** Flow rate through four-way valve E during the 8 inch downstroke
- **(g)** Flow rate through four-way valve E during the return stroke (upstroke)

A free-body diagram of the press piston and ram during the 8 inch downstroke is shown in Fig. 4-66. We see that

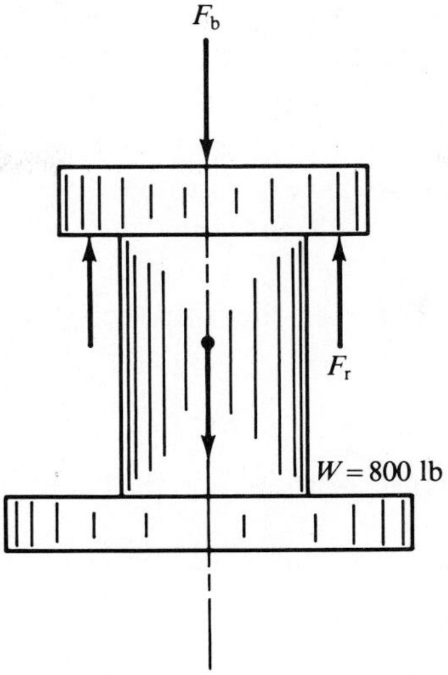

Figure 4-66. Piston free-body diagram.

$$F_r = F_b + W$$

$$p_r A_r = P_b A_b + W$$

$$A_r = \frac{A_b}{2}$$

Combining, we obtain

$$p_r = 2p_b + \frac{2W}{A_b}$$

The pressure drop across valve A is defined as

$$\delta p = p_r - p_b$$

Combining the last two equations gives

$$p_b = \delta p - \frac{2W}{A_b}$$

Substituting given values, we find that

$$\delta p = 320 + \frac{2(800)}{20} = 400 \text{ psi}$$

Then the rod-end pressure is

$$p_r = 2(320) + \frac{2(800)}{20} = 720 \text{ psi}$$

We do not want bypass valve B to be actuated during this 8 inch approach downstroke, so its setting must exceed that of sequence valve A by some margin (say, 200 or 300 psi). Therefore, we set bypass valve B to a spring setting of some 520 to 620 psi.

As soon as the upper platen contacts the forging, the pump outlet pressure will immediately rise above 320 psi. This contact opens bypass valve B, so the rod-end of the cylinder will communicate with the tank through the exit of four-way valve E. Sequence valve A now will be effectively closed since the exhaust path of least resistance is through bypass valve B. The piston speed will be reduced to half its previous value because the regenerative circuit will no longer exist.

A free-body diagram of the press piston during the forging process now produces the equation

$$R = F_b + W$$

where $R = 50{,}800$ lb is the forging force. Then

$$50{,}800 = 20p_b + 800$$

We see that the blank-end pressure has risen to

$$p_b = 2500 \text{ psi}$$

Therefore, we set relief valve C to a setting to exceed this value by a safe margin, say, to 3000 psi.

During the 8 inch downstroke, the volumetric flow rate into the blank-end of the cylinder must be given by

$$Q = \frac{AV}{19.25} = \left(\frac{20}{19.25}\right)\left(\frac{8/12}{1.04}\right)(60) = 40 \text{ gpm}$$

But since the circuit is regenerative during this time, the pump capacity will be only half of this value, that is, 20 gpm (with the other half coming from the rod-end of the cylinder). So, during the 8 inch downstroke there will be 20 gpm flowing through the inlet port and zero through the exit port of four-way valve E.

During the return stroke, the flow into the rod-end of the cylinder will be 20 gpm, and out of the blank-end the flow will be 40 gpm. Thus, the flow will be 20 gpm through the inlet port of the four-way valve E and 40 gpm through the exit port.

If the pump is turned off, the blank-end pressure falls to zero. Then the rod-end pressure will be

$$p_r = \frac{2(800)}{20} = 80 \text{ psi}$$

Neither sequence valve A or bypass valve B will be open and, therefore, the piston cannot fall under its own weight.

Hydraulic Rivet Upsetter. The hydraulic system shown in Fig. 4-67 is for a machine designed to upset (clinch) large rivets. The riveting cycle is initiated by shifting the directional control valve to the left. This shifting will feed the ram upward at 145 fpm until the rivet is contacted, at which time the feed rate will be drastically reduced while the rivet is being clinched (called upsetting). The force required to upset the rivet is 22,500 lb.

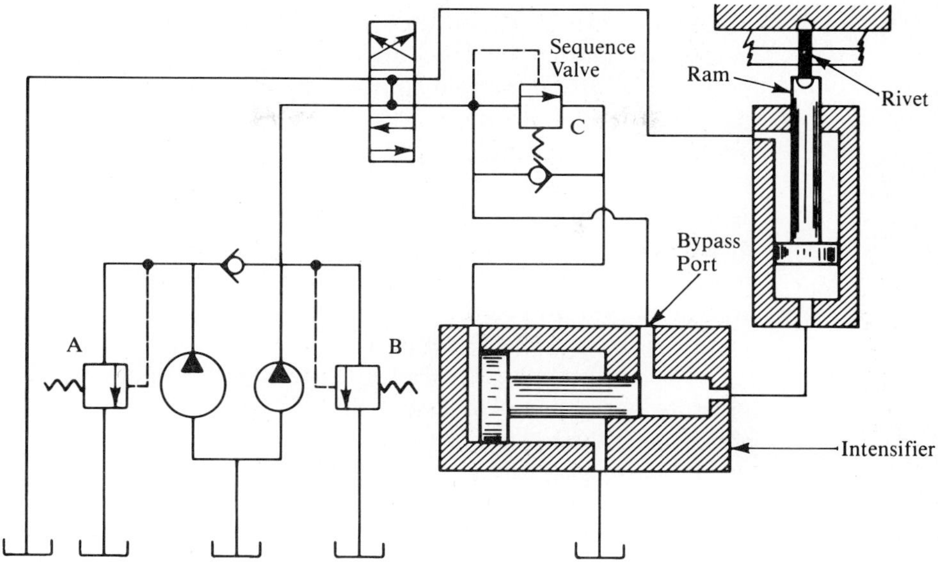

Figure 4-67. Hydraulic riveting machine.

After riveting is completed, the directional control valve is shifted to the right. Shifting will cause the ram to return to its original position at the bottom of its stroke.

The cylinder arrangement shown at the bottom of the figure is called an intensifier; its function is to provide an amplified force for upsetting the rivet once it has been contacted by the ram. The force intensifier ratio is 25:1.

Assume that the setting of relief valve A, for the large pump, is below the common pump outlet pressure required to upset the rivet, and that the setting of relief valve B, for the small pump, is above this required pressure. The setting of the sequence valve is the same as for relief valve A.

The weight of the ram is negligible; the ram piston has an area of 4 square inches, and the rod area is 2 square inches. The large pump has twice the capacity of the small pump. Assume the pressure drop in all fluid lines is negligible.

(a) Describe, in step-by-step fashion, the complete sequence of events that will occur after the directional control valve is shifted to the left.
(b) What is the delivery (gpm) of the smaller pump?
(c) What are the settings of relief valves A and B?
(d) What is the speed of the ram while the rivet is being upset?
(e) What is the speed of the retraction stroke of the ram?

The answers are as follows:

(a) Assume the ram is at the bottom of its stroke and is out of contact with the rivet. Then, when the control valve is shifted to the left, the ram will be fed upward at high speed since both pumps will be supplying fluid at essentially zero pump outlet pressure (since the weight of the ram is negligible). During this time, the force intensifier is bypassed through the port shown. When the ram contacts the rivet, it encounters a force, and the pump outlet pressure rises immediately to the value required to upset the rivet. At this point, relief valve A will open, which diverts the output of the large pump to the reservoir. At the same time, the sequence valve opens and ports the outlet of the small pump to the intensifier. Its piston moves to the right and provides the force required to upset the rivet at slow speed. After riveting is completed, the ram will stall in its upward position and relief valve B will open, which vents all pump outlet flow to the reservoir.

When the control valve is shifted to the right, the pump outlet pressure again drops to essentially zero and the ram is driven down, which shifts the intensifier to the left, uncovering the port, which returns the exhausting oil to the tank. When the piston reaches the bottom of its stroke, the pressure will rise and both pumps will vent, through their relief valves, to the tank.

(b) While the ram is moving upward under no-load, the speed is 145 fpm. Therefore, the capacity of both pumps is

$$Q = \frac{145(4)}{19.25} = 30 \text{ gpm}$$

This means that the capacity of the small pump is 10 gpm.

(c) The pressure on the ram piston during the rivet upsetting is

$$p = \frac{22{,}500}{4} = 5625 \text{ psi}$$

Note that the rod side of the piston communicates directly with the tank and is therefore at zero pressure. Since the intensifier has a 25:1 ratio, the pump outlet pressure is

$$p = \frac{5625}{25} = 225 \text{ psi}$$

Therefore, we set relief valve A at a pressure below 225 psi (say, 100 psi), while relief valve B is set above 225 psi (say, 400 psi).

(d) During the upsetting process, only the small pump is providing fluid to the circuit. With the area of the large piston of the intensifier denoted by A, and that of the small piston by a, the velocity of the intensifier piston is

$$V = \frac{19.25(10)}{A}$$

The velocity of the small piston is the same as for the large piston, and the displacement of the small piston must be the same as for the ram piston. Therefore,

$$A_r V_r = aV = 4V_r$$

Combining these two equations, we find that (since $A/a = 25$),

$$V_r = \frac{19.25(10)}{4(25)} = 1.925 \text{ fpm}$$

(e) On the return stroke, both pumps are in the circuit and the piston area (on the rod side) of the ram is only 2 square inches. Therefore, the return stroke speed is 290 fpm.

4-5-3 Deceleration Control

We have mentioned frequently the necessity for the mechanical systems designer to be concerned with over-running (or overhauling) loads. In hydraulic circuits, in which a cylinder is the actuator, an over-running load may cause the piston to compress the oil between piston and head rather violently. Such hazards must be mitigated by the circuit design.

In the hydraulic circuit sketched in Fig. 4-68(a), imagine that solenoid A of

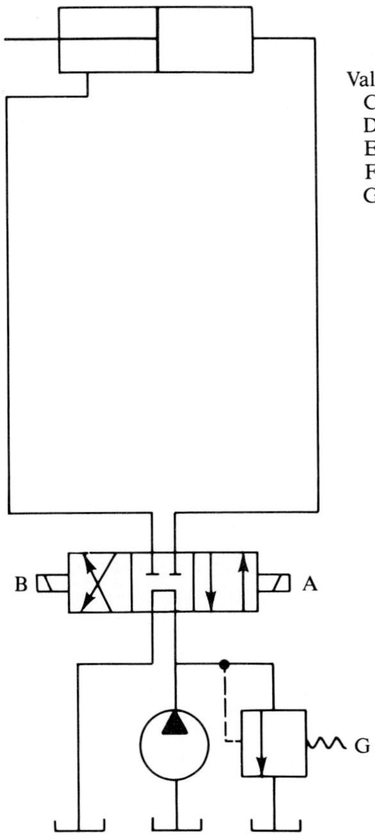

(a)

Valve	Pilot Setting (psi)
C	1500
D	1200
E	1500
F	0
G	1000

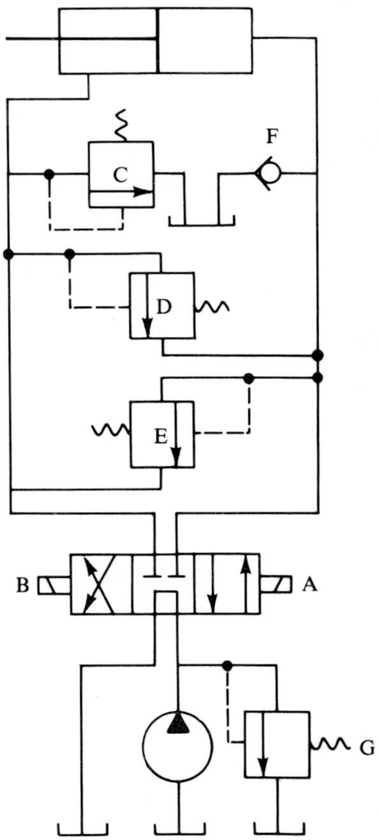

(b)

Figure 4-68. Deceleration controls.

the four-way valve is actuated so that the piston is moving to the left. We suppose that under these conditions the working pressure in the blank-end of the cylinder is 500 psi and that the piston is driving a swinging load of large moment of inertia. To stop this load, solenoid A is de-energized so that the four-way valve returns to center as it is shown in Fig. 4-68(a). Note that in this position the cylinder ports are completely blocked off. Blocking stops the piston immediately but the piston, nevertheless, is still subject to a large over-running load. This load will cause the pressure in the rod-end of the cylinder to rise to dangerously high levels unless relieved.

Similarly, if solenoid B is actuated so that the piston is moving to the right under a 500 psi working pressure in the rod-end, the pressure in the blank-end of the cylinder will rise to a dangerously high level (because of the inertia of the load) when the control valve is returned to center in order to stop the piston.

To relieve these potentially hazardous high pressures, several control valves are incorporated between the lines to the cylinder as shown in Fig. 4-68(b). These valves are deceleration controls.

Assume that the rod-end area is half the blank-end area and that the pilot settings of the various pressure control values are as indicated in Fig. 4-68.

(a) With the piston moving to the right, what will be the maximum pressure that can occur in the blank-end of the cylinder when the four-way valve is returned to center position?

(b) With the piston moving to the left, what will be the maximum pressure that can occur in the rod-end of the cylinder when the four-way valve is returned to center position?

(c) What will be the pressure in the blank-end of the cylinder for the situation described in item (b)?

(d) Assume the pressure drop across the piston is 1200 psi when the attempt is made to stop the piston during its leftward motion. How far will the piston move after the four-way valve is centered if the piston is 6 inches in diameter and 3000 ft-lb of load kinetic energy must be dissipated to bring the inertia load to rest?

The solutions are as follows:

(a) When the four-way valve is centered and the piston begins to over-run, the pressure generated in the blank-end will open deceleration valve E. Therefore, the blank-end pressure will be relieved to 1500 psi. Note that after passing through valve E, half of the oil will be ported into the rod-end of the cylinder and the other half will go to the tank through relief valve C.

(b, c) When the four-way valve is centered and the piston begins to over-run, the pressure generated in the rod-end will open deceleration valve D. Therefore, the rod-end pressure will be relieved to 1200 psi. Now, though, there will not be sufficient flow delivered to the blank-end from valve D. So the blank-end will draw half of its required inflow from the tank through the check valve F at 0 psi.

(d) The area of the blank-end is 28.3 square inches, so the area of the rod-end is 14.15 square inches. Remembering that the pressure in the blank-end is zero, we calculate the force acting on the piston to be

$$1200(14.15) = 17{,}400 \text{ lb}$$

Then the piston displacement will be

$$\frac{3000(12)}{17{,}400} = 2.1 \text{ inches}$$

4-6 ELECTROHYDRAULIC VALVES

An electrohydraulic valve is a valve that is actuated electrically in order to control hydraulic pressure, flow, or position. Electrohydraulic valves make possible the electronic control of hydraulic circuits through the use of electrical analog controls, programmable controllers, or microprocessors.

Electrohydraulic valves contain electrical actuators which are usually solenoids or electrical torque motors. They may also be designed to incorporate an internal feedback in order to accurately position some part of the valve (e.g., a valve spool) or to achieve force amplification. The internal feedback mechanism can be mechanical, hydraulic, or electrical.

Electrohydraulic valves can be classified into two general types: proportional valves and servo-valves. Unfortunately, this nomenclature is not entirely definitive in classifying these valves even though it is used by practitioners of the art. A servo-valve may also be proportional, and a proportional valve may possess internal feedback (which might be interpreted to mean that it is a servo-valve).

Another basis sometimes used to discriminate between proportional valves and servo-valves is the valve frequency response. Broadly speaking, servo-valves are said to have good frequency response, in the 100 Hz range, while proportional valves have a lower frequency response, in the 10 Hz range. This distinction is beginning to disappear, however, as improved proportional valves are developed that have higher frequency response. Since proportional valves are cheaper than servo-valves, there is a great incentive to improve their control characteristics.

4-6-1 Electrohydraulic Proportional Valves

Proportional valves are available in a large variety of electrohydraulic types including direct-actuation and pilot versions (for actuating larger valves). A comprehensive discussion of some proportional valve types can be found in References 4-2, 4-8, and 4-9. Here we shall discuss only two types in order to illustrate the proportional valve concept.

Shown in Fig. 4-69 is a simple solenoid-actuated proportional valve for

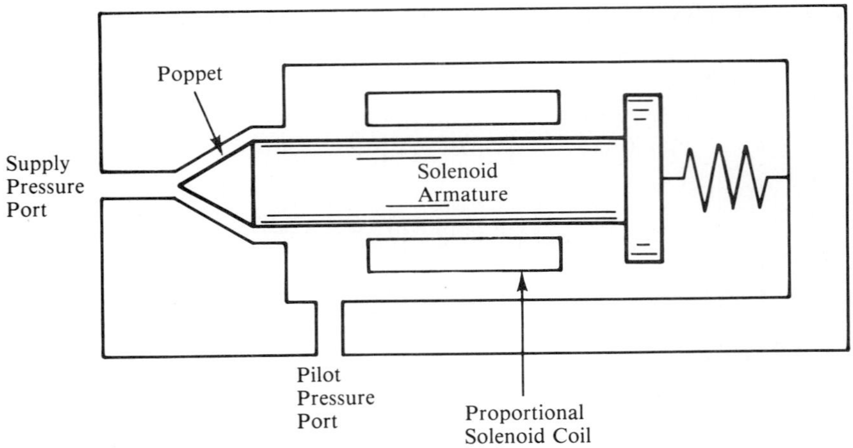

Figure 4-69. Proportional solenoid-actuated pilot stage for a pressure control valve.

actuating the pilot stage of a pressure control valve. The poppet of this valve can be positioned continuously in response to either current or voltage applied to the solenoid. When the poppet is fully seated, the pilot pressure will be zero because the valve supply pressure will not be communicated to the pilot port. As the valve opens in response to solenoid energizing, the pilot pressure will be a fraction of the supply pressure, depending on the position of the poppet.

Note that this solenoid is not a simple on-off solenoid of the type we discussed previously in connection with directional control valves. It is called a proportional solenoid because it is designed to have its armature position proportional to voltage, or current, supplied to it. But note that this valve is an open-loop valve in the sense that poppet position is not measured, with its location being fed back and compared to a position commanded.

If a valve such as that sketched in Fig. 4-69 is used as a pressure relief valve, it could be incorporated into a hydraulic circuit to provide programmed force of the actuator(s) of the circuit. This programmed force would be accomplished by having the voltage, or current, supplied to the relief valve programmed via a remotely located programmable controller or microprocessor.

Flow control can be implemented by using the proportional solenoid actuator to position the spool of a directional control valve. This idea is sketched in Fig. 4-70.

In this design, the proportional solenoid is used to continuously modulate the position of the spool. When used in this fashion, the spool can be made to assume an infinite number of positions between the fully open and fully closed positions. The valve becomes both a directional control valve and a flow control valve because the flow through the valve can be designed to be proportional to the command signal voltage, or current, supplied to the valve. As compared to the simple on-off type of solenoid-actuated directional control valve, the response of the valve spool will be as sketched in Fig. 4-71.

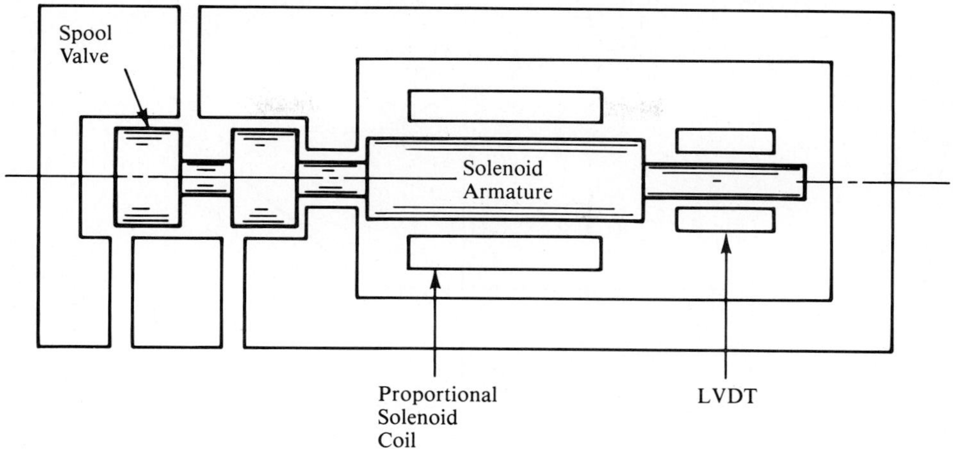

Figure 4-70. Proportional solenoid-actuated directional control valve.

The valve sketched in Fig. 4-70 also contains a linear variable differential transformer (an LVDT transducer) for sensing position of the spool. This feature makes possible the closed-loop feedback control of spool position. In other words, this valve is servo-controlled. Nevertheless, it will not necessarily be called a servo-valve because it has not necessarily been designed and manufactured for the high frequency response and precision associated with the word *servo-valve*. A proportional electrohydraulic valve such as that sketched in Fig. 4-70 is generally manufactured from interchangeable parts, which can result in deadband in the valve response characteristics (see Fig. 4-72).

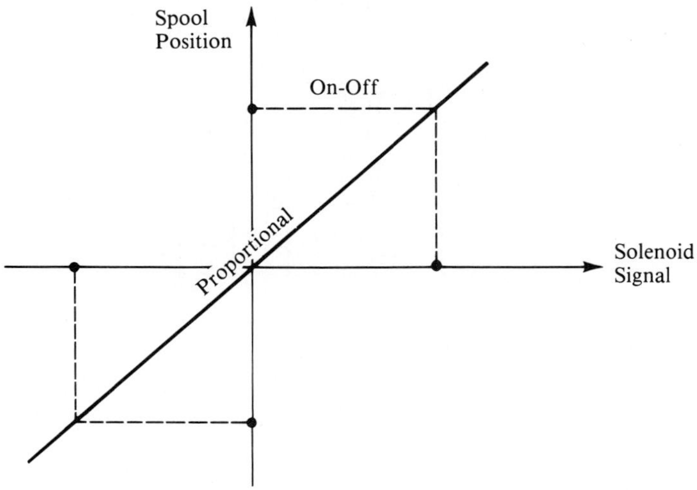

Figure 4-71. Proportional versus on-off valve response.

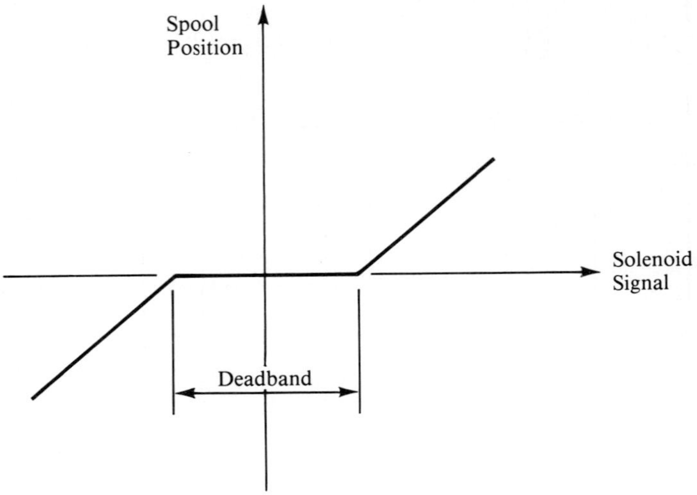

Figure 4-72. Valve deadband due to stiction.

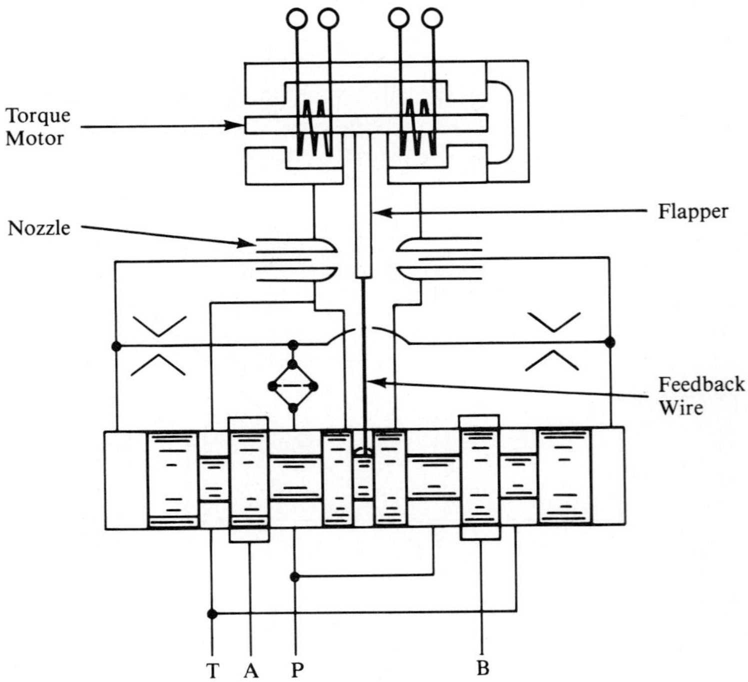

Figure 4-73. Servo-valve in the de-energized state.

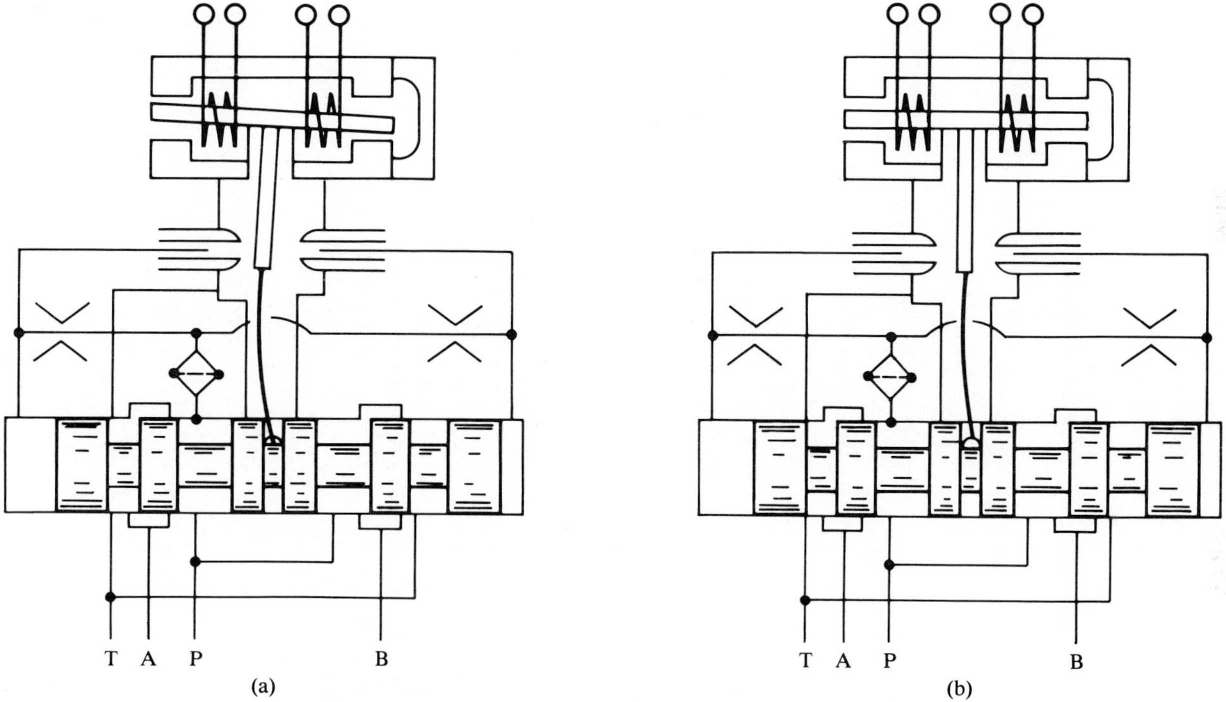

Figure 4-74. Servo-valve response to energization. (a) Torque motor shifts flapper; (b) spool motion re-centers flapper.

When electrohydraulic valves are used in closed-loop hydraulic circuits to obtain, say, the precision feedback control of actuator position, the valves must be precision valves. Proportional valves, although often possessing internal feedback, are generally not precise enough to be used in feedback-controlled hydraulic circuits. Consequently, their use is associated with open-loop hydraulic circuits.

4-6-2 Electrohydraulic Servo-Valves

Where precision valves are to be incorporated into a closed-loop hydraulic circuit, they will be servo-valves. These valves have a dual servo sense. Not only are they designed to contain internal feedback within the valves themselves, but the circuits in which these valves are installed are also servo-controlled (i.e., closed-loop hydraulic circuits).

The precision and response required of servo-valves are obtained by carefully matching spools and valve bodies and by lapping the valves during assembly to minimize deadband. The valves must also be designed to possess high frequency response. These requirements contribute to the high cost of servo-valves.

One common form of servo-valve that utilizes mechanical feedback is the electrohydraulic valve actuated by a torque motor. The operating concept is illustrated schematically in Fig. 4-73.

Actuation of the torque motor causes its armature to rotate slightly so that the flapper attached to it is moved closer to one of the two nozzles which are directing flow against the flapper from opposite sides. This movement will increase the pressure in the side of the valve whose nozzle has been partially blocked off. The increase in pressure causes the main spool of the valve to be shifted. But when the valve spool shifts, the spool will cause a mechanical feedback wire, connecting the spool to the torque motor, to rotate the torque motor back to its centralized position, which again centers the flapper between the opposing jet nozzles. The net result will be to have moved the spool to a new position where it now comes to equilibrium. The sequence just described is sketched in Fig. 4-74.

There are many servo-valve designs that are simply variations of the basic principles shown in Fig. 4-74. These precision valves are used in robotics, mobile equipment, farm equipment, machine tools, and industrial machinery. References 4-9 and 4-10 may be consulted for more information about servo-valves.

4-7 SUMMARY

In this chapter, we have discussed some of the fundamental components and basic concepts of fluid power circuits. The development is not a complete exposé of the many fluid power components and circuits available to the design engineer, but it is sufficient in order for the mechanical systems designer to see how fluid power circuits can be incorporated into machines. Engineering students who master the concepts of this chapter will be able to intelligently discuss fluid power circuits with more experienced designers.

Several important areas of fluid power have been omitted from this chapter. We neglected piping pressure drops in order to facilitate an understanding of the intrinsic properties of fluid power circuits. In real circuit designs, the piping pressure drops must be considered, and this may well require the use or development of computer programs for computer-aided design (CAD). Also, the transient performance of fluid power circuits is sometimes crucial to a system's performance. Again, computer programs may be required to conduct such transient analyses and design.

One of the major problems with fluid power systems is contamination of the system with water or crud. Also, maintenance is often improperly done or neglected. These conditions could easily be the cause of premature system failure or faulty operation. Consequently, it is critical in any fluid power system that filters and a filtration system be designed into the fluid power circuit. This subject, although extremely important to the successful operation of a fluid power circuit, does not contribute to the functional logic of the circuit, and has been omitted for that reason. Reference 4-11 discusses some of the factors involved in filtration practice.

4-8 REFERENCES

4-1. W. Ernst, *Oil Hydraulic Power and Industrial Applications*, McGraw-Hill, New York, 1960.

4-2. R. W. Henke, *Fluid Power Systems and Circuits*, published by the editors of *Hydraulics and Pneumatics Magazine*, Penton/IPC, 1111 Chester Avenue, Cleveland, OH, 1983.

4-3. D. A. Pease, *Basic Fluid Power*, Prentice-Hall, Englewood Cliffs, NJ, 1967.

4-4. G. R. Keller, *Hydraulic System Analysis*, published by the editors of *Hydraulics and Pneumatics Magazine*, Penton/IPC, 614 Superior Avenue West, Cleveland, OH, 1978.

4-5. *Mobile Hydraulics Manual M-2990-A*, published by Sperry-Vickers, Division of Sperry Corporation, Troy, MI, 1981.

4-6. *Industrial Hydraulics Manual*, published by Sperry-Vickers, Division of Sperry Corporation, Troy, MI, 1983.

4-7. *Fluid Power Encyclopedia*, published by Parker-Hannifin Corporation, 17325 Euclid Avenue, Cleveland, OH.

4-8. "Electrohydraulic Valves," *Machine Design*, January 24, 1985, pp. 56–58.

4-9. E. W. Reed and I. S. Lahrman, *Fluid Power with Microprocessor Control*, Prentice-Hall, Englewood Cliffs, NJ, 1985.

4-10. *Electrohydraulic Servo Systems*, published by the editors of *Hydraulics and Pneumatics Magazine*, Penton/IPC, 614 Superior Avenue West, Cleveland, OH, 1979.

4-11. J. R. Pugh, *Filtration for Today's Hydraulics*, available from Parker-Hannifin Corporation, Filter Division, Metamora, OH.

PROBLEMS

4-1. An experimental airflow tunnel is sketched in Fig. P4-1. By applying Bernoulli's equation to this system, the system characteristic can be shown to be closely approximated by

$$P = \left(\frac{120f + 1}{2g}\right)\left(\frac{Q}{A}\right)^2 \text{ feet of air}$$

where $f \cong 0.02$ is the Moody friction factor for flow in smooth ducts, Q is the volumetric flow rate of air in the tunnel in ft^3/sec, and A is the area of the rectangular section ($A = (1.5)(6/144) =$ square foot). Plot the load curve of P versus Q for this system on the plot of Fig. P4-1.

The operating point will occur where the load characteristic intersects the blower characteristic, shown for a 2700 rpm blower. Plot the load curve for the points $Q = 250$, 500, and 750 cfm. Remember that 100 feet of air corresponds to about 1.44 inches of water head. The operating point will occur at a discharge of about 500 cfm.

4-2. The 38 inch pump of Fig. 4-6 is used to pump water at 60°F from a reservoir through 1000 feet of 12 inch ID galvanized iron pipe to a point 200 feet above the reservoir surface. What flow rate and horsepower will result?

4-3. For $Q = 0$, Eq. (4-8) shows that the no-flow head developed by the centrifugal pump is

$$P = \frac{u_2^2}{g}$$

What is the physical interpretation of this result?

4-4. A PDP characteristic is

$$Q = 4 - 0.00002P \qquad (Q \sim \text{gpm},\ P \sim \text{psi})$$

The pump is inserted in a hydraulic circuit whose load characteristic is

$$P = 1000Q$$

Determine the pump internal leakage at the steady-state operating point. If the pump is operated at 1800 rpm, what is its theoretical displacement per revolution?

4-5. Justify Eq. (4-10).

4-6. A centrifugal pump drives a load whose characteristic is

$$P = 1.2Q \text{ kN/m}^2 \qquad (Q \sim \text{m}^3/\text{min})$$

The pump output characteristic is

$$P = \frac{N - 6}{2} - \frac{Q}{5} \text{ kN/m}^2 \qquad (N \sim \text{rev/sec})$$

The pump has a mechanical efficiency of $\eta_o = 0.75$.

(a) Determine the speed at which the pump must be driven such that the system flow rate will be 5 m^3/min.

(b) Display the equilibrium operating condition of this system on the P-Q plane as the intersection of the load and pump characteristics.

4-7. Determine the load of a hydraulic system in which water at 60°F is to be pumped from one reservoir to another which is 150 feet higher in elevation. The pipe used is

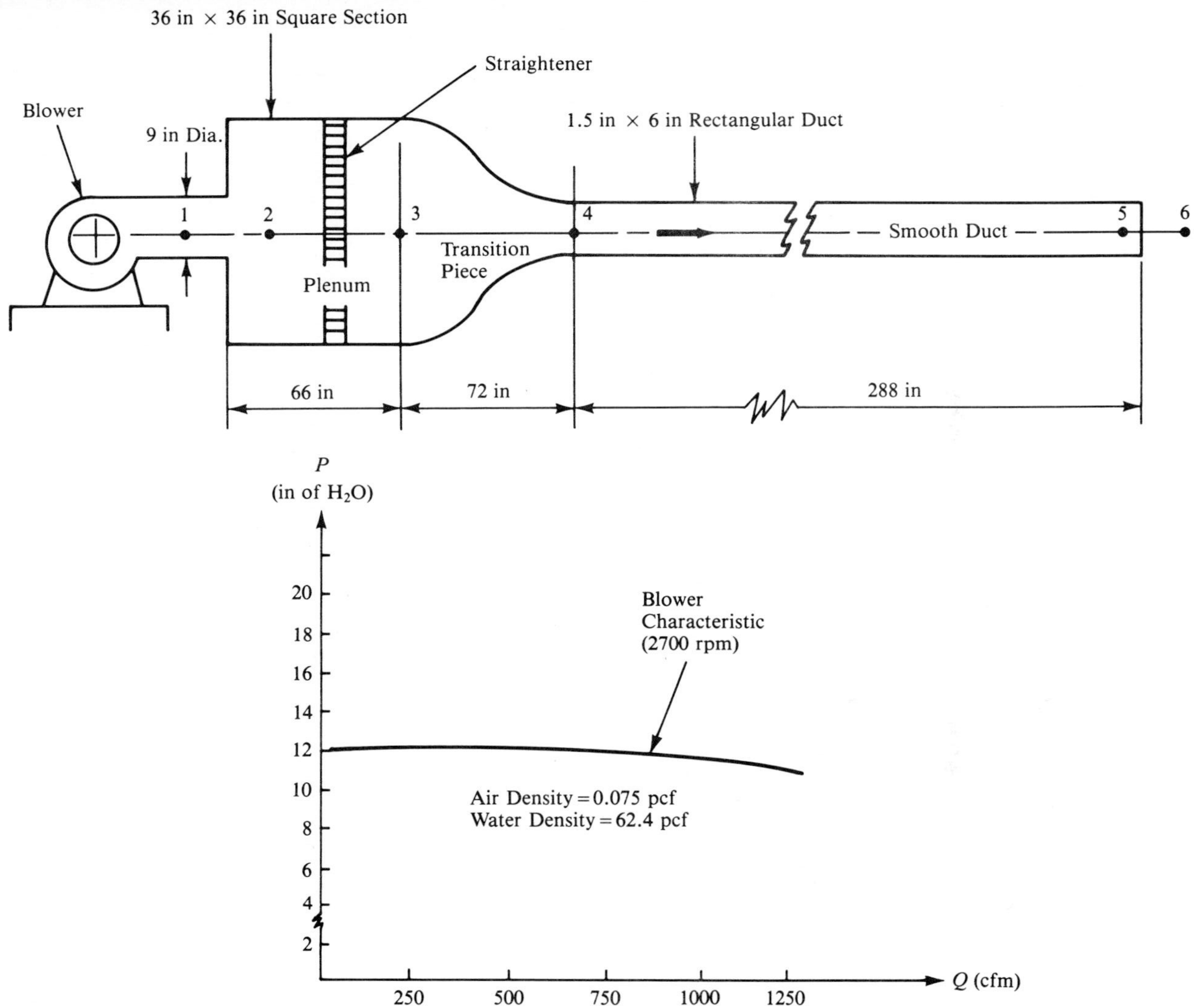

(b)

Figure P4-1. Figures for Problem 4-1.

1000 feet long and 12 inches in diameter. Assume the friction factor is 0.02 and neglect entrance and exit losses. Plot the load characteristic in units of feet of water (head, P) versus gpm (discharge, Q).

4-8. A vertically oriented hydraulic cylinder supports a weight W as it descends (see Fig. P4-8). If the rod area is a and the piston area is A, show that for constant speed downward, the rod-end pressure must exceed the blank-end pressure according to the relationship

$$p_r = \left(\frac{A}{A-a}\right)p_b + \frac{W}{A-a}$$

What resistance (in psi) must be provided in the exhaust line connected to port B to insure that this load W will not fall when the system is static?

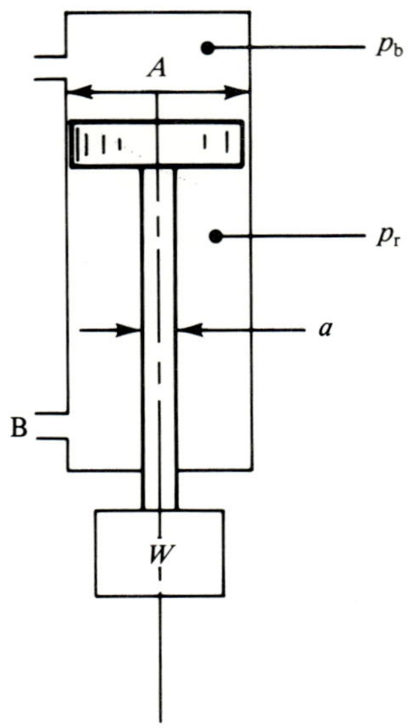

Figure P4-8. Figure for Problem 4-8.

4-9. In the regenerative circuit of Fig. 4-29, show that if the rod area is a and the piston area is A, the relationship between leftward piston speed V_ℓ and rightward piston speed V_r is

$$\frac{V_\ell}{V_r} = \frac{\dfrac{a}{A}}{1 - \dfrac{a}{A}}$$

4-10. As the piston moves to the right in Fig. 4-29, what is the rate of volumetric depletion of oil from the reservoir? What happens to the oil in the reservoir as the piston moves back toward the left?

4-11. A hydraulic intensifier is driven by a hydraulic cylinder B as shown in Fig. P4-11. The region between pistons a and A of the intensifier is vented to the reservoir. The intensifier output drives a hydraulic cylinder C whose piston area is C. The piston area of cylinder B is denoted as B. What is the relationship between the piston forces F_B and F_C and the piston displacements of δ_B and δ_C? Neglect rod-end pressures in cylinders B and C.

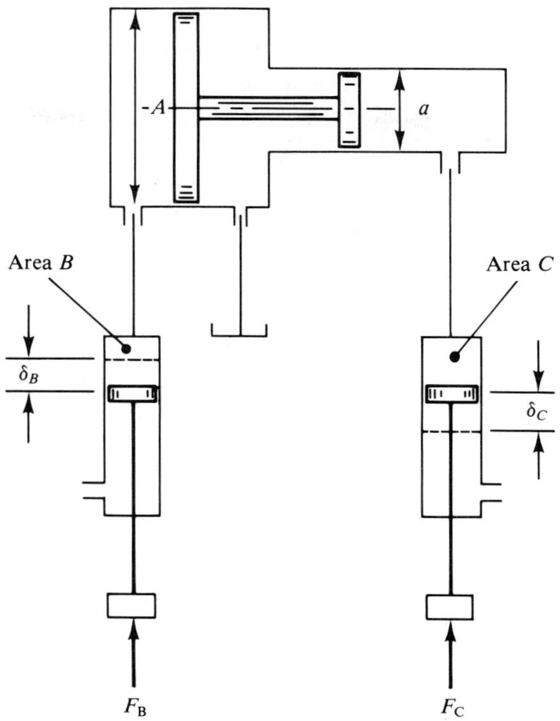

Figure P4-11. Figure for Problem 4-11.

4-12. Assume the axial-piston motor of Fig. 4-34 has a theoretical displacement of 6.8 in^3/rev. What is its volumetric efficiency when it requires 40 gpm to drive it at an inlet pressure of 3000 psi? Assume the motor exhaust pressure is zero.

4-13. Assume the volumetric efficiency of the axial-piston motor of Fig. 4-34 is 100 percent. What will be the motor's torque efficiency when it is driven by 40 gpm at an inlet pressure of 3000 psi? Assume the motor exhaust pressure is zero.

4-14. In Example 4-7, what will dictate the order in which actuators A and B are retracted?

4-15. Develop a hydraulic circuit using two hydraulic cylinders A and B (as in Example 4-7) in which, first, cylinder A is to be extended against its load, followed by cylinder B extended against its load. On the retraction stroke, cylinder A is to be retracted first followed by the retraction of cylinder B. (Hint: You will need to add another sequence valve and another check valve to the circuit of Fig. 4-41.)

4-16. Assume the area of the connecting rod of the spool of Fig. 4-42 is a and the area of the piston(s) is A. Make a force analysis of the counterbalance valve spool and determine the spring setting (in lb) as a function of areas a and A and the pressure threshold for valve opening, p_t. Assume the valve exit pressure is zero.

4-17. With the piston moving toward the right against an external force F in the regenerative circuit shown at the left of Fig. 4-29, explain why the hydraulic fluid delivered by the pump will proceed into the blank-end of the cylinder rather than into the rod-end. Let the rod area be a and the piston area be A. (Hint: Draw a free-body diagram of the piston.)

4-18. With the piston moving toward the left against an external load F in the regenerative circuit of Fig. 4-29, what is the relationship between rod-end pressure p_r and load F?

4-19. It is required that the speed to the left be three times the speed to the right in the circuit of Fig. 4-29. Size the rod, compared to the piston, to achieve this with the regenerative circuit shown.

4-20. The speed requirements for a hydraulic cylinder are the same as for Problem 4-19, but the area ratio is $a/A = 1/2$. How would you design a circuit to achieve the required speed ratio? You can use any combination of sequence, flow control, or directional control valves, but the pump must be of the uni-directional, fixed-displacement type.

4-21. Assume the hydraulic lifting cylinder of Fig. 4-43 has a blank-end diameter of 4.5 inches and a rod-end diameter of 2.25 inches. The cylinder is used to elevate and lower a load of $W = 2$ tons. Assume the counterbalance valve is set so as to provide a 150 psi margin against the load descending under its own weight with the pump off. The relief valve at the pump outlet is set to 150 psi in excess of the maximum pressure ever to be experienced at the pump outlet with this 2 ton load. Determine

(a) The pressure required at the pump outlet to drive the load down

(b) The setting of the pump pressure relief valve

4-22. Figure P4-22 shows a hydraulically operated platen press whose upper platen and piston weigh 10,000 lb. There is a counterbalance valve A in the line connected to the cylinder rod-end. The diameter of the cylinder is 4 inches while that of the rod is 1 inch. The pressure relief valve B is set for 1200 psi. The counterbalance valve is set so that the factor of safety against the platen falling will be 1.15.

(a) Determine the setting of counterbalance valve A.

(b) Determine the pump outlet pressure required to power the platen down when it does not encounter the workpiece.

(c) Determine the maximum working force the platen can provide to the workpiece.

(d) Show how the two-position, two-way, directional control valve C can be connected into this circuit to increase the working force of part (c) by 11,500 lb. Valve C is a pilot-actuated, spring-return, two-position valve.

(e) What should be the spring setting of valve C?

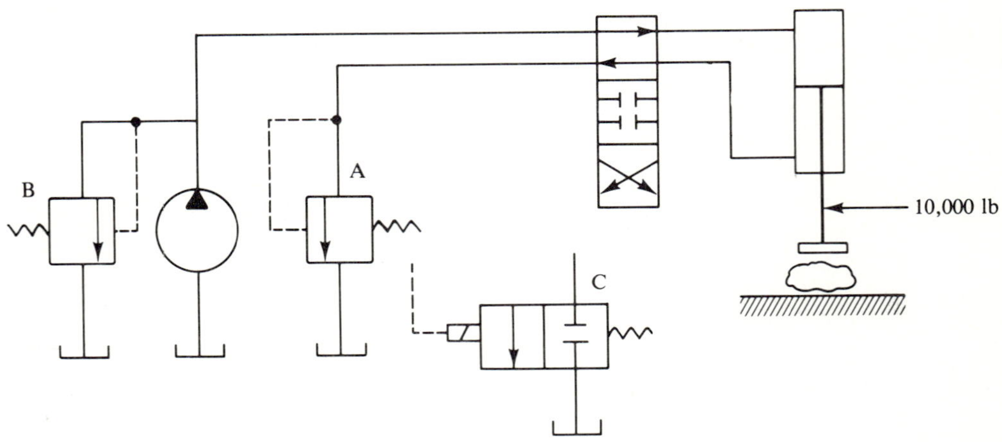

Figure P4-22. Figure for Problem 4-22.

4-23. In Fig. 4-44, why is the position of the valve spool unaffected by the upstream pressure? Why must this valve be externally drained, in contrast to the internal draining of the unloading valve shown in Fig. 4-38?

4-24. Consider the single-acting cylinder whose load is supported by the counterbalance valve as shown in Fig. P4-24.

(a) Suppose the pilot line A of the counterbalance valve is connected to the exhaust line (point C) from the cylinder. What must the minimum value be for the spring setting of the valve in order that the load will not fall?

(b) Suppose pilot line A is connected to the pump supply line (point B) to the cylinder. What will be the minimum value of the counterbalance valve spring setting in order that the load not fall?

(c) When the load is powered down by the pump, what will the pressures be in the rod-end and the blank-end of the cylinder? Assume line A is connected to line C and that the counterbalance valve is set at 1600 psi.

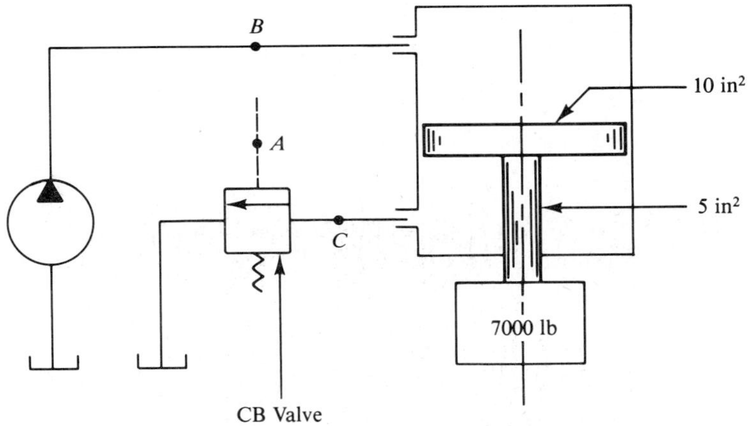

Figure P4-24. Figure for Problem 4-24.

4-25. Suppose a meter-in circuit is used to control the speed of a load that is resistive in nature but that suddenly becomes an over-running load. What can occur to the oil in the line that supplies fluid to the actuator? Explain why.

4-26. Let the pressure drop across a flow control valve be δp. Suppose the load on a cylinder actuator is F, the piston area is A, and the rod area is a. Determine expressions for the pump discharge pressure p_p for (a) meter-in, (b) meter-out, and (c) bleed-off circuits.

4-27. Suppose the rod area is 5 square inches, the piston area is 10 square inches, and the load is 5000 lb on a hydraulic cylinder. The relief valve setting is 1000 psi. Compare the performance of meter-in, meter-out, and bleed-off circuits for these parameters. Will they all work as speed control circuits?

4-28. What is the difference between the closed-center and center bypass (three-position, four-way) directional control valves insofar as the actuator is concerned? What is the difference between these valves insofar as the pump is concerned?

4-29. The hydraulic circuit of Fig. 4-62 is a rapid-advance (RA), slow-feed (F), rapid-return (RR) system for reciprocating a machine tool table. The system is designed such that the F portion of the cycle has a speed of one-fourth (1/4) the RA speed. During the RA part of the cycle, the table does not work against any external force. However, during the F part of the cycle, the table works against an external force of 10,000 lb. The feed (F) will begin when the cam on the piston rod contacts the mechanical actuator of valve B and closes it. This action inserts the metering-out flow control valve D into the cylinder exhaust line.

Assume the capacity of the single PF (fixed-displacement) pump to be 10 gpm and the cylinder blank-end and rod-end areas to be 10 square inches and 5 square inches respectively. The main system relief valve A is set for 3000 psi.

(a) What will be the RA speed of the table?
(b) What will be the F speed of the table?
(c) What is the dumping rate (in gpm) of oil over the relief valve during the F part of the cycle?
(d) What will the pressure drop be across the metering-out valve D during the feed?
(e) Design a circuit that eliminates the metering-out valve D and that provides the same speed program through the use of two pumps (a 2.5 gpm and 7.5 gpm double pump).

4-30. For the circuit of Fig. 4-63, assume that the area of the smaller piston is 10 square inches and that the area of the larger piston is 20 square inches. The connecting rod area is 5 square inches, and the two pistons are sealed from each other around the connecting rod. The sequence valve is set to open at 20 psi.

(a) If the double piston moves to the right with sequence valve A closed, what is the purpose of check valve B?
(b) If the double piston moves to the right, what is the maximum working force it can encounter if the sequence valve is to remain closed?
(c) Suppose the rightward-moving double piston encounters a working force of 4000 lb. What will be the pump outlet pressure?
(d) What is the purpose of check valve C?

4-31. The hydraulic system shown in Fig. P4-31 is a slow-advance, rapid-return circuit. Assume that the directional control valve is actuated by hand. Explain in detail how this system works. The metering valve is adjustable.

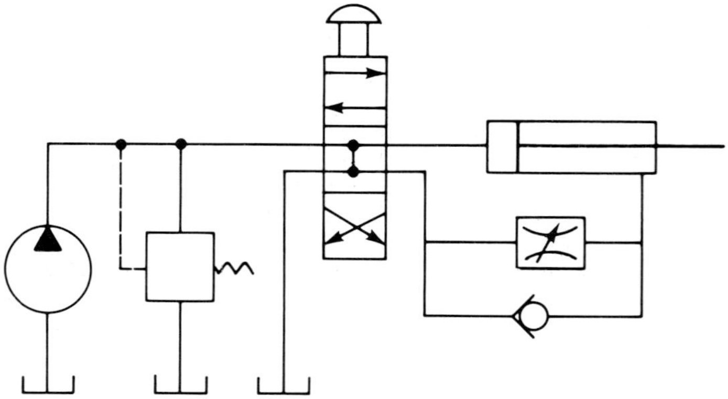

Figure P4-31. Figure for Problem 4-31.

4-32. A rapid-advance, slow-feed, rapid-return circuit that uses a double-pump arrangement is shown in Fig. P4-32. Assume the directional control valve is solenoid-actuated, with the solenoids A and B being energized by the limit switches shown. The on-off valve C is also solenoid actuated by the limit switch(es).

(a) Explain how this system might be made to work as an RA, F, RR circuit. Valve E is a sequence valve that diverts the output of the large pump to the tank when it is actuated. Valve F is simply a relief valve. Valve D is a metering-out valve when it is in the circuit.

(b) Assume that the mechanical load on the piston is negligible compared to the load created by the metering-out valve during the feed stroke. If the rod area (a) is one-half the piston area (A), what will the setting of the sequence valve E be if the pressure drop across the metering valve is 1000 psi?

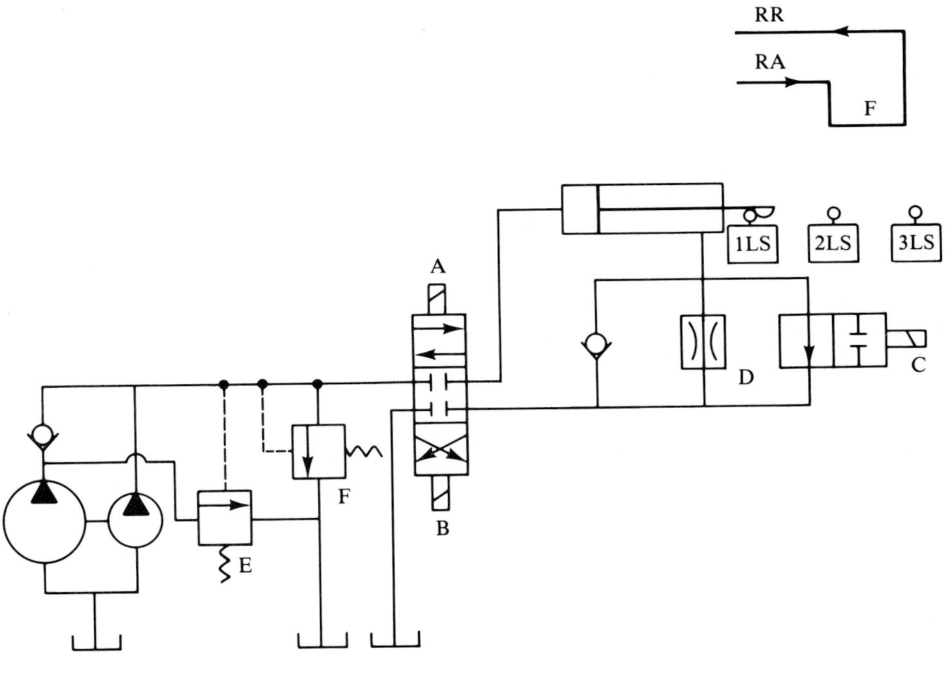

Figure P4-32. Figure for Problem 4-32.

4-33. What is the advantage of a double-pump arrangement for controlling speed versus a single pump and a metering valve?

4-34. In the fluid power circuit sketched in Fig. P4-34, the pump capacity is 10 gpm. The blank-end area of the cylinder is 4 square inches and the rod-end is 2 square inches. The load on the table is 2000 lb when it moves to the right and zero when it moves to the left. Determine

(a) The pressure in the blank-end for motion to the right if the pressure drop in the exhaust line is 60 psi

(b) The speed of the piston as it moves to the right

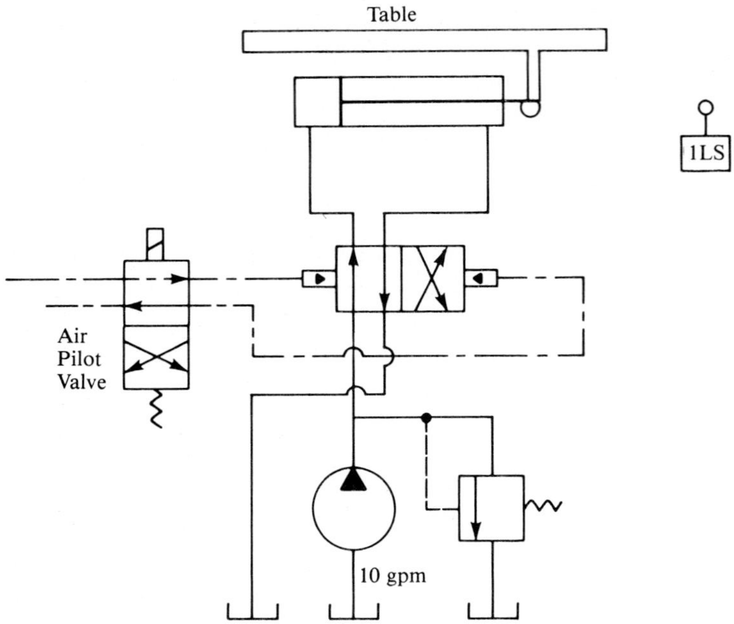

Figure P4-34. Figure for Problem 4-34.

(c) The speed of the piston as it moves to the left
(d) The setting of the relief valve
(e) The horsepower capacity of the pump

4-35. The double-rodded hydraulic cylinder actuator shown in Fig. P4-35 operates intermittently for 5 seconds every minute. During the 5 second work cycle, the working pressure in the cylinder averages 1350 psi and the cylinder requires a flow rate of 5 cfm.

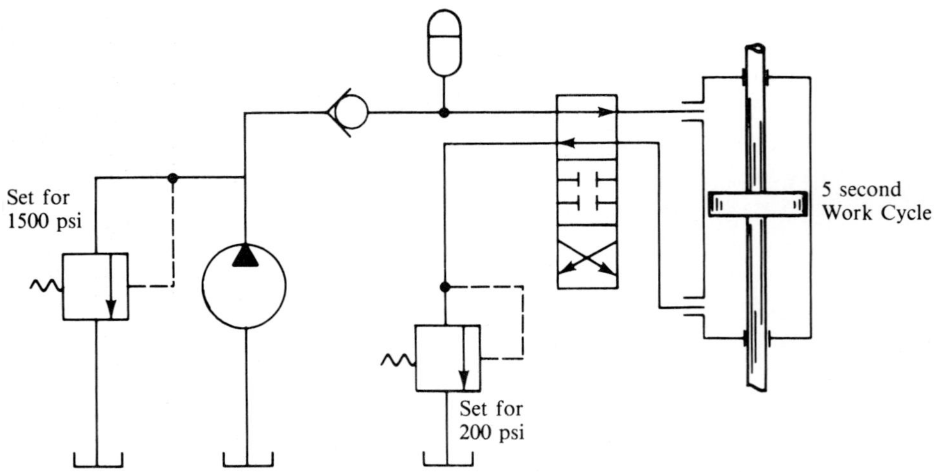

Figure P4-35. Figure for Problem 4-35.

(a) What will be the required horsepower capacity of the pump if no accumulator is provided in the circuit?

(b) If an accumulator is provided (as shown), which will provide the entire cylinder capacity requirements during the 5 second work cycle, what will be the horsepower requirements of the pump? Assume the oil specific gravity is 0.9.

4-36. The ideal relationship between torque T_i developed by a hydraulic motor and the pressure P driving it is given by Eq. (4-33); that is,

$$T_i = P\left(\frac{An\varepsilon}{\pi}\right)$$

We shall define the motor geometry parameter k_m to be

$$k_m = \frac{An\varepsilon}{\pi}$$

The actual torque T_m developed by the motor is somewhat less than T_i due to internal friction in the motor. We shall assume that the torque loss is proportional to motor speed ω_m with proportionality constant f_m. Then the actual torque developed by the motor will be

$$T_m = k_m P - f_m \omega_m$$

In a similar manner, we can develop an approximate relationship for the actual volumetric flow rate Q_m through the motor from the ideal relationship given in Eq. (4-33) for Q_i. This relationship is

$$Q_m = k_m \omega_m + \lambda_m P$$

where λ_m is a leakage constant.

(a) Using the equation for Q_m, determine the constants k_m and λ_m using data from Fig. 4-34. Use data for the speed $N_m = 1400$ rpm ($\omega_m = 8796.46$ rad/min). Give k_m in units of cubic feet and λ_m in units of gpm/psi. (Find $k_m = 0.000517$ cubic foot.)

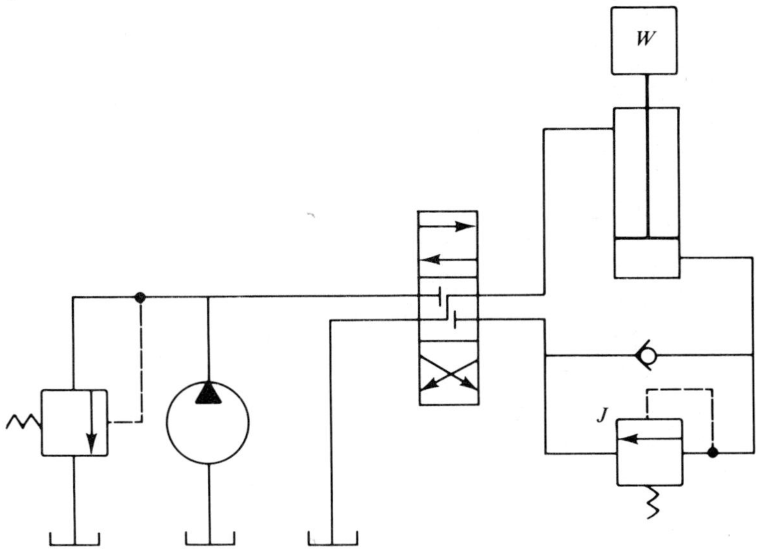

Figure P4-37. Figure for Problem 4-37.

(b) Using the equation for T_m, determine the constants k_m and f_m using data from Fig. 4-34. Use data for the speed $N_m = 1400$ rpm ($\omega_m = 8796.46$ rad/min). Give k_m in units of cubic feet and f_m in units of in-lb/rad min^{-1}. (Find $k_m = 0.0005$ cubic foot.)

4-37. A hydraulic piston is used to elevate a weight of 3 tons (see Fig. P4-37). If the piston area is 3 square inches and the rod area is 1 square inch, determine

(a) The setting of the counterbalance valve J if a margin of 15 percent is to be provided against the weight falling when the system is static

(b) The setting of the system relief valve if a margin of 15 percent is to be provided against it opening

(c) The pump outlet pressure required to drive the weight downward

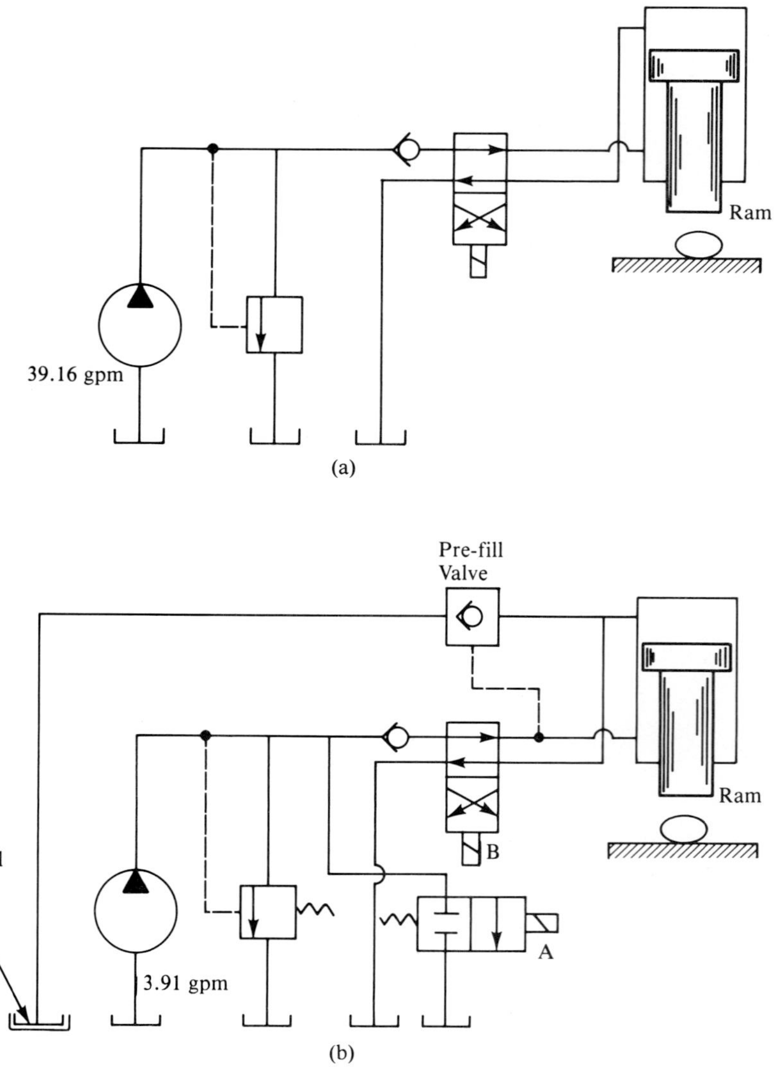

Figure P4-38. Figure for Problem 4-38.

4-38. In the circuit shown in Fig. P4-38(a) the large press ram cylinder is 48 inches in diameter. During the rapid advance toward the workpiece, the downward speed is to be ten times the speed of 0.5 fpm at which the ram will move when it contacts the workpiece and works against an external force. This will require that 35.25 gpm be dumped over the relief valve during the feed stroke if a fixed-displacement pump of 39.16 gpm capacity is used, as shown.

The circuit shown in Fig. P4-38(b) is intended to be used with a fixed-displacement pump of much smaller capacity (3.91 gpm).

With a smaller pump whose capacity is only 39.16 − 35.25 = 3.91 gpm, the feed stroke can be maintained at 0.5 fpm, but the circuit at the left could not provide the rapid advance. The circuit at the right is intended to provide a rapid advance despite the fact that the pump displacement is fixed at 3.91 gpm.

Explain, in words, how this system (on the right) works.

5

Hydrostatic Drives (Transmissions)

In this and the next two chapters, we will discuss several methods of transmitting power using fluid drives. As we shall see, there are basically three ways of transmitting fluid power: by hydrodynamic action, by hydroviscous action, and by hydrostatic action.

In a hydrodynamic drive, torque is developed by fluid momentum exchanges between an impeller and a turbine, and this torque is, therefore, related to fluid vector velocities. These drives are called fluid couplings or torque converters; they are discussed in detail in Chapter 6. Hydrodynamic drives are basically non-positive-displacement devices incapable of providing internal braking or of providing appreciable torque over a broad range of speeds. On the other hand, these hydrodynamic couplings can provide smooth acceleration capability in starting high inertia loads.

Hydroviscous drives develop torque through the mechanism of viscous shearing in thin layers of oil sandwiched between adjacent disks. These drives are subject to the same kind of limitations as are hydrodynamic drives but they do have the virtue of being conveniently servo-controllable. Hydroviscous drives will be discussed in Chapter 7.

A hydrostatic drive (also called a hydrostatic, or hydraulic, transmission) is a fluid power circuit in which a positive-displacement pump (for supplying power) drives a positive-displacement motor (which demands power). Ideally, hydrostatic drives are non-slip transmissions similar to mechanical traction drives. Of course, the small internal leakages that exist in the real hydraulic components cause some potential creep to occur despite the fact that the components are positive-displacement. Practical hydrostatic drives also contain subsidiary fluid power components such as valves and filters in addition to the pump and motor.

The hydrostatic drive is analogous to an electrical drive that consists of a generator (for supplying power) driving an electric motor (which demands power). As is shown in Fig. 5-1, hydraulic pressure is analogous to electrical voltage, and oil discharge rate is analogous to electrical current.

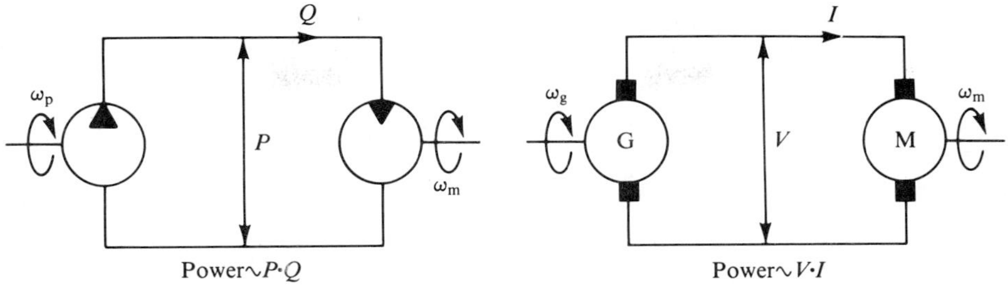

Figure 5-1. Analogous hydraulic and electrical drives.

Because of their similarities, hydraulic and electrical drives may often both be suitable for particular mechanical system applications, and similar technical analyses can be made for them in terms of corresponding analogous parameters. So, what are the advantages of hydrostatic drives? They are listed below.

(a) These drives have high power density, that is, high horsepower transmission in relation to package size and weight.
(b) They are characterized by low inertia of rotating members because of relatively small size. Low inertia permits fast starting, stopping, and reversal with smoothness and precision.
(c) They have flexibility with respect to control of output speed and torque; also, rotation is readily available in either direction.
(d) Accurate speed control can be achieved relatively simply in both servo-controlled and non-servo-controlled systems.
(e) They are capable of providing constant output torque from zero to full-load speed (with small, or zero, creep at zero speed).
(f) They have relatively high mechanical efficiency over a wide range of operating conditions.
(g) They can be stalled, and remain stalled, without damage.
(h) Damping can be provided easily.
(i) Dynamic, or regenerative, braking can be obtained.

Hydrostatic drives are used in a wide range of power transmission applications including the following:

(a) Earth-moving and road-making mobile machinery
(b) Military systems including gun turrets, steering systems, and propulsion of tanks and ships
(c) Machine tools and presses
(d) Industrial and processing machinery
(e) Tension control machinery
(f) Aircraft applications
(g) Agricultural equipment
(h) Transportation vehicles such as automobiles, buses, locomotives, and aerial trams

Recent improvements in the performance of hydrostatic drives, and in their configurations, have opened many new fields of application. These drives are available in power ratings ranging from 5 to 1000 hp.

5-1 CIRCUIT CONFIGURATIONS

The fluid power components from which a hydrostatic drive is synthesized may be arrayed in two basic types of circuit configurations. One configuration, called a closed circuit, is displayed at the left of Fig. 5-1. In this circuit configuration, the exhaust from the motor is piped directly back to the inlet of the pump and no reservoir (tank) is included in the primary circuit shown. Note that the pressure in the low-pressure side of a closed-circuit hydrostatic drive is arbitrary and can be controlled through the use of a small charging pump (see Fig. 5-8). Thus, the level of pressures (or pressure datum) in a closed-circuit hydrostatic drive can be easily controlled.

The other type of hydrostatic drive configuration is the open-circuit configuration. In this circuit, the motor exhaust is discharged directly to the reservoir, and the pump gets its supply from this same reservoir. This means that the low-pressure side of the circuit will always be (open) at atmospheric pressure (zero gage) in an open-circuit configuration.

These two circuit configurations are sometimes called closed-loop and open-loop hydrostatic drives, which can be confusing in light of the fact that these same expressions are also used to distinguish how the drives are controlled. Figure 5-2 will help clarify the nomenclature.

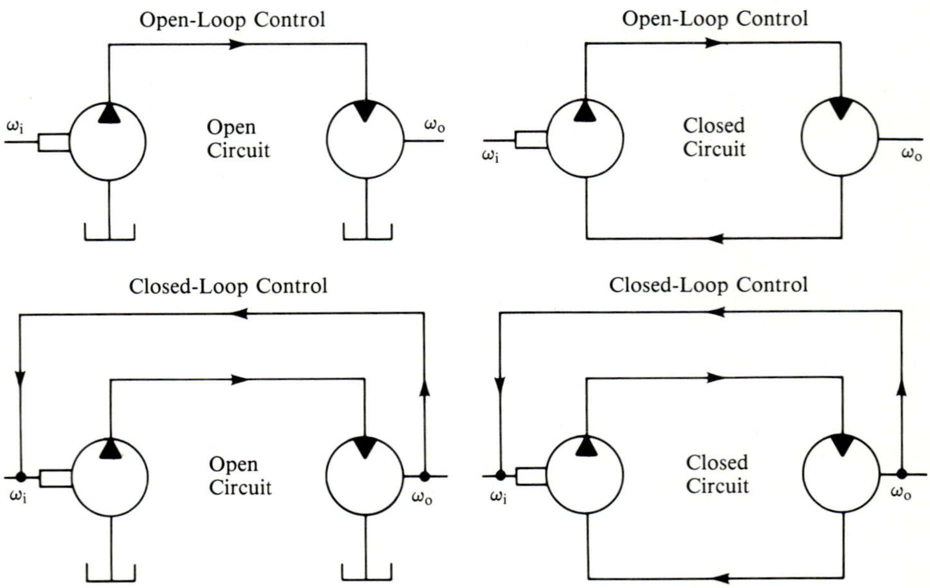

Figure 5-2. Classification of hydrostatic drives.

5-1-1 Closed-Circuit versus Open-Circuit Drives

There are distinct differences, advantages, and disadvantages of closed-circuit and open-circuit hydrostatic drives. One major difference in configuration will be in how the direction of motor rotation is controlled. In the closed-circuit drive, the direction of motor rotation can be simply controlled by using a reversible pump (see Fig. 5-3). This makes possible the reversing of the roles of the high- and low-pressure sides of the circuit. If, additionally, a variable-displacement reversible pump is used, then the hydraulic motor will be provided with infinitely variable speed throughout its entire range in either direction (including zero) without the necessity of having a directional control valve at all.

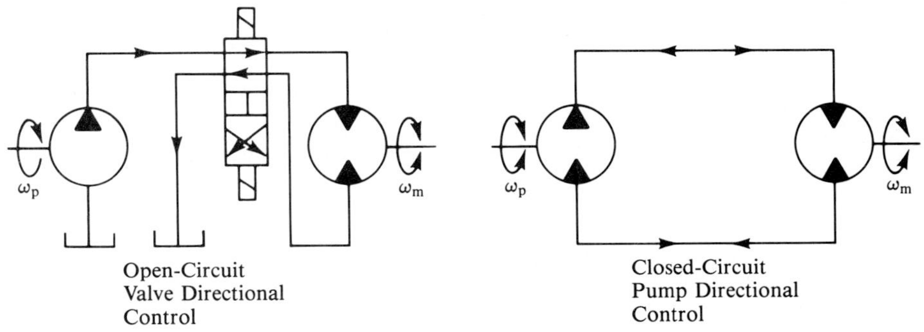

Figure 5-3. Motor directional control.

In an open-circuit drive, a directional control valve must be used to control the motor rotational direction, because reversing the pump will simply pump fluid into a vented reservoir. So the directional valving of an open-circuit drive will be quite different from that of the closed circuit (see Fig. 5-3).

There are additional differences in valving between closed- and open-circuit hydrostatic drives caused by the different ways in which these types of circuits must respond to over-running loads. The problems created in a hydrostatic drive by an over-running load are of two types: potential cavitation and braking.

In a closed-circuit drive, some regenerative, or dynamic, braking is generally inherent in the system. If a load reversal suddenly occurs such as to create an over-running load, the hydraulic motor will then suddenly be operated as a pump with the energy input coming from the kinetic energy of the over-running load. Then, depending on how the circuit is valved, this motor (turned into a pump) may operate the pump of the closed-circuit hydrostatic drive as a motor. The prime mover of the system will be forced to absorb energy, and if this prime mover is an electric motor, the energy can be returned to the electrical supply. This is the regenerative mode of braking. An alternative is to dissipate the energy provided by the motor, turned pump, through a valve and piping arrangement, in which case we call the action dynamic braking.

In a closed circuit, the tendency towards cavitation in the low-pressure line caused by an over-running load can usually be suppressed by the charge pump which is necessary in closed-circuit drives (see Fig. 5-8). We shall discuss this more complete closed-circuit drive in greater detail later.

Although the open-circuit drive seems to be at a disadvantage with respect to over-running loads (which we shall discuss subsequently), it does have one major advantage over the closed-circuit drive. Where more than one output function must be provided to a machine driven by a hydrostatic drive, the closed-circuit configuration will generally require a separate pump for each output function. This can result in a transmission both expensive and bulky which tends to compromise the inherent advantages of fluid power systems.

The open-circuit drive, on the other hand, can effectively drive a multi-function machine using only one (or possibly two) pumps through the use of directional control valves. Installing additional control valves is much more cost effective and can be done more compactly than mounting additional pumps. This approach requires only the addition of a directional control valve for each hydraulic motor that needs to be driven. Advantages of these kinds have led to recent improvements in the design of open-circuit hydrostatic transmissions that have overcome many of the traditional shortcomings such as the inability of the open-circuit drive to provide smooth direction changes at actuators and to deal effectively with over-running loads.

5-1-2 Open-Circuit Hydrostatic Transmissions

One disadvantage of the open-circuit hydrostatic transmission, compared to the closed-circuit transmission, is in how it must be modified to ensure that potential over-running loads do not create hazards. If control in an open-circuit transmission is to be effected using the metering-in principle, the motor will not be prevented from running away under a heavy inertia load because there will be no resistance in the line between the motor outlet port and the tank. Running away could cause a hazard to personnel, and cavitation in the inlet side of the system.

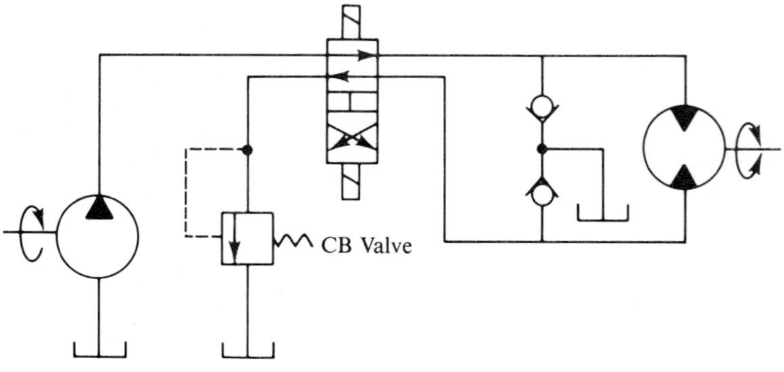

Figure 5-4. Open-circuit drive with a counterbalance valve.

One solution to this problem is shown in the circuit of Fig. 5-4. Braking of the over-running load is provided by a counterbalance valve installed in the exhaust line of the circuit. Also shown in Fig. 5-4 is an arrangement of check valves which are intended to negate the potential cavitation in the line connecting the pump to the motor in the event of an over-running load. Cavitation is avoided by having this line draw fluid from the tank through the low resistance check valve (the two valves are required to take care of both possible directions of rotation of the motor).

Although the arrangement of Fig. 5-4 ameliorates the over-running load problem, its metering-out type of circuit has the disadvantage of pressurizing the motor outlet port (which adds to the motor load pressure) so a limitation on inlet pressure (by the relief valve) will result in not providing the same torque capability that a metering-in type of circuit will provide.

One way out of this dilemma is to use a circuit such as that sketched in Fig. 5-5. This system incorporates a counterbalance circuit in which a pilot-actuated, proportional type of valve is inserted between the inlet and exit lines of the hydraulic motor. During normal forward operation against a positive external load, the inlet line is pressurized (by the load), which causes the counterbalance valve to be in its actuated position. In this position, the motor exit port communicates directly with the tank under no resistance in the exit line so full torque can be extracted by the motor from the available inlet pressure.

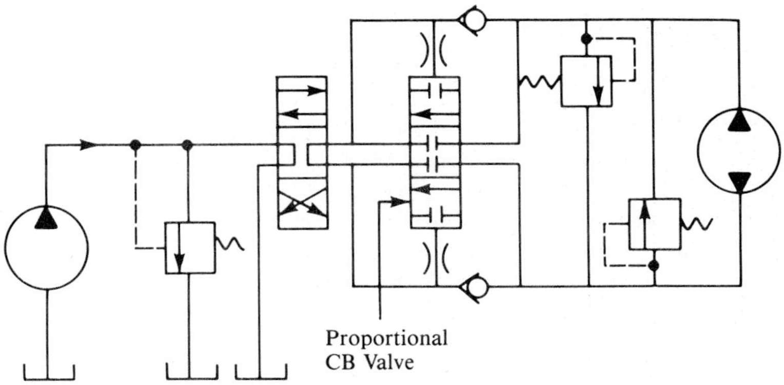

Figure 5-5. Open-circuit drive with a proportional counterbalance valve.

When the motor experiences an over-running load, the inlet pressure will drop drastically, causing the counterbalance valve to move back toward its spring-centered, closed position. Since the valve is proportional, it will not close completely but will introduce a back-pressure in the exit line because of the resistance introduced by the valve between motor exit and the tank. This exit pressure, in turn, prevents an excessive drop in inlet pressure and hence suppresses what would otherwise be cavitation in the inlet line. The magnitude of the resistance created in the exit line balances the over-running load and is proportional to it.

The circuit shown in Fig. 5-5 is symmetrical so that the motor may be operated in either direction. The cross-port relief valves are included to limit maximum circuit pressure so that component integrity is not compromised.

EXAMPLE 5-1

A fixed-displacement (MF) hydraulic motor has a theoretical displacement of $3\,\text{in}^3/\text{rev}$ and a volumetric efficiency of 95 percent. The motor is driven by a variable-displacement (PV) hydraulic pump having a maximum theoretical displacement of $1.9\,\text{in}^3/\text{rev}$ and a volumetric efficiency of 95 percent. The transmission is sketched in Fig. 5-6.

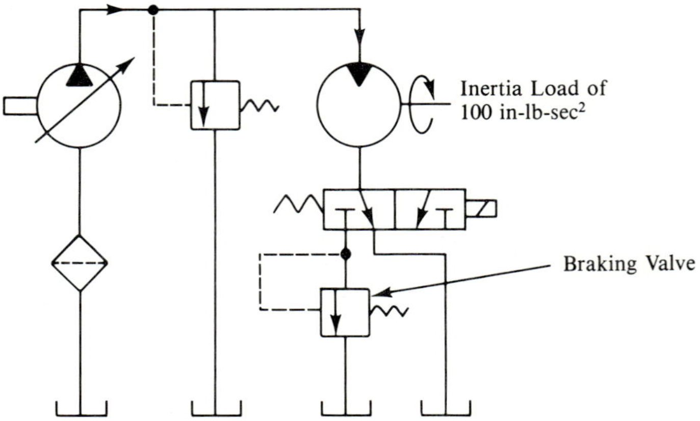

Figure 5-6. Circuit for Example 5-1.

The motor is driving an external torque load of 2000 in-lb at 1000 rpm. The flywheel effect of the load (and motor) is $100\,\text{in-lb-sec}^2$. When the external (resistive) load torque is removed, the hydraulic system is programmed so that the following events are initiated simultaneously:

- **(a)** The pump flow is programmed, by a microprocessor, to diminish linearly to zero in 10 seconds.
- **(b)** A counterbalance valve is inserted in the motor exhaust line by the two-position, three-way valve.

These events are required in order to decelerate the over-running inertia load safely. It is also necessary that cavitation be avoided in the motor inlet line during the braking process. The following items are to be determined:

- **(a)** Motor discharge rate, in gpm, just prior to the removal of the external load torque
- **(b)** Pump rpm just prior to the load removal
- **(c)** Pump outlet pressure just prior to the external load torque removal
- **(d)** Pump horsepower requirement just prior to the removal of the external load torque

(e) Setting of the pump relief valve
(f) Setting of the counterbalance (braking) valve in the motor exhaust line that will just avoid cavitation in the motor inlet line

Solution

We will neglect all frictional dissipation in fluid lines in order to illustrate how the solution is effected. Then, from Eq. (4-34), we can determine the required motor flow rate:

$$Q = \frac{3(1000)}{0.95(231)} = 13.67 \text{ gpm}$$

Since the discharge of the pump must be the same flow rate as at the motor inlet, the pump rpm is given by

$$N = \frac{13.67(231)}{0.95(1.9)} = 1749 \text{ rpm}$$

The pump outlet pressure is obtained by equating pump power to motor power. In other words,

$$\frac{P \cdot Q}{1714} = \frac{N \cdot T}{63{,}025}$$

The numerical constants are conversion factors required to relate P (psi), Q (gpm), N (rpm), and T (in-lb). Then

$$P = \left(\frac{1714}{63{,}025}\right)\left(\frac{1000}{13.67}\right)(2000) = 3979 \text{ psi}$$

It follows that the pump horsepower required will be

$$\text{HP} = \frac{3979(13.67)}{1714} = 31.73 \text{ hp}$$

Since we do not want the relief valve open under normal operating conditions (and since relief valves are generally set in increments of 500 psi), we set the pump pressure relief valve at 4500 psi.

To avoid cavitation in the inlet line to the motor while the pump is slowing down (after the resistive load is removed), we want the motor to be slowed (by the resistance of the counterbalance valve) at the same rate. Consequently, the motor deceleration rate must be

$$\alpha = \frac{\omega}{t} = \frac{2\pi(1000)}{60(10)} = 10.47 \text{ rad/sec}^2$$

The torque against which the over-running motor (now turned pump) will have to work is then

$$T = I\alpha = 100(10.47) = 1047 \text{ in-lb}$$

Then the setting ΔP of the counterbalance valve will be

$$\Delta P = \left(\frac{1714}{63{,}025}\right)\left(\frac{N}{Q}\right)T = \left(\frac{1714}{63{,}025}\right)\left(\frac{1000}{13.67}\right)(1047) = 2082 \text{ psi}$$

Thus, the counterbalance valve must provide a resistance of around 2082 psi. Note that the ratio 1000/13.67 remains constant as the motor decelerates. ■

5-1-3 Closed-Circuit Hydrostatic Transmissions

The primitive closed-circuit drive shown in Fig. 5-1 will require additional fluid power components before it can function as a practical transmission. The pump and motor are not perfect positive-displacement devices. Their volumetric efficiencies will generally be in the vicinity of 95 percent, which means that there are internal leakages in pump and motor that drain to the oil supply reservoir in the system. This lost fluid must be replaced to avoid cavitation.

One possible circuit configuration, called a replenishing network, for keeping the closed-circuit hydrostatic drive full of oil at a specified pressure is shown in Fig. 5-7. A separate, small, charging pump is provided that supplies fluid to the low-pressure side of the circuit, as required, through check valve A or B depending on the direction of pump (and motor) rotation. The level of pressure in the low-pressure side of the circuit is controlled by relief valve E. Thus, the basic datum level of pressure in the system is controlled by relief valve E, and the charge pump continuously replenishes the oil lost by internal leakage in pump and motor.

Relief valve F is provided to limit the pressure in the high-pressure side of the circuit to a safe value. It functions in the same way as the primary system relief valve does in an open circuit. Suppose the motor temporarily encounters a very large resistance. This encounter would demand a very high pressure at the pump outlet. The circuit is protected in the following way. Suppose the flow direction is as indicated in Fig. 5-7. Then relief valve F will open and the pump outlet flow will be bypassed around the motor through check valves D and A. If the motor and pump rotational directions are opposite to those indicated in Fig. 5-7, then when the motor encounters an overload, the pump discharge will again be bypassed around the motor, but now it will be through check valves C and B.

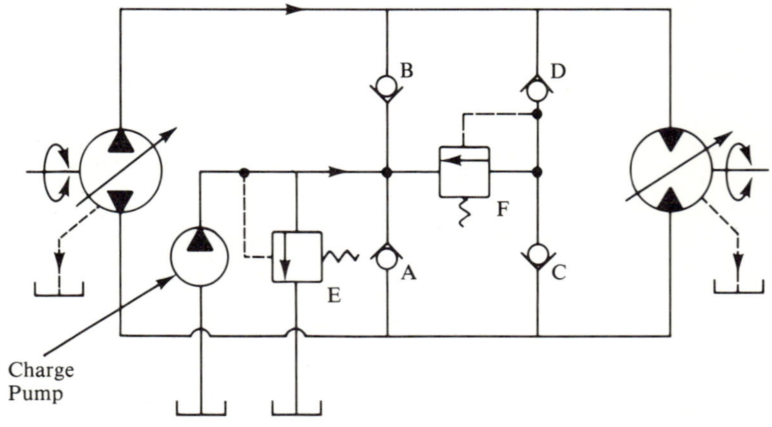

Figure 5-7. Closed-circuit replenishing network.

Relief valve F also serves as a braking valve if, and when, the motor suddenly encounters an over-running load. Suppose the external resistive load that the motor is driving suddenly disappears for the rotational and flow directions indicated in Fig. 5-7 and is replaced by an over-running inertial load. The motor now acts as a pump, and what was previously its exhaust port becomes its pressurized port. Pressurizing can open relief valve F, with the motor flow rate now circulating through check valves C and B. Correspondingly, the pump has an alternative circuit through check valves D and A is necessary. The dissipation through valve F produces braking of the over-running load (although not necessarily well controlled).

An alternative arrangement often used in closed-circuit hydrostatic drives is shown in Fig. 5-8. In this circuit, the two check valves C and D and single relief valve F of Fig. 5-7 are replaced by two, cross-port, relief valves which provide the same kind of overload protection.

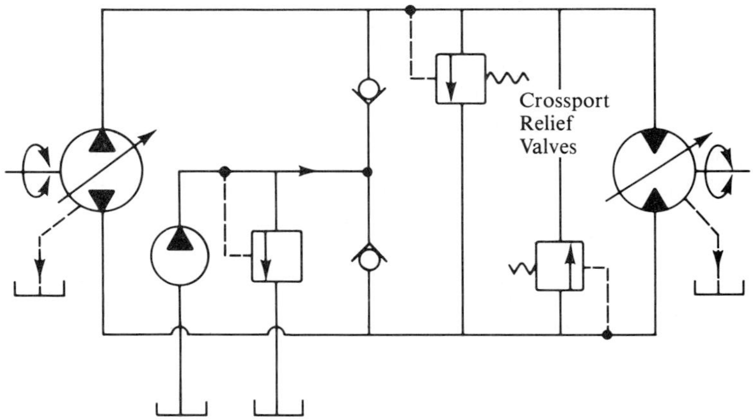

Figure 5-8. Replenishing network with cross-port relief.

5-2 CHARACTERISTICS OF HYDROSTATIC DRIVES

As we shall see, hydrostatic drives can be made to exhibit transmission properties analogous to mechanical drive shafts, gearboxes, and clutches (or combinations of these types of couplings). The desirability of a hydrostatic drive, as compared to these mechanical transmissions, lies in its inherent flexibility and controllability.

To expose the properties of hydrostatic transmissions analytically, it will be necessary to expand on the characteristics of the pumps and motors (discussed in Chapter 4) from which a hydrostatic drive is synthesized. In doing this, we shall limit our discussion to hydrostatic transmissions of the closed-circuit type such as that sketched in Fig. 5-8. The closed-circuit transmission is represented in skeletal form in Fig. 5-9, along with its black box equivalent.

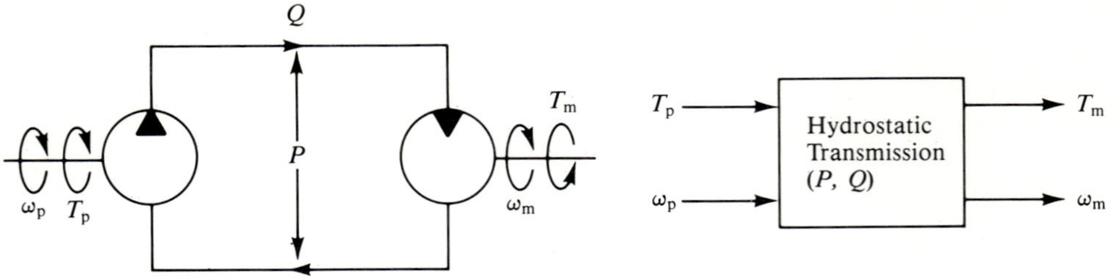

Figure 5-9. Skeletal hydrostatic transmission.

Since the hydrostatic drive is a transmitter simply of torque and angular speed, its characteristics will be the various possible relationships between the input variables (pump) T_p, ω_p and the output variables (motor) T_m, ω_m. Quantities Q and P will be regarded as parameters.

There are two idealized models that we can use in deriving the desired characteristics. In the simpler model, we neglect internal leakage and internal friction in both pump and motor. This model is of only limited value because it cannot produce a useful relationship for the motor torque T_m. But it is useful in classifying hydrostatic drives. In the somewhat more complex model, both internal leakage and internal friction in the hydraulic dynamos (pump and motor) are represented as linear functions. This model is valuable because it exposes how the transmission actually works.

5-2-1 Characteristics Neglecting Leakage and Friction

The ideal relationships of the hydraulic dynamo characteristics, when leakage and friction are neglected, were presented in Chapter 4 as Eqs. (4-22) and (4-33). These relationships can be rewritten as

$$Q_p = k_p \omega_p \qquad T_p = k_p P_p \qquad \text{(pump)} \tag{5-1}$$

$$Q_m = k_m \omega_m \qquad T_m = k_m P_m \qquad \text{(motor)} \tag{5-2}$$

where

$$k_p = \left(\frac{An\varepsilon}{\pi}\right)_p \qquad k_m = \left(\frac{An\varepsilon}{\pi}\right)_m \tag{5-3}$$

Since we are neglecting internal leakage and friction, there is no need to distinguish between Q_p and Q_m or between P_p and P_m. Then, by eliminating Q and P between Eqs. (5-1) and (5-2), we find that

$$\frac{\omega_m}{\omega_p} = \frac{k_p}{k_m} \qquad \frac{T_m}{T_p} = \frac{k_m}{k_p} \tag{5-4}$$

It is a simple matter to show that the geometric constant k is related to the stroke parameter S and maximum displacement (per radian) D according to (see Section 4-1-5)

$$k = \frac{An\varepsilon}{\pi} = S \cdot D \qquad D = \frac{q_{\max}}{2\pi} \tag{5-5}$$

Then Eqs. (5-4) can be written as

$$\frac{\omega_m}{\omega_p} = \left(\frac{S_p}{S_m}\right)\left(\frac{D_p}{D_m}\right) \qquad \frac{T_m}{T_p} = \left(\frac{S_m}{S_p}\right)\left(\frac{D_m}{D_p}\right) \tag{5-6}$$

We define the ratio r of the transmission to be the motor displacement D_m divided by the pump displacement D_p. Then

$$\frac{D_m}{D_p} = r \qquad \frac{\omega_m}{\omega_p} = \left(\frac{S_p}{S_m}\right)\left(\frac{1}{r}\right) \qquad \frac{T_m}{T_p} = \left(\frac{S_m}{S_p}\right)(r) \tag{5-7}$$

Recall that for a fixed-displacement hydraulic dynamo, the stroke S is not variable. It is fixed at the value $S = 1$.

Pump and Motor Both of Fixed Displacement. If the hydrostatic drive is one in which both the pump and motor displacements are fixed at their values D_p and D_m, then the strokes S_p and S_m are unity. For such a drive, Eqs. (5-7) reduce to

$$\omega_p = r\omega_m \qquad T_m = rT_p \tag{5-8}$$

These equations show that this particular hydrostatic transmission has precisely the same properties as a mechanical gearbox of ratio r. It will be a positive drive (with no slip) because we have assumed internal leakage to be absent.

For the case $r = 1$, the hydrostatic drive is nothing more than a simple drive shaft insofar as its properties are concerned. Note, however, that this is a much more flexible drive shaft than the mechanical drive shaft because the pump and motor can be connected by flexible hoses which can bend around corners.

Motor Displacement Fixed, Pump Displacement Variable. If the motor displacement is held fixed at the value D_m, then $S_m = 1$ and Eqs. (5-7) reduce to

$$\omega_m = \left(\frac{S_p}{r}\right)\omega_p \qquad T_m = \left(\frac{r}{S_p}\right)T_p \tag{5-9}$$

Imagine that the pump speed ω_p is held fixed. Then the left of Eqs. (5-9) shows that the motor speed ω_m can be controlled linearly between zero and a maximum value ω_p/r as the pump stroke is varied between zero and unity. This relationship is shown in dimensionless form in Fig. 5-10.

This hydrostatic drive has the same properties as a mechanical transmission consisting of a gearbox plus clutch. The clutch is engaged by stroking the pump away from $S_p = 0$. But note that there is no dissipation of power in the hydraulic arrangement as there is in a simple mechanical clutch.

With the motor displacement fixed, the constant k_m is fixed and Eq. (5-2) shows that the motor torque is proportional to the hydrostatic head P_m. If P_m is

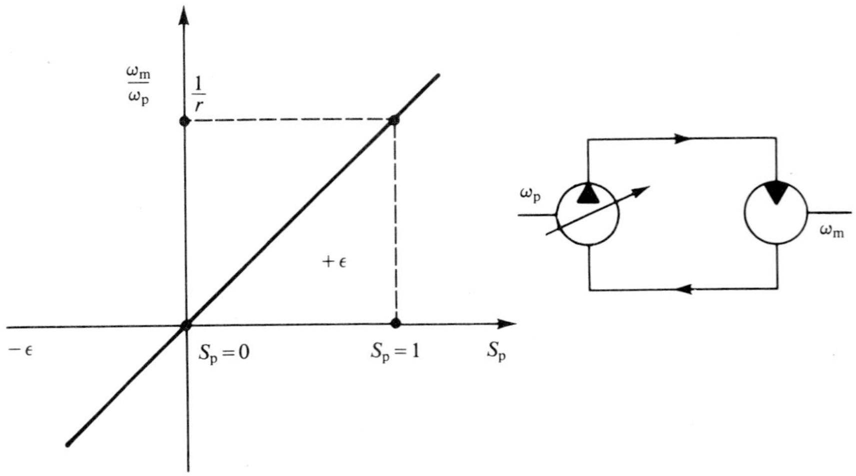

Figure 5-10. Constant-torque transmission (no leakage).

held fixed, as is often assumed, then the motor torque T_m will also be fixed. For this reason, the hydrostatic transmission consisting of a fixed-displacement motor and variable-displacement pump is often called a constant-torque transmission. This kind of language can be misleading and should be regarded merely as a convenient way of classifying hydrostatic drives.

The correct way of viewing hydrostatic transmissions is to recognize that the load torque placed on the hydraulic motor will dictate what the system pressure P_m shall be. If the load torque remains constant (say, as in the case of a non-accelerating lifting hoist), then the transmission pressure will remain fixed. But it will be the constancy of the load torque, in this instance, rather than the transmission itself, that will dictate that this be a constant-torque system.

Ordinarily, we would think of a lifting hoist system as being a constant-torque system because its external torque is fixed. But, it must be recognized that during the start-up, or braking, there will be inertia torques in addition to the constant steady-state torque. Therefore, the lifting hoist system cannot be viewed as a constant-torque system during transient operation. During acceleration, the hydraulic motor torque (and hence hydrostatic drive pressure) will exceed its steady-state value unless the inertia torque is negligible.

In systems having resistive loads, the steady-state load torque will not be constant, as in a hoist, but will ascend with speed. In this case, a constant-torque drive will be possible, but it will require a control scheme for maintaining the transmission pressure constant.

There are basically two ways of controlling transmission pressure constant during acceleration or deceleration periods for a resistive load system. The first scheme uses a variable-displacement pump with its stroking mechanism adjusted to maintain transmission pressure constant. The adjustment is made with a feedback link that ties system pressure to pump displacement, as shown in Fig. 5-11. In this fashion, the motor torque can be held constant, say, during the

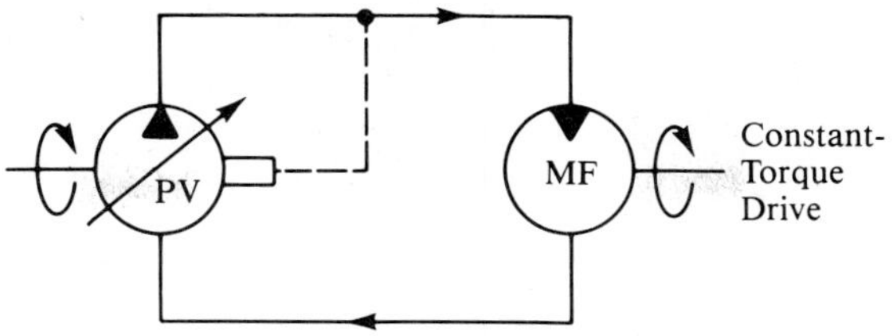

Figure 5-11. Pressure-stroked variable-displacement pump.

start-up period when a resistive load is being accelerated to its operating point value (see Fig. 5-12).

The second method of maintaining system pressure constant is to place a suitably set pressure relief valve in parallel with the pump so that some of the pump output will be bypassed around the motor in order to keep system pressure constant (see Fig. 5-13). In this scheme, the pump is of fixed displacement, and

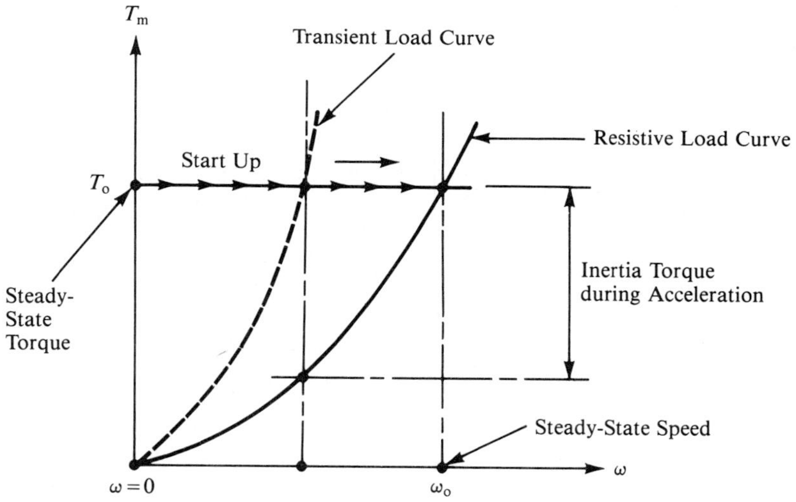

Figure 5-12. Constant-torque system for a resistive load.

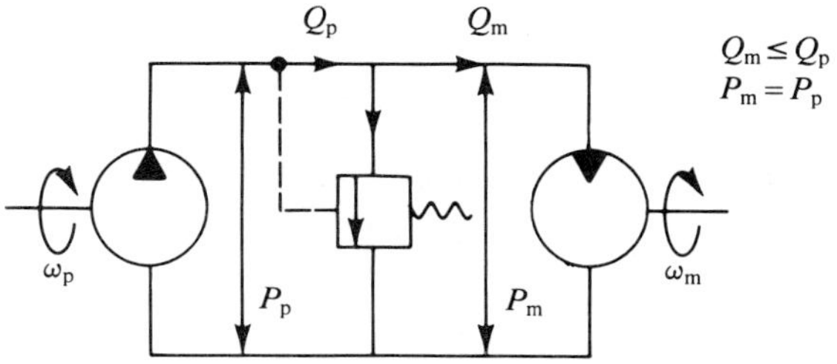

Figure 5-13. Bypass type constant-torque control.

the volumetric flow that the motor receives is varied because some of the pump output bypasses the motor. During start-up, when the motor speed is zero, all of the pump output passes through the relief valve. As the motor accelerates, less and less fluid is bypassed until, at steady-state operating speed, all of the pump output passes through the motor. The system pressure, and hence motor torque, remains constant as dictated by the pressure relief valve setting.

Suppose the pump is designed in such a way that it is capable of producing negative eccentricity as well as positive eccentricity ε. Then, if the pump is stroked from maximum displacement (with positive ε) toward, and through, zero to maximum displacement (with negative ε; see Fig. 5-10, quadrant III), its outlet and inlet ports will become reversed and the pump will become capable of driving its motor in either direction without changing the pump rotational direction. In other words, this type of transmission can provide reverse in addition to clutching.

Finally, we note that if the ratio r is unity, the constant-torque hydrostatic drive is analogous to a simple mechanical clutch.

EXAMPLE 5-2

A constant-torque hydrostatic drive is to be used to drive a resistive load whose load characteristic is

$$T_\ell = 3.5N_\ell \text{ in-lb} \qquad (N_\ell \sim \text{rpm})$$

A variable-displacement pump is to be used which is driven at 1750 rpm and whose discharge at full stroke is 50 gpm.

The motor is of fixed displacement and is to drive the load at a speed of 1000 rpm in the steady state. Neglect both leakage and friction losses.

(a) Determine the steady-state transmission pressure.

(b) Determine the pump maximum displacement and the motor fixed displacement.

(c) Assuming the motor is to be started by stroking the pump from $S_p = 0$ to $S_p = 1$, at constant transmission pressure, determine the time required to start it if the motor plus load rotational inertia is 25 slug-ft^2.

Solution

The steady-state load (and motor) torque is determined from

$$T_o = 3.5(1000) = 3500 \text{ in-lb}$$

Then the steady-state horsepower is given by

$$\text{HP} = \frac{2\pi(1000)(3500/12)}{33{,}000} = 55.5 \text{ hp}$$

Since all losses are being neglected, the hydraulic and mechanical powers are equal. Therefore,

$$\frac{P \cdot Q}{1714} = \frac{T \cdot N}{63{,}025}$$

The hydrostatic transmission pressure in the steady state will then be

$$P = \frac{(1714)(3500)(1000)}{(50)(63{,}025)} = 1904 \text{ psi}$$

The pump discharge at full stroke is

$$Q = 50 \text{ gpm} = 11{,}550.8 \text{ in}^3/\text{min}$$

Since the pump speed is 1750 rpm, the pump maximum displacement is

$$D_p = \frac{11{,}550.8}{2\pi(1750)} = 1.0505 \text{ in}^3/\text{rad}$$

Ratio r of this constant-torque transmission is given by Eq. (5-9). At full stroke, we find

$$\omega_m = \left(\frac{1}{r}\right)\omega_p$$

$$r = \frac{1750}{1000} = 1.75$$

Then the motor displacement is given by

$$D_m = rD_p = 1.75(1.0505) = 1.838 \text{ in}^3/\text{rad}$$

The relationship between motor and load torques during the starting period is

$$T_m - T_\ell = I\frac{d\omega_m}{dt}$$

The motor torque T_m will be held constant at 3500 in-lb, because the system pressure is to be held constant at 1904 psi (either by a pressure relief valve or by a pump-stroking mechanism). Then the torque equation becomes

$$\frac{3500 - 3.5N_m}{12} = 25\left(\frac{2\pi}{60}\right)\frac{dN_m}{dt}$$

where $N_m = N_\ell$ is speed in rpm and t is time in seconds. Integrating this equation, we obtain

$$\left[\frac{\ln|3500 - 3.5N_m|}{-3.5}\right]_0^{1000} = \frac{60t}{(12)(25)(2\pi)}$$

Solving, we find that the time t required to accelerate the load from rest to 1000 rpm is $t = 73.25$ seconds. The system properties are sketched in Fig. 5-14. ■

EXAMPLE 5-3

Re-examine Example 5-2 for a hoisting load that represents a constant external load of 3500 in-lb. Assume the transmission is a constant-torque system with the same properties as in Example 5-2. The load is to be accelerated from $\omega_m = 0$ to $\omega_m = 1000$ rpm in 73.25 seconds by stroking the variable-displacement pump from $S_p = 0$ to $S_p = 1$ such that the acceleration will be constant. There is no pressure relief valve in the system.

(a) Determine the maximum and minimum motor torques during start-up.
(b) Determine the maximum and minimum pump torques during start-up.
(c) Determine the maximum and minimum transmission pressures during start-up.
(d) Is this a constant-torque drive?

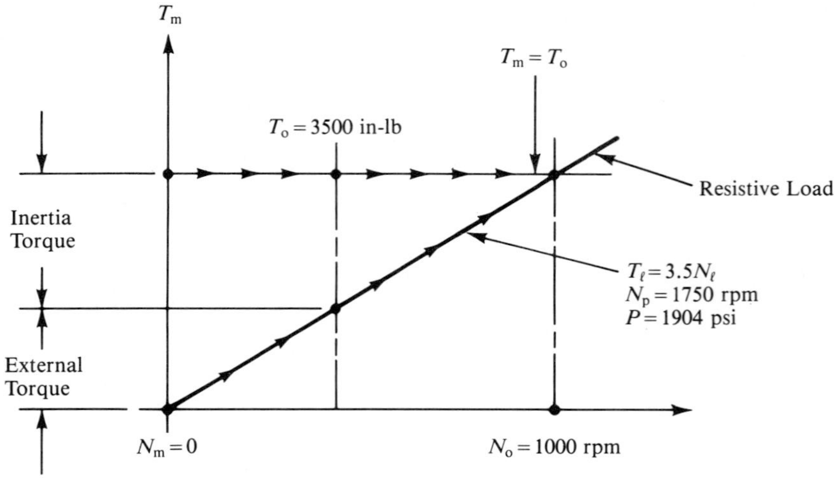

Figure 5-14. Conditions for Example 5-2.

Solution

During start-up, the torque equation is

$$T_m - T_\ell = I\,\frac{d\omega_m}{dt}$$

For the conditions stated, the maximum motor torque is given by

$$\frac{T_m - 3500}{12} = 25\left(\frac{2\pi}{60}\right)\left(\frac{1000}{73.25}\right)$$

$$T_m = 3928 \text{ in-lb}$$

This motor torque remains constant during the start-up period and then drops instantaneously to the value 3500 in-lb when the speed reaches 1000 rpm. The discontinuity is caused by the assumption of constant acceleration during start-up which suddenly goes to zero acceleration when the load is fully accelerated to 1000 rpm.

The pump torque is given by Eq. (5-9). Ratio r is 1.75, so we have the relationship

$$T_p = \frac{S_p T_m}{1.75}$$

The minimum value of pump torque is seen to be $T_p = 0$ when the stroke is $S_p = 0$. The pump torque increases to the maximum value of

$$T_p = \frac{(1)(3928)}{1.75} = 2244.6 \text{ in-lb}$$

When the load is fully accelerated, this value will drop instantaneously to

$$T_p = \frac{(1)(3500)}{1.75} = 2000 \text{ in-lb}$$

The transmission pressure corresponding to the motor torque of 3928 in-lb is given by Eq. (5-2), with $k_m = D_m = 1.838\ \text{in}^3/\text{rad}$. Then

$$P_m = \frac{T_m}{k_m} = \frac{3928}{1.838} = 2137\ \text{psi}$$

This system pressure will drop instantaneously to the value 1904 psi when the speed of 1000 rpm is reached.

We see that although this is called a constant-torque transmission, the motor torque actually varies between the values 3928 in-lb and 3500 in-lb. The system properties are sketched in Fig. 5-15. ■

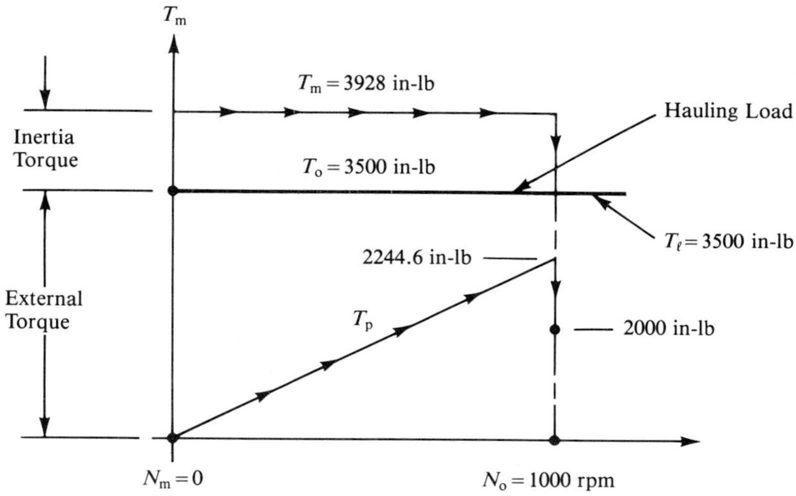

Figure 5-15. Conditions for Example 5-3.

Pump Displacement Fixed, Motor Displacement Variable. If the pump displacement is held fixed at the value D_p, then $S_p = 1$ and Eqs. (5-7) reduce to

$$\omega_m = \left(\frac{1}{rS_m}\right)\omega_p \qquad T_m = (rS_m)T_p \tag{5-10}$$

Again imagine that the pump speed ω_p is held fixed. Then the minimum motor speed, as dictated by $S_m = 1$, will be ω_p/r. As the motor stroke S_m is reduced toward zero, we see that very high motor speeds can be obtained and are limited only by the centrifugal stresses developed in the motor by high rotational speeds. This relationship is shown in dimensionless form in Fig. 5-16.

With pump displacement fixed, the constant k_p is fixed, and Eq. (5-1) shows that the pump torque will be fixed if the transmission pressure P is held fixed. With both ω_p and T_p fixed, the pump horsepower is fixed and, for this reason, this transmission is called a constant-horsepower drive. As we discussed earlier, this is simply a convenient way of classifying hydrostatic drives since P is regarded as a parameter of the drive rather than as a variable.

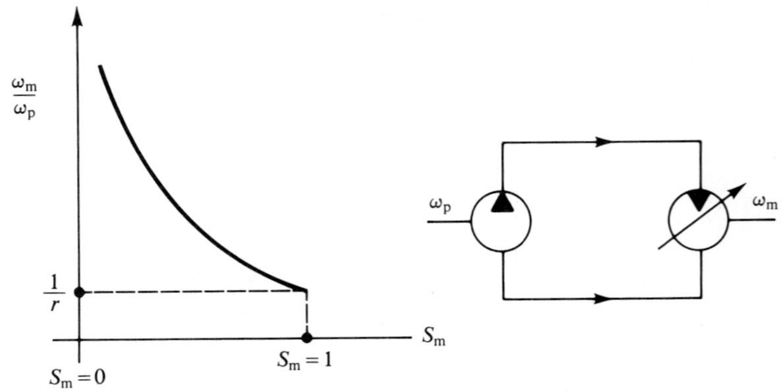

Figure 5-16. Constant-horsepower transmission.

Generally, the constant-horsepower drive is designed with $r > 1$ so that the motor minimum speed is slower than the pump speed.

EXAMPLE 5-4

A constant-horsepower hydrostatic transmission is to provide 25 hp for a motor speed range of 900 to 1800 rpm. Assume that the motor is fully stroked at 900 rpm and that the transmission pressure is held fixed at 2000 psi. Displacements D_m and D_p are to be taken equal. Neglect leakage and friction and determine

(a) The pump (and hence motor) displacement D_p
(b) The motor torques at the motor speeds of 900 and 1800 rpm
(c) The pump torque

Solution

Ratio r of the transmission is $r = 1$ because $D_m = D_p$. Then Eq. (5-10) shows that the pump speed will be $N_p = 900$ rpm, because $S_m = 1$ when $N_m = 900$.

The horsepower relationships are

$$\text{HP} = \frac{P \cdot Q}{1714} = \frac{T \cdot N}{63{,}025}$$

Therefore, the pump discharge will be

$$Q_p = \frac{25(1714)}{2000} = 21.425 \text{ gpm} = 4949.5 \text{ in}^3/\text{min}$$

The pump displacement D_p is then

$$D_p = \frac{Q_p}{2\pi N_p} = \frac{4949.5}{2\pi(900)} = 0.875 \text{ in}^3/\text{rad}$$

The motor maximum displacement is then also $0.875 \text{ in}^3/\text{rad}$.

To obtain the higher speed of 1800 rpm, the motor must be stroked to $S_m = 1/2$ (see Eq. [5-10]). The pump torque is obtained from

$$T_p = \frac{63{,}025}{900}(25) = 1750 \text{ in-lb}$$

Consequently, the motor torque will be 1750 in-lb when the motor speed is 900 rpm and 875 in-lb when the motor speed is 1800 rpm. The pump torque remains fixed. ∎

Pump and Motor Both of Variable Displacement. The properties of both the constant-torque drive and the constant-horsepower drive can be obtained in a single hydrostatic transmission by synthesizing it from a variable-displacement pump driving a variable-displacement motor. When the motor displacement is held fixed, the drive will be constant-torque, and when the pump displacement is held fixed, the drive will be constant-horsepower. The properties of this transmission are usually presented in the form of the torque-speed curve shown in Fig. 5-17.

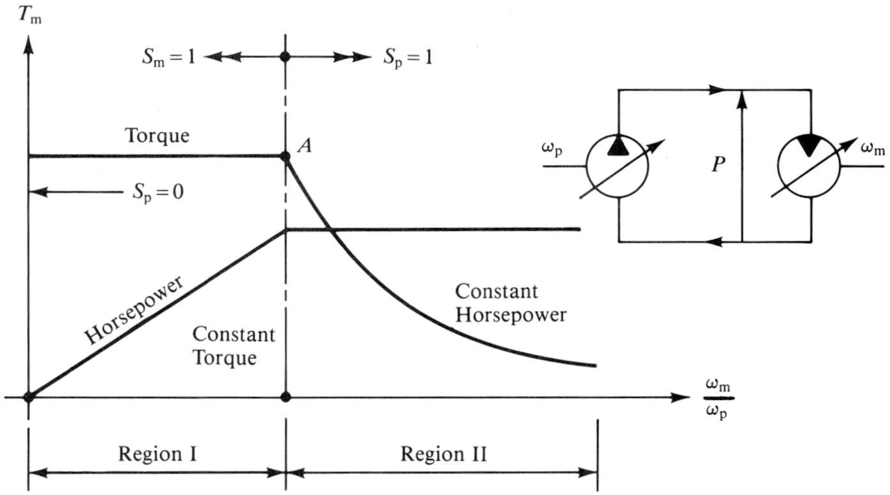

Figure 5-17. Traction drive characteristics.

In region I, the motor stroke is held at the maximum value $S_m = 1$ while the pump stroke increases from $S_p = 0$ to $S_p = 1$. This is the constant-torque region (if P is fixed), in which both speed ω_m and horsepower increase linearly with S_p. At the breakpoint A, where pump displacement is a maximum ($S_p = 1$), the motor displacement must be decreased toward zero from the value $S_m = 1$ in order to obtain further motor speed increases. This is region II, where horsepower remains constant but motor torque diminishes hyperbolically as the motor speed is increased. The total characteristic shown in Fig. 5-17 is called the traction drive characteristic. We shall encounter this type of torque-speed characteristic again in Chapter 8 where it is exposed for an analogous DC electrical drive.

5-2-2 Characteristics with Linear Leakage and Friction

Suppose we have a hydrostatic transmission that is being driven at constant pump speed ω_p. Then Eq. (5-4) predicts that the motor speed ω_m depends only on the pump and motor geometric constants k_p and k_m. Consequently, Eq. (5-4) predicts that ω_m will be fixed at the value ω_p if both pump and motor geometries are fixed and equal. This is the case of the rigid drive shaft we discussed earlier.

In particular, Eq. (5-4) predicts that the motor speed ω_m will be independent of the torque being transmitted, which is unrealistic. The transmission, in fact, is not perfectly rigid because of three factors. First of all, the oil is not perfectly incompressible; second, there is internal leakage in both motor and pump; and, third, there is internal friction in both motor and pump. What will actually happen, as load torque is increased, is that the motor speed will diminish even though the pump speed is held fixed. The reason is that the transmission pressure will increase with load torque which will cause increased leakage and oil compression. With internal leakage, all of the pump displacement will not appear as motor displacement because some of it will be lost due to leakage.

In the following analyses, we will still assume the hydraulic fluid to be incompressible, but the internal leakage and friction will be represented linearly by the relationships

$$Q_p = k_p \omega_p - \lambda_p P_p \qquad T_p = k_p P_p + f_p \omega_p \tag{5-11}$$

$$Q_m = k_m \omega_m + \lambda_m P_m \qquad T_m = k_m P_m - f_m \omega_m \tag{5-12}$$

Equations (5-11) and (5-12) state that the internal leakage in pump and motor is proportional to pressure P. The proportionality constants are λ_p and λ_m. Equation (5-11) for the pump discharge Q_p states that the actual pump discharge Q_p is less than the ideal value $k_p\omega_p$ by the amount of internal leakage $\lambda_p P_p$. Equation (5-12) for the motor volumetric flow rate Q_m states that the flow Q_m into the motor exceeds the theoretical value $k_m\omega_m$ by the amount of internal leakage $\lambda_m P_m$.

In the torque relationships, the torque lost through friction is represented by $f_p\omega_p$ for the pump and by $f_m\omega_m$ for the motor. The actual torque T_p required to drive the pump exceeds the theoretical value $k_p P_p$ by the amount lost in friction, which is proportional to pump speed ω_p. Similarly, the torque T_m appearing at the motor output shaft is less than the theoretical amount $k_m P_m$ by the amount lost through friction.

If there are no leaks in the piping connecting the pump and motor and if there is negligible pressure drop in this piping, then we have the conditions

$$Q_p = Q_m = Q \qquad P_p = P_m = P \tag{5-13}$$

Then the pressure can be eliminated between the two equations for Q to obtain

$$Q = \frac{k_p \omega_p + L k_m \omega_m}{1 + L} \tag{5-14}$$

where

$$L = \frac{\lambda_p}{\lambda_m} \tag{5-15}$$

And if Q is eliminated from these same two equations, an expression can be obtained for P, namely,

$$P = \frac{k_p \omega_p - k_m \omega_m}{\lambda_m (1 + L)} \tag{5-16}$$

Suppose the motor is stalled ($\omega_m = 0$). Then Eqs. (5-14) and (5-16) become

$$Q = \frac{k_p \omega_p}{1 + L} \tag{5-17}$$

$$P = \frac{k_p \omega_p}{\lambda_m (1 + L)} \tag{5-18}$$

Equation (5-17) shows that there will be flow in the system despite the fact that the motor is stalled. This flow is the leakage through the motor $\lambda_m P$. Equation (5-18) shows that when the motor is stalled, the head P developed in the transmission is proportional to the pump speed ω_p.

More generally, Eq. (5-16) shows that the head developed in the transmission depends on the weighted difference between pump speed and motor speed. This equation can be written in a different form which provides a comparison with Eq. (5-4), namely,

$$\frac{\omega_m}{\omega_p} = \frac{k_p}{k_m} - \frac{\lambda_m (1 + L) P}{k_m \omega_p} \tag{5-19}$$

Equations (5-11) and (5-12) show that the total leakage Q_l in pump and motor is $\lambda_m (1 + L) P$. Therefore,

$$\frac{\omega_m}{\omega_p} = \frac{k_p}{k_m} - \frac{Q_l}{k_m \omega_p} \tag{5-20}$$

For a constant-torque drive, Eq. (5-20) can be simplified to

$$\frac{\omega_m}{\omega_p} = \frac{S_p}{r} - \frac{Q_l}{k_m \omega_p} \tag{5-21}$$

This equation is plotted in Fig. 5-18.

The curve shown dashed in Fig. 5-18 is the same as the one shown in Fig. 5-10, where leakage was neglected. With leakage, we see that the motor shaft will not turn until the pump (having positive ε) is stroked by the amount δ. For a pump having negative ε (stroked past center), there will be a similar deadband to the left of the origin, in which the pump will have to be stroked by the amount $-\delta$ before the motor will turn. This deadband is due to internal leakage in pump and motor and must be recognized when designing the controls for a system using such a transmission. As Eq. (5-19) shows, displacement δ is proportional to transmission pressure and, hence, load torque. The greater the load torque, the greater will be the deadband.

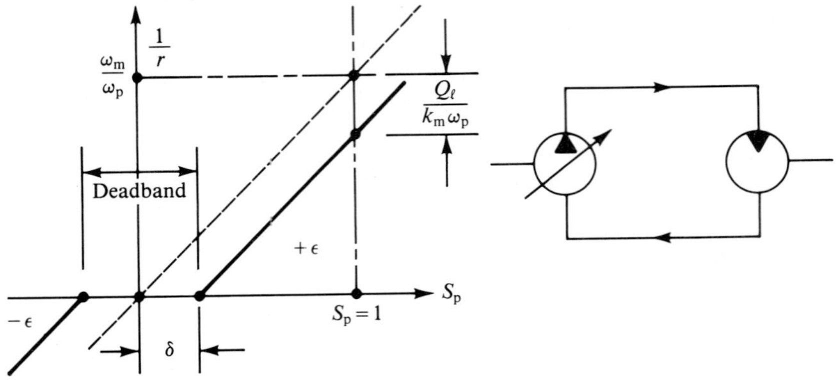

Figure 5-18. Constant-torque transmission (with leakage deadband).

All of the above analysis is based on fluid flow continuity considerations along with the fact that leakage is linearly proportional to transmission head P. In order to relate motor speed ω_m to motor torque T_m, we will have to introduce the torque equation (5-12). If Eq. (5-12) is solved for P and the result is substituted into Eq. (5-19), we obtain

$$T_m = A\frac{k_p}{k_m}\omega_p - (A + f_m)\omega_m \tag{5-22}$$

where

$$A = \frac{k_m^2}{\lambda_m(1+L)} \tag{5-23}$$

Equation (5-22) shows that the hydrostatic transmission has its own characteristic output torque-speed curve when leakage is included. The pump speed ω_p can be regarded as a parameter for this drive that is generally held constant. We see that when leakage is included, the hydrostatic drive is tantamount to a prime mover that has energy flowing in at the pump and energy flowing out at the motor.

EXAMPLE 5-5

A constant-torque hydrostatic drive has the following parameters:

Leakage constants: $\lambda_m = \lambda_p = 0.002$ gpm/psi

Ratio: $r = \dfrac{D_m}{D_p} = 1.75$

Motor displacement: $D_m = 1\ \text{in}^3/\text{rad}$

Motor friction: negligible; $f_m = 0$

Pump speed: $N_p = 1750$ rpm

The drive is directly coupled to a resistive load whose characteristic is

$$T_\ell = 3.5N_\ell \text{ in-lb} \qquad (N_1 \sim \text{rpm})$$

Determine

(a) The equilibrium (steady-state) operating speed of the motor (and load) when the pump is fully stroked
(b) The transmission pressure at the equilibrium operating point with the pump fully stroked
(c) The no-load speed of the motor when the pump is fully stroked

Solution

The constant A of Eq. (5-23) is

$$A = \frac{k_m^2}{\lambda_m(1+L)} = \frac{D_m^2}{\lambda_m + \lambda_p} = \frac{(1)^2}{0.004}\ \frac{\text{in}^6}{\text{gpm/psi}}$$

In a more useful set of units,

$$A = \frac{7.48(60)(1)^2}{0.004(1728)} = 64.93\ \text{in-lb-sec}$$

For a constant-torque drive, $S_m = 1$ and $k_p/k_m = S_p/r$. Consequently, Eq. (5-22) becomes

$$T_m = 64.93\left(\frac{S_p}{1.75}\,\omega_p - \omega_m\right)$$

In a set of units consistent with the above equation, the load is

$$T_\ell = 33.42\omega_\ell$$

For direct coupling, we have $\omega_\ell = \omega_m$, $T_\ell = T_m$. Then the following equation is obtained for the motor speed when the pump is fully stroked ($S_p = 1$):

$$64.93\left(\frac{\omega_p}{1.75} - \omega_m\right) = 33.42\omega_m$$

Solving this relationship for a pump speed of 1750 rpm ($\omega_p = 183.26$ rad/sec), we find that

$$\omega_m = 69.14\ \text{rad/sec} \qquad N_m = 660.2\ \text{rpm}$$

The equilibrium operating torque is then

$$T_m = 3.5(660.2) = 2310.7\ \text{in-lb}$$

And since friction has been neglected in the motor,

$$P = \frac{T_m}{k_m} = \frac{T_m}{D_m} = \frac{2310.7}{1} = 2310.7\ \text{psi}$$

When the load torque is zero, Eq. (5-22) reduces to

$$N_m = \frac{N_p}{1.75} = \frac{1750}{1.75} = 1000\ \text{rpm}$$

We see that as this transmission is loaded from zero to 2310.7 in-lb of torque, its output shaft slows down from 1000 rpm to 660.2 rpm. ■

EXAMPLE 5-6

A constant-torque hydrostatic drive with ratio $r = 1$ is used to drive an elevator by means of a cable drum 36 inches in diameter and a single wire rope. The elevator and load weigh 3000 lb and are counterweighted by a 2000 lb weight. The motor displacement is 6 in^3/rad and the pump is driven at 875 rpm. During acceleration and deceleration between stops, the acceleration is constant at ± 6 ft/sec^2. When running at constant speed, the elevator moves at 400 fpm. The distance between stops is 20 feet. Neglect internal friction but assume the leakage constants λ_m and λ_p are each 0.002.

- **(a)** What distance is required to accelerate the elevator from rest to its steady 400 fpm speed?
- **(b)** What time is required to accomplish part (a)?
- **(c)** What is the motor speed when the elevator moves at 400 fpm?
- **(d)** What transmission pressure is required to just hold the elevator static when it is not braked?
- **(e)** What pump stroke S_p is required to just hold the elevator static or to move it at constant speed?
- **(f)** What will the transmission pressure be during the acceleration period?
- **(g)** Write the equation for the stroke S_p during the acceleration period in terms of ω_m/ω_p.
- **(h)** Write the equation for the stroke S_p during the acceleration period in terms of position s from the starting position $s = 0$. This equation will be needed if the stroke is to be controlled by a feedback from elevator position. Plot the equation.

Solution

The acceleration can be written in the form

$$d\left(\frac{V^2}{2}\right) = a\,ds$$

Integrating with constant acceleration, we find that

$$\frac{V^2}{2} = as$$

Now

$$V = \frac{400}{60} = 6.667 \text{ fps} \qquad \frac{V^2}{2} = 22.22 \text{ ft}^2/\text{sec}^2$$

Then

$$s = \frac{22.222}{6} = 3.7 \text{ feet}$$

The time required is given by

$$t = \frac{V}{a} = \frac{6.667}{6} = 1.111 \text{ seconds}$$

The motor speed when the elevator moves at 400 fpm is given by

$$\omega = \frac{V}{R} = \frac{400}{1.5} = 266.67 \text{ rad/min} = 42.44 \text{ rpm}$$

In order to just hold the elevator static (or to move it at constant 400 fpm speed), the transmission pressure must be (neglecting motor friction)

$$P = \frac{T_m}{k_m} = \frac{T_m}{D_m} = \frac{1.5(1000)(12)}{6} = 3000 \text{ psi}$$

Then the total leakage in pump and motor when the elevator is static is

$$Q_l = (\lambda_m + \lambda_p)P = 0.004(3000) = 12 \text{ gpm} = 2772 \text{ in}^3/\text{min}$$

The leakage constant is therefore

$$\frac{Q_l}{k_m \omega_p} = \frac{2772}{6(2\pi)(875)} = 0.084$$

The pump stroke to just hold the elevator static will then have to be given by

$$\frac{\omega_m}{\omega_p} = 0 = S_p - 0.084$$

or

$$S_p = 0.084$$

During the acceleration period, the motor acceleration will be

$$\alpha = \frac{a}{R} = \frac{6}{1.5} = 4 \text{ rad/sec}^2$$

The angular inertia is given by

$$I = mR^2 = \frac{1000}{32.2}(1.5)^2 = 69.875 \text{ slug-ft}^2$$

Then the inertia torque is

$$I\alpha = 69.875(4) = 279.5 \text{ ft-lb} = 3354 \text{ in-lb}$$

And the total motor torque during acceleration must be

$$T_m = 1.5(1000)(12) + 3354 = 21{,}354 \text{ in-lb}$$

The corresponding transmission pressure is

$$P = \frac{21{,}354}{6} = 3559 \text{ psi}$$

Then the leakage constant becomes

$$\frac{Q_l}{k_m \omega_p} = \frac{(0.004)(3559)(1728)}{6(2\pi)(875)(7.48)} = 0.0997$$

Therefore, the pump stroke during the acceleration period must be given by

$$S_p = \frac{\omega_m}{\omega_p} + 0.0997$$

But during the acceleration period,

$$\omega_m = \alpha t = 4t \text{ rad/sec}$$

or

$$\omega_m = \frac{4(60)t}{2} = 38.197t \text{ rpm} \qquad (t \sim \text{seconds})$$

With $\omega_p = 875$ rpm, we see that

$$S_p = \frac{38.197t}{875} + 0.0997 = 0.04356t + 0.0997$$

But the relationship between elevator position and time is

$$s = \frac{1}{2}at^2$$

Then

$$s = \frac{1}{2}(6)t^2 = 3t^2$$

or

$$t = \frac{s^{1/2}}{1.732}$$

Then

$$S_p = 0.0251s^{1/2} + 0.0997$$

The stroke profile, as the elevator is accelerated through the distance of 3.7 feet, is sketched in Fig. 5-19. ■

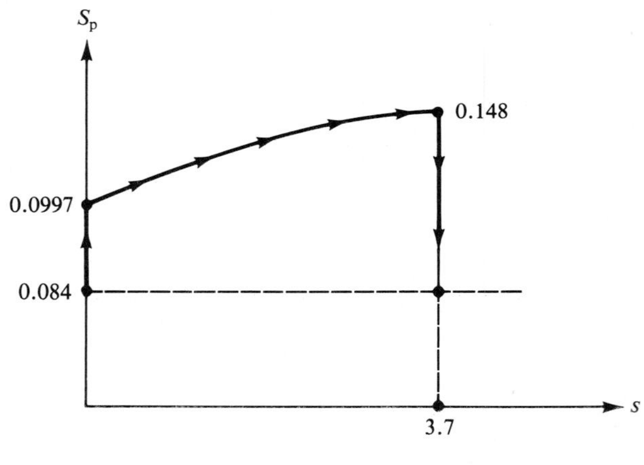

Figure 5-19. Stroke profile for Example 5-6.

5-3 INTEGRATED HYDROSTATIC DRIVES

In the larger horsepower sizes, the most common type of hydrostatic drive is one synthesized from off-the-shelf pumps, motors, and valves connected together with high-pressure hoses or tubing. However, in the lower horsepower capacities (below about 50 hp), these fluid drives do not compete well with their electrical or mechanical counterparts.

To be competitive in the lower power range, manufacturers of hydrostatic drives have developed packaged drives in which pump, motor, and valving are designed and manufactured to be integrated within a single housing enclosure. These integrated hydrostatic drives resemble gearboxes because only the mounting surface(s) and input and output shafts appear externally.

Integrated hydrostatic drives provide all of the features of conventional hydrostatic drives. In addition, they are compact and leak-free. Their advantages, compared to analogous electrical and mechanical drives, are a broad range of speed reduction, high starting torque, and stepless operation. They are particularly well suited to mobile drive applications.

Most integrated hydrostatic drives are of the constant-torque type. That is, they incorporate a variable-displacement pump with a fixed-displacement motor. The hydraulic dynamos may be of the vane type or of either the radial-piston or axial-piston type. Speed ratios are available over a wide range. Typically, the speed range is around 40:1 with a 1750 rpm input speed. The slip in these transmissions depends on torque output and is of the same order of magnitude as in an induction (electric) motor (see Chapter 8).

A typical design is shown in Fig. 5-20. This transmission incorporates a fixed-displacement (MF) radial-piston motor, a variable-displacement (PV) radial-piston pump, valves, plumbing, and a pump controller, all within the same housing. This arrangement minimizes the oil flow paths, leaks, and valving problems attendant to a design fabricated from individual off-the-shelf components.

A typical output torque-speed characteristic curve is shown in Fig. 5-21 for a 5 hp transmission of this design. The curve corresponds to a fixed transmission pressure. Although this is theoretically a constant-torque transmission (since

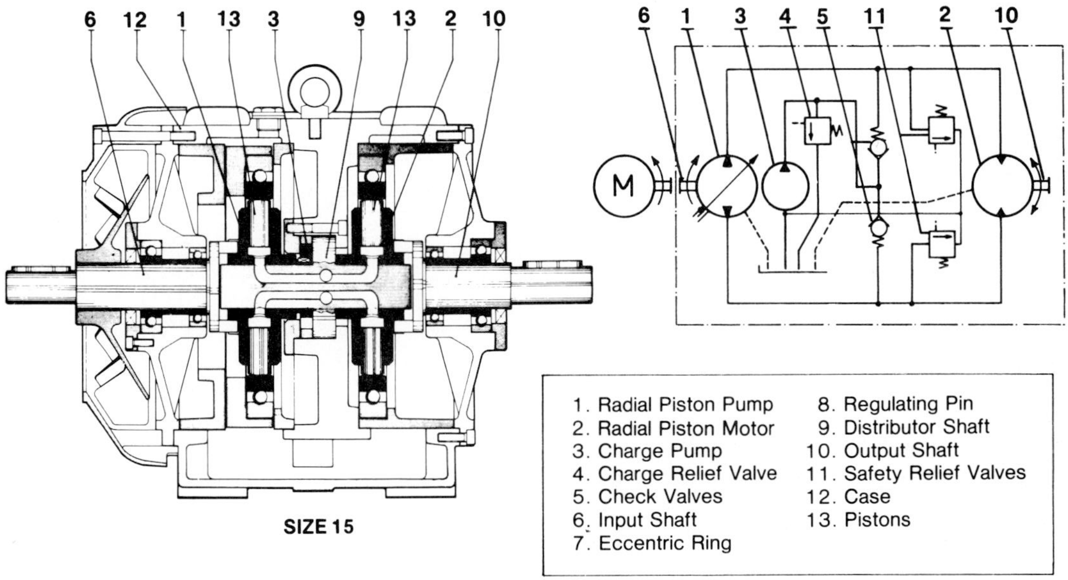

Figure 5-20. Radial-piston integrated hydrostatic drive. (Courtesy of Formsprag, Dana Corporation.)

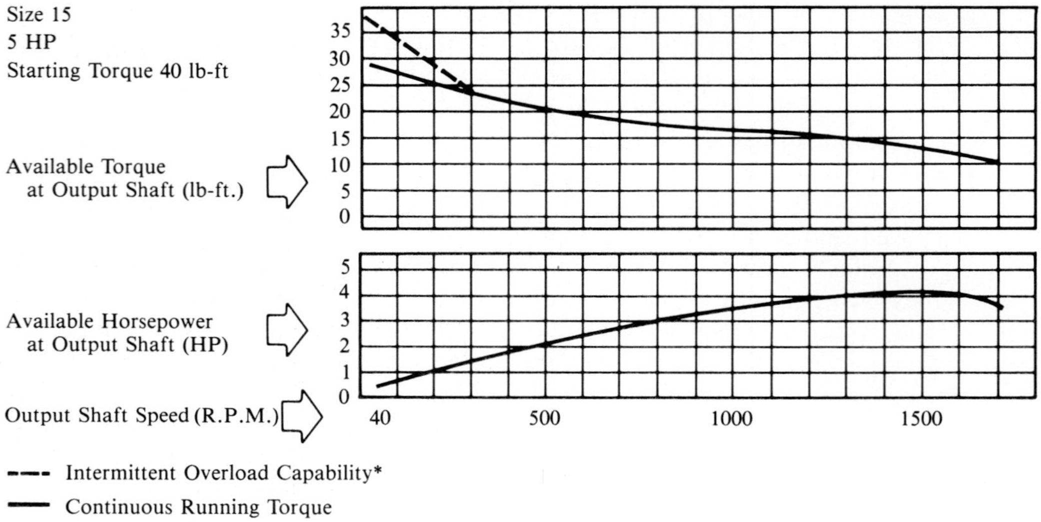

Figure 5-21. Typical characteristics of an integrated hydrostatic drive. (Courtesy of Formsprag, Dana Corporation.)

TABLE 5-1. Service Factors for Hydrostatic Drives. (Courtesy of Formsprag, Dana Corporation.)

Duty, C_1

Running Hours per Day	Service Factor C_1
8	1.0
8–15	1.1
15–24	1.2

Shock Loading, C_2

Shock Load in Driven Equipment	Service Factor C_2
Low	0.0
Medium	0.1
Heavy	0.3
Very heavy	Consult factory

Temperature, C_3

Ambient Temperature	Service Factor C_3
Up to 80°F	0.0
From 85°F to 105°F	0.4
Higher than 105°F	Consult factory

Notes: The service factor is the sum of C_1, C_2, and C_3:

$$SF = C_1 + C_2 + C_3$$

If high loads are to be cycled through the speed range more than 40 times per hour, consult the factory.

pressure is fixed), there is some drooping in the torque characteristic due to internal leakage and frictional losses which increase for the higher speeds. Note the wide speed range (from 40 to 1700 rpm with a 1750 rpm input).

Service factors are applied to integrated hydrostatic drives in much the same way as for gearboxes. Typical values are shown in Table 5-1. The total service factor is the sum of the individual values for duty, shock loading, and temperature.

EXAMPLE 5-7

A conveyor is to be driven with an existing integrated hydrostatic transmission of the constant-torque type. The transmission is rated at 2000 psi, 10 hp, 1750 rpm input speed, $r = 1$, with torque density 25 in-lb/100 psi. The volumetric efficiency of pump and motor is 0.95 each.

The conveyor is to operate 24 hours per day with a shock loading estimated as medium in a temperature environment of 100°F. The steady-state operating torque is 600 in-lb in a speed range of 100 to 600 rpm. Determine if the use of this transmission is feasible.

Solution

With service factors applied, the output torque required of the hydrostatic drive must be

$$T_m = (1.2 + 0.1 + 0.4)(600) = 1020 \text{ in-lb}$$

The maximum power required will be

$$\text{HP} = \frac{2\pi(600)(1020/12)}{33{,}000} = 9.7 \text{ hp}$$

At rated transmission pressure, the drive is capable of producing a torque of

$$T_m = \frac{25}{100}(2000) = 500 \text{ in-lb}$$

The maximum output speed of the drive is given by

$$\frac{\omega_m}{\omega_p} = \left(\frac{S_p}{r}\right)(\eta_v)^2 = \left(\frac{1}{1}\right)(0.95)^2$$

Therefore,

$$\omega_m = (0.95)^2(1750) = 1579.4 \text{ rpm}$$

We see that the maximum speed is beyond what is required but that the torque capacity of 500 in-lb is insufficient. So, as it stands, the transmission does not have sufficient output torque capacity. However, if we connect the transmission to the conveyor headshaft with a geared drive having a ratio of around 2:1, then the maximum speed of the output can be reduced to $1579.4/2 = 789.7$ rpm while the output torque is amplified to $500(2) = 1000$ in-lb.

Thus, this transmission is feasible if we use the 2:1 geared drive as a coupling. To achieve the 600 rpm conveyor speed, the pump of the transmission will have to be stroked at the value

$$S_p = \frac{600}{789.7(0.95)^2} = 0.84$$

Since the horsepower capacity of the transmission is 10 hp, it exceeds the required value of 9.7. ■

5-4 LOW-SPEED, HIGH-TORQUE DRIVES

In many applications, the hydrostatic transmission will be integrated with a gearbox in order to achieve the torque and speed relationships required. The possible arrangements are diverse and may include planetary gear trains, split-power arrangements, shunt transmissions, and other complex transmission schemes. Reference 5-2 contains a varied assortment of such drives.

Hydraulic power systems are admirably suited to large earth-moving and similar construction equipment. Some of these mobile machines require enormous amounts of power transmission capability. Hydraulic systems permit the concentration of this power into highly compact components. An additional virtue of fluid power in these applications is the ability of the hydraulic system to easily direct the power around corners and joints which, in a mechanical drive system, might require a complicated set of linkages.

Many of these mobile machines will be wheel driven. Often, the traction wheels may be powered directly by hydraulic motors. But in some very large machines, the torque and speed requirements of the driving wheels would require inordinately large hydraulic motors. It would not be unusual for one of these machines to require wheel torques of a quarter million ft-lb at the low speed of 10 rpm. These large torques may produce very high overhung loads on the transmission driving the wheels.

In applications calling for high torques at low speeds, there is the choice to be made between a low-speed, high-torque hydraulic motor (which tends to be large in size) and a higher speed hydraulic motor coupled to a reduction gearbox. Although it is generally more economical to use a hydraulic motor alone, the combination of gearbox and hydraulic motor can be more competitive when the output speeds are very low.

In some applications, the integration of a planetary gear reduction unit with a hydraulic motor will provide a compact transmission capable of handling the torque and overhung load requirements at low speeds. One such arrangement is shown schematically in Fig. 5-22. Here, the traction wheel would be attached directly to the low-speed output drive flange shown.

EXAMPLE 5-8

The transmission illustrated in Fig. 5-22 is to transmit 500 hp with an output speed of 5 rpm. What will be the rpm of the barrel of the hydraulic motor? What will be the overhung load on the output shaft if the gears are 25 degree pressure angle and the traction wheel is 6 feet in diameter? Assume the following information: $N_1/N_2 = 13/19$; $N_3/N_4 = 20/61$; $N_7/N_5 = 80/17$; $N_{10}/N_8 = 88/21$.

Solution

For the planetary part of the gear reduction unit, we have the relationships

$$\frac{n_5 - n_8}{n_7 - n_8} = -\frac{N_7}{N_5} = -\frac{80}{17}$$

$$\frac{n_8 - n_9}{n_{10} - n_9} = -\frac{N_{10}}{N_8} = -\frac{88}{21}$$

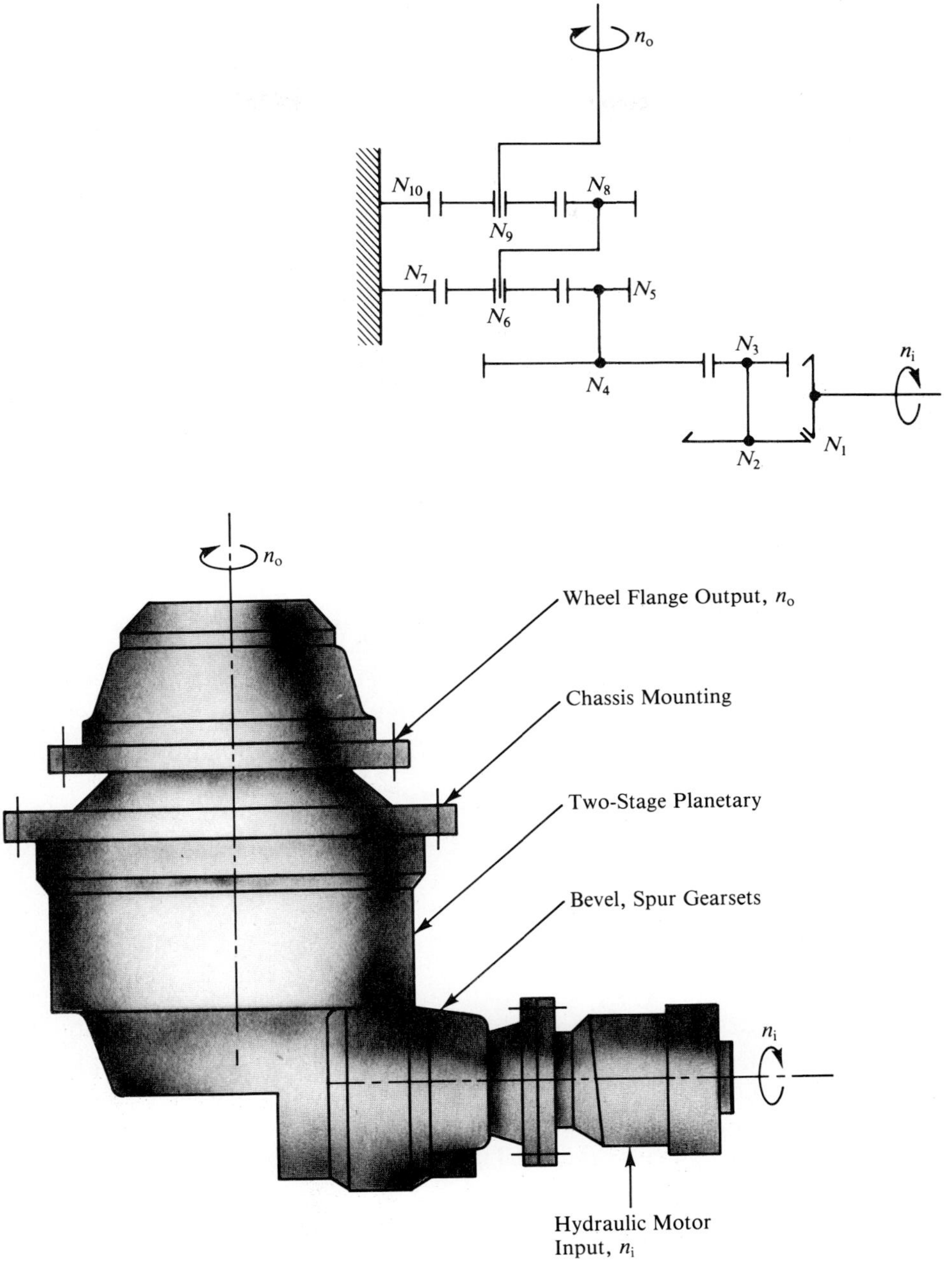

Figure 5-22. Low-speed, high-torque transmission.

Recognizing that n_9 is the output speed n_o, the above equations will produce the result

$$\frac{n_o}{n_5} = \frac{1}{\left(1 + \frac{80}{17}\right)\left(1 + \frac{88}{21}\right)} = 0.03388$$

For the other parts of the gear reduction, we have

$$\frac{n_2}{n_1} = \frac{N_1}{N_2} = \frac{13}{19} = 0.6842$$

$$\frac{n_4}{n_3} = \frac{N_3}{N_4} = \frac{20}{61} = 0.3279$$

Therefore,

$$\frac{n_o}{n_i} = \left(\frac{n_o}{n_5}\right)\left(\frac{n_4}{n_3}\right)\left(\frac{n_2}{n_i}\right) = (0.03388)(0.6842)(0.3279) = 0.0076$$

or

$$\frac{n_i}{n_o} = 132$$

Then

$$n_i = 132(5) = 660 \text{ rpm}$$

The output torque is

$$T_o = \frac{33{,}000(500)}{2\pi(5)} = 525{,}211 \text{ ft-lb}$$

The overhung load on the output shaft will be

$$F = \frac{525{,}211}{3} = 175{,}070 \text{ lb}$$ ■

5-5 SUMMARY

There are certain mechanical system designs for which hydrostatic transmissions are particularly well suited. Early applications were in agricultural and transportation equipment. Hydrostatic drives are still used for tractors, buses, and locomotives. Other important vehicle applications are found in earth-moving and military fighting equipment. And the high power density of the hydrostatic drive makes it attractive in aircraft applications.

Industrial applications are found in machine tools, in fan drives, in robotics, in winch and hoist systems, and in equipment where control of tension is important. Information about some of these applications can be found in the references listed at the end of this chapter, particularly References 5-2 and 5-3.

In this chapter, we have introduced the fundamental principles of hydrostatic transmissions which are prerequisite to the integration of such transmissions in more complex mechanical systems such as the ones mentioned above. However, there is much more to be learned about the system load characteristics to which these transmissions are to be coupled. A good introductory discussion of this topic can be found in the chapter on hydrostatic transmissions in Reference 5-1.

5-6 REFERENCES

5-1. R. W. Henke, *Fluid Power Systems and Circuits*, published by Penton/IPC, 1111 Chester Avenue, Cleveland, OH, 1983.

5-2. *Hydrostatic Transmissions for Vehicle Applications*, Institute of Mechanical Engineers Conference Publication ISBN 0-85298-476-6, 1981, available from Society of Automotive Engineers (SAE), 400 Commonwealth Drive, Warrendale, PA.

5-3. *Transmissions and Transmission Controls for Off-Highway Equipment*, Society of Automotive Engineers (SAE) Publication SP-549 (ISBN 0-89883-320-5), 1983, available from Society of Automotive Engineers (SAE), 400 Commonwealth Drive, Warrendale, PA.

5-4. *Mobile Hydraulics Manual*, Sperry-Vickers Manual M-2990-A, available from Sperry Corporation, Troy, MI, 1981.

PROBLEMS

5-1. How would you add an on-off, two-position, two-way control valve to the circuit of Fig. 5-4 so as to remove (bypass) the resistance of the counterbalance valve in the exhaust line except when there are over-running loads?

5-2. Explain why the metering-out flow control principle reduces the torque available at the output of an open-circuit hydrostatic drive when compared with a system using the bypass type flow control.

5-3. Show that the parameter $k = An\varepsilon/\pi$ is the same as $k = SD$, where S = stroke and D = maximum displacement (per radian). You will need to review Section 4-1-5.

5-4. In Example 5-2, assume that a pressure relief valve set at 1904 psi is used to hold the transmission pressure constant during start-up. Assume a fixed-displacement pump and determine how many gpm are being dumped over this relief valve 5 seconds after the acceleration of the load is begun. What power loss is associated with this dumping rate?

5-5. In Example 5-3, determine a start-up acceleration profile such that the starting torque profile will merge smoothly into the lifting hoist torque (3500 in-lb) at the beginning speed $N_m = 0$ and the ending speed $N_m = N_o$.

5-6. Suppose the pump outlet pressure P_p of an open-circuit hydrostatic drive exceeds the motor inlet pressure P_m by the amount of pressure drop in the piping connecting the pump to the motor. Also, suppose that the volumetric flow rate leaving the pump, Q_p, exceeds that coming into the inlet of the motor, Q_m, because of leaks in the piping connecting the pump to the motor. Then the assumptions made in Section 5-2 will be invalid, because $P_p \neq P_m$ and $Q_p \neq Q_m$. Show that the overall efficiency of the hydrostatic transmission that contains these piping leaks and friction will be

$$\eta = \eta_p \eta_m \left(\frac{Q_m}{Q_p}\right)\left(\frac{P_m}{P_p}\right)$$

where η_p and η_m are the overall efficiencies of the pump and motor respectively. Remember that overall efficiency η is defined as

$$\eta = \eta_v \eta_T$$

5-7. The volumetric efficiency η_v of a hydrostatic drive having no leaks in the piping connecting the pump to the motor is defined as

$$\eta_v = \eta_{v(p)} \eta_{v(m)}$$

where $\eta_{v(p)}$ and $\eta_{v(m)}$ are the volumetric efficiencies of pump and motor respectively. Show that

$$\eta_v = 1 - \frac{\lambda_m(1+L)P}{k_p \omega_p} = 1 - \frac{Q_l}{k_p \omega_p}$$

5-8. Show that the torque efficiency of a hydrostatic drive having no friction (or leaks) in the piping connecting the pump to the motor is

$$\eta_T = \frac{k_p T_m}{k_m T_p}$$

5-9. Assume that the volumetric efficiency of a hydrostatic drive is $\eta_v = 90$ percent. The drive is to produce an output speed and torque of 1500 rpm and 800 in-lb respectively. Neglect internal friction in motor and pump and determine

(a) The required speed of the pump if the drive is to operate at a pressure of 2000 psi with a ratio $r = 2$

(b) The input horsepower neglecting frictional losses

5-10. Rework Example 5-5 for the case in which the pump is stroked at one-half ($S_p = 1/2$) of its full value ($S_p = 1$). Determine the steady-state operating speed and output torque.

5-11. Assume that the load of Example 5-5 is to be controlled such that the speed is to be maintained constant at $N_m = 330.1$ rpm independent of load torque. Devise a closed-loop feedback control scheme for achieving this constant-speed control.

5-12. An elevator is connected to a cable drum 36 inches in diameter by a single wire rope. The cable drum is connected directly to the output shaft of a hydrostatic transmission that is driven (at the pump) at 530 rpm. The displacement of the motor is fixed at 20 in^3/rev, while the pump is variable-displacement (a constant-torque drive) with maximum value, at full stroke, equal to the motor displacement.

(a) What is the motor speed when the elevator moves at a speed of 500 fpm?

(b) Neglecting leakage, what is the volumetric discharge (in in^3/min) flowing through the motor?

(c) What is the value of the stroke at which the pump is being operated? Neglect leakage.

5-13. Suppose the torque on the motor of Problem 5-12 is 663 ft-lb. Neglect internal friction in the motor and pump but assume that they each have internal leakage defined by the constants

$$\lambda_m = \lambda_p = 0.002 \text{ gpm/psi}$$

Recalculate Problem 5-12 accounting for leakage with the assumption that the pump is stroked at the value $S_p = 0.1$. Determine the motor speed assuming the pump operates at 530 rpm. What does your answer imply about the system behavior?

5-14. In the fluid power circuit shown in Fig. P5-14, the PF (fixed-displacement) pump operates continuously and supplies 0.5 gpm. The MF (fixed-displacement) motor has a displacement of 3 in^3/rev.

The motor is being cycled in the following way. First, the motor turns 20 revolutions cw in 5 seconds. This is followed by a 55 second dwell (no rotation). Then the motor turns 20 revolutions ccw, followed by a 55 second dwell. Then the cycle repeats. The cycling is accomplished by the solenoid-actuated directional control valve.

During the 5 seconds that the motor turns, its supply pressure drops from 1500 psi to 1000 psi. The unloading valve B is set for 1500 psi, so with the directional control valve centered, the pump output will be bypassed to the tank (through valve B) once the accumulator D has been charged to 1500 psi. The unloading valve will close immediately when the supply line pressure falls below 1500 psi. Neglect all internal leakages and friction and determine

(a) The motor rpm when the motor turns
(b) The flow rate (in gpm) through the motor when it is turning
(c) The flow rate out of the accumulator when the motor turns
(d) How long (in seconds) it takes to recharge the accumulator
(e) The average external torque against which the motor works when it is turning

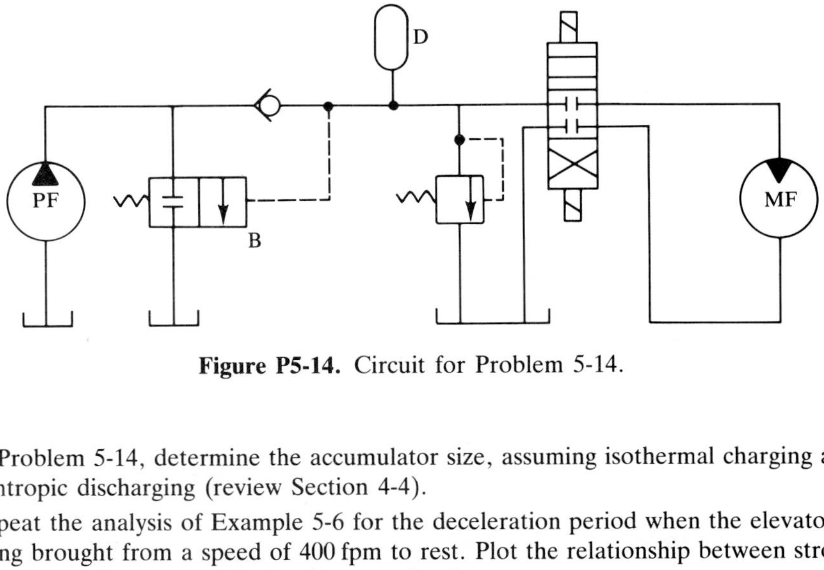

Figure P5-14. Circuit for Problem 5-14.

5-15. In Problem 5-14, determine the accumulator size, assuming isothermal charging and isentropic discharging (review Section 4-4).

5-16. Repeat the analysis of Example 5-6 for the deceleration period when the elevator is being brought from a speed of 400 fpm to rest. Plot the relationship between stroke and elevator position while the elevator decelerates.

5-17. A constant-torque hydrostatic transmission circuit for driving a winch drum is sketched in Fig. P5-17. The main pump A in this drive is a pressure-compensated, variable-displacement dynamo that provides full delivery up to a pressure of 1250 psi and then reduces the delivery linearly until it is zero at the deadhead pressure of 2250 psi.

This winch system is intended to be used in mooring ships tethered to the cable drum by a wire rope. The motor displacement is 153.6 in^3/rad, and the winch drum is 24 inches in diameter. The no-load pump discharge is 15 gpm.

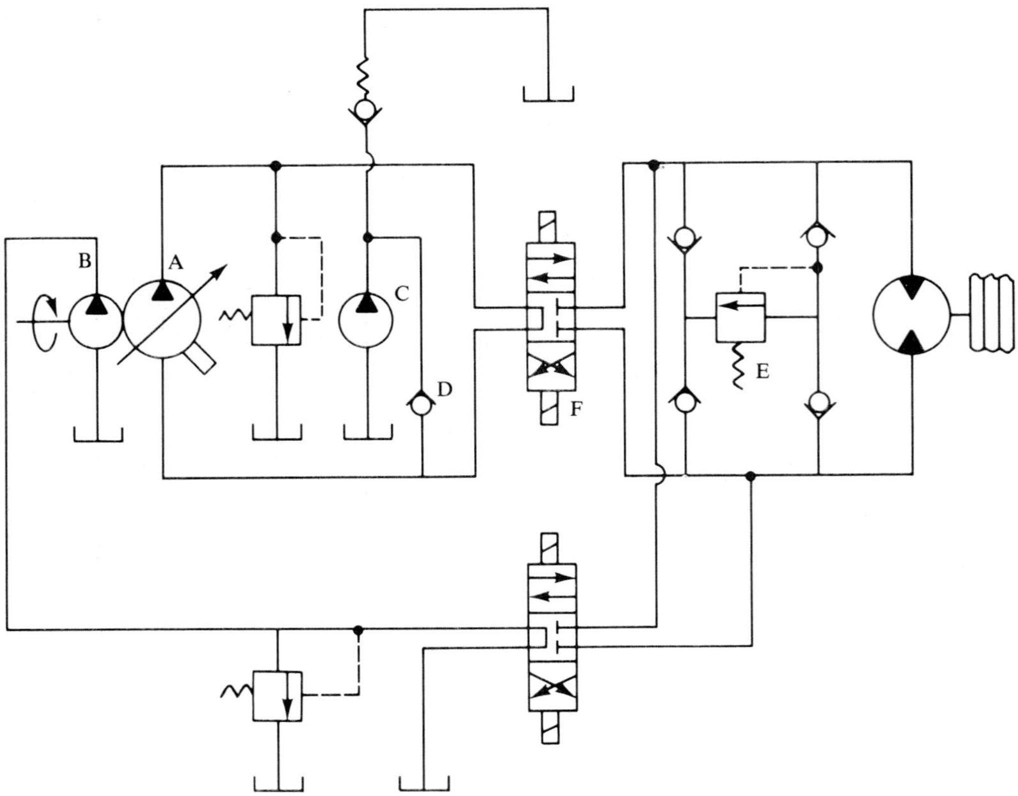

Figure P5-17. Circuit for Problem 5-17.

(a) What is the function of pump C and its associated check valve D?
(b) What are the circumstances under which valve E will be actuated?
(c) Draw the characteristic *P-Q* curve for pump A.
(d) What is the purpose of the fixed-displacement pump B?
(e) Assume valve F is centered. What will be the retrieval rate of this winch system if the line tension (in the wire rope) is below 8 tons? Neglect friction and leakage.
(f) What will happen if the ship pulls on this winch with a line tension exceeding 14.4 tons?
(g) Why are the directional control valves three-position rather than two-position (on-off) valves?

5-18. Show that Eq. (5-21) can be written in the form

$$\frac{\omega_m}{\omega_p} = \left(\frac{S_p}{r}\right)(\eta_v)^2$$

where η_v is the volumetric efficiency of both pump and motor. Comparing this formulation with Eq. (5-21), does it predict a deadband for S_p?

5-19. A tandem-wheel road roller is to be capable of being driven up a grade of 15 degrees at 8 mph (see Fig. P5-19). The vehicle has a gross weight of 8000 lb, and it has rollers 40 inches in diameter. A hydrostatic drive is used in which a single pump, whose

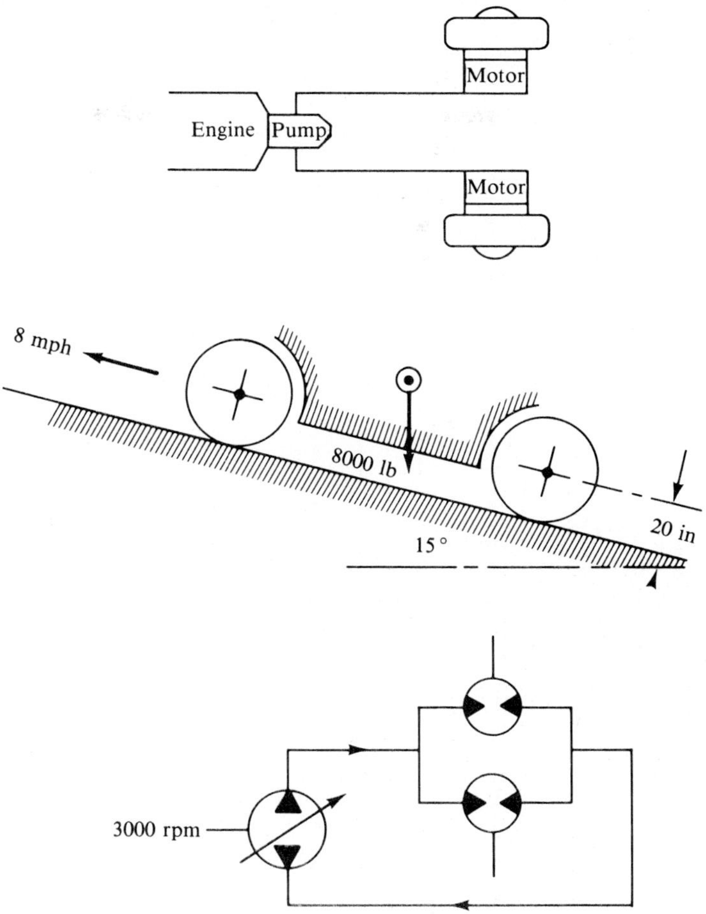

Figure P5-19. Figures for Problem 5-19.

displacement is 0.375 in^3/rad, drives two wheel-mounted motors each of which has a displacement of 0.375 in^3/rad. The pump is driven at 3000 rpm. Assume the pump is fully stroked when the roller speed is 8 mph. Neglect all inertia effects.

(a) Determine the pump output (in gpm) when it is fully stroked (neglecting leakage).

(b) The hydraulic motors shown are integral and concentric with planetary gearboxes located between the motor output shafts and the wheel hubs. What is the speed ratio of these gearboxes if pump and motor leakages are neglected?

(c) Assume the overall torque efficiency of the system (including gearboxes, motors, and pump) is $\eta_T = 0.90$. Determine the hydrostatic transmission pressure if the planetary gear ratio is assumed to be 22.3:1.

6

Hydrodynamic Drives

In a hydrostatic drive, we have a positive-displacement pump (PDP) driving a positive-displacement motor (PDM). In such a drive, the fluid pressure, and hence torque capacity, is independent of the volumetric transfer of fluid. The transmission pressure is dictated by the load torque or is otherwise controlled through the use of suitable circuit components, such as relief valves, or by using the principle of pump pressure compensation.

A hydrodynamic drive is one in which a non-positive-displacement pump (NPDP) drives a non-positive-displacement motor (NPDM) (see Fig. 6-1). In a hydrodynamic drive, the torque capacity is not independent of fluid volumetric transfer. It is derived from the dynamic conversion of fluid kinetic energy rather than from hydrostatic pressure (as in the hydrostatic drive). The torque capacity of the hydrodynamic drive is determined by the design of the device itself rather than through the use of circuit components.

To avoid the piping frictional losses attendant to the separation of pump and motor, as shown in Fig. 6-1, the pump and motor are integrated together in a manner similar to the integrated hydrostatic drive, which we discussed in Chapter 5.

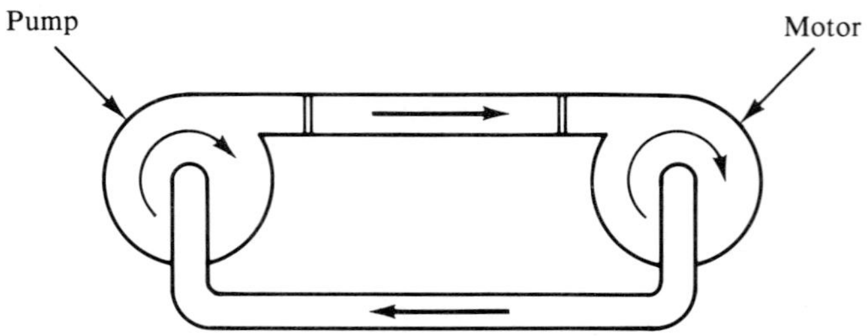

Figure 6-1. Hydrodynamic drive.

6-1 THE FLUID COUPLING

One version of an integrated hydrodynamic drive is the fluid coupling shown in Fig. 6-2. The fluid coupling consists of two half-toroidal rotors, placed face-to-face, in such a way that one rotor can serve as the pump while the other serves as the motor. The pump is called the impeller and the motor is called the turbine, as shown in Fig. 6-3.

Figure 6-2. Fluid coupling. (Courtesy of Renold PTC.)

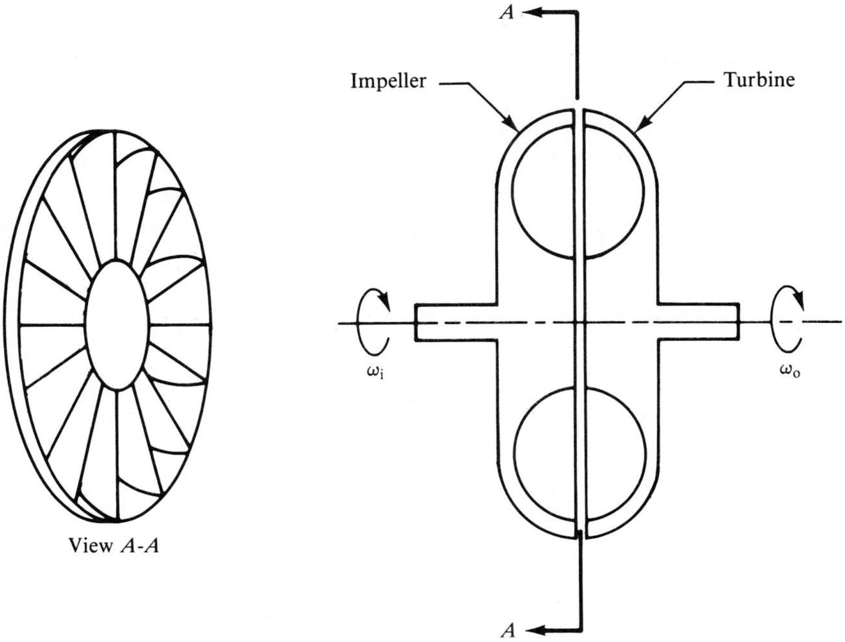

Figure 6-3. Schematic diagram of a fluid coupling.

Each half of the fluid coupling takes on the form of a half-toroid containing radial vanes similar to a half-grapefruit with its fruit removed. Combined together, the two half-toroids form a sealed toroidal volume filled with hydraulic fluid. The halves of the fluid coupling are not connected together mechanically in any way. Each half has its own shaft and can be rotated independently of the other. The impeller shaft is connected to the prime mover while the turbine shaft is connected to the load.

The mechanism of torque transfer involves changes in fluid angular momentum as fluid transfers to the turbine from the impeller. Suppose the turbine is at rest (stalled) while the impeller is rotated. Centrifugal forces on the oil in the impeller will cause a circulation of the hydraulic fluid between the impeller and turbine. Since the angular momentum of the rotating impeller fluid will be brought to rest by the static turbine, a torque will be exerted on the turbine as dictated by the angular momentum theorem of fluid mechanics. This torque on the turbine will be balanced by a counter-torque exerted on the impeller because the net torque on the whole device must be zero (for steady-state operation).

The torque exerted on the turbine by the impeller fluid will cause the turbine to rotate if it exceeds the external torque of the load connected to the turbine shaft. The turbine will then accelerate to the steady state, in which the fluid torque acting on the turbine is just balanced by the load torque. In the steady state, there will be a slip between the two halves of the fluid coupling (if a torque is transmitted), because if the impeller and turbine rotate at the same speeds (zero slip), there can be no transfer of angular momentum (and, hence, no torque transmission). Fluid couplings are selected and applied such that the normal operating slip for rated torque transmission will be around 3 to 5 percent. As we shall see later, the slip represents dissipation of power and must be kept relatively small.

Integrated hydrodynamic drives such as the one we have just discussed are classified as either fluid couplings or torque converters. They are available as fixed-fill or variable-fill devices. A variable-fill device is one in which the amount of fluid in the coupling can be externally controlled.

Some advantages of hydrodynamic drives are listed below:

(a) During start-up, smooth acceleration of large inertial loads can be obtained. The result is a soft-start for the motor, making possible the use of squirrel cage induction motors if the prime mover is electrical.

(b) Similarly, an external load torque can be started smoothly by accelerating the coupling impeller until the requisite torque is produced by the coupling.

(c) Hydrodynamic drives used to couple diesel engines to loads will attenuate the intrinsic torque fluctuations produced by these engines. This protects gearboxes in the power train from deleterious torque fluctuations.

(d) Similarly, hydrodynamic couplings minimize and damp shock loads and torsional vibrations without seriously impeding the operation of the transmission.

(e) The driving side of the transmission will not stall even if the output is stalled, thus protecting the prime mover.
(f) Torque transmission can be controlled by changing the amount of oil in the coupling (variable-fill couplings).
(g) Couplings will withstand rugged use conditions and require only minimal maintenance.
(h) There are no meshing parts to wear, so long life is obtained.

The disadvantages of hydrodynamic drives are as follows:

(a) They are slip devices and hence dissipate energy (about 3 to 5 percent under normal operating conditions).
(b) They are not positive drives and cannot be used where synchronization of drives is required.
(c) Brakes must be used to hold constant-torque loads during start-up because hydrodynamic drives do not develop torque at zero impeller speeds.

Despite these limitations, hydrodynamic drives have been found useful in a range of applications, including:

(a) Conveyor drives to provide smooth starts and to minimize belt stretch
(b) Large induced-draft fans
(c) On diesel engines to attenuate torque fluctuations
(d) Automatic transmissions and other automotive propulsion drives
(e) Marine transmissions
(f) In conjunction with epicyclic gear trains, to effectively amplify the torque that a given coupling can safely transmit
(g) Earth-moving machines, elevators, hoists, forging machines, mining applications, and pumps

Of the various types of hydrodynamic drives that have been developed, we will discuss only two: the fluid coupling and the single-stage torque converter. The fundamental fluid mechanical principle of operation of both of these devices is the angular momentum theorem.

6-2 FLUID COUPLING ANGULAR MOMENTUM RELATIONSHIPS

A schematic diagram of a fluid coupling is shown in Fig. 6-4. This configuration is slightly different from Fig. 6-3 in that a core has been inserted in the toroidal volume to promote a smoother circulation of flow between impeller and turbine. The fluid motion occurring between impeller and turbine is a complicated three-dimensional flow. The component of this flow in a plane containing the fluid coupling shaft axis can be visualized to be a vortex as indicated in Fig. 6-4.

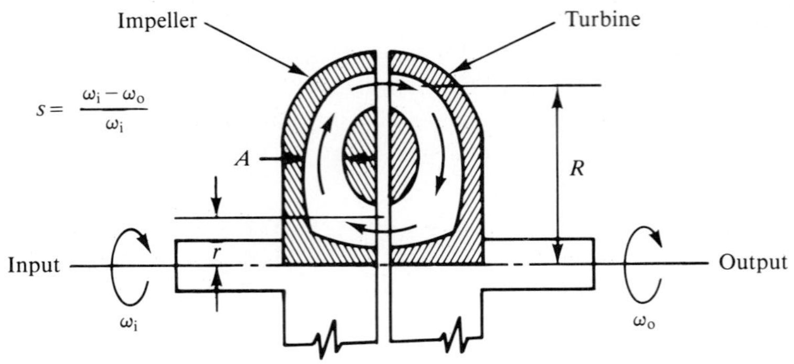

Figure 6-4. Flow velocities in a fluid coupling.

In the steady state, the net concentric torque acting on this device must be zero. Only two torques (input and output) exist, since the device is not externally fixed (or grounded). Consequently, we see that the output torque is always equal to the input torque (in magnitude). This is the distinguishing feature of the fluid coupling.

To derive the desired angular momentum relationships, we surround the impeller portion of the fluid coupling by a control volume, as shown in Fig. 6-5. The angular momentum theorem states that the torque applied to this control volume (by the impeller input shaft) is equal to the net outward flow of angular momentum through the control volume surface.

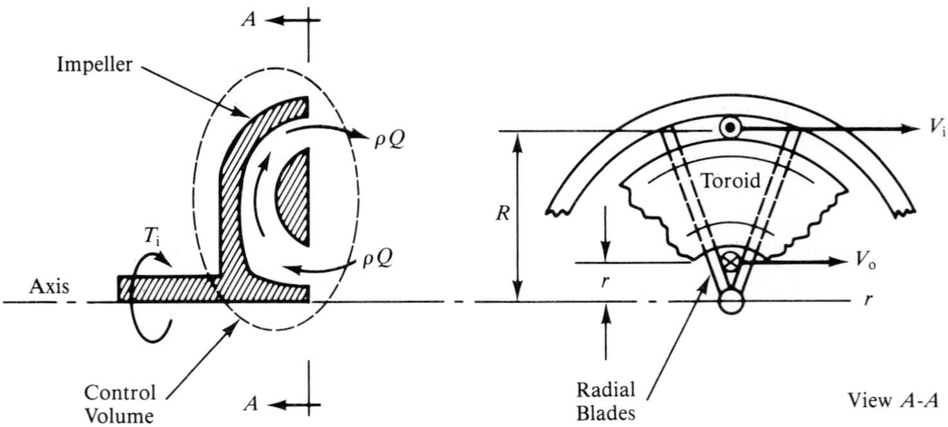

Figure 6-5. Flow velocities in a fluid coupling impeller.

The diagram of Fig. 6-5 indicates the nomenclature to be used. The impeller (input) speed and torque are ω_i and T_i respectively, while the turbine (output) speed and torque are ω_o and T_o respectively. There is no positive connection between impeller and turbine; fluid simply transfers from impeller to turbine at the outer periphery and from turbine to impeller at the inner periphery. There is a slip s between impeller and turbine, defined by

$$s = \frac{\omega_i - \omega_o}{\omega_i} = 1 - \frac{\omega_o}{\omega_i} \tag{6-1}$$

Seals prevent the hydraulic fluid from leaking at the interface between impeller and turbine.

With the blading assumed to be simply radial plates, the fluid transfer is as indicated in Fig. 6-5. Fluid flows axially out of the impeller at its outer periphery at the mass flow rate defined as ρQ, where ρ is fluid density. Although there is a small component of axial linear momentum associated with this axial mass flow, the predominant momentum effect is peripheral rather than axial. The value of the linear peripheral momentum, per unit mass of the fluid, is V_i. Since the fluid has the peripheral momentum of the impeller where it transfers from the impeller, we see that

$$V_i = \omega_i R \tag{6-2}$$

At the inner periphery, fluid returns to the impeller from the turbine at the same mass flow rate ρQ. Again, the peripheral momentum dominates the axial momentum, so we can assume that the linear peripheral momentum entering the impeller is simply the linear speed of the turbine inner periphery. The value is

$$V_o = \omega_o r \tag{6-3}$$

The difference between the moments of these two linear momenta is simply the net outflow of angular momentum from the impeller and, hence, is the torque T_i required to drive the impeller. In other words,

$$T_i = \rho Q(RV_i - rV_o) \tag{6-4}$$

or

$$T_i = \rho Q(\omega_i R^2 - \omega_o r^2) \tag{6-5}$$

Equation (6-4) is of the same form as Eq. (4-3) for the centrifugal pump as discussed in Chapter 4. The fundamental difference between the centrifugal pump impeller and the fluid coupling impeller is that in the centrifugal pump the mass flow rate ρQ is always essentially radial, while in the fluid coupling it is axial (i.e., along the axis of the coupling). We could just as well have called the impeller of the fluid coupling a pump instead of an impeller.

The mechanical efficiency of the fluid coupling is defined as

$$\eta = \frac{\omega_o T_o}{\omega_i T_i} = (1 - s)\frac{T_o}{T_i} \tag{6-6}$$

In the steady state, when both ω_o and ω_i are constant, the input torque must equal the output torque. Therefore, in the steady state we have

$$T_o = T_i \tag{6-7}$$

$$\eta = 1 - s \tag{6-8}$$

If Eq. (6-1) is substituted into Eq. (6-5), we find that

$$T_i = \rho Q \omega_i R^2 \left[1 - (1 - s)\left(\frac{r}{R}\right)^2 \right] \tag{6-9}$$

It would be desirable to eliminate the unknown value Q from Eq. (6-9). The value of Q will depend on the resistance it encounters as it makes the circuit through the coupling from impeller to turbine and then back again.

Recall that the volumetric flow rate Q has the form

$$Q = VA \tag{6-10}$$

In trying to assess, qualitatively, what characteristic velocity V and area A to insert in Eq. (6-10) for the purpose of estimating Q, we should keep in mind that Q will be zero when the slip is zero and will be a maximum when the slip is unity (turbine stalled). When the turbine is stalled ($s = 1$), we would expect Q to increase as the impeller speed increases. Therefore, it is plausible to assume that

$$V = C_1 V_i s \tag{6-11}$$

The simplest assumption we could make for area A would be

$$A = C_2 R^2 \tag{6-12}$$

This is tantamount to assuming that all fluid couplings would be geometrically similar. Then, upon substituting Eqs. (6-2), (6-11), and (6-12) into Eq. (6-10), we obtain

$$Q = C_3 \omega_i R^3 s \tag{6-13}$$

where the constant C_3 equals $C_1 C_2$. Then, if Eq. (6-13) is substituted into Eq. (6-9), we obtain

$$T_i = C_3 \rho \omega_i^2 R^5 s \left[1 - (1 - s)\left(\frac{r}{R}\right)^2 \right] \tag{6-14}$$

Equation (6-14) reveals the intrinsic input properties of the fluid coupling:

(a) Torque capacity is proportional to the square of impeller speed.
(b) Torque capacity is proportional to the fifth power of impeller radius (or diameter).
(c) For small slips, the torque capacity is proportional to slip.

Note the important effect that coupling size has on torque capacity. Large increases in torque capacity can be obtained with only small increases in size. These properties are borne out by experimental studies on fluid couplings.

EXAMPLE 6-1

A certain fluid coupling has a characteristic diameter of 1 foot and transmits 200 ft-lb of torque at a slip of 5 percent when it is driven at a speed of 1750 rpm. A geometrically similar fluid coupling is to be selected to drive a 10,000 hp load at a load speed of 500 rpm and a slip of 3 percent. What will be the characteristic diameter of the 10,000 hp coupling if both couplings are filled with the same kind of oil? How much energy will be dissipated in the larger fluid coupling?

Solution

The torque capacity of the 10,000 hp coupling will have to be

$$T_2 = \frac{33{,}000(10{,}000)}{2\pi(500)} = 105{,}000 \text{ ft-lb}$$

Its input speed will be

$$\omega_2 = \frac{500}{0.97} = 515.5 \text{ rpm}$$

Assuming at these small slips that

$$T_i = C_3 \rho \omega_i^2 R^5 s$$

then

$$D_2 = \left[\left(\frac{T_2}{T_1}\right)\left(\frac{\omega_1}{\omega_2}\right)^2\left(\frac{s^1}{s_2}\right)\right]^{1/5} D_1 = \left[\left(\frac{105{,}000}{200}\right)\left(\frac{1750}{515.5}\right)^2\left(\frac{0.05}{0.03}\right)\right]^{1/5}(1) = 6.32 \text{ feet}$$

The energy dissipated will be 3 percent of 10,000 hp, or 300 hp. ■

EXAMPLE 6-2

Assume that a fluid coupling obeys the relationship

$$T = C\rho\omega_i^2 D^5 s$$

where C is a constant. This coupling is used to connect a torque load of 1500 ft-lb to an electric motor running at 1760 rpm. The slip of the coupling is 5 percent.

(a) Determine the efficiency of the coupling and the speed of the load.
(b) If the load rises temporarily to 3000 ft-lb while the electric motor speed drops to 1740 rpm, what will be the new speed of the load?

Solution

The efficiency of the coupling is

$$1 - s = 0.95$$

The speed of the load is given by

$$\frac{1760 - \omega_o}{1760} = 0.05$$

$$\omega_o = 1672 \text{ rpm}$$

When the load is increased, the new slip, s_2, will be given by

$$s_2 = \left(\frac{T_2}{T_1}\right)\left(\frac{\omega_1}{\omega_2}\right)^2 s_1 = \left(\frac{3000}{1500}\right)\left(\frac{1760}{1740}\right)^2(0.05) = 0.1023$$

Then the new speed of the load is given by

$$\frac{1740 - \omega_o}{1740} = 0.1023$$

$$\omega_o = 1562 \text{ rpm}$$ ■

6-2-1 Fluid Coupling Torque Coefficient

Equation (6-14) can be rewritten in non-dimensional form as

$$C_T = \frac{T_{\rm i}}{\rho \omega_{\rm i}^2 D^5} = f\left(s, \frac{r}{R}\right) \qquad \textbf{(6-15)}$$

where

$$D = 2R \qquad \textbf{(6-16)}$$

The non-dimensional quantity C_T is called the torque coefficient of the fluid coupling.

The function $f(s, r/R)$ can be determined by performing tests on families of geometrically similar fluid couplings. Results similar to Fig. 6-6 are obtained. For small slips ($s < 0.1$), the torque coefficient is proportional to slip, but as the slip increases beyond about 0.1, the slope of the curve begins to diminish with slip.

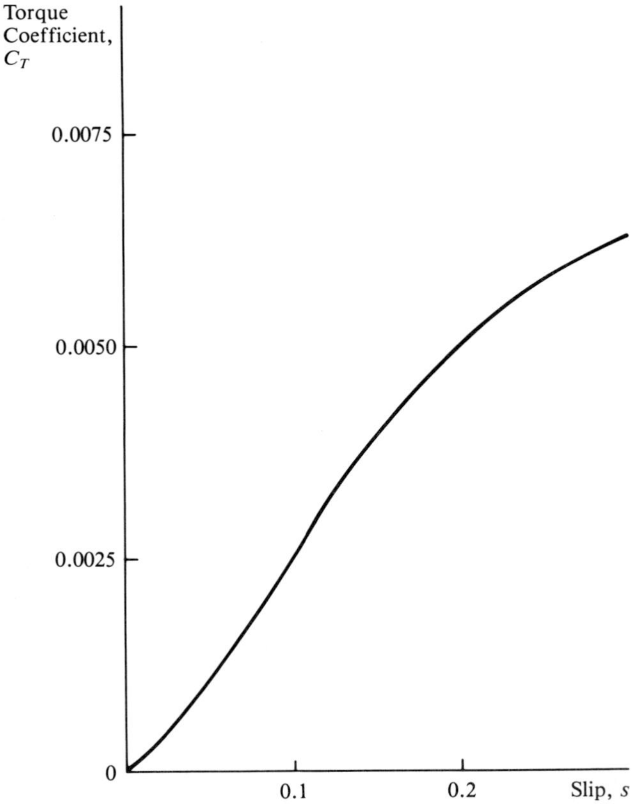

Figure 6-6. Torque coefficient versus slip for a fluid coupling.

EXAMPLE 6-3

Suppose the torque coefficient curve of Fig. 6-6 represents a certain family of geometrically similar fluid couplings. If the oil density is 1.65 slug/ft^3 in a 1 foot diameter fluid coupling, determine the prime mover speed required for the coupling to deliver 15 hp to the load with a slip of 5 percent.

Solution

The torque to be transmitted is given by

$$T_o = \frac{550(15)}{\omega_o} = \frac{8250}{\omega_o}$$

But since $T_o = T_i$ and $\omega_o = 0.95\omega_i$, we obtain

$$T_i = \frac{8684.2}{\omega_i}$$

From Fig. 6-6, we find that at a slip of 0.05, the torque coefficient $C_T = 0.001$. Consequently,

$$\frac{T_i}{\rho\omega_i^2 D^5} = 0.001$$

Solving these last two equations simultaneously for $D = 1$ foot and $\rho = 1.65$ slug/ft^3, we find that

$$\omega_i = 174 \text{ rad/sec} = 1661 \text{ rpm}$$

It follows that

$$\omega_o = 0.95(1661) = 1578 \text{ rpm}$$

and

$$T_o = T_i = 49.9 \text{ ft-lb}$$

■

6-3 MATCHING A FLUID COUPLING TO ITS PRIME MOVER

The universal properties of the fluid coupling can be displayed in non-dimensional form as in Fig. 6-6. However, when we are considering the problem of matching the coupling to a prime mover, the convenient way of displaying the coupling input characteristics will be in dimensional form as shown in Fig. 6-7. Here, the input characteristic is shown as a family of parabolic torque-speed curves with slip s as parameter. For zero slip, the characteristic curve is simply the x axis. As slip increases, the curves ascend toward the y axis until the limiting stall curve ($s = 1$) is reached.

The fluid coupling input characteristic presented in Fig. 6-7 exhibits the same type of system load characteristic that was discussed earlier in this book.

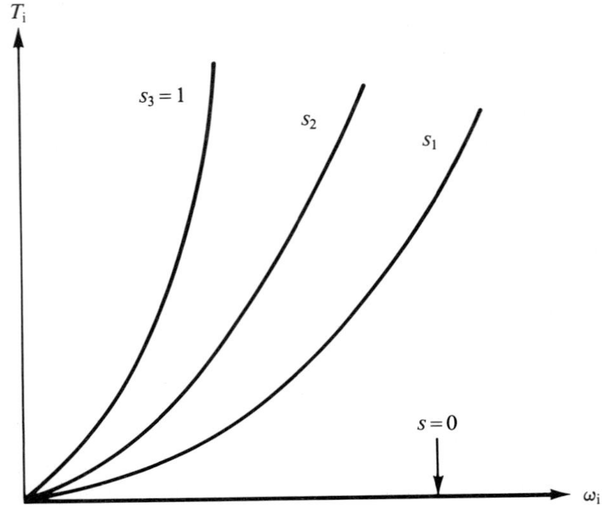

Figure 6-7. Typical input characteristics for a completely filled fluid coupling.

Here we have the torque-speed characteristic curves of the fluid coupling viewed as a system to be driven.

In Fig. 6-8, we have superimposed typical internal combustion engine characteristics upon the coupling input characteristics. The points of intersection represent potential steady-state operating points. Two engine characteristics are shown for two different throttle conditions.

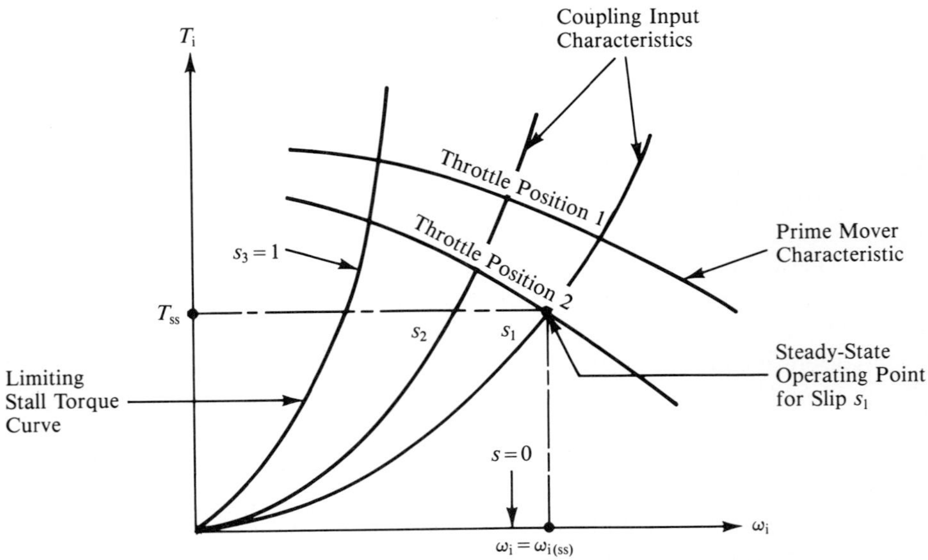

Figure 6-8. Matching a prime mover to a fluid coupling.

EXAMPLE 6-4

The torque-speed curve of a certain automobile engine at half-throttle is a parabola having the value of 125 ft-lb at 3600 rpm and a maximum value of 150 ft-lb at a speed of 2400 rpm. This engine is the prime mover of a 4000 lb automobile having a 28 inch tire diameter and a 4:1 gear reduction in the back axle. The engine is coupled to the back axle by a fluid coupling having a characteristic diameter of 12 inches. Assume this coupling belongs to the family represented by the curve of Fig. 6-6.

(a) What automobile speed is attained at the engine speed of 3600 rpm at half-throttle (on a horizontal surface)?
(b) What is the coupling efficiency for the conditions of part (a)?
(c) Assume that the vehicle drag varies as the square of vehicle speed and that the vehicle is ascending a 4 percent grade. What is the speed of ascent at engine half-throttle conditions? Assume the specific gravity of the oil in the coupling is 0.89.

Solution

The oil density is

$$\rho = \frac{0.89(62.4)}{32.2} = 1.725 \text{ slug/ft}^3$$

Then, for a torque of 125 ft-lb (at 3600 rpm engine speed) and a coupling diameter of 12 inches, the torque coefficient will be

$$C_T = \frac{125}{1.725(1)^5\left(\dfrac{2\pi(3600)}{60}\right)^2} = 0.00051$$

Reading from Fig. 6-6, we find the slip to be

$$s = 0.026$$

Thus, the output speed of the coupling is

$$\omega_o = 0.974(3600) = 3506.4 \text{ rpm} = 367 \text{ rad/sec}$$

The back axle speed is then

$$\omega_a = \frac{367}{4} = 91.8 \text{ rad/sec}$$

The tire surface speed is then (relative to the axle)

$$V_s = R\omega_a = \frac{28}{2(12)}(91.8) = 107 \text{ fps}$$

With no skidding, the vehicle speed will be

$$V_s = \frac{60}{88}(107) = 73 \text{ mph}$$

The coupling efficiency under the conditions stated is

$$\eta = 1 - s = 0.974$$

The vehicle speed is 73 mph when the coupling torque is 125 ft-lb. This torque is amplified 4:1 through the gear ratio of the back axle, so the wheel torque is

500 ft-lb. The tractive force at the wheels is then

$$F_t = \frac{500}{28}(2)(12) = 428.6 \text{ lb}$$

This value is the same as the drag on the vehicle for horizontal travel at constant speed. Therefore, the proportionality constant between drag and speed is

$$C = \frac{\text{drag}}{(\text{speed})^2} = \frac{428.6}{(107)^2} = 0.0374 \frac{\text{lb-sec}^2}{\text{ft}^2}$$

Ascending a 4 percent grade, the component of the vehicle weight along the negative direction of motion is

$$W_c = \frac{4}{100}(4000) = 160 \text{ lb}$$

So, on a 4 percent grade, the wheel tractive force must provide for both drag and gravity effects, and

$$F_t = 0.0374V_s^2 + 160$$

The velocity of this automobile is related to the output speed of the fluid coupling by the relationship

$$V_s = \frac{28}{2(12)}\left(\frac{\omega_o}{4}\right) = 0.282\omega_o$$

Therefore,

$$F_t = 0.003189\omega_o^2 + 160$$

The load torque is related to F_t by the relationship

$$T_o = \frac{28}{2(12)}\left(\frac{F_t}{4}\right) = 0.292F_t$$

Then the load torque can be expressed as

$$T_o = 0.00093\omega_o^2 + 46.72$$

The load as seen by the prime mover at the coupling input side is then

$$T_i = 0.00093(1-s)^2\omega_i^2 + 46.72 \qquad \textbf{(1)}$$

For slips of less than around 0.1, we can approximate the coupling input characteristic of Fig. 6-6 by the linear relationship

$$\frac{C_T}{s} = \frac{0.0025}{0.1}$$

or

$$T_i = 0.025\rho\omega_i^2D^5s$$

With $D = 1$ foot and $\rho = 1.725 \text{ slug/ft}^3$, this equation reduces to

$$T_i = 0.0432\omega_i^2s \qquad \textbf{(2)}$$

If a parabolic curve is fitted to the engine characteristic, the result is found to be

$$T_i = -0.00158\omega_i^2 + 0.795\omega_i + 50 \qquad \textbf{(3)}$$

Equations (1) through (3) can be solved simultaneously by trial and error, to determine that

$$T_i = 140 \text{ ft-lb}, \ s = 0.03$$

$$\omega_i = 328 \text{ rad/sec} = 3132 \text{ rpm}$$

$$\omega_o = 318.2 \text{ rad/sec} = 3038.6 \text{ rpm}$$

The steady-state operating point is shown in Fig. 6-9. Note that the tractive load of the vehicle as seen by the engine, Eq. (1), and fluid coupling input characteristic, Eq. (2), converge at the point of intersection with the engine characteristic as dictated by the slip of $s = 0.03$. This is an interesting problem in that it shows how the engine sees the actual tractive load and how it sees the fluid coupling as a load. Of course, these must be one and the same thing at the steady-state operating point.

The speed of ascent of the automobile under these equilibrium conditions is given by

$$V_s = 0.292\omega_o = 0.292(318.2) = 92.9 \text{ fps} = 63.35 \text{ mph}$$ ■

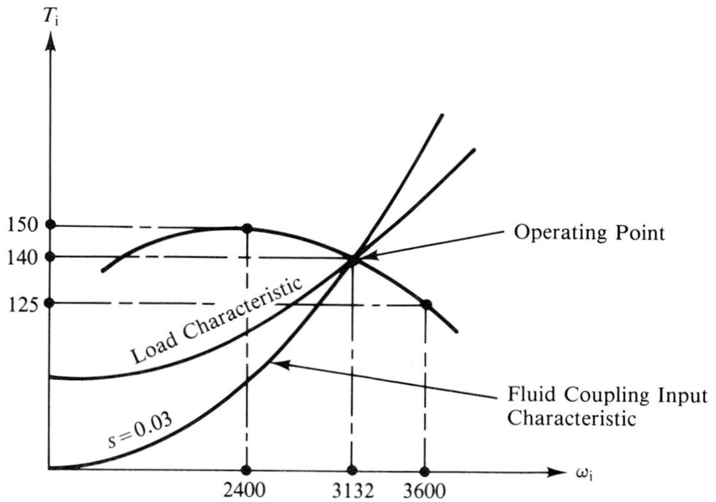

Figure 6-9. Operating point for Example 6-4.

6-4 FLUID COUPLING OUTPUT CHARACTERISTICS

The characteristic curve of a fluid coupling is shown in Fig. 6-7 from the point of view of impeller requirements (T_i versus ω_i, with slip as parameter). Viewed from the input side, the fluid coupling has characteristics of a system to be driven (in Fig. 6-8, the fluid coupling is shown driven by a typical prime mover).

The characteristic of a fluid coupling as viewed from the output will be seen to exemplify behavior typical of a driving system. We can see this behavior by substutiting Eq. (6-1) into Eq. (6-14). Recalling that $T_i = T_o$, we obtain

$$T_o = C_3 \rho R^5 \left(1 - \frac{\omega_o}{\omega_i}\right)\left[\omega_i^2 - \omega_o \omega_i \left(\frac{r}{R}\right)^2\right] \tag{6-17}$$

This equation shows that $T_o = 0$ when $\omega_o = \omega_i$, and that $T_o = T_{max}$ when $\omega_o = 0$, where

$$T_{max} = C_3 \rho R^5 \omega_i^2 \tag{6-18}$$

If we plot Eq. (6-17) with ω_i as a parameter, the family of output characteristic curves will be as sketched in Fig. 6-10. The output characteristic is the typical drooping behavior of torque versus speed that we have come to identify as a driver characteristic. Note that the slip increases along each curve from $s = 0$ at the right to the value $s = 1$ at the stall torque at the left.

Also plotted in Fig. 6-10 is a demand characteristic of a typical driven system. The curve shown is representative of an automobile's torque-speed requirement when ascending a grade as was discussed in Example 6-4. The intercept of the demand characteristic on the ordinate represents the torque needed to provide for the vehicle weight, while the remaining portion provides for friction and drag when the vehicle is in motion. The intersection of the fluid coupling output characteristic with the load characteristic defines the speed and torque at which this vehicle will operate for a given engine speed and coupling slip.

So, when a fluid coupling is used as a clutch in a vehicle, we see that the engine drives the fluid coupling impeller, the impeller drives the fluid coupling turbine, and the turbine drives the vehicle.

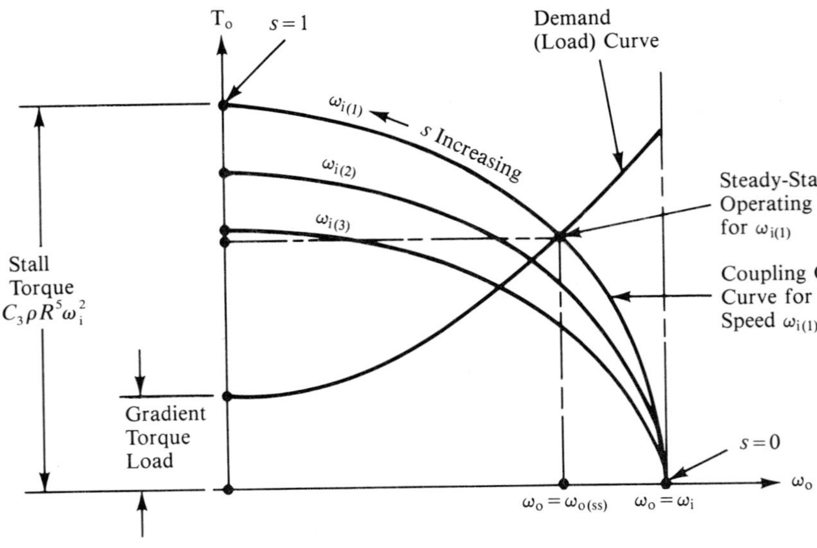

Figure 6-10. Typical output characteristics for a completely filled fluid coupling.

EXAMPLE 6-5

The straight lines of Fig. 6-11 represent the torque versus impeller speed characteristics of a fluid coupling on a log-log plot with slip as parameter. Also shown is the output torque versus shaft speed of a 50 hp internal combustion engine which drives the fluid coupling at a specific throttle setting. A load, as sketched in Fig. 6-12, is connected to the turbine of the fluid coupling.

(a) Plot the torque versus output speed characteristic of the engine, as seen by the load, on the plot of Fig. 6-12.
(b) Determine the speed and torque at which the load will be driven.
(c) What are the coupling slip and engine speed at the equilibrium operating point?

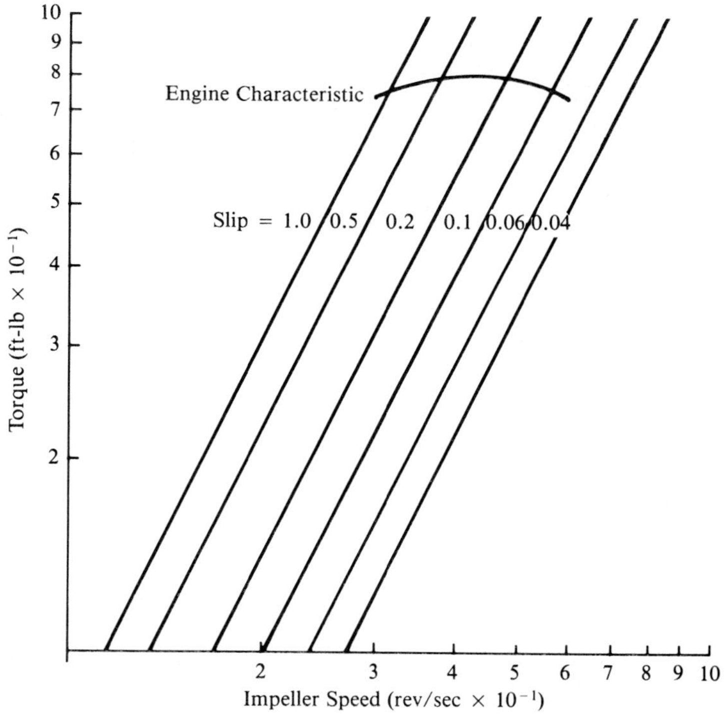

Figure 6-11. Input characteristic curves for Example 6-5.

Solution

At the intersection of the engine and fluid coupling characteristic curves of Fig. 6-11, we can determine the impeller speed and torque for a slip of $s = 0.1$. These values are

$$T_{i(s=0.1)} = 75.5 \text{ ft-lb}$$

$$\omega_{i(s=0.1)} = 56.5 \text{ rev/sec}$$

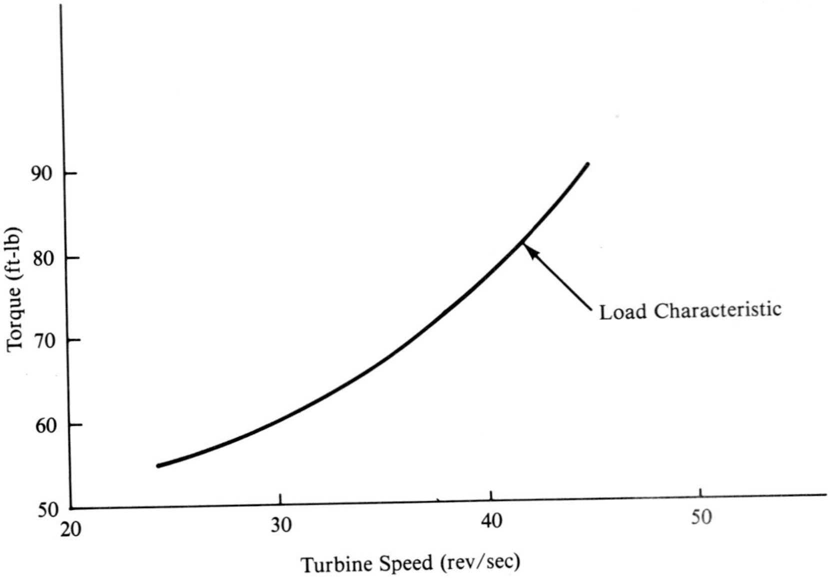

Figure 6-12. Load characteristic for Example 6-5.

The output speed and torque corresponding to these values are

$$T_{o(s=0.1)} = 75.5 \text{ ft-lb}$$

$$\omega_{o(s=0.1)} = 0.9(56.5) = 50.85 \text{ rev/sec}$$

Similarly, for the intersection points corresponding to slips of 0.2 and 0.5, we can determine that

$$T_{o(s=0.2)} = 79 \text{ ft-lb}$$

$$\omega_{o(s=0.2)} = 38.4 \text{ rev/sec}$$

$$T_{o(s=0.5)} = 79 \text{ ft-lb}$$

$$\omega_{o(s=0.5)} = 19 \text{ rev/sec}$$

If these three points are plotted on Fig. 6-12, a curve drawn through them will represent the engine curve as seen by the load. The point of intersection will be found to occur at

$$T_o = 78 \text{ ft-lb}$$

$$\omega_o = 40.6 \text{ rev/sec}$$

This is the steady-state operating point. It lies at a slip between $s = 0.1$ and $s = 0.2$ close to $s = 0.2$. We estimate the slip to be $s = 0.19$. Then the engine speed will be

$$\omega_i = \frac{40.6}{0.81} = 50.12 \text{ rev/sec}$$

■

6-5 VARIABLE-FILL FLUID COUPLINGS

Equation (6-14) is the basic angular momentum equation for a fluid coupling. It can be rewritten in the form

$$T_i = C_4 M \omega_i^2 R^2 s\left[1 - (1 - s)\left(\frac{r}{R}\right)^2\right] \tag{6-19}$$

where

$$M \sim \rho R^3 \qquad \text{(mass of hydraulic fluid)} \tag{6-20}$$

We see that for a given slip, the torque capacity of a given coupling is proportional to the mass of hydraulic fluid inside the coupling. This implies a method of speed or torque control by using couplings whose fill can be varied while they are in use. Speed control can be obtained in various ways, the most common of which is to use a scoop tube whose location inside the coupling can be varied.

Variable-fill couplings serve two important coupling functions:

(a) *Starting control.* By varying the fluid mass inside the coupling in inverse proportion to the slip, the coupling is made to function as a clutch and can be used to accelerate constant-torque loads smoothly from rest to steady-state speed. This technique is used to start long, loaded belt conveyors. More generally, machinery can be started through the use of pre-set programs that control the fill of the coupling as a function of time, or of speed, during starting.

(b) *Speed control.* By varying the fill of the coupling during machinery operation, speed control can be obtained in either the open-loop or the closed-loop mode. In some designs, a stepper motor is used to control the scoop tube position via an electronic controller. This technique has been used to control speed of centrifugal rotating equipment such as pumps and blowers. It also makes possible the use of simple squirrel cage induction motors as prime movers (see Chapters 8 and 9).

One method of controlling the fill of a fluid coupling by using a scoop tube is sketched in Fig. 6-13. In this device, a fluid reservoir is designed integral with, and adjacent to, the fluid coupling impeller and rotates with it. Small leak holes connect the impeller chamber to the reservoir chamber so all the oil will eventually be in the reservoir when the coupling is static. When the impeller and connecting reservoir are driven by the prime mover, the oil will accumulate at the outer periphery of the reservoir owing to centrifugal force. By extending a non-rotating but radially movable scoop tube into this oil pool, the oil is pumped into the impeller through the scoop tube by a ram effect. The amount of oil pumped into the impeller is controlled by the position of the scoop tube. If it is desired to decrease the amount of oil in the impeller-turbine chamber, the scoop tube is retracted toward the coupling axis. Then oil will leak through the leak holes back into the reservoir, until the pool of oil is again in contact with the scoop tube.

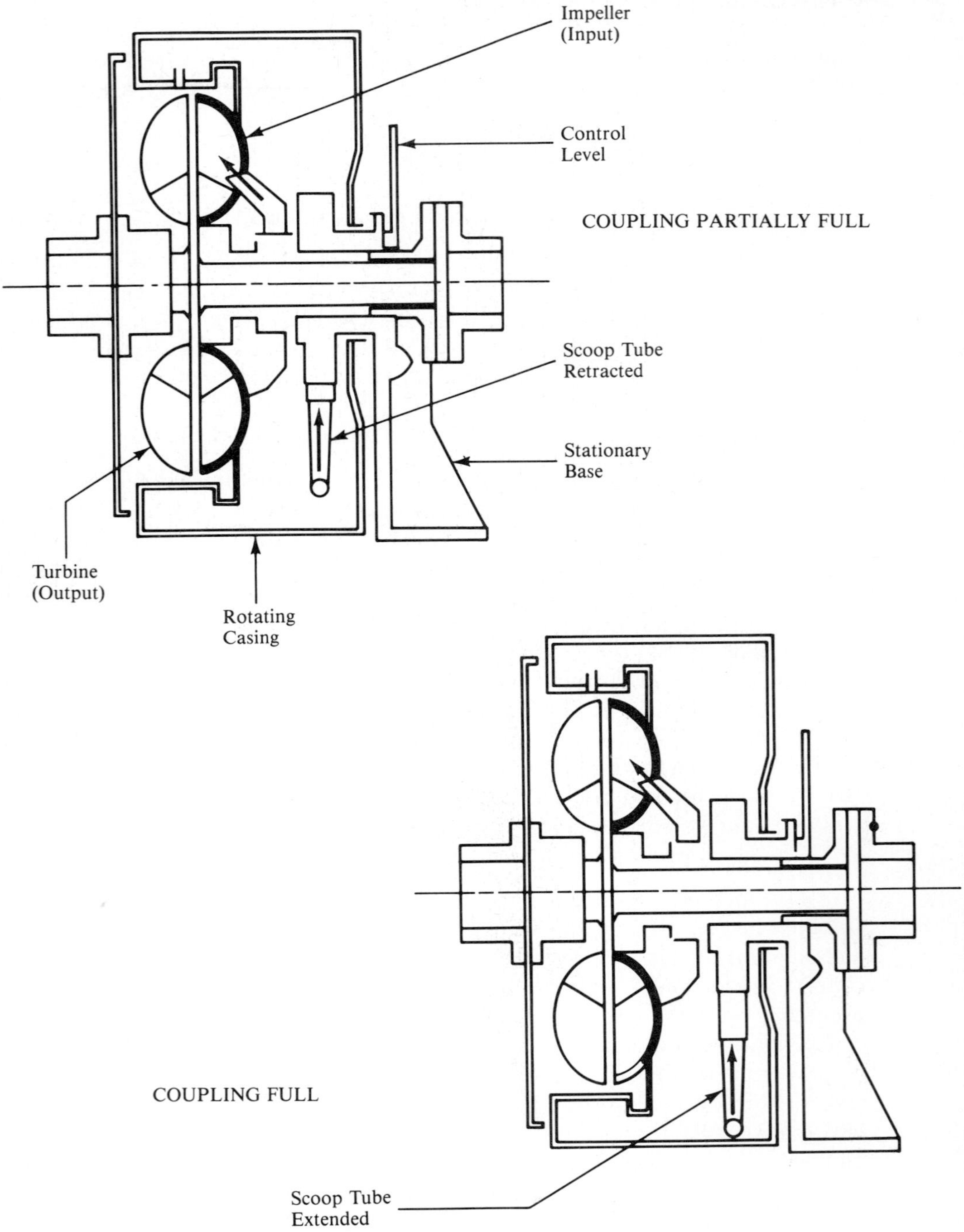

Figure 6-13. Variable-fill fluid coupling with a scoop tube.

6-5-1 Variable-Fill Coupling Output Characteristics

The effect of reduced fill on the output characteristic of a fluid coupling is to flatten the characteristic for the larger slips (see Fig. 6-14). Comparing Figs. 6-10 and 6-14, we see that changing fill at constant impeller speed has an effect similar to changing impeller speed at maximum (constant) fill. This means that control of either the impeller speed or the fill of the fluid coupling offers an opportunity for speed control in either the open-loop or the closed-loop mode. Generally, there is not as much flexibility of control over impeller speed (e.g., when an induction motor is the prime mover) as there is over the amount of fill in the fluid coupling.

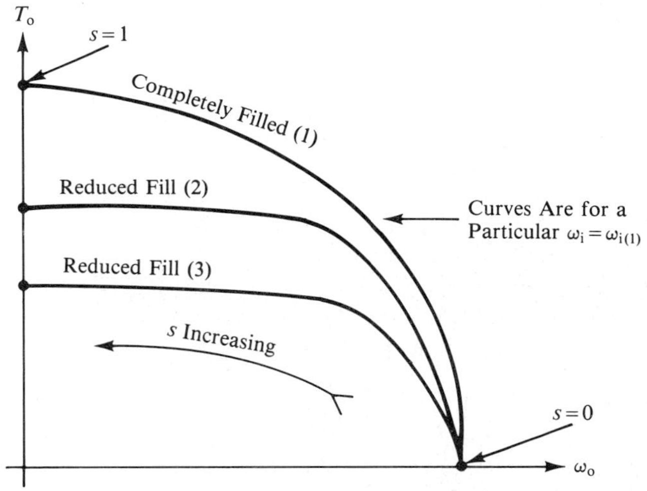

Figure 6-14. Typical output characteristics for a partially filled fluid coupling.

As we shall see in Chapter 10, the fluid coupling is the hydrodynamic analog of the electrical eddy current coupling. Furthermore, the amount of fill in the fluid coupling is the parameter analogous to the electrical excitation of the armature of the eddy current coupling. This means that fluid couplings can provide speed control, by changing the fill of the coupling, in the same way that an eddy current coupling provides speed control by changing its armature excitation. The manner in which this is accomplished is shown in Fig. 10-4, and an application that illustrates how this method of control can save power in a centrifugal pump application is illustrated in Example 10-1.

EXAMPLE 6-6

The rated torque and horsepower of one manufacturer's fluid coupling are defined to be its torque and horsepower capacity with the output (turbine) stalled when the coupling is 100 percent full of oil. The torque capability of one such coupling as a function of oil fill is sketched in Fig. 6-15.

This fluid coupling is to be used to drive a conveyor system that requires 1880 ft-lb of frictional load torque independent of speed. The system is to be started from rest, at constant acceleration, in 15 seconds. After the system reaches its steady-state speed, the load will be driven at 1715 rpm and the coupling will have a 2 percent slip.

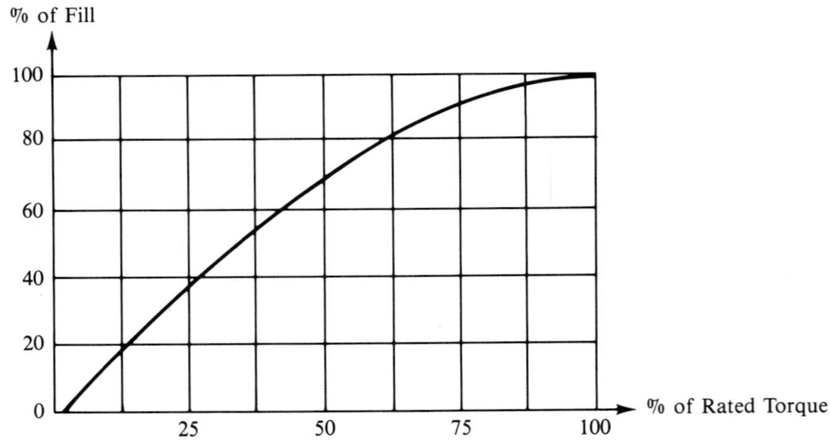

Figure 6-15. Variable-fill fluid coupling characteristic for Example 6-6.

The inertia of the conveyor system referred to the fluid coupling output shaft is 1000 lb-ft^2. The rated horsepower of the coupling is 1000 hp. Assume the impeller speed is constant.

(a) How full must this coupling be to be able to start this load?
(b) At what input horsepower will the coupling operate after the load has been accelerated to its equilibrium speed?
(c) How much horsepower is dissipated when the load has reached its equilibrium speed?

Solution

The torque required to accelerate the load is

$$T_a = I\alpha = I\,\frac{\omega}{t} = \left(\frac{1000}{32.2}\right)\left(\frac{1715}{60}\right)\left(\frac{2\pi}{15}\right) = 372 \text{ ft-lb}$$

The total starting torque is then

$$T_t = T_a + T_f = 372 + 1880 = 2252 \text{ ft-lb}$$

The impeller speed at steady state is

$$\omega_i = \frac{\omega_o}{1 - s} = \frac{1715}{0.98} = 1750 \text{ rpm}$$

Then the rated torque of this fluid coupling is

$$T_r = \frac{33{,}000(1000)}{2\pi(1750)} = 3001 \text{ ft-lb}$$

The percentage of rated torque required is therefore

$$\frac{T_t}{T_r} = \frac{2252}{3001} = 0.75$$

From Fig. 6-15, we see that the percent fill required will be 90 percent.

At full-load under steady-state conditions, the input horsepower will be

$$\text{HP}_i = \frac{2\pi(1750)(1880)}{33{,}000} = 626 \text{ hp}$$

The horsepower dissipated under these conditions is

$$\text{HP}_{diss} = 0.02(626) = 12.5 \text{ hp}$$

■

6-6 FLUID COUPLING BRAKING CHARACTERISTICS

Imagine that the slip is zero between the impeller and turbine of a fluid coupling; both impeller and turbine have the same angular speed. Under these circumstances, there will be no transfer of fluid from impeller to turbine, nor will there be any transfer of fluid angular momentum. Consequently, the torque capability is zero.

Another way of viewing this situation is to see the centrifugal head developed by the turbine to be the same as that developed by the impeller and in opposition to it. These opposing heads balance each other and prevent the transfer of both mass and momentum when impeller and turbine have the same speed (see Fig. 6-16[a]).

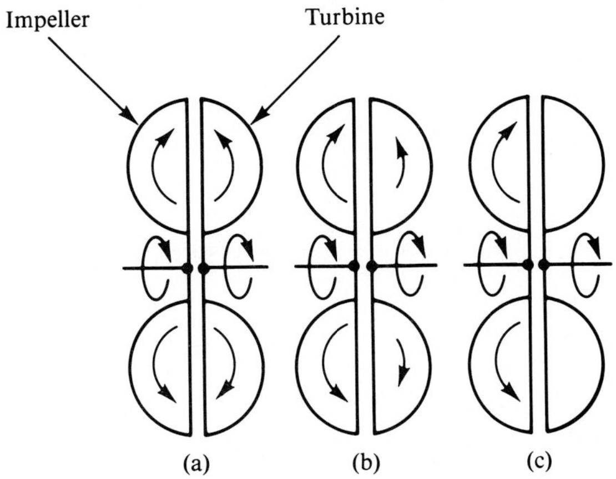

Figure 6-16. Impeller versus turbine-developed heads.

The situation sketched in Fig. 6-16(a) is analogous to that of a DC motor (see Chapter 8) in which the counter-emf developed by the armature as a result of its rotation is in opposition to the emf applied to the terminals of the motor. When the counter-emf is equal and opposite to the emf, there is no current flow in the armature and, hence, no torque developed by the motor.

As the turbine of the fluid coupling is slowed by a torque load on its shaft, its developed head will be diminished compared to that of the impeller which is being rotated, say, at constant speed (see Fig. 6-16[b]). The impeller head will overcome the opposing turbine head, and fluid mass and angular momentum will be forced from impeller to turbine to provide for the torque requirement. The slip between impeller and turbine will be dictated by this balance between torque and angular momentum.

When the turbine is stalled, its centrifugal head will disappear completely and the impeller head will be completely unopposed (see Fig. 6-16[c]). Maximum mass and angular momentum will then be transferred, and the (stall) torque developed will be at a maximum.

Except for the difference in shearing forces at the impeller-turbine interface, the effects we have described above are independent of the direction of rotation of the turbine! In other words, if the turbine is slowed from a value equal to impeller speed ($s = 0$) to zero speed ($s = 1$), and it then begins to rotate in a direction opposite to that of the impeller ($s > 1$), the turbine will again begin to develop a head in opposition to the impeller head even though it is rotating in the opposite direction! When the slip reaches the value $s = 2$, the turbine will be rotating opposite to the impeller but at the same angular speed. Its head will now again equal that of the impeller so the torque developed will be zero. The characteristic of these circumstances is sketched in Fig. 6-17.

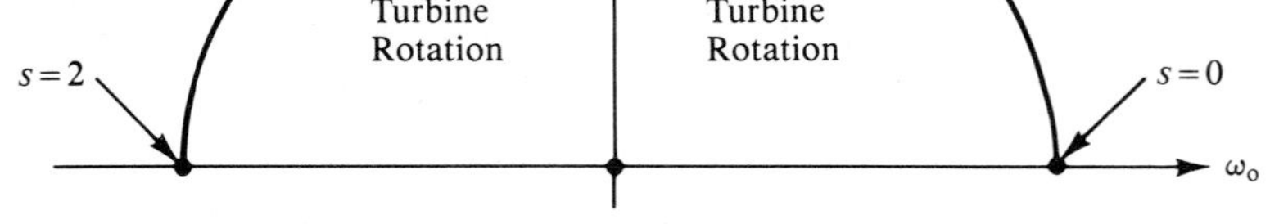

Figure 6-17. Operating domains of a fluid coupling.

The domain ($1 \leq s \leq 2$) of Fig. 6-16 represents braking of the load since the speed of the load will now be in opposition to torque applied to it. Note that the braking torque vanishes if, and when, the load is rotating at the same speed as the impeller but in the opposite direction!!

If a fluid coupling is used in an application such as hoisting, extreme care must be taken in control if the coupling is to be used for hoisting and also as a brake for lowering. The load could run away upon lowering if the stall torque for a given impeller speed is exceeded by the load torque! This can produce a hazard to both personnel and property.

If the slip should exceed $s = 2$, the direction of torque transfer will change and the falling load will try to run the prime mover backward, which could damage it.

For negative slips ($s < 0$), the turbine has the same direction of rotation as the impeller but over-runs it. Again, the torque will be reversed such as to tend to overspeed the prime mover (but it will act as a brake for the over-running load).

EXAMPLE 6-7

Suppose a fluid coupling is being used in a hoisting application where the impeller is being driven by a 1750 rpm induction motor. A certain wieght is being lifted at 5 percent slip. What will be the speed of the coupling output for hoisting, and what will be the speed at which this same load can be lowered at constant speed?

Solution

We assume that the characteristic curve of Fig. 6-17 is symmetrical about a slip of $s = 0$. The hoisting speed will be

$$\omega_o = 0.95(1750) = 1662 \text{ rpm}$$

This same torque will occur, during braking, when the slip is $s = 1.95$. Consequently,

$$\omega_o = (1 - 1.95)(1750) = -1662 \text{ rpm}$$

■

6-7 THE FLUID COUPLING TORQUE DIVIDER

An arrangement is shown in Fig. 6-18 in which a fluid coupling is combined with an epicyclic gear train. This transmission, called a torque divider or split-power

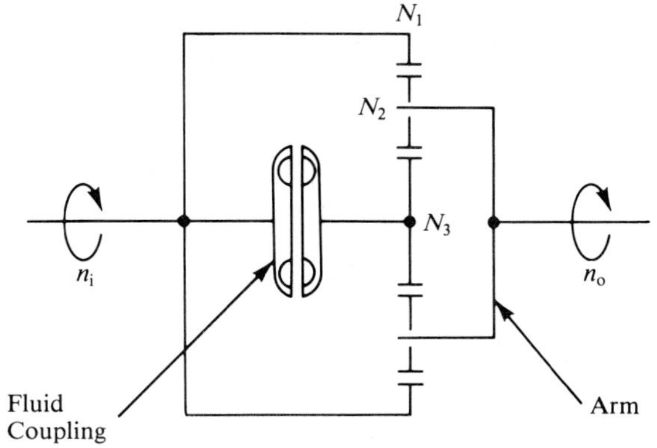

Figure 6-18. Torque divider.

transmission, has been used in automobile power trains to provide smooth acceleration and to amplify the torque that can be effectively carried by a small fluid coupling in a variable-speed application.

If the slip s of the fluid coupling and gear ratio α of the epicyclic are defined as

$$s = 1 - \frac{\omega_3}{\omega_i} \qquad \alpha = \frac{N_1}{N_3} \tag{6-21}$$

then it is possible to show (see Chapter 2) that the ratio of output speed to input speed is

$$\frac{\omega_o}{\omega_i} = \frac{(1-s)+\alpha}{1+\alpha} \tag{6-22}$$

Note that when $s = 0$, the device simply rotates as a rigid body, with $\omega_o/\omega_i = 1$. On the other hand, if the sun gear (N_3) is stalled so that $s = 1$, then

$$\frac{\omega_o}{\omega_i} = \frac{\alpha}{1+\alpha} \tag{6-23}$$

The slip of the overall transmission is defined as

$$S = 1 - \frac{\omega_o}{\omega_i} \tag{6-24}$$

By combining Eqs. (6-22) and (6-24), it is possible to show that

$$S = \frac{s}{1+\alpha} \tag{6-25}$$

So we see that the slip of the overall transmission is less than that of the fluid coupling.

It is possible to show further that the torque and power that the fluid coupling transmits as fractions of the output torque and power are

$$\frac{T_3}{T_o} = -\frac{1}{1+\alpha} \qquad \frac{P_3}{P_o} = -\frac{(1-s)}{(1-s)+\alpha} \tag{6-26}$$

The power dissipated in the fluid coupling as a fraction of the input power is

$$\frac{P_{diss}}{P_i} = -\frac{s}{1+\alpha} \tag{6-27}$$

The minus signs in Eqs. (6-26) and (6-27) indicate that the quantities have opposite directions. In Eq. (6-27), for example, the power lost in the coupling represents power flow out of the system.

EXAMPLE 6-8

Consider a torque divider whose planetary train has the gear ratio $\alpha = 3$. Determine the performance of this transmission for a fluid coupling slip of 5 percent ($s = 0.05$).

Solution

The ratio of output speed to input speed is

$$\frac{\omega_o}{\omega_i} = \frac{(1-0.05)+3}{1+3} = 0.9875$$

The speed ratio limits of this transmission are

$$(s=1)\ 0.75 \leq \frac{\omega_o}{\omega_i} \leq 1.0\ (s=0)$$

The overall transmission slip is

$$S = \frac{0.05}{1+3} = 0.0125 = 1.25 \text{ percent}$$

The fluid coupling torque compared to the output torque is

$$\frac{T_3}{T_o} = -\frac{1}{1+3} = -0.25$$

The power transmitted by the coupling compared to the output power is

$$\frac{P_3}{P_o} = -\frac{(1-0.05)}{(1-0.05)+3} = -0.24$$

Finally, the power dissipated as a fraction of the input power is

$$\frac{P_{diss}}{P_i} = -\frac{0.05}{1+3} = -0.0125$$ ■

6-8 THE TORQUE CONVERTER

If we plot the output to input torque ratio and mechanical efficiency as functions of the output to input speed ratio, the results for a fluid coupling will be as shown in Fig. 6-19. Since the torque ratio is constant, the fluid coupling, by itself, will not be a suitable power transmission component for machines requiring torque

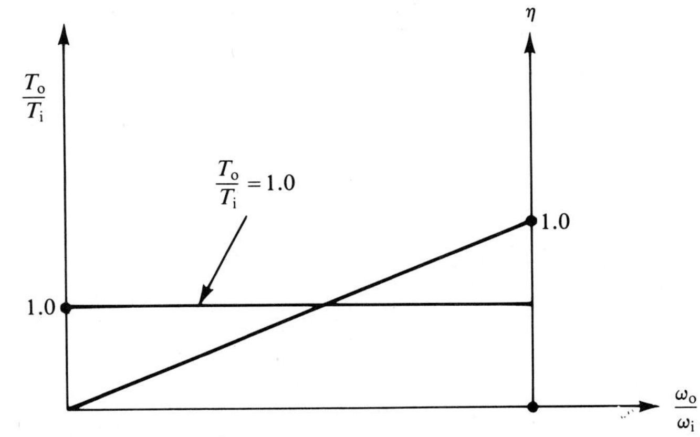

Figure 6-19. Fluid coupling dimensionless characteristics.

amplification at some point in their operation (typically during start-up or other periods of acceleration).

The torque converter (which would be better named torque amplifier) is a transmission device somewhat similar to a fluid coupling but with a modification that permits torque amplification to be achieved (particularly at high slips, that is, low ratios of ω_o/ω_i). A sketch of a torque converter configuration is shown in Fig. 6-20. Note that in this version of a torque converter, both the impeller and turbine lie on one side of the casing while on the other side is a fixed guide vane device called the stator (or reactor). Other types of internal configurations are described in Reference 6-1.

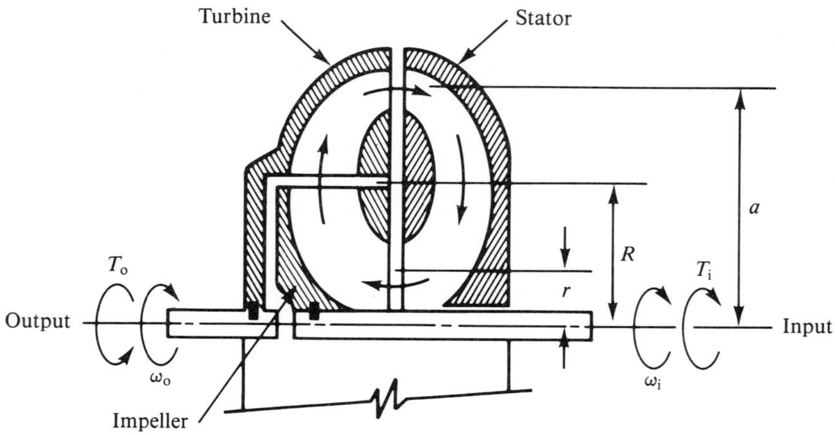

Figure 6-20. Schematic diagram of a torque converter.

In contrast to the fluid coupling, a free-body diagram of the torque converter illustrated in Fig. 6-20 will now show that the steady-state (equilibrium) torque relationship is

$$T_i + T_o + T_s = 0 \tag{6-28}$$

or

$$\left|\frac{T_o}{T_i}\right| = 1 - \frac{T_s}{T_i} \tag{6-29}$$

where T_s is the torque required to hold the stator at rest. We see that the magnitude of the output/input torque ratio is not necessarily unity. At higher slips, T_s will be in a direction opposite to T_i, and torque amplification will be obtained.

The mechanical efficiency now becomes

$$\eta = \frac{\omega_o T_o}{\omega_i T_i} = (1 - s)\frac{T_o}{T_i} \tag{6-30}$$

Since T_o/T_i is not necessarily constant (unity) as it is for a fluid coupling, the mechanical efficiency of the torque converter is not a simple linear function of slip as is shown in Fig. 6-19. The efficiency will generally be a maximum in the slip interval of approximately $0.2 \le s \le 0.4$, and the maximum value will not generally exceed $\eta = 0.85$ (see Fig. 6-21). So, for the maximum efficiency, torque converters are operated at higher slips than are fluid couplings, which operate at high efficiencies with small slips of the order of 0.05 and less.

The blading in a torque converter is not simple radial blading as it usually is in the fluid coupling. Therefore, development of the angular momentum principle as applied to the torque converter is considerably more complex. It is not our purpose here to become involved in the actual detailed fluid mechanics analysis appropriate for torque converter blading. This has been done in Reference 6-2. It will be sufficient for us to simply write the following three torque equations for the impeller, turbine, and stator:

$$T_i = \rho Q(RV_i - rV_o) \tag{6-31}$$

$$T_o = \rho Q(RV_i - aV_s) \tag{6-32}$$

$$T_s = \rho Q(rV_o - aV_s) \tag{6-33}$$

These equations are similar in form to Eq. (6-4) developed earlier for the fluid coupling.

Quantities V_i, V_o, and V_s are simply characteristic velocities which could be determined from the actual fluid velocity vector polygons associated with particular blading configurations. Similarly, R, r, and a are regarded as characteristic lengths appropriate to the analysis.

A development for the factor ρQ similar to that made previously for the fluid coupling will show that Eq. (6-31) can be written in the form

$$C_T = \frac{T_i}{\rho \omega_i^2 D^5} = f\left(s, \frac{r}{R}\right) \tag{6-34}$$

Function f is a slight function of Reynolds number as well. But, basically, we find that the input characteristic of the torque converter is of the same form as for the fluid coupling, namely, parabolic for a given slip. This function can be determined by performing experiments on families of geometrically similar torque converters.

Equations (6-31) and (6-32) can be combined to produce

$$\frac{T_o}{T_i} = \frac{1 - \left(\frac{a}{R}\right)\left(\frac{V_s}{V_i}\right)}{1 - \left(\frac{r}{R}\right)\left(\frac{V_o}{V_i}\right)} \tag{6-35}$$

A detailed analysis of the torque converter fluid kinematics will show that ratio V_o/V_i increases and ratio V_s/V_i decreases as the speed ratio ω_o/ω_i decreases (slip s increases). The result will be a torque amplification curve that is essentially a function of slip only, as sketched in Fig. 6-21. The solid curves of Fig. 6-21 refer to the torque converter characteristics whereas the dashed curves represent the fluid coupling characteristics.

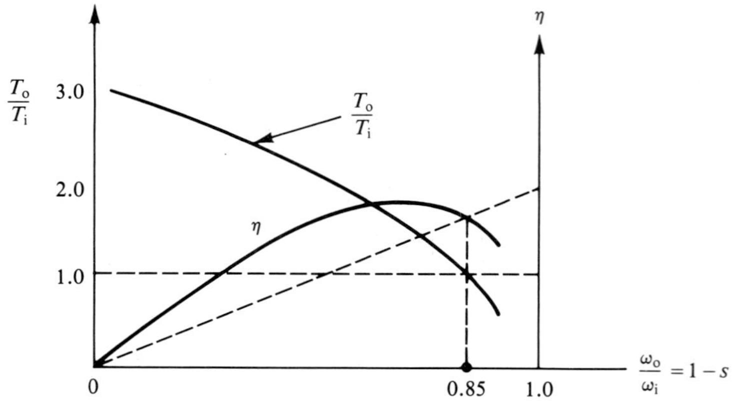

Figure 6-21. Torque converter dimensionless characteristics.

Note that the ratio T_o/T_i for the torque converter becomes unity, for this particular converter, around the value $\omega_o/\omega_i = 0.85$. This means that the torque T_s, required to restrain the stator from rotating diminishes as the torque amplification falls, and changes sign after the amplification ratio falls below unity (see Eq. [6-29]). Above the value $\omega_o/\omega_i = 0.85$, the direction of the torque on the stator will tend to cause the stator to turn in the same direction as the turbine.

6-8-1 The Two-Phase Torque Converter

The torque converter of Fig. 6-20, whose characteristics are displayed in Fig. 6-21, is called a single-stage, single-phase torque converter. It is a single-stage converter because it has only one stator element. It is a single-phase converter because, as described in Fig. 6-20, it can operate only in the torque converter phase and not in the fluid coupling phase.

Note in Fig. 6-21 that for speed ratios ω_o/ω_i greater than 0.85, the output torque of the torque converter falls below the input torque. The efficiency also begins to fall off rather drastically in this interval. This type of behavior tends to undo all of the advantages of the converter as a torque amplifier for the smaller speed ratios. This limitation can be avoided by incorporating an over-running clutch on the stator so that the stator is prevented from rotating for speed ratios below 0.85 and permitted to turn for ratios greater than 0.85.

Since the direction of torque T_s exerted on the stator changes direction when the torque ratio falls below unity (at $\omega_o/\omega_i = 0.85$ in this example), incorporating a one-way clutch permits the stator to turn in the same direction as the turbine for speed ratios exceeding 0.85. The free-wheeling of the stator will result in its rpm becoming essentially the same as that of the turbine. Thus, in this phase of operation, the torque converter will have the same configuration as a fluid coupling.

The phase of operation for speed ratios below 0.85 is called the torque converter phase, while that above 0.85 is called the fluid coupling phase. The combination is called a two-phase torque converter. The combined dimensionless characteristics of the two-phase converter are sketched in Fig. 6-22. When an engine is used as a prime mover to drive a torque converter such as that defined in Fig. 6-22, a converter size is usually selected that will absorb maximum net engine horsepower when the converter is operating somewhere between peak efficiency and the point where the torque ratio is unity.

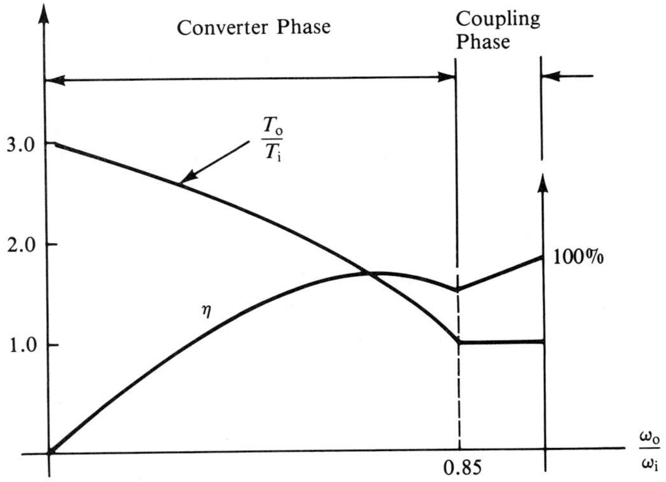

Figure 6-22. Two-phase torque converter.

In addition to the two-phase, single-stage torque converter discussed above, more complex geometries have been designed for special applications, particularly for vehicle transmissions. More than one stator element can be used, each of which has its over-running clutch. These elements are called polyphase torque converters. In the larger sizes, torque converters have been built with torque amplifications of up to 5:1 and with capacities up to 20,000 hp.

EXAMPLE 6-9

The sprockets of the propelling tracks of a bulldozer have a pitch diameter of 24 inches. They are driven by a concentric-gear reducer having a speed ratio of 12:1. The gear reducer input shaft is coupled to an internal combustion engine by a single-phase, single-stage torque converter.

The full-throttle torque available from the engine is a function of engine speed as specified in Table 6-1. Also shown in Table 6-1 are the results of tests on the torque converter for an input speed of 1400 rpm.

(a) What is the maximum tractive effort that the bulldozer can develop?

(b) If the bulldozer is running at a constant speed such that the engine is running at 1800 rpm at full-throttle, what is the bulldozer's speed?

TABLE 6-1. Data for Example 6-9.

Full-Throttle Engine Characteristic

Output torque, T_i (ft-lb)	430	440	425	405	375
Engine speed (rpm)	1000	1200	1400	1600	1800

Torque Converter Characteristic for $\omega_i = 1400$ rpm

Input torque, T_i (ft-lb)	600	545	490	415	325	230
Output torque, T_o (ft-lb)	2870	2060	1410	810	400	130
Output speed, ω_o (rpm)	0	200	400	600	800	1000

Solution

The torque ratio T_o/T_i of a torque converter is substantially independent of input speed ω_i; it depends only on the speed ratio ω_o/ω_i as indicated in Fig. 6-21. The maximum torque ratio occurs when the turbine is stalled. Therefore, we can determine the maximum torque ratio from the first column of the bottom portion of Table 6-1:

$$\frac{T_o}{T_i} = \frac{2870}{600} = 4.783 \qquad (\text{for } \omega_o = 0)$$

The engine torque is a maximum at the speed of $\omega_i = 1200$ rpm. The value is $T_i = 440$ ft-lb. Consequently, the maximum output torque that can be provided by the fluid coupling is

$$T_{o(\max)} = 4.783(440) = 2104 \text{ ft-lb}$$

The torque that can be provided to the sprocket by the 12:1 reducer is then

$$T_{s(\max)} = 12(2104) = 25{,}248 \text{ ft-lb}$$

This is also the maximum bulldozer tractive effort because the sprocket radius is 1 foot. So

$$\text{TE}_{\max} = 25{,}248 \text{ lb}$$

For a given slip (or speed ratio ω_o/ω_i), Eq. (6-34) shows that the input torque developed by the converter is proportional to the square of the input speed. Therefore,

$$T_i = C\omega_i^2 \qquad (\text{for a specific } \omega_o/\omega_i)$$

The constant C will depend on the slip (or speed ratio ω_o/ω_i). For example, at zero output speed ($s = 1$, $\omega_o/\omega_i = 0$), we can determine the constant C from the first column of the torque converter data of Table 6-1. In other words,

$$600 = C(1400)^2$$

$$C = 0.000306$$

$$T_i = 0.000306\omega_i^2 \qquad (\text{for } \omega_o/\omega_i = 0)$$

Proceeding in this fashion, we can determine a constant C for each input torque curve corresponding to a given slip (or ratio ω_o/ω_i). For the second column, for example, the ratio ω_o/ω_i is 200/1400, or 0.143. The corresponding constant C is $545/(1400)^2$, or 0.000278.

Then the input torque corresponding to these ω_o/ω_i ratios for an input speed of $\omega_i = 1800$ rpm can be calculated from the various speed-torque curves. The values are presented in Table 6-2. As an illustration, for $\omega_o/\omega_i = 0.143$, we find that

$$T_i = 0.000278(1800)^2 = 900 \text{ ft-lb}$$

TABLE 6-2. Converter Characteristics for $\omega_i = 1800$ rpm.

Input torque, T_i, at $\omega_i = 1400$	600	545	490	415	325	230
Output speed, ω_o	0	200	400	600	800	1000
Ratio ω_o/ω_i	0	0.143	0.286	0.429	0.571	0.715
Constant $C = T_i/\omega_i^2$	0.000306	0.000278	0.00025	0.000211	0.000166	0.000117
Input torque, T_i, at $\omega_i = 1800$	991	900	810	683	538	379

For an engine speed of 1800 rpm, the engine torque developed is $T_i = 375$ ft-lb. From Table 6-2, we see that this value corresponds to a speed ratio of about $\omega_o/\omega_i = 0.715$. Consequently, the converter output speed will be

$$\omega_o = 0.715(1800) = 1287 \text{ rpm}$$

The corresponding bulldozer sprocket speed will be

$$\omega_{\text{sprocket}} = \frac{1287}{12} = 107.25 \text{ rpm}$$

The track speed is then

$$V_t = \frac{2\pi(1)(107.25)}{60} = 11.23 \text{ fps}$$

The bulldozer speed will then be

$$V_t = \frac{60}{88}(11.23) = 7.66 \text{ mph}$$

■

EXAMPLE 6-10

The requirements of a load and the availability of a prime mover are such that a load is to be coupled to the prime mover by a shiftable three-speed gearbox in series with a torque converter. The three gearbox speed ratios are 3.0, 1.8, and 1.0.

The prime mover speed is governed at 1800 rpm, and the maximum speed of the load is 1500 rpm, at which speed the load requires 500 ft-lb of torque. At this maximum speed, the load is directly connected to the prime mover through the torque converter (i.e., the gearbox ratio is 1.0). Shifting of the gearbox occurs at load speeds of 240 and 600 rpm.

The torque converter characteristics are displayed in Fig. 6-23. Assume that the mechanical efficiency of the gearbox is 100 percent and that the torque capability of the prime mover is constant. Determine

(a) The maximum starting torque available to the load
(b) The maximum torque available to the load at the mean load speed provided by the transmission when the gearbox is in its high ratio (i.e., when the transmission is in "low gear")
(c) The efficiency of the torque converter under the conditions of part (b)
(d) The efficiency of the torque converter of a transmission that will provide the load speed of part (b) but that contains only the torque converter (i.e., without the gearbox)
(e) The torque capability of a prime mover to provide the torque availability to the load as determined in part (b) but without the gearbox
(f) The maximum torque available to the load at the mean load speed provided by the transmission when the gearbox is in its intermediate ratio (1.8:1)
(g) The efficiency of the torque converter at the conditions of part (f)
(h) The efficiency of the torque converter of a transmission that will provide the load speed of part (f) but that has no gearbox

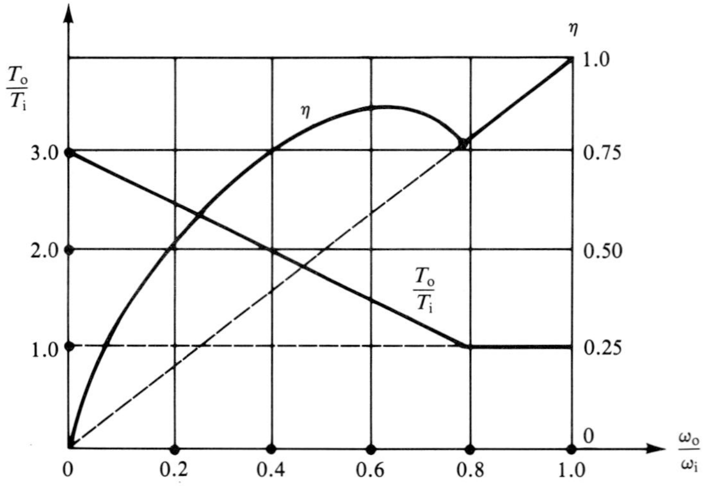

Figure 6-23. Torque converter characteristics for Example 6-10.

Solution

(a) The maximum starting torque will occur when the gearbox ratio is 3.0 and the load is stalled. Therefore,

$$T_s = 3(3)(500) = 4500 \text{ ft-lb}$$

Note that the maximum prime mover torque is 500 ft-lb, as dictated by the condition of direct connection.

(b) The mean speed in low gear is $\omega_o = 120$ rpm. The torque converter input at this speed is $T_i = 3(500) = 1500$ ft-lb. The torque converter input speed under these conditions is $\omega_i = 1800/3 = 600$ rpm. Therefore,

$$\frac{\omega_o}{\omega_i} = \frac{120}{600} = 0.2$$

We see that $T_o/T_i = 2.5$ at $\omega_o/\omega_i = 0.2$. Consequently,

$$T_o = 2.5(1500) = 3750 \text{ ft-lb}$$

(c) The efficiency of the coupling at $\omega_o/\omega_i = 0.2$ is seen to be $\eta = 0.5$.

(d) Without the gearbox, the speed ratio would have to be $\omega_o/\omega_i = 120/1800 = 0.067$. At this speed ratio, we see that $\eta = 0.25$.

(e) We require $T_o = 3750$ ft-lb at a speed ratio of $\omega_o/\omega_i = 0.067$. The amplification at this ratio is $T_o/T_i = 2.67$. Therefore,

$$T_i = \frac{3750}{2.67} = 1406 \text{ ft-lb}$$

(f) The mean output speed will be $\omega_o = (600 + 240)/2 = 420$ rpm. The input speed is $\omega_i = 1800/1.8 = 1000$ rpm. Therefore, $\omega_o/\omega_i = 0.42$. At this speed ratio, the torque amplification of the coupling is $T_o/T_i = 2.0$. The gearbox torque amplification is a factor of 1.8. Therefore,

$$T_o = (2.0)(1.8)(500) = 1800 \text{ ft-lb}$$

(g) At $\omega_o/\omega_i = 0.42$, we see that $\eta = 0.75$.

(h) Without the gearbox, we would have $\omega_o/\omega_i = 420/1800 = 0.233$, for which $\eta = 0.5$. ■

6-9 REFERENCES

6-1. *A Guide to Hydrodynamic Drive Applications*, Twin Disc Clutch Co., Rockford, IL, 1961. (This is, in part, a collection of articles appearing in *Machine Design* in 1960.)

6-2. A. H. Burr, *Mechanical Analysis and Design*, Elsevier, New York, 1982.

6-3. J. C. Hunsaker and B. G. Rightmire, *Engineering Applications of Fluid Mechanics*, McGraw-Hill, New York, 1947.

PROBLEMS

6-1. How does the power transmitted by a fluid coupling vary with impeller speed and impeller radius? If a fluid coupling is doubled in size, what is the percent increase in power and torque capacity?

6-2. Assume the function $f(s, r/R)$ of Eq. (6-15) is

$$f\left(s, \frac{r}{R}\right) = 0.006[1 - (1 - s)^4]$$

This torque coefficient function applies to a class of geometrically similar fluid couplings.

(a) What will be the characteristic diameter of a fluid coupling that has an input power of 125 hp at an input speed of 600 rpm when the turbine is stalled? Assume the hydraulic fluid density is 51.5 lb/ft^3.

(b) What will be the efficiency of this coupling if it has an input power of 198 hp with an input speed of 1100 rpm? What will be the turbine speed under these conditions?

6-3. Suppose a family of geometrically similar fluid couplings has its input characteristics represented by the equation

$$T_i = 0.008\rho\omega_i^2 D^5 s^{1/2}$$

If the oil density is to be 1.65 slug/ft^3, determine the coupling characteristic diameter D required to deliver 25 hp to the load with a prime mover speed of 1600 rpm and a slip of 3 percent.

6-4. Assume that the stall torque of the engine of Example 6-4 is linearly proportional to throttle position. What will the throttle position have to be to just hold the automobile static on the 4 percent grade?

6-5. Show how, conceptually, speed can be controlled as torque changes by controlling fluid coupling fill in a closed-loop mode. See Chapter 10.

6-6. What are the magnitude and direction of turbine speed ω_o compared to impeller speed ω_i for $s = 1.5$? Repeat this calculation for $s = -0.5$.

6-7. A hoist load can be elevated at a slip of $s = 0.04$ with an 1800 rpm prime mover. What is the speed at which the load can be lowered without changing the speed of the prime mover? Assume the torque-speed curves are symmetrical about $s = 1$ and $s = 2$.

6-8. An existing fluid coupling is to be redesigned to increase its efficiency from 94 percent to 97 percent under the same operating conditions. The new coupling is to be geometrically similar to the existing one. What will be the percent change required in the characteristic diameter?

6-9. A machine load is coupled to a 1760 rpm induction motor by a fluid coupling. The motor delivers 2000 ft-lb of torque when the coupling slip is 3.5 percent. What are the output speed, torque, and horsepower delivered to the load by the coupling? Under temporary overload conditions, the torque rises by 50 percent while the motor speed drops to 1740 rpm. What are the new values of output speed, torque, and horsepower?

6-10. The torque-speed curve of a class of aircraft engines, at half-throttle, droops by 100 percent as the speed is increased from 2400 rpm to 3600 rpm. The engine is coupled to the propeller of a small blimp by a fluid coupling. The drag coefficient of the blimp is $C_d = 0.5$, and the blimp's projected area is 200 square feet. If the blimp is traveling at 10 mph when the engine speed is 3000 rpm and the coupling slip is 5 percent, what will be the engine speed and coupling slip when the blimp travels at 15 mph? Assume the engine can provide the requisite torque at half-throttle.

6-11. Derive Eqs. (6-22), (6-25), (6-26), and (6-27).

6-12. Assume the ring gear in a torque divider has $N_1 = 49$ teeth while the sun gear has $N_3 = 30$ teeth. How much of the coupling output torque is transmitted by the fluid coupling?

6-13. A fluid coupling is being considered to provide a torque T_o directly to a load at a coupling slip of $s = 0.05$. An alternative arrangement is to use a torque divider (using this same coupling) to provide the same torque. Assume the coupling torque is proportional to slip for the range of slips that will occur. Additionally, assume the planetary gear train is frictionless, with $N_1 = 49$ and $N_3 = 30$. What are the respective mechanical efficiencies of these two alternative arrangements?

6-14. If the torque of Problem 6-13 is 146 ft-lb and the load operates continuously at a speed of 3600 rpm, what is the annual savings in energy (in kWh) provided by the torque divider alternative over the direct coupling arrangement?

6-15. A prime mover supplies 180 ft-lb of torque to a single-stage torque converter at an input speed of 2800 rpm. The torque converter is operating at a torque ratio of 3.0 when the slip is 0.75.

(a) What is the torque acting on the stator?

(b) What is the efficiency of the torque converter?

6-16. A load requires a torque of 600 ft-lb at a speed of 600 rpm. A single-stage torque converter drives the load by means of a prime mover operating at 1750 rpm. What power will be required from the prime mover, and what torque and speed ratios will result, if the converter is operated at an efficiency of 0.75?

6-17. Suppose the torque converter of Example 6-9 is a two-phase, single-stage converter. Determine the bulldozer speed at which the torque converter will revert to fluid coupling operation.

7

Hydroviscous Drives

The fundamental characteristic of a hydrostatic drive is that the static pressure developed at the positive-displacement pump outlet will automatically rise to meet the torque demand encountered by the positive-displacement motor. In the absence of a pressure relief valve, or pressure compensation in the pump, the outlet speed of the hydrostatic drive will be relatively independent of torque demand, provided the prime mover, driving the pump, has sufficient power.

In a hydrodynamic drive, such as a fluid coupling, torque is developed by the transfer of angular momentum rather than from static pressure. However, being a non-positive-displacement device, torque is developed by slip between the impeller and turbine halves of the fluid coupling. Consequently, for a given input speed (held constant), the output speed will be an inverse function of output torque.

Yet another way of developing torque in a fluid drive is to use the property of viscousness (viscosity) of the fluid. If a shearing action of the fluid can be developed, say, between two adjacent rotating disks, it will be possible to transmit a torque from one of the disks to the other. This is the fundamental principle of the hydroviscous drive (also called an oil-shear drive) (see Fig. 7-1). Although this type of drive also requires a slip in order to transmit torque hydraulically, it differs from the hydrodynamic drive in several respects.

Under steady conditions of operation, a Couette flow is established in the oil film located between the two rotating disks when a slip exists. This shearing flow of the oil will result in a transmitted hydraulic torque proportional to the disk relative speed $(\omega_i - \omega_o)$. Consequently, we might be led to expect that the transmitted torque of a hydroviscous drive will vanish as the slip falls to zero just as is the case with a hydrodynamic drive. However, in all practical hydroviscous drives the torque-transmitting mechanism transitions from hydroviscous action to the more conventional surface contact friction for small slips (range of zero to 10 percent), because the disks will be in close proximity at small slips (if a torque is being carried). Transmitted torque is then determined by the coefficient of friction between rubbing plates and the pressure between them. Thus, by combining hydroviscous and friction drive characteristics, the hydroviscous drive can provide both variable-speed and synchronous (lock-up) -speed characteristics.

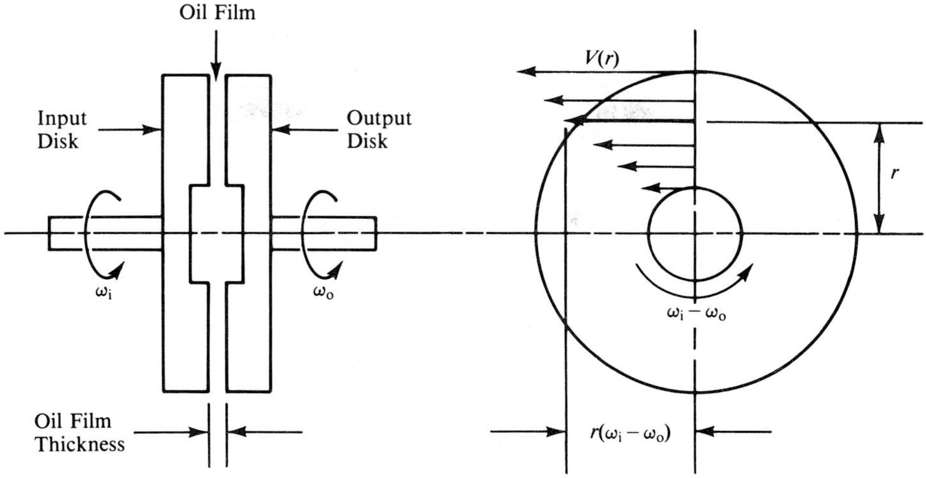

Figure 7-1. Sketch of a simple hydroviscous drive.

This is beyond the capability of the fluid coupling which requires a slip (generally around 3 percent) in order to transmit torque along with the associated dissipation of power.

The principle of torque transmission by using an oil-shear film has been known for many years. But the development of practical heavy-duty oil-shear drives is relatively new for industrial applications, despite the early use of oil-shear clutches in automotive transmissions. The first industrial applications were on large stamping presses where the combination of high rotating inertia and high-frequency cycling was not compatible with the capability of conventional dry clutches and brakes.

Some advantages of hydroviscous drives over hydrodynamic drives are listed below.

- **(a)** Hydroviscous drives have a high torque to inertia ratio. Torque diminishes as the cube of the diameter, while inertia shrinks as the fifth power of the diameter. This behavior is particularly advantageous in clutching applications that involve machine cycling.
- **(b)** The characteristic generally used to control hydroviscous drives is the relationship between torque and disk clamping pressure, which is linear. This characteristic is in contrast to the variable-fill fluid coupling, in which the control characteristic is a non-linear relationship between torque and scoop-tube position.
- **(c)** The response characteristics of the hydroviscous drive are better than those of the hydrodynamic drive; the hydroviscous drive has a faster reaction time in response to actions initiated by the controller.
- **(d)** The hydroviscous drive is amenable to integration with gearing and brakes, all contained within a single housing. Fail-safe brakes can be integrated conveniently with a clutch disk-pack co-axially.

(e) Integrated hydroviscous drives (clutch plus gearing) tend to be more compact and require fewer components than the equivalent hydrodynamic torque divider, or split-power, arrangements.

Disadvantages of the hydroviscous drive are as follows:

(a) Where control of the hydroviscous drive is by means of electrohydraulic servo-valves, there is the common problem of oil contamination and crud build-up in some application environments, which affects valve operation.
(b) Initial capital costs of the hydroviscous drive are generally higher than those of the fluid coupling.
(c) There is the potential for degradation of the disk surfaces if the hydroviscous drive is operated for extended periods of time at small slips (5 percent and less). Provisions must be made for adequate cooling under these circumstances.

Applications of hydroviscous drives cover a broad range, including:

(a) *Cycling drives*. Examples are stamping presses, packaging machines, machine tools, indexing mechanisms, and conveyors.
(b) *Constant slip*. These include unwind brakes, rewind clutches, constant-tension drives, dynamometers, and drilling equipment brakes.
(c) *Adjustable-speed drives*. These must compete with adjustable-speed electric drives and with hydrostatic drives.
(d) *Large power applications*. These include power plant feed pumps and both forced- and induced-draft fans. Fans with rotary inertias up to 400,000 lb-ft^2 and powers up to some 4000 hp have been driven by means of hydroviscous drives. Pumps of 6000 hp capacity have been driven using hydroviscous couplings.

Wherever possible, the hydroviscous drive component in these power transmission arrangements should be placed at the point of low torque in order to reduce cost (since cost is proportional to torque capability in a hydroviscous drive).

7-1 HYDROVISCOUS DRIVE CHARACTERISTICS

To learn how a hydroviscous drive (or coupling) works, consider the simple flow configuration shown in Fig. 7-2. In the steady-state mode of operation, a Couette flow is established in the gap, of thickness h, between the two rotating disks. The velocity distribution established between the disks will then be linear, as shown in the figure.

The relative velocity between the disks at radius r is

$$V_{\mathrm{r}} = r(\omega_{\mathrm{i}} - \omega_{\mathrm{o}}) \tag{7-1}$$

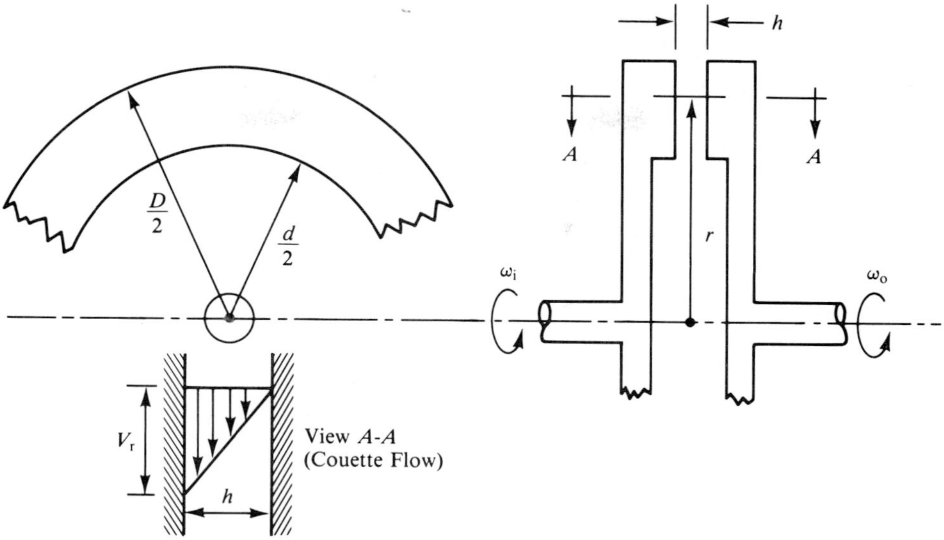

Figure 7-2. Basic hydroviscous drive configuration.

For the Couette flow established in the gap under these circumstances, the shear stress developed in the oil film is

$$\tau = \mu \frac{V_r}{h} = \frac{\mu r(\omega_i - \omega_o)}{h} \tag{7-2}$$

The torque contribution to one of the disks by a ring of fluid of thickness dr at radius r will then be

$$dT = r \cdot (2\pi r\, dr) \cdot \tau \tag{7-3}$$

or

$$dT = \frac{2\pi\mu(\omega_i - \omega_o) r^3\, dr}{h} \tag{7-4}$$

Integrating this last expression between the limits $d/2$ and $D/2$, we obtain the torque on a single disk as

$$T = \frac{\pi\mu(\omega_i - \omega_o)(D^4 - d^4)}{32h} \tag{7-5}$$

Defining the slip s as

$$s = \frac{\omega_i - \omega_o}{\omega_i} \tag{7-6}$$

Equation (7-5) can be written in the form

$$T = \frac{\tau\mu\omega_i D^4}{32h}(s)\left(1 - \frac{d^4}{D^4}\right) \tag{7-7}$$

We find that the torque developed by a single pair of disks is proportional to the slip, s, proportional to the input speed, ω_i, and proportional to the fourth power of diameter, D^4. Recall that hydrodynamic fluid coupling torque is proportional to the square of the input speed, ω_i^2, proportional to the fifth power of diameter, D^5, and for the smaller slips the torque is proportional to the slip, s.

If there are N total disks with $N/2$ of them connected to the output and $N/2$ of them connected to the input, there will be $N-1$ oil film gaps established, and the theoretical torque developed will then be

$$T = \frac{\pi\mu(N-1)\omega_i D^4}{32h}(s)\left(1 - \frac{d^4}{D^4}\right) \tag{7-8}$$

One possible arrangement, for $N = 8$, is shown in Fig. 7-3.

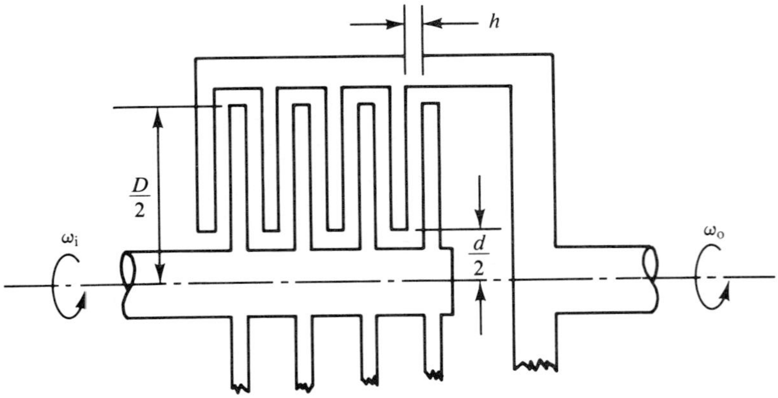

Figure 7-3. Arrangement of multiple disks.

For a particular geometry, Eq. (7-8) can be written in the form

$$T = C\,\frac{\omega_i s}{h} \tag{7-9}$$

where C is a constant that depends on the fluid used, the number of disks, and the diameters d and D. Note that in contrast to the hydrodynamic fluid coupling, whose torque must vanish when slip s goes to zero, the hydroviscous coupling can still provide an appreciable torque, even for small slips, if the film thickness h is allowed to become small. For the special case $h = 0$, we have the disks in direct contact, and it will be possible to transmit torque even for $s = 0$. This is called disk lock-up.

As the disks of the hydroviscous coupling rotate, the oil will move out radially between the disks due to centrifugal action and a pressure will be developed between adjacent disks. As dimension h is diminished, this pressure will be increased, so a relationship will be established between the oil pressure and torque (torque will increase as h is diminished). Viewed as a potential brake, the braking torque developed will depend on braking pressure in a manner similar to the ordinary dry friction brake. So, the hydroviscous coupling can be used as a brake or clutch by controlling the dimension h separating adjacent disks. It will be necessary with such a brake or clutch to incorporate a pump to supply the continuous flow of oil through the oil film gaps between the disks.

7-2 THE HYDROVISCOUS CLUTCH DISK-PACK

Theoretically, the hydroviscous type of coupling could be used in a manner similar to the hydrodynamic fluid coupling by simply maintaining the oil film thickness h constant. Then, for a given input speed ω_i, the coupling will provide a torque proportional to slip, with torque vanishing as slip goes to zero and with maximum torque developed when the output shaft is stalled ($s = 1$). However, used in this mode, the hydroviscous coupling would not have any advantages over the hydrodynamic fluid coupling.

The really useful applications of the hydroviscous coupling are obtained with designs in which the oil film thickness h between adjacent disks is externally controllable. And, since the torque capacity of the coupling is multiplied by using multiple disks, practical hydroviscous drives consist of disk-packs in which the gap thickness h can be controlled by having the disks splined, alternately, to an input and an output shaft. The disks can be shifted axially by using a hydraulic (or pneumatic) piston arrangement in such a way as to diminish, or increase, the film thickness h. This process, called clamping, is shown schematically in Fig. 7-4.

Note that one set of disks is splined to the inside of the outer barrel connected to the input shaft. The other set of disks is splined to the inner barrel

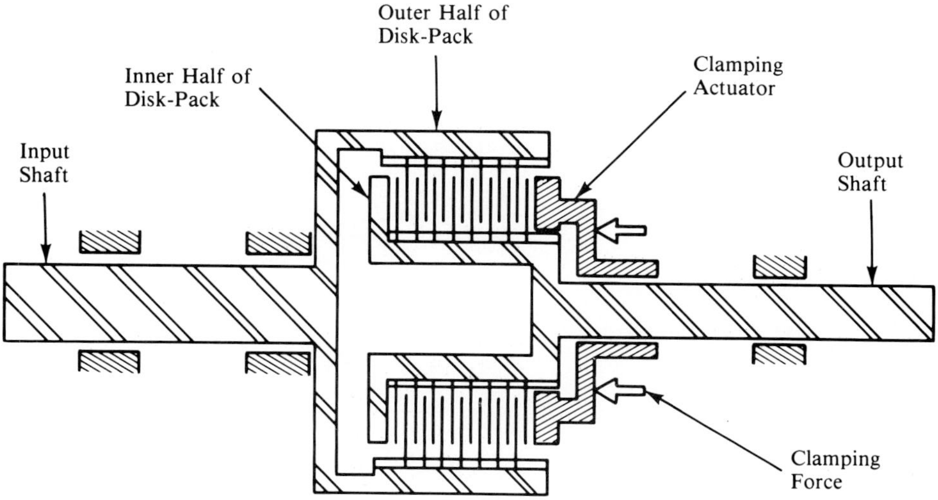

Figure 7-4. Schematic diagram of a hydroviscous disk-pack.

connected to the output shaft. All disks are free to move axially along their splines. The complete system of disks is called a disk-pack.

Also shown in Fig. 7-4 is an actuator arrangement whereby the disk-pack can be clamped together (diminishing dimension h between adjacent disks) by using an annular type of hydraulic piston. As the clamping force (piston pressure) becomes larger, dimension h becomes smaller and the torque transfer between input and output shafts increases.

EXAMPLE 7-1

An internal combustion engine is coupled to a load by a hydroviscous disk-pack clutch. Assume that the disk-pack relationship between torque (ft-lb), input speed (rpm), slip, and disk spacing (inches) is given by

$$T = \frac{0.26\omega_i s}{h}$$

Also assume that the hydrodynamic pressure necessary to clamp the disk-pack is inversely proportional to the disk spacing h.

The full-throttle engine characteristic is sketched in Fig. 7-5. Assume the engine is initially operating at a load torque of 4400 ft-lb under full-throttle. Suddenly the load torque increases by 9.1 percent. The initial clutch slip of 0.03 is to be maintained. Determine

(a) The final load speed
(b) The percent increase required in hydraulic clamping pressure
(c) The disk-pack spacing if the load is to be stalled at 5200 ft-lb of torque

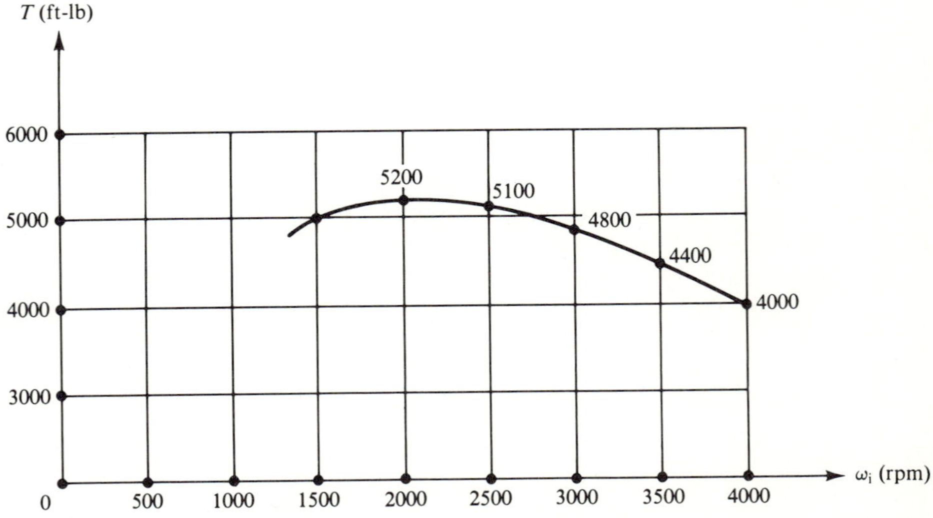

Figure 7-5. Engine characteristic for Example 7-1.

Solution

(a) The new (final) torque is $T = 1.091(4400) = 4800$ ft-lb. From Fig. 7-5, the new engine speed is then seen to be $\omega_i = 3000$ rpm. The new load speed is therefore

$$\omega_o = (1 - s)\omega_i = 0.97(3000) = 2910 \text{ rpm}$$

(b) The initial disk-pack spacing was

$$h = \frac{0.26(3500)(0.03)}{4400} = 0.0062 \text{ inch}$$

The final disk-pack spacing is

$$h = \frac{0.26(3000)(0.03)}{4800} = 0.0049 \text{ inch}$$

Now

$$\frac{p_f}{p_i} = \frac{h_i}{h_f} = \frac{0.0062}{0.0049} = 1.265$$

where p_f = final pressure, p_i = initial pressure, h_i = initial spacing, and h_f = final spacing. Therefore, the clamping pressure must be increased by 26.5 percent.

(c) The slip is $s = 1$ under stalled conditions. Therefore,

$$h = \frac{0.26(2000)(1)}{5200} = 0.10 \text{ inch}$$

■

7-3 APPLICATIONS

The three major kinds of applications of the hydroviscous disk-pack having controllable clamping force are:

(a) Viscous friction brakes or drag devices
(b) Soft-start clutches
(c) Variable-speed drives

In the brake application, the outer set of disks (see Fig. 7-4) is prevented from rotating because the disks are splined to an outer barrel which is connected rigidly to the non-rotating housing of the brake; the housing is connected rigidly to the frame of the machine for which it is being used as a brake. The inner set of disks is splined to the rotating shaft which is to be braked.

In a spring-set brake, the disks are held apart by a hydraulic or a pneumatic piston acting against the closing force of the spring(s). The brake is actuated by interrupting the hydraulic or pneumatic pressure so that the spring will then set (clamp) the disk-pack together. It is called a fail-safe brake because the brake will be set if there is a failure that interrupts the hydraulic or pneumatic pressure holding the brake open. Then a braking torque will be exerted on the rotating disks by the stationary disks which are fixed to the housing.

In a hydraulically set or pneumatically set brake, the disk-pack is actually clamped together by supplying a pressure to overcome the springs which, in this application, hold the disks apart. In either the spring-set or pressure-set version, the hydroviscous coupling can be used as a brake to either bring a shaft to rest or supply a controllable drag via a feedback control system.

Hydroviscous disk-packs have been used as soft-start clutches in power plant, pump, and conveyor applications of up to 20,000 hp. In these applications, the clutch input shaft is connected to the prime mover while the clutch output shaft is connected to the pump, conveyor, etc. The prime mover is started with the disk-pack unclamped so only a small torque (due to parasitic effects such as windage and bearing losses) will be transmitted by the clutch. Then, by clamping the disk-pack in a controlled manner (which is done by controlling the hydraulic pressure of the clamping cylinder), a controlled torque is used to accelerate the shaft of the pump, conveyor, etc., in a controlled manner. This process of starting, called a soft-start, permits the prime mover to be sized for the steady-state operating condition rather than for the overloads that would occur with an uncontrolled start. Consequently, smaller prime movers are made possible, with the resulting economies in capital outlays.

A closed-loop control arrangement whereby a soft-start can be obtained is shown in Fig. 7-6. The type of start desired is determined by how the command signal is programmed. For example, in some instances a ramp function command signal will provide the starting characteristics required.

A unique feature of this type of soft-start transmission is that it can be made to operate, at full-load torque, with zero slip. Zero slip occurs at the end of the soft-start when the clamping pressure forces the disks of the disk-pack together in

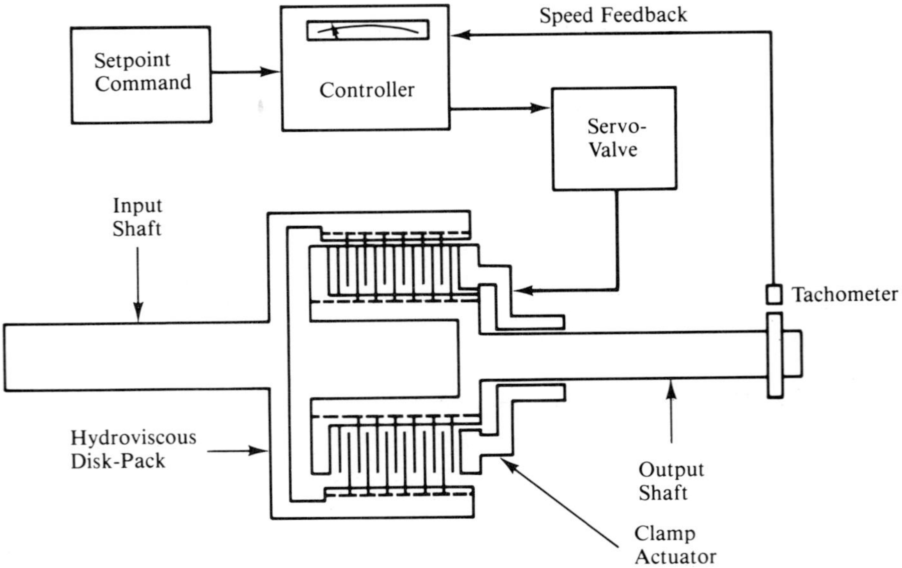

Figure 7-6. Servo-control of a hydroviscous disk-pack.

direct physical contact ($h = 0$). This is called disk lock-up. Operated in this mode, the hydroviscous clutch has an advantage over the other soft-start clutches (hydrodynamic fluid couplings, eddy current clutches, frequency-controlled induction motors) in that the latter all require a slip in order to transmit full-load torque. As we have seen, power dissipation is directly proportional to slip, and this dissipation is eliminated in the locked-up hydroviscous drive. In some high-power transmission applications, savings of up to $20,000 annually have been projected by the elimination of the power dissipation that occurs at the small 2 to 5 percent slips associated with these other types of soft-start clutches.

The hydroviscous disk-pack drive can be used as a variable-speed drive (also called an adjustable-speed drive). Figure 7-6 shows the closed-loop control arrangement that makes this possible. The speed of the output shaft will be determined by the command signal of the controller. This, in effect, determines the correct oil film thickness h required to provide the necessary slip s between specified output and input speeds. The closed-loop system provides the additional benefit of maintaining a set output speed constant independent of load demand torque. If the load increases, the output shaft will tend to slow down. Slow down is detected by the magnetic speed pick-up, which then causes the servo-valve to produce more clamping pressure. In turn, the torque transmitted by the disk-pack increases, which will restore the output speed to its setpoint (commanded) value.

7-4 INTEGRATED HYDROVISCOUS DRIVES

Used as a variable-speed drive, both the hydrodynamic and hydroviscous drives are inherently inefficient compared to a hydrostatic drive (or a DC electric motor drive). The reason is that a greater slip is required to provide the lower speed range, and this greater slip means a greater dissipation of the input power. In a hydrostatic drive, variable speed is obtained by stroking either the pump or the motor (or both), and this does not, inherently, require greater slip for the lower output speeds.

To surmount this disadvantage and yet be able to implement the advantages of the hydroviscous coupling, integrated hydroviscous drives have been designed in which the hydroviscous disk-pack is incorporated into a gearbox. An example appears in Fig. 7-7.

7-4-1 The Hybrid Drive

The drive of Fig. 7-7 consists of a hydroviscous clutch coupled integrally to a high-low shiftable geared transmission. So-called hot-shift clutches are used to engage either the high gear ratio or the low gear ratio of this integrated drive. The shifting of the transmission can be done under power.

A schematic diagram of this integrated drive (also called a hybrid drive) is shown in Fig. 7-8. For low-speed operation of the load, the low-ratio gearset, shown on the left of Fig. 7-8, is engaged so the slip required in the hydroviscous

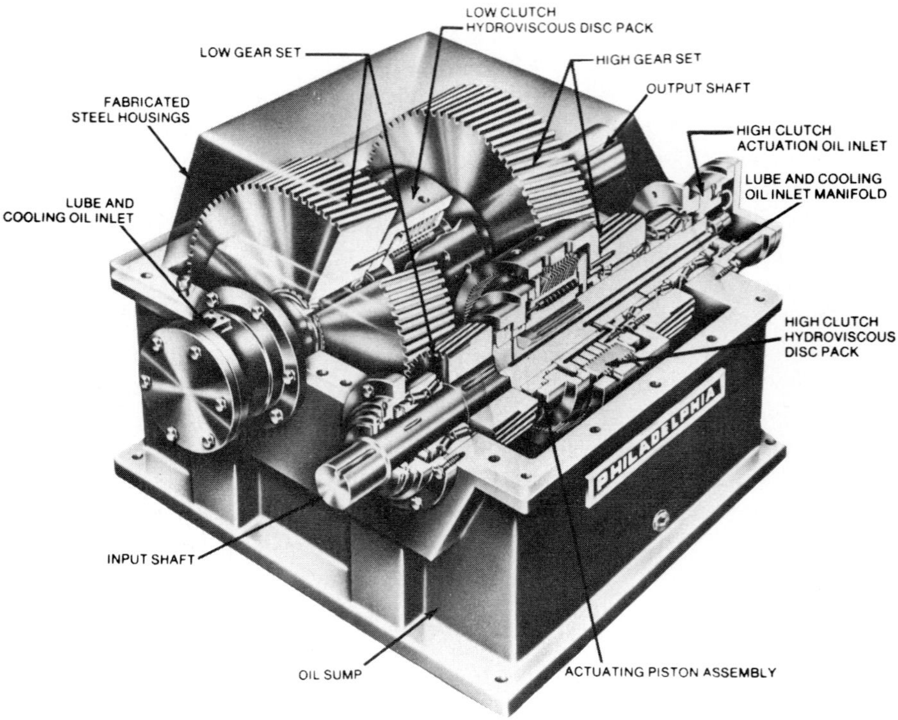

Figure 7-7. Integrated (hybrid) hydroviscous drive. (Courtesy of Philadelphia Gear Corporation.)

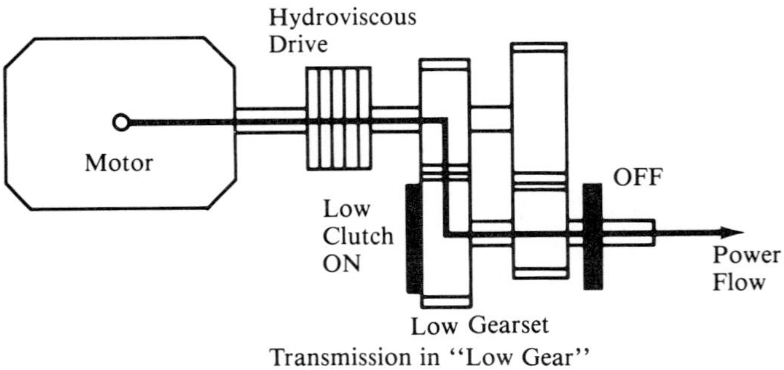

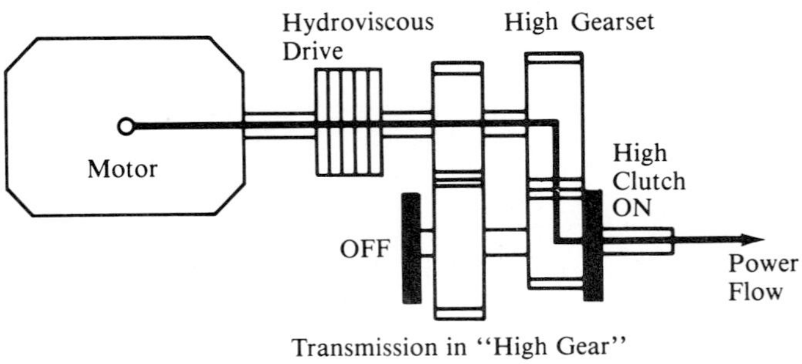

Figure 7-8. High-low shiftable hybrid drive.

coupling is minimized for variable-speed operation in this low-speed range. Engagement of the "low" gearset is by means of a shiftable double-clutch arrangement as shown (the low clutch is ON while the high clutch is OFF). For operation in the high-speed range, the high-ratio gearset is engaged by shifting the clutches so that the high clutch is ON and the low clutch is OFF. This provides variable-speed operation in the high-speed range of the load.

The manufacturer of this type of integrated hydroviscous drive will provide the optimum gear ratios of the low and high gear transmissions in order to match the operating requirements of the load; if a broader range of shifting is required, it is possible to provide a three-speed transmission as well as the two-speed device sketched in Fig. 7-8.

EXAMPLE 7-2

A centrifugal boiler-fan is to be driven by a high-low hydroviscous drive like that shown in Fig. 7-8. In "high gear," the fan is to be driven at a maximum speed of 1000 rpm when the hydroviscous disk-pack is locked up ($s = 0$). The high-speed range of this fan load is 800 to 1000 rpm.

At the lower speed of the high-speed range (800 rpm), the transmission is to be shifted into "low gear" such that the fan operates at 800 rpm with the disk-pack locked up. The low-speed range is 600 to 800 rpm.

The prime mover of this power transmission is a 1200 rpm (synchronous speed) AC induction motor. The fan requires 4000 hp when operating at 1000 rpm. Assume that the head developed by the fan is proportional to the square of fan speed and that the fan discharge is proportional to fan speed.

(a) What are the gear ratios of the high-gear and low-gear gearsets? Assume the induction motor operates with a 3 percent slip.
(b) What power is dissipated by the disk-pack when the fan is operated at 900 rpm?
(c) What power is dissipated by the disk-pack when the fan is operated at 700 rpm with the high-gear pair engaged?
(d) What power is dissipated by the disk-pack when the fan is operated at 700 rpm with the low-gear pair engaged?
(e) If the duty cycle is 95 percent and the cost of electricity is 5¢/kWh, what are the annual savings in operating expense provided by the arrangement of part (d) as compared to part (c)?

Solution

(a) With 3 percent slip, the prime mover operates at $0.97(1200) = 1164$ rpm. Therefore, the gear ratio of the high gearset is

$$r_h = \frac{1000}{1164} = 0.859$$

The gear ratio of the low gearset will be

$$r_l = \frac{800}{1164} = 0.687$$

(b) When the fan is operated at 900 rpm, the high gearset is engaged. The slip in the disk-pack will then have to be

$$1 - s = \frac{900/0.859}{1164}$$

$$s = 0.10$$

In Chapter 4, we showed that the power developed by a centrifugal fan is given by Eq. (4-6), namely,

$$\text{HP} = \gamma QP$$

where Q is the discharge and P is the head. From the assumptions made, we see that the power is proportional to the cube of fan speed. The fan power at 900 rpm is then

$$\text{HP}_{900} = \left(\frac{900}{1000}\right)^3 (4000) = 2916 \text{ hp}$$

With a slip of 10 percent, the efficiency is 0.9 and the input power will have to be

$$\text{HP}_i = \frac{2916}{0.9} = 3240 \text{ hp}$$

The dissipation in the disk pack is then

$$\text{HP}_{\text{diss}} = 3240 - 2916 = 324 \text{ hp}$$

(c) A calculation similar to the one given in part (b) will show that at 700 rpm (in high gear), the power dissipated in the disk-pack will be

$$\text{HP}_{\text{diss}} = 588 \text{ hp}$$

(d) With the transmission in low gear, the disk-pack slip for a fan speed of 700 rpm will be

$$1 - s = \frac{700/0.687}{1164} = 0.875$$

$$s = 0.125$$

The fan power at 700 rpm will be

$$\text{HP}_{700} = \left(\frac{700}{1000}\right)^3 (4000) = 1372 \text{ hp}$$

The corresponding input horsepower is given by

$$\text{HP}_i = \frac{1372}{0.875} = 1568 \text{ hp}$$

The dissipation in the disk-pack is then

$$\text{HP}_{\text{diss}} = 1568 - 1372 = 196 \text{ hp}$$

(e) The difference in horsepower dissipation between the schemes of part (c) and part (d) is

$$\text{HP savings} = 588 - 196 = 392 \text{ hp} = 292 \text{ kW}$$

Under the assumption of a duty cycle of 95 percent, the fan will operate for 0.95(365)(24) = 8322 hours annually. Therefore, the dollar savings will be

$$\text{Dollar savings} = \frac{5}{100}(8322)(292) = \$121{,}501/\text{year}$$ ■

7-4-2 The Hydroviscous Clutch/Brake

In high-production manufacturing and other industrial operations, it is often necessary to cycle a variable-speed power transmission output shaft between ON and OFF without stopping the input shaft. This can be done by integrating a hydroviscous clutch and a hydroviscous brake within the same housing, as shown in Fig. 7-9.

Figure 7-9. Hydroviscous clutch/brake drive. (Courtesy of Force Control Industries, Inc.)

The clutch/brake package of Fig. 7-9 is operated with a single hydraulic (or pneumatic) piston actuator which clamps the brake disk-pack at the same time that the clutch disk-pack is unclamped, or vice versa. This is done by a shuttle that shifts back and forth as the piston is extended or retracted. This type of operation permits the input shaft to operate continuously while the output is cycled between ON and OFF. The input shaft is connected to clutch disks attached to an outer barrel, while the fixed disks of the brake are attached to a fixed outer barrel connected to the clutch/brake housing. The output shaft contains two separate sets of disks; one set engages the clutch disks connected to the input, and the other set engages the brake disks.

7-4-3 Integrated Epicyclic Hybrid Drive

An interesting variable-speed transmission can be obtained by integrating an epicyclic gear train with a hydroviscous brake. One set of disks of the disk-pack is attached to the ring gear of the epicyclic while the other set is attached to the non-rotating housing of the gearbox (see Fig. 7-10).

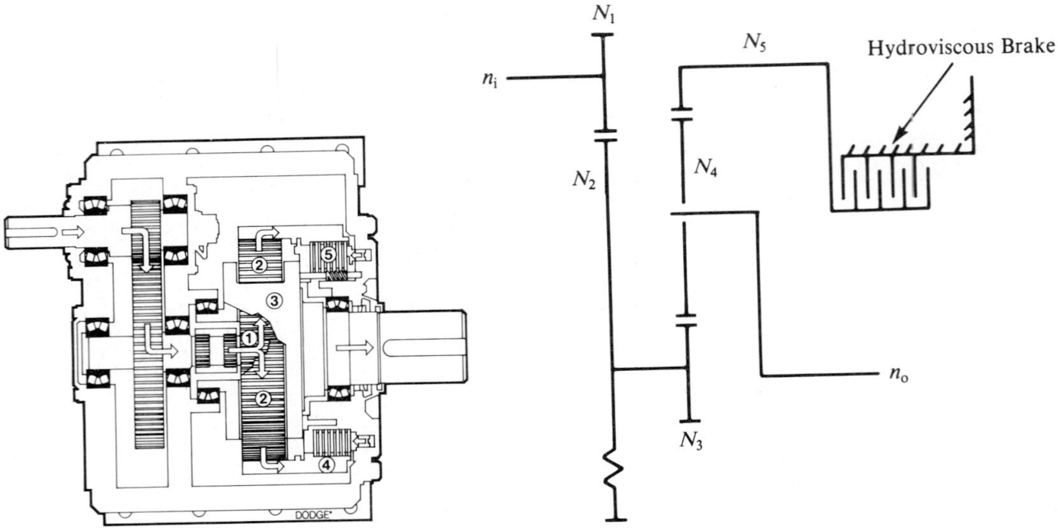

Figure 7-10. Integrated epicyclic hydroviscous drive. (Courtesy of Dodge/Reliance Electric, Greenville, S.C.)

With the ring gear clamped so that it remains at rest, the device becomes simply a two-terminal gearbox of fixed ratio n_i/n_o. When the ring gear is unclamped, the torque capability is zero and the ring gear has a free-wheeling rotation while the planet carrier (and, hence, output shaft) remains at rest. When the output shaft demands torque, the hydraulic piston actuator of the brake is caused to apply pressure, which begins to clamp the ring gear to the housing. As the ring gear slows at a controlled rate, the planet carrier will begin to rotate, with output speed increasing as the ring gear speed decreases. Full output speed is obtained when the ring gear is totally clamped (stopped). The result is a controlled soft-start.

7-5 HYDROVISCOUS COUPLING PERFORMANCE WITH A QUADRATIC LOAD

We have already seen that one common type of load characteristic is the variation of load torque T_o proportional to the square of load speed ω_o. This load characteristic is typical of fans, blowers, and pumps. Therefore, assume that

$$T_o = \alpha \omega_o^2 \tag{7-10}$$

Input and output power are defined by

$$P_i = T_i \omega_i \qquad P_o = T_o \omega_o \tag{7-11}$$

The power dissipation in the hydroviscous coupling is

$$P_s = P_i - P_o = T_i \omega_i - T_o \omega_o \tag{7-12}$$

Since input torque is substantially equal to output torque (neglecting parasitic losses such as bearing friction and windage), we see that the power dissipated is

$$P_s = T_o(\omega_i - \omega_o) \tag{7-13}$$

Combining Eqs. (7-10) and (7-13), we obtain

$$P_s = \alpha\omega_o^2(\omega_i - \omega_o) \tag{7-14}$$

It is easy to show that maximum power dissipation occurs at a slip of 1/3 (or output speed of $\frac{2}{3}\omega_i$).

With Eq. (7-10) and Eqs. (7-11) combined, the input and output power are given by

$$P_i = \alpha\left(\frac{\omega_o}{\omega_i}\right)(\omega_i^2\omega_o) \qquad P_o = \alpha\left(\frac{\omega_o}{\omega_i}\right)(\omega_o^2\omega_i) \tag{7-15}$$

Efficiency and slip are given by

$$s = 1 - \frac{\omega_o}{\omega_i} \qquad \eta = 1 - s \tag{7-16}$$

These various relationships are plotted versus ω_o/ω_i in Fig. 7-11 for a quadratic load.

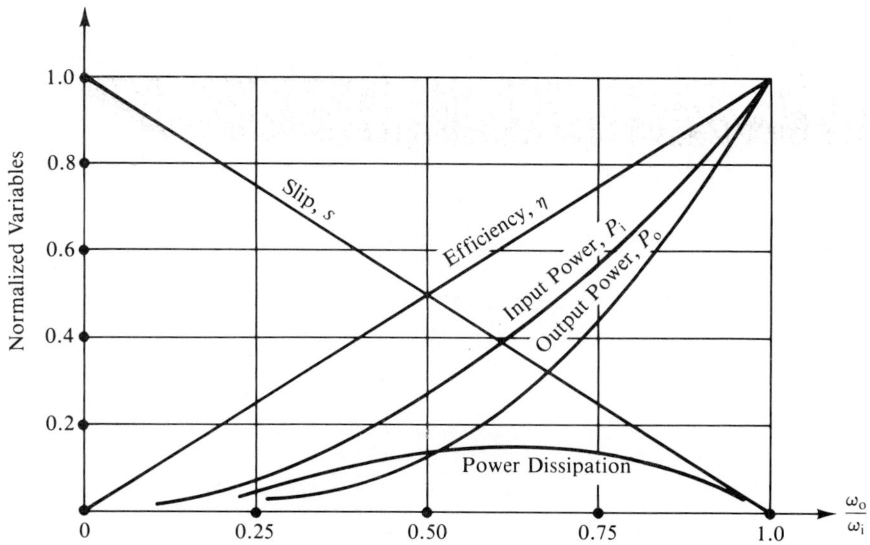

Figure 7-11. Characteristics of a quadratic load.

7-6 SUMMARY

The hydroviscous (or oil-shear) coupling can be utilized as a brake, as a clutch, as a clutch/brake, or as a variable-speed drive. This device competes with dry friction clutches and brakes, with the hydrodynamic coupling, and with a variety of electrical couplings (eddy current clutches, adjustable-frequency AC drives, and DC drives). When integrated with a gear reduction unit in a single housing, the hydroviscous drive provides a smooth-running, low-inertia, closed-loop-controllable, variable-speed drive having certain advantages over competitive drives.

In the smaller sizes, the hydroviscous drive is ideal in machine applications involving high-frequency cycling operations or in tension control applications. The oil-shear drive is intrinsically amenable to cooling (by circulating the oil through a heat exchanger), which is necessary in cycling operations which can wear out ordinary clutches rapidly. And the oil-shear drive has an advantage over the hydrodynamic coupling insofar as closed-loop control and potential lock-up are concerned.

In the larger sizes, the hydroviscous drive has been used frequently in power plants to drive feed water pumps, fans, and compressors. Other large industrial applications are to conveyor drives, pipeline pumps, and industrial fans.

A major advantage of the hydroviscous drive is that it can be locked up to eliminate the slip (and associated dissipation) inherent in hydrodynamic and eddy

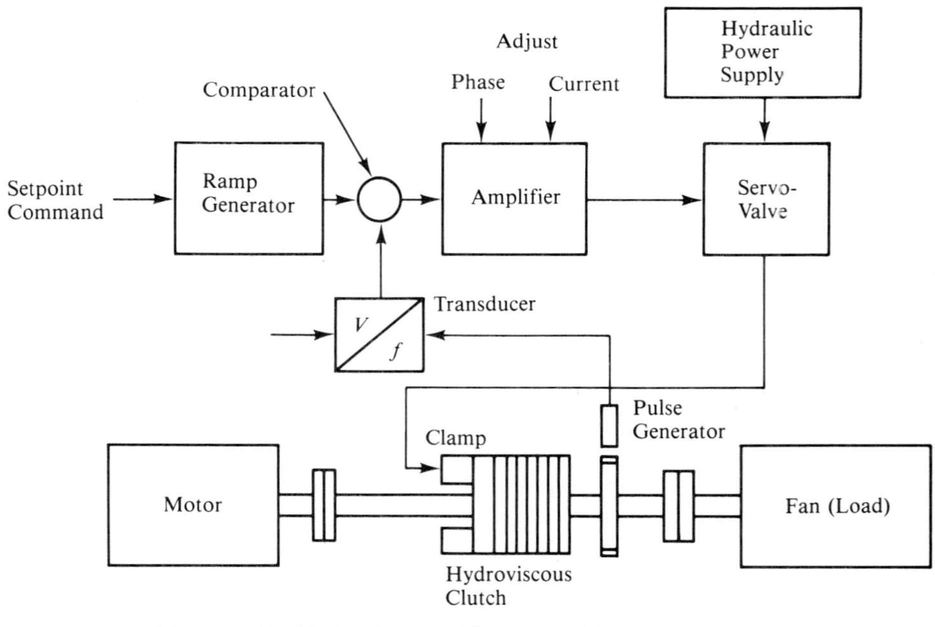

Figure 7-12. Hydroviscous drive control loop components.

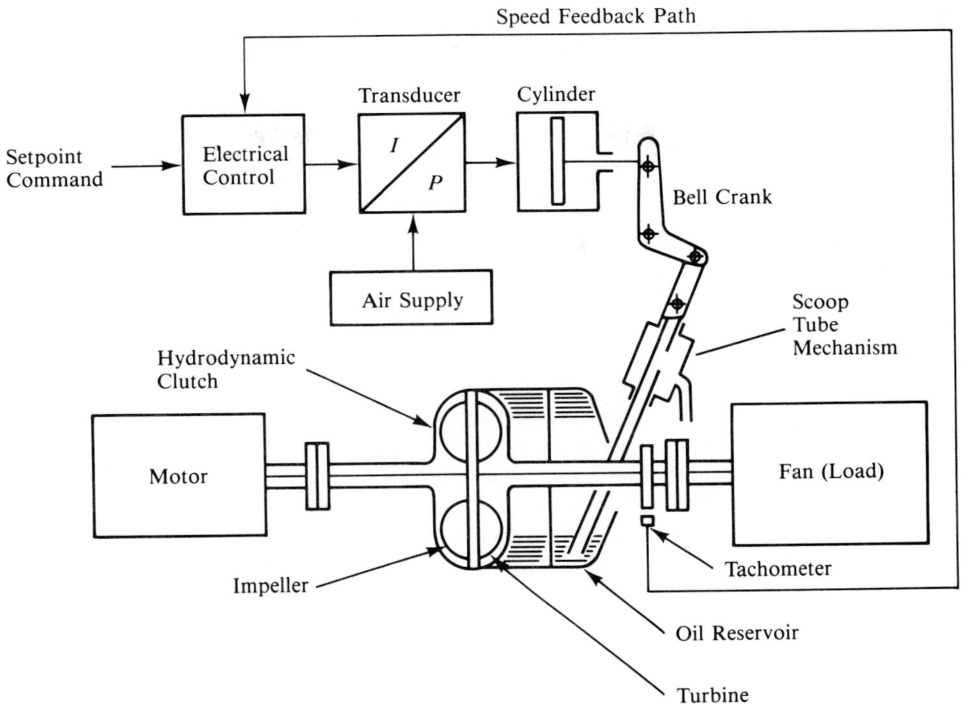

Figure 7-13. Hydrodynamic drive control loop components.

current couplings (see Chapter 10). Thus, the hydroviscous drive can reduce operating costs substantially in the larger horsepower applications.

In regard to capital cost, the hydroviscous drive is at a disadvantage with respect to competitive hydrodynamic drives, but it is more versatile and responsive from a controls point of view. Hydrodynamic drives usually contain more control elements and depend on the insertion or withdrawal of a scoop tube to accomplish control. This operation is about an order of magnitude slower than the disk-pack clamping or unclamping that provides the control of the hydroviscous drive. Figures 7-12 and 7-13 show the control loop elements of these two competing drives.

The hydroviscous drive enjoys a capital cost advantage over its competing electric motor drives that derive variable-speed control by means of solid-state rectifiers or frequency controllers. Controllers will be discussed in Chapter 11.

7-7 REFERENCES

7-1. J. K. Liu, *Recent Developments in Belt Conveyor Drives*, Paper No. 111579L, presented at the National Conference on Power Transmission, Chicago, November, 1979.

7-2. J. K. Liu, *High Efficiency Hydroviscous Drives for Mainline Pumps*, ASME Paper 80-Pet-80, presented at the Energy Technology Conference and Exhibition, New Orleans, February, 1980.

7-3. *Variable Speed Synchrodrives*, Catalog SD-2, published by Philadelphia Gear Corporation, King Of Prussia, PA.

7-4. *Posidyne Clutch/Brake Drives*, Catalog 204D, published by Force Control Industries Inc., Hamilton, OH.

PROBLEMS

7-1. In the hydroviscous coupling configuration shown in Fig. P7-1, the oil film gaps are sandwiched between concentric cylinders. One set of cylinders is connected to the input shaft of the coupling while the other set is connected to the output shaft. If the oil film gap between adjacent cylinders is h and the effective length of the cylinders is L, determine the characteristic equation relating torque transmission to slip s, oil film thickness h, input speed ω_i, diameters D and d, and number of cylinders (N). Assume each cylinder is of thickness t.

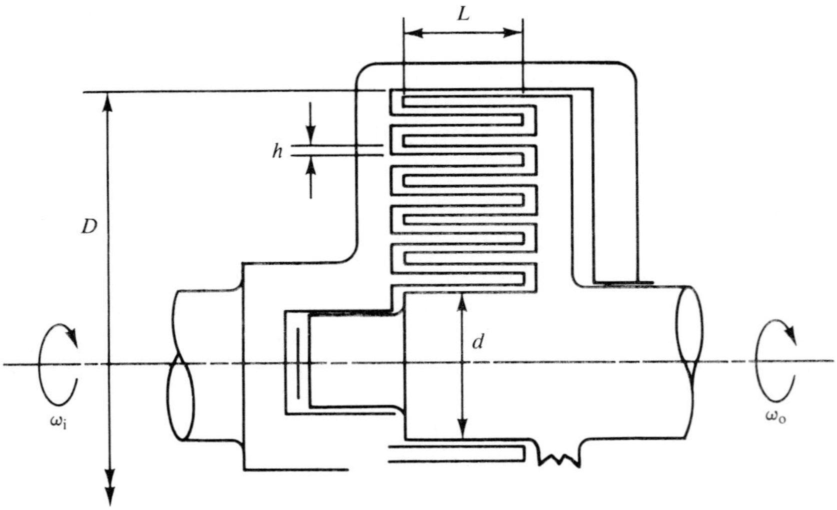

Figure P7-1. Figure for Problem 7-1.

7-2. An epicyclic hybrid drive is sketched in Fig. P7-2 in which the hydroviscous brake is applied to the ring gear of the planetary portion of the gear train.

(a) Assuming the input speed n_1 is 1760 rpm, what is the output speed n_o when the brake holds the ring gear fixed?

(b) Suppose the hydroviscous brake is totally disengaged such that the ring gear torque is zero. Analyze the transmission under these circumstances and show that the torque transmission capability of the transmission vanishes.

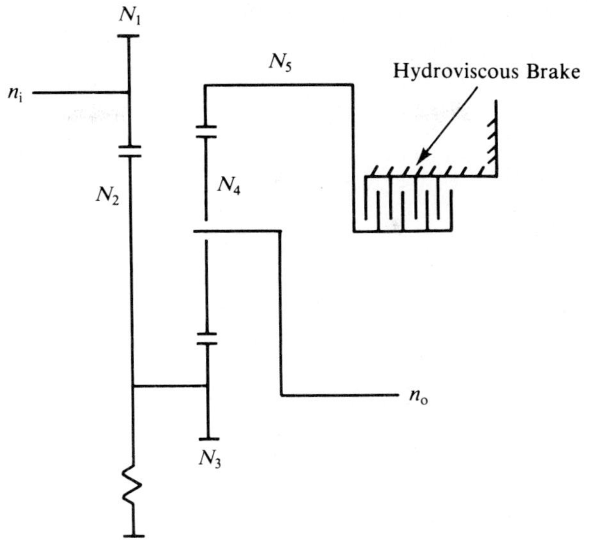

Figure P7-2. Figure for Problem 7-2.

7-3. An integrated hydroviscous drive is shown in Fig. P7-3 in which the numbers of teeth on the gears are $N_1 = 52$ and $N_2 = 19$. Assume the input shaft is driven at 1160 rpm.
(a) What is the output speed with the disk-pack locked up?
(b) What is the output speed when the disk-pack experiences a slip of 0.5?

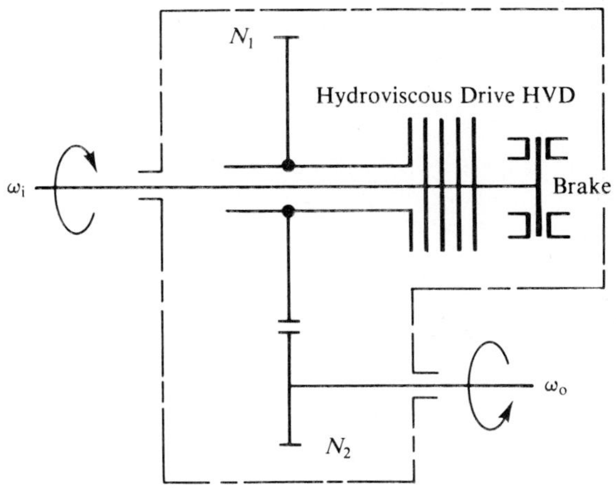

Figure P7-3. Figure for Problem 7-3.

7-4. How does the power transmitted by a hydroviscous coupling vary with input speed and disk-pack radius? If a hydroviscous coupling is doubled in size, what are the percent increases in power and torque capacity (assume d is small)?

7-5. Assume that a family of geometrically similar hydroviscous couplings has its input characteristics represented by the equation

$$T_i = \frac{5.66\omega_i D^4 s}{h}$$

with D and h in inches, ω_i in rad/sec, and T in in-lb. Determine the coupling diameter D required to deliver 25 hp to the load with a prime mover speed of 1600 rpm and a slip of 3 percent.

7-6. An existing hydroviscous coupling is to be redesigned to increase its efficiency from 94 percent to 97 percent under the same operating conditions. The new coupling is to be geometrically similar to the existing one. What will be the percent change required in characteristic diameter D? Assume d is small.

7-7. Design a high-medium-low integrated hydroviscous drive (choose the gears and shaft configuration) that will provide lock-up output speeds of 1200, 900, and 600 rpm for an input speed of 1800 rpm. Plot the mechanical efficiency of this drive as a function of the ratio of output speed to maximum output speed (1200 rpm).

Part C

ELECTRICAL COMPONENTS

Almost all mechanical systems will incorporate some electrical components. The prime mover will often be an induction motor. Even in the most elementary application, the induction motor will require a motor starter (a sophisticated switch). The motor starter, in turn, may involve electromechanical relays, electronic circuits, or microelectronic devices. In more sophisticated mechanical systems, the electric motor might be a direct-current (DC) dynamo capable of providing precise control of position or speed (a DC servo-drive). In this case, an electrical amplifier and electrical signal converters may also be required.

Electromechanical relays are also used in fluid power circuits to actuate directional control valves. And many of the fluid power and mechanical components discussed previously have their electrical or magnetic analogs. A good example is the Ward-Leonard electrical drive which is the electrical analog of the fluid power hydrostatic drive. Another example is the eddy current coupling which is the electrical analog of the fluid coupling discussed earlier. Thus, the mechanical designer may, at some point, have to make a choice among these different types of drives and couplings in designing the power train of a mechanical system.

Obviously, we cannot cover all of the electrical components about which the compleat mechanical designer will be knowledgeable. These components would include all kinds of motors and their electrical controllers, transformers, rectifiers, amplifiers, power supplies, special circuits, instrumentation, and many more.

In the next several chapters, we will discuss a few electrical devices with which all mechanical system designers should be familiar. We shall begin with electric motors.

ELECTRIC MOTORS

Electric motors are available in a diverse variety of types which are a result of the ingenuity of scientists and electrical engineers during the past century and a half. Practically all electric motor designs are based on the principle of the creation of a force (or a torque) when two or more magnetic fields interact with one another. In some cases, it is simpler to view this interaction as being one between a magnetic field and a current-carrying conductor that finds itself in the magnetic field.

Perhaps the simplest manifestation of the electric motor principle is provided by two interacting permanent magnets where one magnet is moved and the other magnet then attempts to remain aligned with the moving magnet by virtue of the force exerted on it by that moving magnet. Technically, we would say that the system of two magnets attempts to configure itself so as to minimize the reluctance of the magnetic system.

For many years, the only way to produce a substantial magnetic field was by using an electromagnet which is simply an iron core with a current-carrying coil wound concentrically about it. Most of the electric motors that have been developed represent the many possible combinations of alternating-current (AC) or direct-current (DC) electromagnets in a variety of geometric and kinematic configurations. For example, a synchronous motor is one in which a three-phase AC power source is used to create a rotating magnetic field in a fixed stator. A DC power source is used to create a DC electromagnet in a rotating rotor that attempts to remain aligned with the rotating field (in order to minimize the reluctance in the dynamo).

Recently, powerful permanent magnets have been developed. They will make possible new types of motors that utilize not only these permanent magnets but also the ingenious electronic circuits that microelectric technology makes possible. In fact, some of these motors have already been developed, and we can expect more in certain kinds of mechanical systems not requiring large torques (e.g., in small robotic devices).

Many of the electric motors to be discussed can also be operated as generators. Consequently, we call such a device a dynamo, which is a generic name describing electromagnetic-mechanical rotating devices that convert electrical energy into mechanical energy or vice versa. When operated as a generator, a dynamo requires a mechanical torque as an input and provides a voltage as an output. The direction of the electromagnetic torque created in the dynamo will be counter to the direction of rotation of the input shaft of the generator. When operated as a motor, a dynamo requires a voltage as an input and provides a torque as an output. For a motor, the direction of the electromagnetic torque created in the dynamo will be in the same directional sense as the direction of rotation of the output shaft.

8

Direct-Current (DC) Dynamos

A direct-current (DC) dynamo is an electromagnetic-mechanical device consisting of an armature, a field, and a commutator configured in such a way as to make the force interaction, between the electrical current of the armature (the rotor) and the magnetic flux of the field (the stator), manifest itself as a net torque on the armature. Normally, the armature rotates at some angular speed ω_r.

A schematic diagram of an elementary dynamo is presented in Fig. 8-1. The magnetic flux Φ of the field is created either by permanent magnets or by a DC

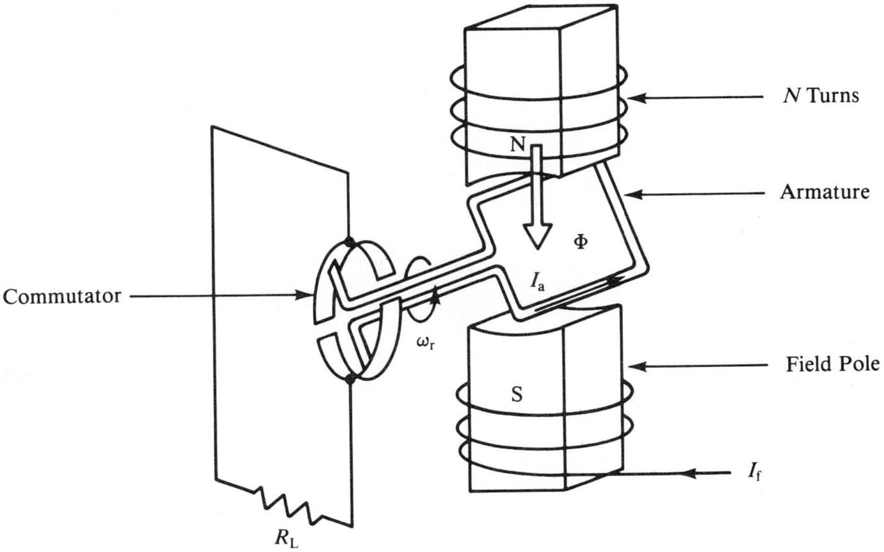

R_L = load resistance

Figure 8-1. Schematic diagram of a DC dynamo.

electromagnet. If an electromagnet is used, there will be a field winding wound about field poles and carrying the direct current I_f. Magnetic flux Φ is proportional to the ampere-turns of the field winding(s) NI_f, where N is the number of turns.

The external circuit to which the armature is electrically connected has a direct current flowing in it. If this external circuit were to be directly connected electrically to the armature (using slip rings), there would be no net torque associated with the rotation of the armature, because each conductor of the armature would experience a reversal of the direction of force as it passed successively the north and south poles of the field Φ. In other words, the instantaneous torques would be self-canceling in a small time interval.

This problem can be solved if the electrical connection between the armature and the external circuit is accomplished using a commutator and brushes rather than by using slip rings. The function of the commutator is to mechanically reverse the polarity of the voltage on the armature in synchronization with angular position of the armature so that there will be constancy of direction of force acting on the conductors by the field Φ. There will then be a net torque associated with the DC dynamo rotor.

The conductors of the armature will be carrying a current I_a, and it is the interaction of this armature current with the magnetic flux that creates the torque on the armature. Also, as the armature rotates, its conductors will cut the magnetic field lines, with the result that an electromotive force E will be induced in the armature.

We will develop some of the relationships between these dynamo electrical parameters later. For our present purposes, it will be sufficient to discuss the dynamo from the point of view of a black box having four characteristic power parameters: voltage V, current I, torque T, and angular speed ω_r. The dynamo black box has two inputs and two outputs, as shown in Fig. 8-2.

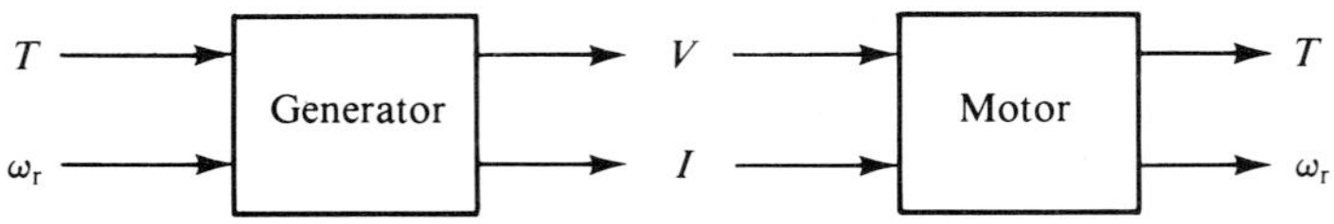

Figure 8-2. DC dynamos as black boxes.

When the input parameters of the dynamo are torque T and speed ω_r, the outputs will be voltage V and current I. This configuration is called a generator. When the input parameters are voltage and current and the outputs are torque and speed, we call the dynamo a motor.

There are no significant differences in the essential features of dynamos as would be dictated by their use as motors rather than as generators; the same dynamo can be used both as a motor and as a generator, depending on the direction of armature torque for a given direction of rotation. The four possible modes of operation of a dynamo can be defined in terms of the four quadrants of the torque-speed plane (see Fig. 8-3).

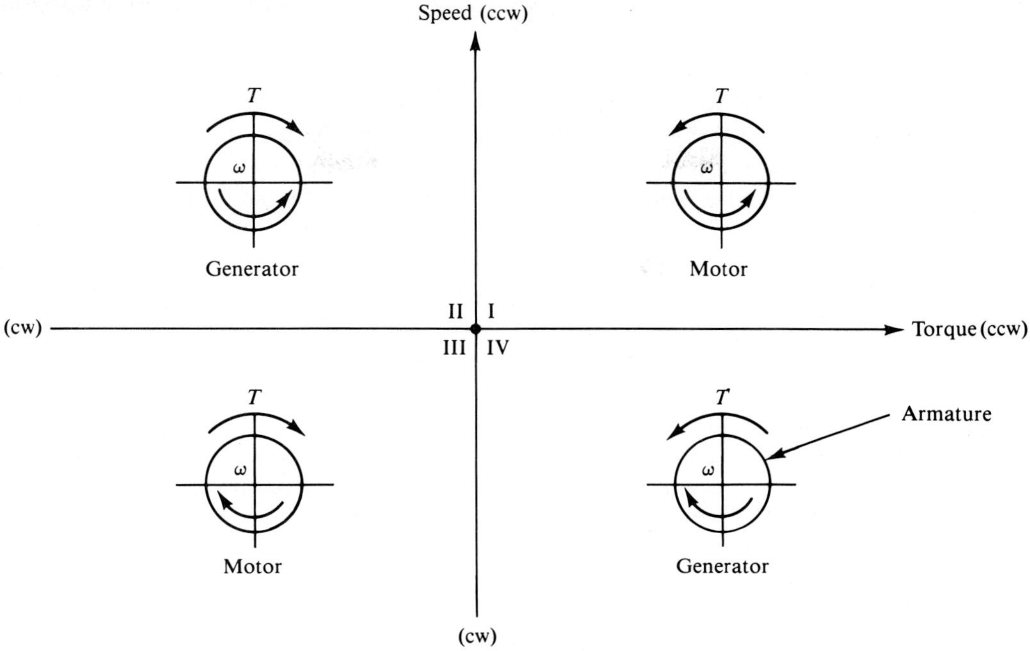

Figure 8-3. Four-quadrant operation of DC dynamos.

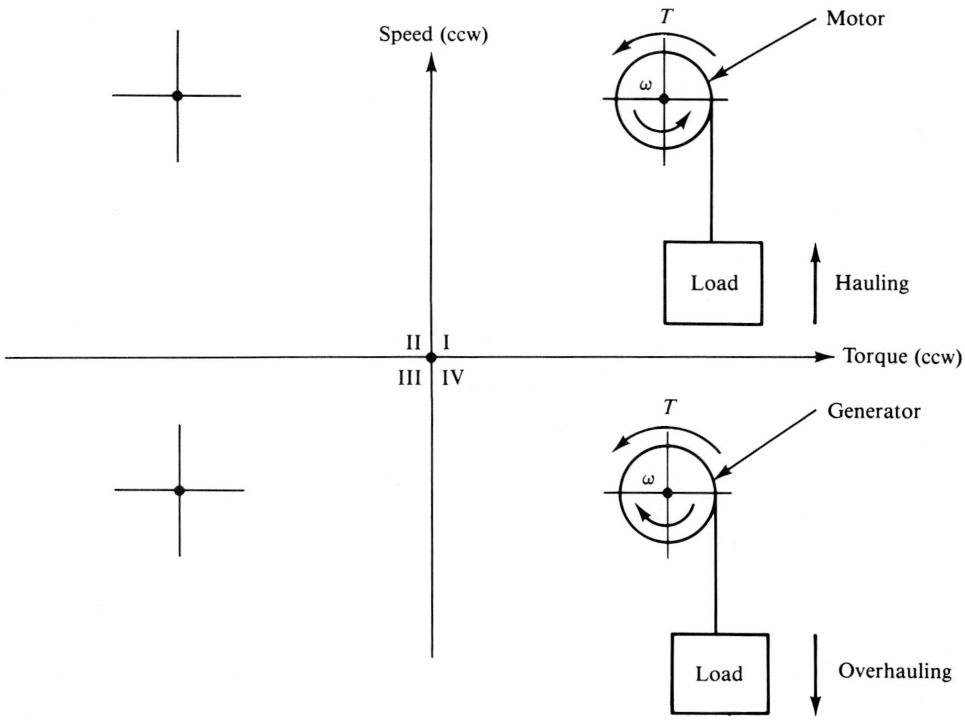

Figure 8-4. DC dynamo used for hoisting.

The fact that the same device can be used as both motor and generator can be used to advantage in certain applications. Consider, for example, a mechanical system consisting of a DC dynamo for hoisting a load (see Fig. 8-4).

In quadrant I, we have a dynamo (motor) lifting (hauling) a load. This lifting will require power from the electrical supply, and the electromagnetic torque on the armature will be in the same direction as the rotation. In quadrant IV, we have a load driving the dynamo (overhauling) as the load is lowered and acting as a power source (now the load is a generator). This power will be returned to the electrical supply lines, and the armature torque will be in opposition to the rotation. This process is called regenerative braking. Similar examples can be provided for quadrants II and III.

Although our interest is primarily in the dynamo as a motor, its operation as a generator does not introduce any significantly different concepts. Also, the combination of a generator and motor has transmission properties analogous to the hydrostatic transmission. This will be discussed later (the Ward-Leonard drive).

8-1 ELECTROMAGNETIC INTERACTIVE EFFECTS

In deriving the steady-state characteristic equations for DC dynamos, we shall assume that the magnetic flux Φ is proportional to the ampere-turns of the field winding when the magnetic flux is created by an electromagnet. Therefore,

$$\Phi = K_1(NI_f) \tag{8-1}$$

where K_1 is a constant.

This assumed linearity will be appropriate provided the field current I_f lies below the magnetic saturation value. Beyond the saturation region, the magnetization curve becomes non-linear, and eventually the dependence of flux Φ on current will cease. This could be a problem in analyzing some DC dynamos, but it will not be for the shunt and series motors to be considered here.

The conductors in the armature will carry current I_a. It is the interaction of this armature current with the magnetic field flux that creates the torque on the armature. In other words,

$$T = K_2 I_a \Phi \tag{8-2}$$

Combining Eqs. (8-1) and (8-2), we find that

$$T = K_3 I_a I_f \tag{8-3}$$

where

$$K_3 = K_1 K_2 N$$

As the armature conductors rotate, they will cut the magnetic field lines and an electromotive force E will be induced in the armature. This electromotive force (emf) is proportional to the speed at which the flux lines are cut and also to the magnetic flux itself. Therefore,

$$E = K_4 \omega_r \Phi \tag{8-4}$$

Substituting Eq. (8-1) into (8-4), we find that

$$E = K_5 I_f \omega_r \tag{8-5}$$

where

$$K_5 = K_1 K_4 N$$

These characteristic equations for the DC electrical dynamo are analogous to the characteristic equations we developed for hydrostatic dynamos in Chapters 4 and 5. They can be put in the forms

$$\begin{aligned} T &= k_e I_a \\ E &= k_e \omega_r \end{aligned} \qquad \text{(electrical dynamos)}$$

where k_e is a constant. These forms are analogous to

$$\begin{aligned} T &= k_h P \\ Q &= k_h \omega \end{aligned} \qquad \text{(hydrostatic dynamos)}$$

where k_h is a constant. We see that armature current I_a in the electrical dynamo is analogous to hydrostatic pressure P in the hydraulic dynamo and that electrical rotor speed ω_r is analogous to hydraulic pump (or motor) rotor speed ω.

The constant k_e of the electrical dynamo is dependent upon the number of poles p, the number of armature conductors N, and the magnetic flux Φ (see Reference 8-1) according to

$$k_e = \frac{pN\Phi}{\pi}$$

This constant is analogous to the constant k_h of the hydraulic dynamo (see Eqs. [5-3]), that is,

$$k_h = \frac{An\varepsilon}{\pi}$$

Consequently, we see that magnetic flux Φ in the electrical dynamo is analogous to eccentricity ε of the hydrostatic dynamo.

The implication that the same constant k_e applies to both the torque and emf equations means that

$$K_3 = K_5 \qquad K_2 = K_4$$

Indeed, this can be shown to be the case, similar to the result we found earlier for hydrostatic dynamos.

8-2 IDEALIZED CHARACTERISTICS OF DC MOTORS

The idealized DC motor is considered to consist of field windings and their resistances R_f, and an armature and its resistance R_a. The type of DC motor (series, shunt, compound, separately excited) is determined by how the field winding is related, electrically, to the armature. Equivalent circuit diagrams for these various electrical configurations are shown in Fig. 8-5.

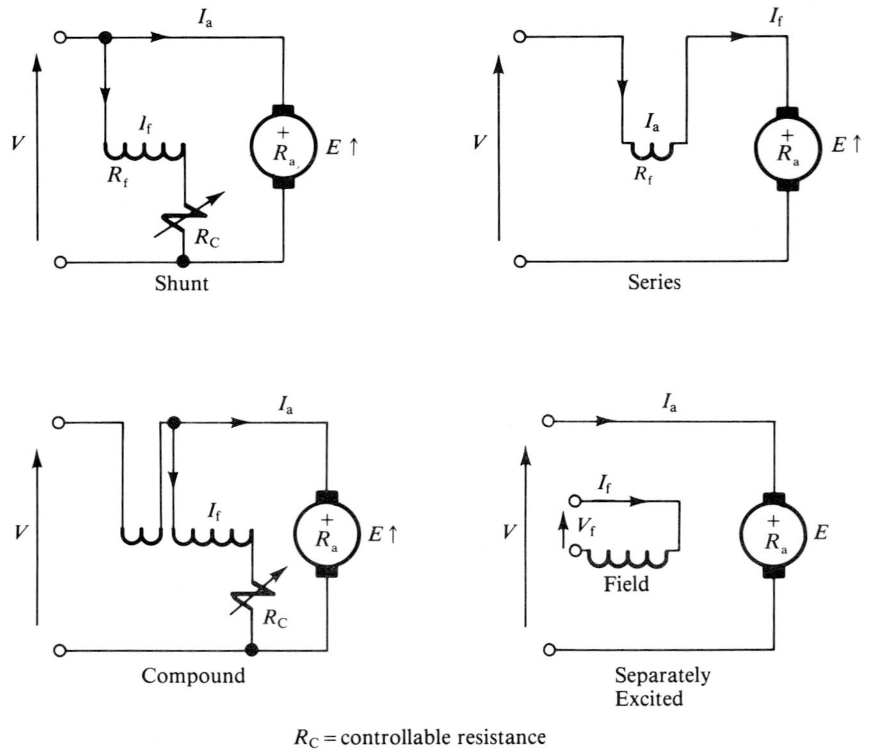

Figure 8-5. DC motor circuit configurations.

In the shunt motor, the field winding is in parallel with the armature, while in the series motor, it is in series with the armature. The compound motor is a shunt-series motor with windings in both series and parallel with the armature. Note that in these first three configurations, there is no independent control of current in the field winding(s). In the separately excited motor, the field winding has its own separate power source and supply voltage V_f.

Voltage V is the external voltage impressed on the motor which causes it to operate. Current I is the current drawn from the power supply lines. Voltage E is the counter-electromotive force (counter-emf) developed by the armature as a result of its rotation in the field. Both the field windings and the armature have characteristic inductances and resistances. The resistances of field and armature are identified as R_f and R_a respectively.

The armature resistance is small in all of the DC motor configurations of Fig. 8-5 (a typical resistance in a 75 hp motor is 0.10 Ω). The resistance of the field R_f, however, depends on how it is connected electrically. In the series motor, the resistance of the field is kept low to minimize the *IR* drop across it because it is in series with the armature which carries substantial current. So, in a series motor, the flux of the magnetic field is derived primarily from high current rather than from number of turns. In the shunt motor, the resistance of the field is kept high to insure that most of the supply current will pass through the armature. In this case, the magnetic flux of the field is derived primarily from numbers of turns of the shunt field rather than from current in it.

The shunt motor operates somewhat similarly to the separately excited motor in that the field is relatively independent of what is happening in the armature branch of the circuit. And in all of these DC motor configurations, the inductances are of no interest to us because we simply want to determine the characteristic steady-state relationships between V, I, T, and ω_r.

The externally applied voltage V must balance the counter-emf E and provide the *IR* drop in the armature branch of the equivalent circuit. Therefore,

$$V = E + I_a R \qquad \textbf{(8-6)}$$

where R is the resistance in the armature branch. Equations (8-3), (8-5), and (8-6) make possible the determination of the characteristic relationships among the power parameters for the various motor configurations of Fig. 8-5.

EXAMPLE 8-1

A DC motor runs at a speed of 1200 rpm when the resistance in the armature branch is 0.2 Ω and the armature current is 25 A. The applied voltage is 115 V. Determine the speed at which this motor will run if the field excitation is maintained constant and

(a) The load is increased until the armature current rises to 30 A
(b) The load is decreased until the armature current drops to 15 A

Solution

The original counter-emf is, from Eq. (8-6),

$$E_o = 115 - 25(0.2) = 110 \text{ V}$$

With the increase in load, the counter-emf becomes

$$E_i = 115 - 30(0.2) = 109 \text{ V}$$

From Eq. (8-4), we see that

$$\frac{E_i}{\omega_i} = \frac{E_o}{\omega_o}$$

Therefore,

$$\omega_i = \frac{109}{110}(1200) = 1189 \text{ rpm}$$

In a similar way, we can find that the decrease in load increases the speed to

$$\omega_d = \frac{112}{110}(1200) = 1222 \text{ rpm}$$

■

8-3 DC MOTOR STARTING CONDITIONS

If Eqs. (8-5) and (8-6) are combined, we see that the armature current is

$$I_a = \frac{V - K_5 I_f \omega_r}{R} \tag{8-7}$$

At starting conditions, the armature is at rest, and Eq. (8-7) shows that

$$I_a = \frac{V}{R} \quad \text{(at starting)} \tag{8-8}$$

Since the resistance of the armature branch R is always low, Eq. (8-8) shows that application of full line voltage V would result in a very large inrush of current (perhaps twenty times higher than normal operating current). This situation occurs because, at starting, the armature cannot develop an emf to counter the applied voltage. For this reason, DC motors are started by placing a starting rheostat (a variable resistance) in series with the armature to limit the current to a safe value until the armature speed provides a countering emf. The resistance is a maximum at starting and is diminished as the armature comes up to speed. The maximum rheostat resistance at starting is chosen so as to limit the starting current to one and a half to two times normal operating current. In a shunt motor, the rheostat should be placed between the field and the armature so that full field voltage will be available for the field winding.

DC motors can be stopped simply by disconnecting the motor from the power supply lines. Various methods of braking the armature are available.

EXAMPLE 8-2

What would be the inrush of current to the motor of Example 8-1 if the motor were to be started without a rheostat? What should be the rheostat maximum resistance in order to limit the starting current to two times its armature current at 1200 rpm?

Solution

Without the rheostat,

$$I_a = \frac{115}{0.2} = 575 \text{ A}$$

This is twenty-three times the normal value of 25 A. If the starting current is to be limited to 2(25) = 50 A, then the rheostat maximum resistance is determined to be

$$50 = \frac{115}{0.2 + R_r}$$

$$R_r = 2.1\ \Omega$$

■

8-3-1 Base Speed

The DC motor does not have a reference synchronous speed, as does the AC induction motor, to which the other speeds of its characteristic torque-speed curve can be referred. Instead, we refer to the base speed of a DC motor, which is its speed at full field excitation and full (rated) armature voltage. The base speed for a motor is a value provided by the manufacturer for each particular type of motor.

8-4 DC MOTOR SPEED CONTROL

One of the great advantages of DC motors over AC motors is the intrinsic versatility and flexibility they possess with respect to speed control possibilities. From Eqs. (8-5) and (8-6), we see that the speed of the DC motor is

$$\omega_r = \frac{V - I_a R}{K_5 I_f} \tag{8-9}$$

Equation (8-9) shows that there are three basic methods of speed control: (a) adjustment of the field current I_f using rheostats in the field winding, (b) adjustment of the armature current I_a using rheostats in the armature circuit, and (c) adjustment of the applied voltage V.

By the use of one or more of these basic methods of speed control, the DC motor can provide a versatile adjustable-speed drive. For many years, the basic modes of speed control were through the use of variable resistances (rheostats) in series with either the field or armature circuits. The more modern electronic mode of speed control is by controlling the applied voltage V using the principle of phase-controlled rectification (see Chapter 11).

There is no great problem in slowing a DC motor down to zero speed and then having it change direction; this is done simply by changing the polarity of either the armature or the field.

The idealized basic speed ranges available for shunt and series motors are sketched in Fig. 8-6. The droop of the characteristic torque-speed curves of these DC motors may also be readily modified by simply making resistance changes. However, in viewing these characteristics, it is important to understand the distinction between speed control and speed regulation. Speed regulation refers to change in motor speed in response to change in load (torque) when other electrical parameters (resistances R_f or R_a or applied voltage V) of the motor are held fixed. The speed regulation curve is simply the output characteristic curve of the type we have been discussing throughout this book. When speed is on the abscissa and torque is on the ordinate, we call the curve the characteristic curve. When speed is on the ordinate and torque is on the abscissa, the curve is called a speed regulation curve.

Speed control, on the other hand, represents changes in speed of the motor because of changes in the operating electrical parameters overtly made by an operator or control system. As an example, suppose a motor slows down because

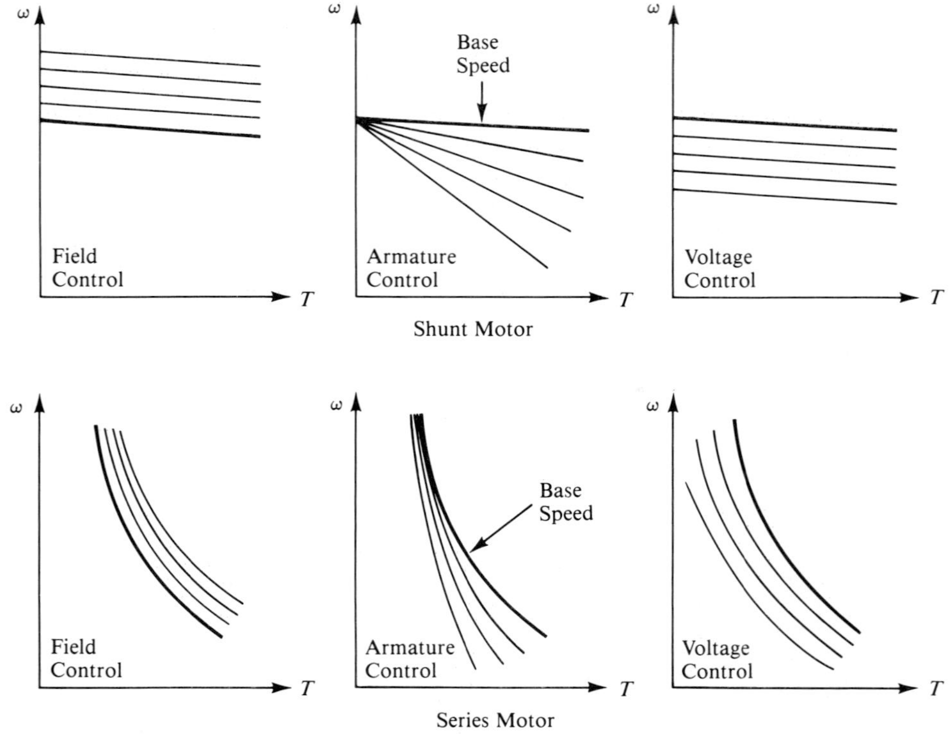

Figure 8-6. Modes of DC dynamo speed control.

of an increase in load torque acting upon it. The amount of speed slowdown as a result of a given increase in load torque is a measure of the motor's speed regulation. The speed can be restored to its original value, despite the increased torque, by effecting a speed control action which could be simply to increase the applied voltage.

8-5 SHUNT MOTOR SPEED CONTROL

In the shunt motor, the resistance of the shunt field is related simply to the applied voltage and field current. Also, the resistance of the armature branch is the same as the resistance of the armature itself. Therefore,

$$R_f = \frac{V}{I_f} \qquad R = R_a \tag{8-10}$$

If Eqs. (8-9) and (8-10) are combined, the shunt motor speed ω_r can be found to be

$$\omega_r = \frac{R_f}{K_5}\left[1 - \left(\frac{I_a}{V}\right)R_a\right] \tag{8-11}$$

Equation (8-11) shows that there are two basic ways in which the speed of the shunt motor can be controlled by using rheostats.

Consider the field resistance R_f first. As R_f is increased, Eq. (8-11) shows that the motor speed will increase. The physical reason is that an increase in field resistance causes a decrease in magnetic flux Φ. Consequently, the counter-emf drops, causing the armature current to increase. Although the decreased current in the field tends to cause the electromagnetic (motor) torque to drop somewhat, it is more than compensated for by the increase in torque because of increased armature current. Also, the drop in speed (see Eq. [8-11]) due to increased armature current does not overcome the speed increase due to increased R_f. So the torque increases and the speed rises. But the increase in speed causes the counter-emf to be restored which ultimately reduces armature current and torque. When the initial torque is restored, the net result will be to supply the same torque as was supplied initially, but at a higher speed.

If the resistance of the armature R_a were to be increased by using a rheostat in series with the armature, then Eq. (8-11) shows that the motor speed will decrease. The reason is that the decreased armature current will cause a decrease in motor torque (see Eq. [8-3]). The temporary superiority of load torque over motor torque causes the speed to drop. This, in turn, decreases the counter-emf, which reflects itself in increased armature current. This increased current will restore the motor torque, and the net result will be to supply the load torque but at a lower speed.

EXAMPLE 8-3

A 220 V shunt motor has a speed of 1000 rpm when its armature current is 40 A. The armature resistance is 0.2 Ω and the field resistance is 200 Ω. Suppose the field resistance is suddenly increased to 400 Ω. What will be the effect on counter-emf, armature current, torque, and speed? Assume the torque is 50 ft-lb.

Solution

The initial counter-emf is

$$E = 220 - 40(0.2) = 212 \text{ V}$$

When the field resistance is doubled, this counter-emf becomes (temporarily)

$$E = \frac{1}{2}(212) = 106 \text{ V}$$

The drop in counter-emf causes the armature current to increase to

$$I_a = \frac{220 - 106}{0.2} = 570 \text{ A}$$

If we assume this temporary inrush of current does not burn out the armature, the motor torque will now rise to (see Eq. [8-3])

$$T = 50\left(\frac{570}{40}\right)(0.5) = 356 \text{ ft-lb}$$

This will cause the motor speed to increase until equilibrium is restored at the original torque of 50 ft-lb. The new equilibrium armature current must be double the original 40 A, because the field current was halved (see Eq. [8-3]). So the new equilibrium value is $I_a = 80$ A. The new counter-emf will then be

$$E = 220 - 80(0.2) = 204 \text{ V}$$

The new speed will be determined by Eq. (8-4). Its value will be

$$\omega_r = 1000\left(\frac{204}{212}\right)(2) = 1924 \text{ rpm}$$ ■

EXAMPLE 8-4

Continuing Example 8-3, what resistance would have to be inserted into the armature branch in order to reduce the speed to 500 rpm?

Solution

Since the field flux is unchanged, Eq. (8-4) shows that the new counter-emf will be

$$E = \frac{1}{2}(212) = 106 \text{ V}$$

The armature current will not change, because the final torque will be the same as it was initially (see Eq. [8-3]). Therefore, the new armature resistance will have to be

$$R_a = \frac{220 - 106}{40} = 2.85\ \Omega$$

Since the original armature resistance was 0.2 Ω, the amount of resistance that will have to be inserted is

$$R = 2.85 - 0.2 = 2.65\ \Omega$$ ■

8-6 SHUNT MOTOR SPEED REGULATION (THE CHARACTERISTIC CURVE)

Equation (8-11) also exposes the characteristic relationship between speed and torque of the shunt motor if we substitute Eq. (8-3) into it. The result is

$$\omega_r = \frac{R_f}{K_5}\left[1 - \left(\frac{R_a R_f}{K_3 V^2}\right)T\right] \quad \textbf{(8-12)}$$

Equation (8-12) is a straight line on the $\omega_r - T$ plane with the slope determined by the constants K_3 and K_5 and the parameters R_a, R_f, and V. Speed regulation curves are shown in Fig. 8-7 for a typical shunt motor.

Recall that normally the resistance R_a of the armature is small. Ideally, it would be zero, and then Eq. (8-12) would show that the idealized shunt motor is a constant-speed machine obeying the relationship

$$\omega_r = \frac{R_f}{K_5} = \omega_B \quad \textbf{(8-13)}$$

For the idealized shunt motor, we see that the base speed is determined by field resistance R_f and magnetizing constant K_5. For the shunt motor of Fig. 8-7, note

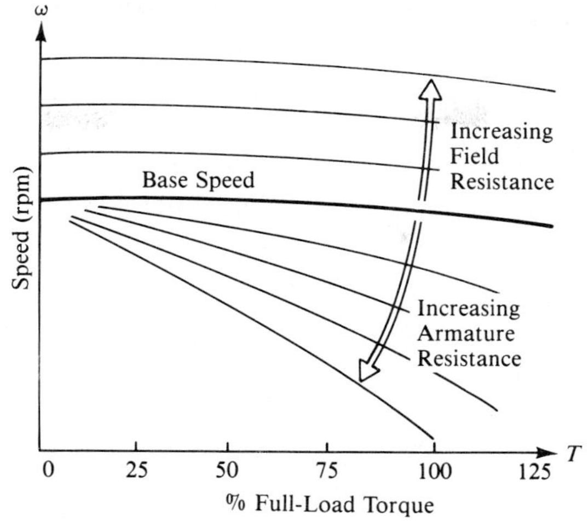

Figure 8-7. Speed regulation curves of the DC shunt motor.

that the curves for negligible armature resistance are practically constant-speed lines (very small droop), and the motor operates close to its base speed at all torques up to 125 percent of full-load torque.

As armature resistance is increased, the droop in the speed regulation increases because of the increasing influence of the second term of Eq. (8-12).

If we plot Eq. (8-13) with speed on the abscissa, we get the characteristic curve of the idealized shunt motor, shown in Fig. 8-8. Note that the idealized shunt motor has the same characteristic on the torque-speed plane as does the idealized positive-displacement pump on the head-discharge plane. As the idealized shunt motor is loaded up, it will provide whatever torque is necessary (and, hence, armature current) to meet the load without a speed change. Consequently, this type of motor needs overload protection analogous to the relief valve in the hydraulic circuit. Protection can be accomplished mechanically (shear pin) or electrically (fuses).

Figure 8-8. Idealized DC shunt motor characteristic.

EXAMPLE 8-5

Suppose the base speed of the motor of Example 8-3 is 1200 rpm. What is the slope of the speed regulation curve?

Solution

Differentiating Eq. (8-12), we find the slope to be

$$\frac{d\omega_r}{dT} = -\frac{R_f^2 R_a}{K_5 K_3 V^2} \frac{\text{rpm}}{\text{ft-lb}}$$

Substituting Eq. (8-13),

$$\frac{d\omega_r}{dT} = -\omega_B \frac{R_f R_a}{K_3 V^2}$$

Constant K_3 is found from Eq. (8-3) to be

$$K_3 = \frac{50}{40(220/200)} = 1.136 \text{ ft-lb/A}^2$$

Substituting this into the expression for the slope gives

$$\frac{d\omega_r}{dT} = -\frac{1200(200)(0.2)}{1.136(220)^2} = -0.973 \text{ rpm/ft-lb}$$ ■

8-7 SERIES MOTOR SPEED CONTROL

In the series motor, the field current is the same as the armature current:

$$I_f = I_a \tag{8-14}$$

Equation (8-3) then becomes

$$T = K_3 I_a^2 \tag{8-15}$$

Since the armature current is high under starting conditions, we see that the series motor has a very powerful starting torque characteristic (in the shunt motor, the torque is proportional to I_a).

If Eqs. (8-5), (8-6), and (8-14) are combined, the speed ω_r of the series motor is found to be

$$\omega_r = \frac{1}{K_5}\left(\frac{V}{I_a} - (R_a + R_f)\right) \tag{8-16}$$

Equation (8-16) shows that speed control of the series motor can be obtained by varying the resistance (e.g., with a rheostat) in the armature circuit.

If Eq. (8-15) is substituted into Eq. (8-16), the characteristic speed-torque relationship is obtained for the series motor:

$$\omega_r = \frac{K_3^{1/2} V}{K_5 T^{1/2}} - \frac{R}{K_5} \tag{8-17}$$

Equation (8-17) is plotted in Fig. 8-9 for a typical series motor.

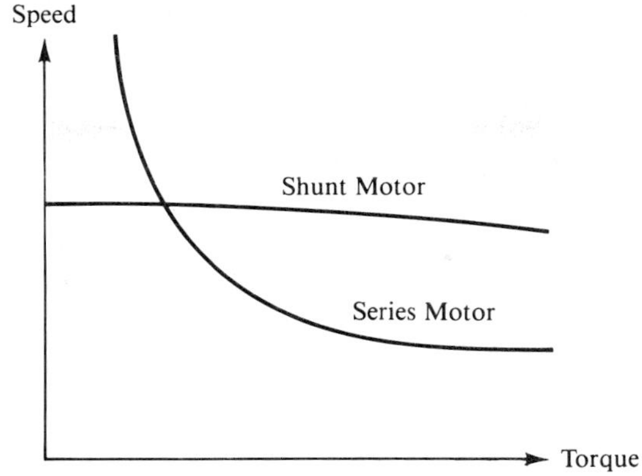

Figure 8-9. Speed regulation of the DC series motor.

Figure 8-9 shows that the speed regulation of the series motor is poor, making it unsuitable for use in many mechanical systems. This motor has been widely used, however, as a traction motor on cranes and in rail applications involving heavy loads and where speed is under the manual control of an operator.

Since the armature circuit resistance R is small, the series motor speed-torque curve can be approximated by

$$\omega_r = \frac{K_3^{1/2}V}{K_5T^{1/2}} \tag{8-18}$$

On a torque versus speed graph, Eq. (8-18) shows the characteristic curve of the idealized series motor (see Fig. 8-10).

Notice the dangerous runaway propensity of the series motor as the load torque is reduced; this is why speed variation is best controlled by an operator rather than by depending on the intrinsic speed-torque characteristic.

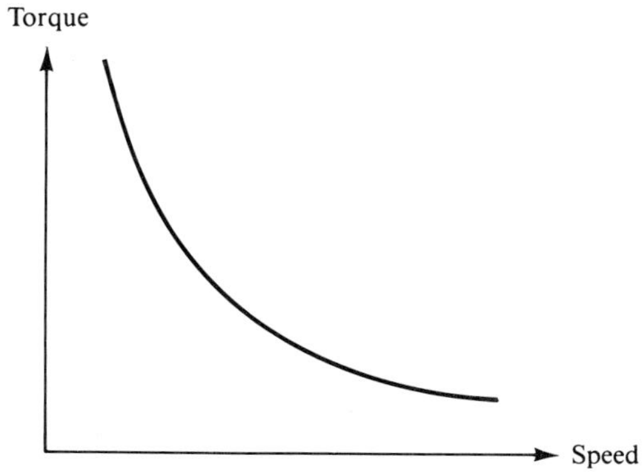

Figure 8-10. Idealized DC series motor characteristic.

The speed regulation of a typical shunt motor is also shown in Fig. 8-9. As the torque is reduced on a shunt motor, the motor simply approaches its base speed asymptotically; there is no danger of a large speed increase as load is reduced.

EXAMPLE 8-6

A 220 V series motor has a speed of 1000 rpm when its armature current is 40 A. The armature resistance is 0.2 Ω and the field resistance is 0.1 Ω. Suppose there is a rheostat in series with the armature circuit to limit the starting torque and current. If the starting torque is to be twice the torque at the 1000 rpm running speed, what resistance must be placed in the circuit by the rheostat?

Solution

From Eq. (8-15), we find the starting current to be

$$I = \sqrt{2}(40) = 56.6 \text{ A}$$

The total resistance in the armature circuit must then be

$$R = \frac{220}{56.6} = 3.89\ \Omega$$

Subtracting the armature and field resistances, the rheostat resistance required is

$$R = 3.89 - (0.2 + 0.1) = 3.59\ \Omega$$

Suppose this resistance is left in the circuit while the motor comes up to its steady-state speed. This speed can be determined as follows. The 1000 rpm counter-emf is given by

$$E = 220 - 40(0.2 + 0.1) = 208 \text{ V}$$

The new steady-state counter-emf with the rheostat in the circuit will be

$$E = 220 - 40(3.89) = 64.4 \text{ V}$$

Then Eq. (8-5) shows that

$$\omega_r = 1000\left(\frac{64.4}{208}\right) = 309 \text{ rpm}$$

As the rheostat resistance diminishes to zero, the motor will speed up to its normal speed of 1000 rpm. ■

8-8 THE COMPOUND MOTOR

The compound DC motor is an attempt to gain the benefits of both the shunt and series motor in one machine. Consequently, it has both series and shunt windings from which the magnetic flux of the motor is derived. The relative strengths of the two windings will dictate the shape of the torque-speed characteristic of the compound motor. Generally, the relative field strengths are chosen so that the characteristic is similar to that of the shunt motor but with more droop in the curve.

8-9 APPLICATIONS OF DC MOTORS

The speed control and speed regulation potential of DC motors make their use desirable in a wide variety of mechanical systems. Motor speeds can be easily controlled smoothly, and even reversed. DC motors are also extremely responsive to control signals because of their high torque to inertia qualities.

Another desirable feature of the DC motor is that the droop of the torque-speed characteristic can be varied easily by simply changing values of internal motor resistances. These machines are highly adaptable compared to induction motors.

The DC shunt motor with only a slight droop in its speed regulation curve represents high-speed stiffness and good transient performance. Therefore, it is highly suitable as an electrical servo-motor in machines such as robots. Its speed range can also be easily adjusted simply by controlling the shunt field strength. Field control makes the shunt motor ideal in mechanical systems used to control speed or tension of a continuous web (film, textiles, etc.).

The series motor can provide very high starting torque, albeit at the expense of creating certain intrinsic control problems. The high starting torque makes it suitable in heavy-duty applications under the manual control of an operator (cranes, hoists, etc.).

The fact that a DC motor can be operated as a generator makes possible regenerative or dynamic braking. These types of braking are very desirable in certain applications where there may be large overhauling loads, such as in large

TABLE 8-1. DC Motor Properties and Applications.

Motor Type	Properties	Applications
Shunt	Essentially constant speed, or Controllable droop Speed control via field resistance and/or armature resistance Good starting torque	Servo-drives Robotics Compressors Pumps Machine tools Many industrial applications
Series	Poor speed regulation Superior starting torque Speed control via armature circuit resistance Runaway propensity at low torque	Traction devices Cranes Hoists Electrified rail applications High inertia loads
Compound	High starting torque Good speed regulation Speed control via shunt and/or series field resistance or armature resistance Controllable droop	Machine tools Punch presses Pumps Compressors Many industrial applications

traction apparatus. The need for a mechanical brake will not be obviated, but the size of the mechanical brake can be greatly reduced.

The disadvantages of DC motors are that they are complex in construction, compared to induction motors, and require commutators. To avoid commutation problems, good maintenance is required.

A list of DC motor properties and applications is presented in Table 8-1.

8-10 THE DC GENERATOR

When the electromagnetic armature torque of the DC dynamo is in the same direction as its rotation, the dynamo acts as a motor. When the armature torque and rotation rates are in opposite directions, the dynamo acts as a generator.

Relationships (8-3) and (8-5) are also valid for the DC generator. However, the relationship between emf, line voltage, and *IR* drop in the dynamo becomes

$$E = V + I_a R \tag{8-19}$$

In a generator, the electromotive force E is developed as a consequence of an input torque provided to the dynamo in order to drive it. The output voltage V that the generator supplies at its terminals is somewhat less than the emf because of the *IR* drop in the armature circuit.

All of the configurations we discussed for DC motors are applicable to the DC generator. But our interest in the DC generator in this book will be confined to just two types of application: its use as a brake and its use as a component in the Ward-Leonard drive.

8-11 THE DC DYNAMO AS A BRAKE

When a dynamo operates as a generator, the electromagnetic torque is in opposition to the direction of rotation. It therefore is acting as a brake insofar as its applied torque is concerned (i.e., the torque applied to the generator at its input side).

In applications of DC motors where the driven load may tend to overhaul, as in elevators, cranes, and rail applications, the DC motor can very effectively be made to provide some braking of the overhauling load. This can be done in two ways: by regenerative braking and dynamic braking.

In regenerative braking, the overhauling torque is used to drive the motor as a generator by suddenly reducing the resistance in the field circuit. The increased magnetic flux of the field will then increase the motor counter-emf above line voltage, and the motor will now act as a generator delivering power back into the electrical supply. However, at some point, as the speed falls because of braking, the counter-emf will again fall below the line voltage, and the dynamo will again be acting as a motor. So braking by the regenerative method will not be able to reduce the speed all the way to zero. Regenerative braking is excellent on long electrified railway downhill runs where the objective is not to stop the train but merely to slow it down.

In dynamic braking, the armature circuit is disconnected from the electrical supply and a dissipative resistor is placed across it. The field is kept connected to the electrical supply. The emf is now used to drive current through the dissipating resistor rather than to oppose the applied voltage of the electrical supply. Thus, the motor acts as a generator supplying a dissipative load, and braking can be continuously provided down to low speeds. Eventually, of course, a small mechanical brake must be used to stop the armature, because the dynamic braking torque disappears as the speed approaches zero.

8-12 THE WARD-LEONARD DRIVE

A highly versatile and flexible DC drive arrangement that can control speed, position, and direction of rotation of a drive motor is sketched in Fig. 8-11. The arrangement is called the Ward-Leonard drive. It consists of a separately excited DC generator driving a separately excited DC motor. The generator can be driven by any prime mover, but most frequently it is driven by an induction motor (see Chapter 9).

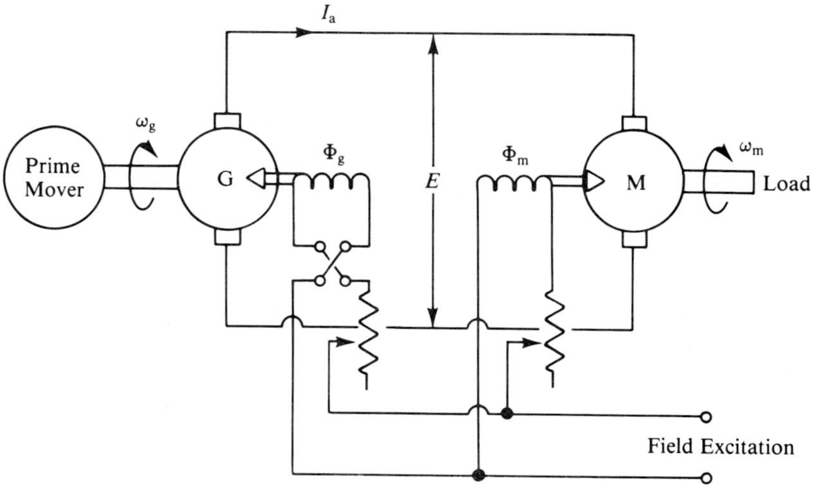

Figure 8-11. Ward-Leonard drive (transmission).

The Ward-Leonard drive is particularly well suited to drives that have a high horsepower requirement (such as are found in steel mill roll drives involving powers of up to 5000 hp). With the Ward-Leonard drive, speed control is effected by changing only field resistances; this has the advantage over a simple shunt motor speed control, where resistance is placed in series in the armature circuit. The disadvantage of the Ward-Leonard drive is that three dynamos are involved rather than just one variable-speed motor, making the drive not only costly but bulky as well.

The characteristics of the Ward-Leonard drive can be obtained by applying Eqs. (8-3), (8-5), and (8-6) to the DC generator and DC motor respectively. For the DC generator, we have the ideal governing equations

$$E_g = K_{5(g)} I_{f(g)} \omega_g \quad \textbf{(8-20)}$$

$$V_g = E_g - I_g R_g \quad \textbf{(8-21)}$$

Combining Eqs. (8-20) and (8-21) gives

$$V_g = K_{5(g)} I_{f(g)} \omega_g - I_g R_g \quad \textbf{(8-22)}$$

For the DC motor, we have

$$E_m = K_{5(m)} I_{f(m)} \omega_m \quad \textbf{(8-23)}$$

$$V_m = E_m + I_m R_m \quad \textbf{(8-24)}$$

$$T_m = K_{3(m)} I_m I_{f(m)} \quad \textbf{(8-25)}$$

Combining Eqs. (8-23) and (8-24) gives

$$V_m = K_{5(m)} I_{f(m)} \omega_m + I_m R_m \quad \textbf{(8-26)}$$

Now, if Eqs. (8-22), (8-25), and (8-26) are combined, the result is

$$T_m = \left(\frac{K_{3(m)} K_{5(g)} I_{f(g)} I_{f(m)}}{R_g + R_m} \right) \omega_g - \left(\frac{K_{3(m)} K_{5(m)} I_{f(m)}^2}{R_g + R_m} \right) \omega_m \quad \textbf{(8-27)}$$

In deriving Eq. (8-27), the conditions for the Ward-Leonard drive were applied, namely,

$$V_g = V_m = V \qquad I_g = I_m = I_a \quad \textbf{(8-28)}$$

Equation (8-27) has the same form as does Eq. (5-22) for the hydrostatic drive discussed in Chapter 5, that is,

$$T_m = A\omega_g - B\omega_m \quad \textbf{(8-29)}$$

The Ward-Leonard drive provides a wide and sensitive range of speed control with both directions of rotation of the motor. All that is required is manipulation of rheostats in the separately excited fields of the generator and motor. Note, too, that this drive is capable of providing a torque when the motor speed is zero. This capability is an advantage in traction applications such as for electrified train prime movers.

Speed control of the Ward-Leonard drive can be effected in two ways. The first mode is by changing the resistance of the generator field winding in order to provide the motor with a voltage from zero to a maximum at, say, maximum motor field excitation. Additional speed control can be obtained in a second mode (at maximum generator voltage) by varying the field excitation of the motor.

8-13 SUMMARY

The DC dynamo as a generator, as a motor, or as a generator/motor drive is analogous to its positive-displacement hydraulic counterpart. Compared to a hydraulic drive, the DC dynamo is generally bulkier and limited to much lower power densities. Commutation is also a maintenance problem when it is done mechanically. With modern electronic commutation schemes, this disadvantage will begin to disappear. Also, the DC dynamo does not suffer from leaking oil, leaking seals, and many of the messier problems of hydrostatic drives. Its versatility in speed control means that it will remain an important component in mechanical systems for many years to come.

8-14 REFERENCES

8-1. A. B. Carlson and D.G. Gisser, *Electrical Engineering*, Addison-Wesley, Reading, MA, 1981.

8-2. M. G. Say and E. O. Taylor, *Direct Current Machines*, Wiley, New York, 1980.

8-3. R. H. Engelmann, *Static and Rotating Electromagnetic Devices*, Marcel-Dekker, New York and Basel, 1982.

PROBLEMS

8-1. A series motor is operating at 1800 rpm with a torque of 20 ft-lb, an armature current of 50 A, and an applied voltage of 110 V. The armature resistance is 0.3 Ω and the field resistance is 0.05 Ω. Suddenly the mechanical load on the motor drops to 10 ft-lb. What will be the new value of the armature current and speed? What are the original horsepower and mechanical efficiency? What is the final horsepower?

8-2. A shunt motor is operating at 1200 rpm, an applied voltage of 220 V, and a total input current of 160 A. The armature resistance is 0.1 Ω and the shunt field resistance is 100 Ω. Assume the efficiency of the motor is 0.85 and determine how much additional resistance must be inserted in the field to increase the speed to 1500 rpm (this additional resistance is not part of the flux-producing field coil). With the additional resistance in the field circuit, what will the speed become if the load is reduced until the armature current falls to 80 A?

8-3. A 250 V separately excited motor has an armature resistance of 0.05 Ω and a speed of 1200 rpm. The external load torque on the motor is 144 ft-lb. Assume that constants K_3 and K_5 of Eqs. (8-3) and (8-5) are equal. Determine the percent increase in field current required to regeneratively brake the load (and motor) inertia down to a speed of 1000 rpm if the external load is removed.

8-4. A 220 V DC series motor draws 24 A when its speed is 950 rpm and its load torque is 38 ft-lb. The armature and field resistances are 0.19 and 0.11 Ω respectively. Determine

(a) The efficiency of the motor

(b) The load torque and motor speed if the motor current increases to 45 A

(c) The motor speed if the torque drops to 5 ft-lb

8-5. A DC shunt motor has an armature current of 200 A, a field current of 2.5 A, a speed of 1800 rpm, and a torque of 78.4 ft-lb when the applied voltage is 110 V. The armature resistance is 0.03 Ω. Suddenly the load drops and the armature current decreases to 120 A. Determine

(a) The motor efficiency when the armature current is 200 A
(b) The new speed after the load drops
(c) The percent change in speed based on the original speed
(d) The percent change in torque based on the original torque
(e) The speed regulation

8-6. A 220 V DC series motor has an armature resistance of 0.25 Ω and a field resistance of 0.08 Ω. The rated horsepower is 10 hp when the motor speed is 1500 rpm.

(a) If the motor efficiency is 0.847, what resistance value must be placed in series with the motor if the starting current is to be limited to a value one and one-half times the rated load current?
(b) If the starting resistance is left in the motor circuit, what will be the resulting steady-state speed under rated torque?

8-7. A DC shunt motor and a DC series motor both run at 600 rpm and develop a 5 hp output when operating with an applied voltage of 220 V and an armature current of 22 A. Determine

(a) The speeds of these motors when the load drops such that the armature currents in both motors becomes 11 A
(b) The torques of the two motors when the currents are 11 A

8-8. A 220 V, 20 hp DC shunt motor runs at 840 rpm and draws 76 A from the power supply. The field current is 2.1 A and the armature resistance is 0.1 Ω. The speed is reduced by placing a 0.8 Ω resistance in the armature circuit. Assuming the total motor current does not change, determine

(a) The new motor speed
(b) The ratio of the new torque to the original torque

8-9. A DC shunt motor has a rated output horsepower of 7.5 hp at a rated speed of 1800 rpm. The applied voltage is 250 V, and the rated input current is 26 A. The armature resistance is 0.5 Ω and the field resistance is 250 Ω.

The motor is to be started from rest in three stages by cutting out a total rheostat resistance of 6 Ω in two steps (3 Ω/step). The rheostat is in series with the armature. At each stage of the starting process, the motor will be permitted to come up to an intermediate speed such that the armature current will be equal to its rated value before the next 3 Ω of rheostat resistance are cut out (see Fig. P8-9).

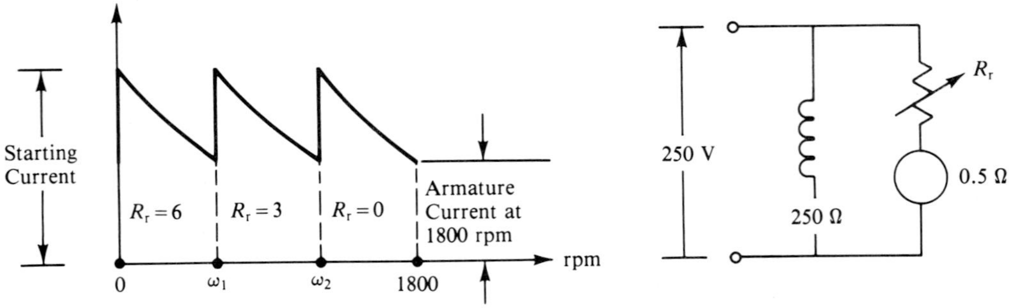

Figure P8-9. Figure for Problem 8-9.

Determine

(a) The mechanical efficiency at rated load conditions
(b) The field current
(c) The armature current at rated speed
(d) The counter-emf at rated speed
(e) The starting current
(f) The intermediate speeds ω_1 and ω_2
(g) The base speed of the motor

8-10. A centrifugal pump drives a load whose characteristic is

$$P = 1.2Q \text{ kN/m}^2 \qquad (Q \sim \text{m}^3/\text{min})$$

The pump output characteristic is

$$P = \frac{N-6}{2} - \frac{Q}{5} \text{ kN/m}^2 \qquad (N \sim \text{rev/sec},\ Q \sim \text{m}^3/\text{min})$$

The pump has a mechanical efficiency of $\eta = 0.75$. It is driven by a 110 V DC shunt motor having an armature resistance of 0.1 Ω and a field resistance of 110 Ω.

(a) At what speed must the pump be driven such that the system discharge will be $5\text{ m}^3/\text{min}^2$?
(b) What input current will the motor draw at this discharge?
(c) Display the operating point of this system, in the steady state, on the *P*-*Q* plane.

8-11. The relationship between terminal voltage and load current of a DC generator is

$$V = 230 - 0.5I$$

The generator is supplying power to a circuit consisting of a pure resistance of 6 Ω in parallel with a battery which is being charged. The battery has an open-circuit emf of 215 V and an internal resistance of 0.5 Ω. Determine

(a) The current *I* and power *P* delivered by the generator to this load
(b) The terminal voltage and generator current if the load is changed to a pure resistance such that 9 kW are delivered by the generator
(c) The maximum power that this generator can deliver
(d) The resistance the load must have for maximum power to be delivered

9

Alternating-Current (AC) Motors

An alternating-current (AC) motor is an electromechanical dynamo consisting of a rotor and a stator. The rotor is a cylinder located inside a concentric hollow cylinder called the stator (see Fig. 9-1).

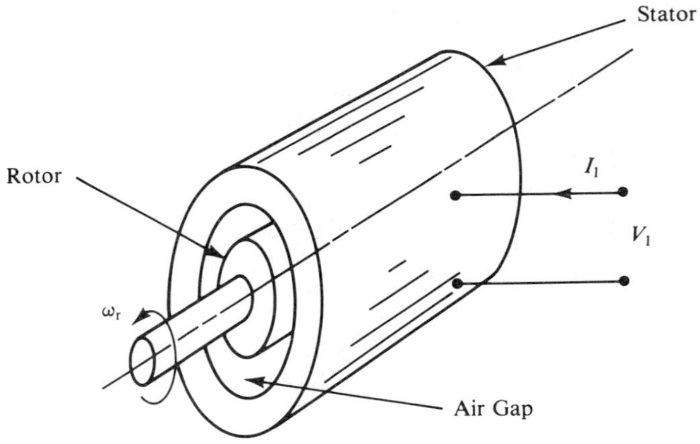

Figure 9-1. AC motor configuration.

The stator is an iron core containing windings arranged in such a way as to constitute one or more electromagnets which produce magnetic flux. The number of poles of the stator depends on how the windings are configured. There may be two, four, six, or more poles.

The stator is a fixed component supplied with an externally applied AC voltage V_1. The resultant current flow in the stator windings will depend on the resistance and reactance of the stator windings.

The rotor is also an iron core. It contains integral conductors that may be arranged on it in a variety of ways. The rotor rotates at angular velocity

ω_r (rad/sec). As it does so, its conductors may cut across the magnetic field produced by the stator.

The fundamental principle of operation of the AC motor is the force interaction between current existing in the rotor and the magnetic field of the stator which produces an electromagnetic torque in the direction of rotation of the rotor. The principle of operation is the same as for the DC motor, but it is implemented in a different way.

If the current in the rotor is produced by electromagnetic induction, we call the AC motor an induction motor. Induction motors are of two basic types: the squirrel cage (or simply cage) motor and the wound rotor motor. These motors can be either three-phase or single-phase devices.

The cage type of motor is called a brushless motor because its rotor is a self-contained electrical circuit isolated mechanically and electrically from the surroundings. The cylindrical rotor consists simply of an iron core fabricated from sheet steel laminations that has axial slots on its periphery. Imbedded in these slots are large conductors short-circuited by end rings in a cagelike configuration (see Fig. 9-2).

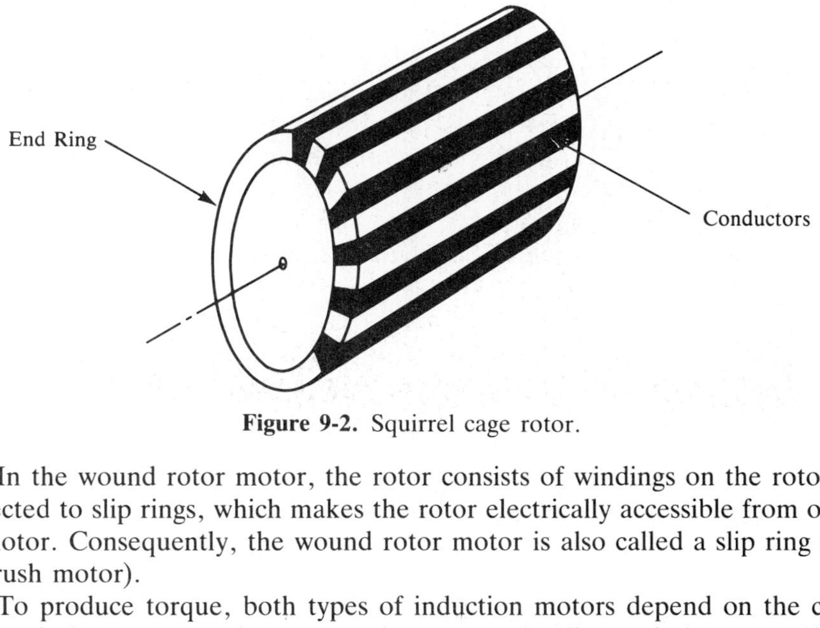

Figure 9-2. Squirrel cage rotor.

In the wound rotor motor, the rotor consists of windings on the rotor core connected to slip rings, which makes the rotor electrically accessible from outside the motor. Consequently, the wound rotor motor is also called a slip ring motor (or brush motor).

To produce torque, both types of induction motors depend on the cutting action of the rotor conductors as they cross the lines of the magnetic flux produced by the stator. This cutting action requires a relative slip s between the motions of the stator field and the rotor conductors. In some respects, the induction motor is similar to an AC transformer when we view the stator as the primary coil and the rotor as the secondary coil. However, whereas the transformer has an iron core linking the fluxes of the primary and secondary coils, the AC induction motor achieves this linkage via the air gap at the interface between the rotor and stator.

The cutting action of the rotor conductors across the stator magnetic field induces a voltage in the rotor which, in turn, causes currents to flow in the rotor. Because of inductive reactance in the rotor, there will be some phase shift between the induced rotor emf and the resulting current.

There is a third type of AC motor of interest to us that works on the principle of interaction between DC currents supplied to the rotor and AC fields produced by the stator. This hybrid DC-AC motor is called a synchronous motor because it does not depend on slip between the rotor and stator field to develop torque. The speed of the rotor remains constant at the synchronous speed of the stator's rotating magnetic field. Synchronous motors will be discussed later. The physical construction of the synchronous motor is somewhat similar to the wound rotor motor in that the rotor is made accessible to the outside DC supply by using slip rings.

9-1 THREE-PHASE INDUCTION MOTORS

The three-phase induction motor is the workhorse of industry. It is simple in concept and rugged in construction. The electromagnetic induction occurring in the rotor is a consequence of an ingenious method of using a three-phase power supply to create a rotating magnetic field in an armature (the stator) without the necessity for the armature itself to rotate, as we will now see.

Figure 9-3 is a schematic representation of a three-phase power supply in which each phase current is represented by a rotating vector (a phasor). The three vectors representing the three phases are separated by 120 temporal degrees, and each results in an equivalent sinusoidal phase current variation as shown at the upper right.

Suppose the three phases of the power supply are used to energize three separate two-pole windings spaced equally around the periphery of the stator (120 spatial degrees apart), as shown in the lower part of Fig. 9-3. Let us now examine the magnetic fluxes these three windings will create in the stator for the three instants specified by 0, 120, and 240 degrees on the time axis.

For the instant corresponding to 0 degrees, currents I_1, I_2, and I_3 are zero, negative, and positive respectively. Therefore, the phase 1 winding will produce no flux while the phase 2 and phase 3 windings produce fluxes Φ_2 and Φ_3 in the directions shown. Φ_2 and Φ_3 have a resultant vector, $\mathbf{\Phi}_{23}$, situated at 90 degrees, as shown. For the instant corresponding to 120 degrees, currents I_1, I_2, and I_3 are positive, zero, and negative respectively. They produce fluxes Φ_1 and Φ_3, and the resultant vector, $\mathbf{\Phi}_{13}$, is situated at -30 degrees. Finally, at the instant corresponding to 240 degrees, we see that the resultant flux vector, $\mathbf{\Phi}_{12}$, is situated at -150 degrees.

The result of this three-phase supply excitation is production of a magnetic flux vector rotating cw around the armature (stator) axis at the same angular velocity as that of the original three-phase supply phasors. The angular velocity of the rotating magnetic field is called the synchronous speed ω_s. If the frequency of the power supply is designated as f (cps), then we see that for this two-pole stator,

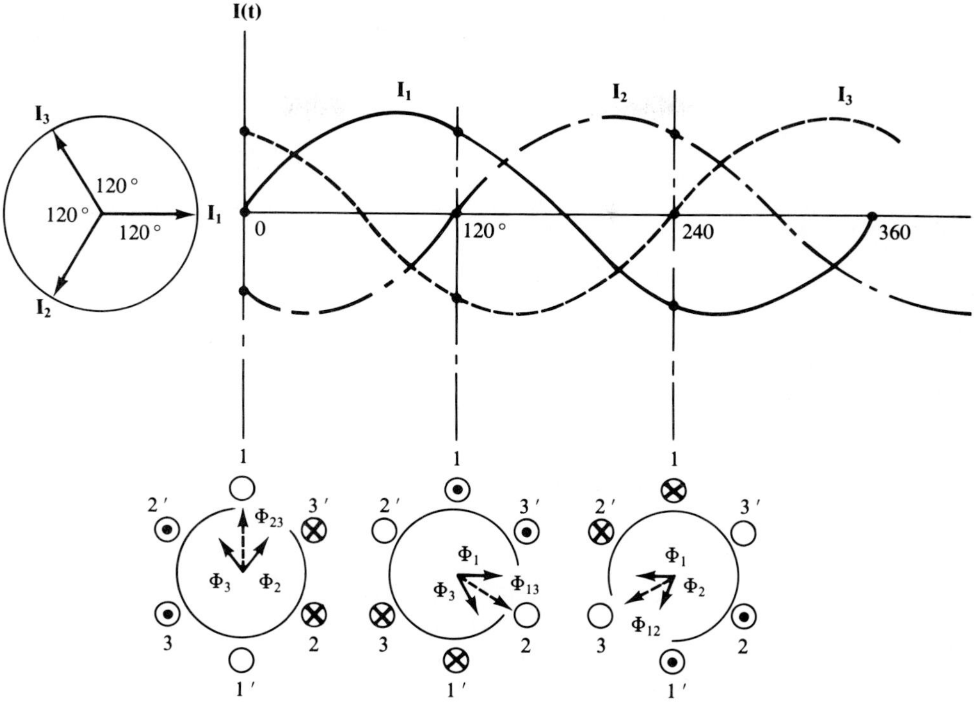

Figure 9-3. Three-phase rotating magnetic field.

the synchronous speed is

$$\omega_s = 2\pi f \text{ rad/sec} \tag{9-1}$$

If we designate the synchronous speed in rpm as n_s, then

$$n_s = 60f \tag{9-2}$$

For a 60 cps supply frequency, Eq. (9-2) shows that the synchronous speed of the two-pole stator field is 3600 rpm.

A more detailed analysis of the rotating magnetic field of the three-phase stator will show that the relationship between synchronous speed n_s and number of poles p is

$$n_s = \left(\frac{120}{p}\right) f \tag{9-3}$$

Equation (9-3) indicates that the synchronous speed of a four-pole winding is 1800 rpm. This type of motor is used often in industrial applications.

Unless the angular speed of the rotor is the same as the synchronous speed of the rotating field, the rotor conductors will be cut by the rotating field (see Fig. 9-4). As discussed earlier, it is the cutting of the rotor conductors by the magnetic field that induces an emf in the rotor and produces the associated electromagnetic torque.

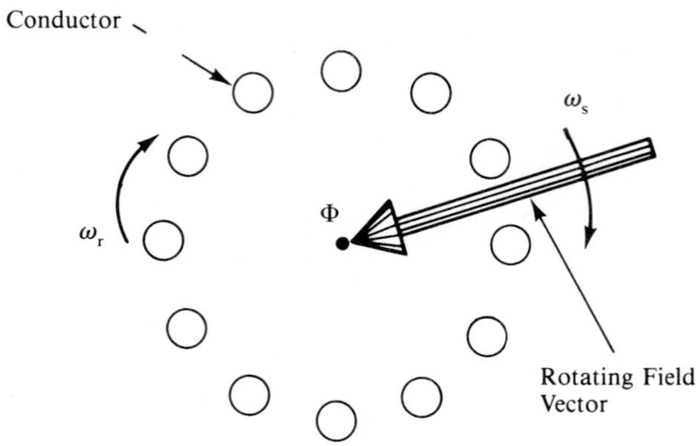

Figure 9-4. Rotor conductors cut by the rotating field.

Let the slip s between the rotor speed and the rotating field speed be defined by

$$s = \frac{\omega_s - \omega_r}{\omega_s} = \frac{n_s - n_r}{n_s} \tag{9-4}$$

The maximum cutting rate will occur when the rotor is locked (at rest), which corresponds to a slip of $s = 1$. For a slip of $s = 0$, there is no cutting of the rotor conductors (and, hence, no emf or torque developed), because $\omega_r = \omega_s$.

The frequency at which AC voltage and current are induced in the rotor is defined as f_r (also called the slip frequency). It seems intuitively obvious that the relationship between slip frequency and the frequency of the three-phase power supply is

$$f_r = sf \tag{9-5}$$

This relationship can be proved easily.

Based on the concept of a three-phase induction motor as it has been described up to this point, it seems plausible that the induction motor would have a characteristic output torque-speed relationship like that sketched in Fig. 9-5. Figure 9-5 indicates that the induction motor torque is a function of slip increasing from zero at a slip of $s = 0$ to a maximum at a slip of $s = 1$. When the load torque on the motor is zero, the motor rotor will rotate at synchronous speed (1800 rpm for a four-pole rotor). Of course, this assumes that the parasitic drag torque on the rotor, due to bearing friction and rotor windage, is negligible.

With the rotor rotating at synchronous speed under no-load, suppose we apply a load torque T_L to the rotor. What happens is that the load torque decelerates the rotor until the proper slip s is developed that will result in the creation of an electromagnetic motor torque T_m tending to equal the load torque. When the motor torque becomes equal to the load torque, the deceleration of the rotor will cease and the motor will be in an equilibrium operating state, with $T_m = T_L$.

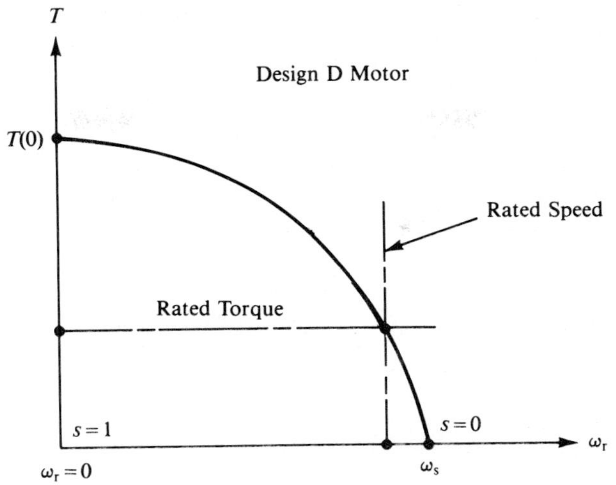

Figure 9-5. Torque-speed characteristic of the Design D motor.

Figure 9-5 shows that as the load torque is continuously increased, the slip will increase until the rotor eventually stalls in producing the stall torque $T_m = T(0)$. This is the maximum torque capability of this particular motor, which has the same classical drooping type of characteristic curve as does the centrifugal pump with backward-curved blading.

As a matter of fact, induction motors with the type of characteristic curve shown in Fig. 9-5 can be purchased. They are called Design D motors.

A more detailed analysis of the electromagnetic behavior of induction motors will show that the Design D type of characteristic curve results when the resistance of the rotor is relatively high (and of the same order of magnitude as the rotor's inductive reactance). For relatively lower rotor resistances, the maximum torque of the motor will be experienced at some slip less than 1, perhaps around $s = 0.25$. This type of motor is called a Design A motor, having a characteristic as sketched in Fig. 9-6. The maximum torque is called the breakdown torque, and the torque at zero rotor speed ($s = 1$) is called the starting torque.

Since slip represents dissipation of energy, induction motors are designed such that their rated torques (normal operating torques) are to occur at small slips of order of magnitude $s = 0.03$. For a motor having a synchronous speed of 1800 rpm, this corresponds to a rotor speed of about 1750 rpm.

We see a fundamental distinction between the Design D and Design A motors sketched in Figs. 9-5 and 9-6. With the Design D motor, the rated torque is less than the starting torque, whereas in the Design A motor, the rated torque is greater than the starting torque. Not only can the Design A motor not be started under rated load, but there is also the possibility that the motor can be stalled if the load torque exceeds the breakdown torque at some point during operation.

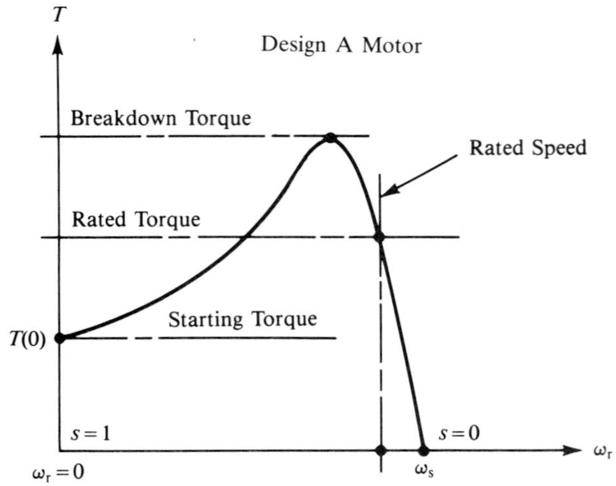

Figure 9-6. Torque-speed characteristic of the Design A motor.

9-2 CHARACTERISTICS OF THE IDEAL THREE-PHASE INDUCTION MOTOR

We can give some analytical substance to our previous intuitive deductions by making several simplifying assumptions about the rotor and stator electrical circuits. An equivalent circuit for the rotor is sketched at the left of Fig. 9-7. Note that the rotor is simply a closed circuit having both inductive reactance x_2 and resistance r_2. The induced emf caused by the cutting of the rotor conductors by the rotating magnetic field is designated as e_2.

For the equivalent circuit shown at the left of Fig. 9-7, the current I_2 in the rotor is given by

$$I_2 = \frac{e_2}{\sqrt{r_2^2 + x_2^2}} \tag{9-6}$$

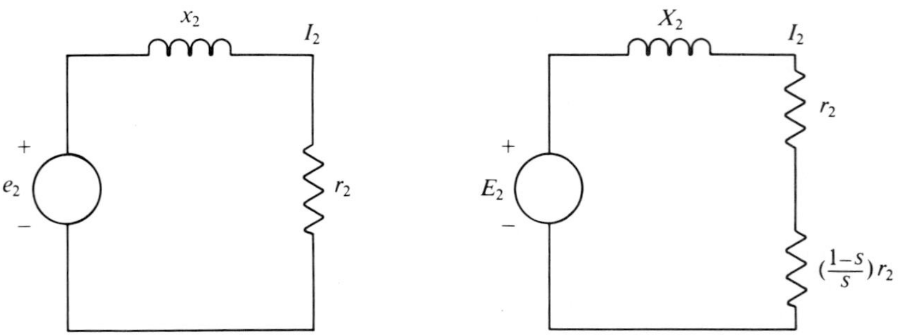

Figure 9-7. Induction motor rotor equivalent circuits.

We now define E_2 and X_2 to be the emf and inductive reactance respectively that would be developed in the rotor if it were to be locked ($s = 1$). Since reactance x_2 is proportional to the frequency f_r at which currents are generated in the rotor, and since reactance X_2 is proportional to frequency f, we conclude that

$$x_2 = sX_2 \tag{9-7}$$

Similarly, the electromotive force e_2 (at some slip s) is proportional to the locked rotor electromotive force according to

$$e_2 = sE_2 \tag{9-8}$$

Substituting Eqs. (9-7) and (9-8) into Eq. (9-6), we find that

$$I_2 = \frac{sE_2}{\sqrt{r_2^2 + (sX_2)^2}} \tag{9-9}$$

Equation (9-9) can be rewritten as

$$I_2 = \frac{E_2}{\sqrt{\left(\frac{r_2}{s}\right)^2 + X_2^2}} \tag{9-10}$$

The equivalent circuit corresponding to Eq. (9-10) is sketched at the right of Fig. 9-7. Note that the resistance term r_2/s of Eq. (9-10) is represented by the two resistances r_2 and $\left(\frac{1-s}{s}\right)r_2$ in series in the circuit on the right. This circuit is preferable to its equivalent on the left because it indicates how much of the power associated with current I_2 is apportioned to dissipation in the rotor circuit ($I_2^2 r_2$) and how much is apportioned to the external load on the rotor $I_2^2\left(\frac{1-s}{s}\right)r_2$. This circuit also has the virtue that electromotive force E_2 and inductive reactance X_2 are fixed values defined by the locked rotor conditions.

Suppose the power flowing into the rotor from the stator is designated as P_2. Using the equivalent circuit at the right of Fig. 9-7, we define P_2 as

$$P_2 = E_2 I_2 \cos\theta_2 \tag{9-11}$$

where θ_2 is the phase angle between E_2 and I_2. Note that

$$\cos\theta_2 = \frac{r_2}{\sqrt{r_2^2 + (sX_2)^2}} \tag{9-12}$$

The power P_m available at the output shaft of the rotor is given by

$$P_m = P_2 - I_2^2 r_2 \tag{9-13}$$

The power dissipation $I_2^2 r_2$ in the rotor circuit is zero for zero slip and a maximum when the rotor is locked ($s = 1$). Therefore, we set

$$I_2^2 r_2 = sP_2 \tag{9-14}$$

In other words, power loss is proportional to slip.

The available output power P_m is then

$$P_m = (1 - s)P_2 \tag{9-15}$$

where

$$P_m = T_m \omega_r \tag{9-16}$$

The output rotor torque is T_m, and the rotor angular speed is ω_r.

Now, if Eqs. (9-4), (9-9), (9-11), (9-12), and (9-15) are substituted into Eq. (9-16), the result is

$$T_m = \frac{sE_2^2 r_2}{\omega_s[r_2^2 + (sX_2)^2]} \tag{9-17}$$

This is the output characteristic $T_m(s)$ of our idealized induction motor. Since E_2 is assumed to be constant, Eq. (9-17) can be written in the form

$$T_m = K\left(\frac{s\alpha}{s^2 + \alpha^2}\right) \tag{9-18}$$

where

$$K = \frac{E_2^2}{\omega_s X_2} \qquad \alpha = \frac{r_2}{X_2} \tag{9-19}$$

Three-phase induction motor characteristic curves are plotted as the solid curves in Fig. 9-8 for several values of parameter α. Note that torque is plotted against slip increasing to the right on the abscissa. Slip increasing is equivalent to rotor angular velocity decreasing to the right on the abscissa from synchronous speed ω_s toward zero. The slip scale is shown above, and the speed scale below, the horizontal axis.

The motor domain of Fig. 9-8 is confined between the slip values $0 \leq s \leq 1$. In this domain, the torque T_m and rotor angular velocity ω_r have the same

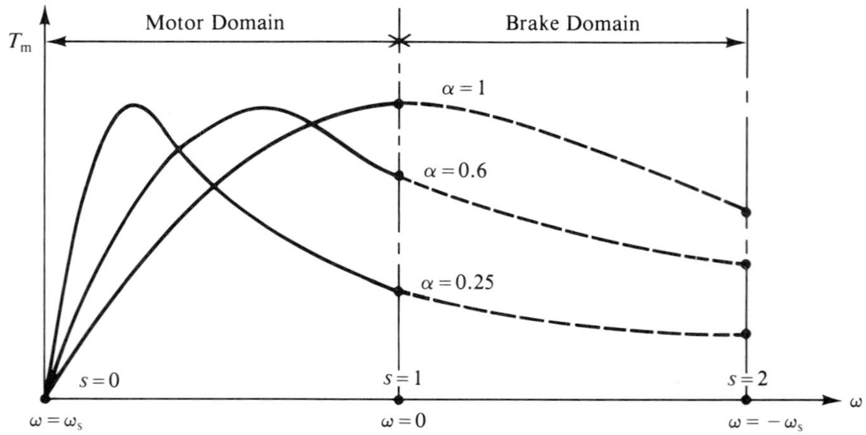

Figure 9-8. Idealized three-phase dynamo characteristics.

direction. Recall that the three-phase induction motor is a dynamo designed to operate at very small slip (0.03) under rated load. For these conditions, the motor characteristic curve of Eq. (9-18) can be closely approximated by

$$T_m = \left(\frac{K}{\alpha}\right)s \tag{9-20}$$

So we see that for small slips, the torque is proportional to slip and the motor characteristic curve is similar to that of the DC shunt motor.

Equation (9-20) shows that rated torque (torque at low slips) can be increased by making the resistance of the rotor smaller. However, doing this tends to reduce the starting torque, as Eq. (9-18) will show. So we see that there is an incentive to try to control the rotor resistance during motor starting. The wound rotor motor has the capability of providing this control.

The slip inverval $1 \le s \le 2$ shown in Fig. 9-8 corresponds to the braking domain of the three-phase induction dynamo. As s increases beyond unity, the direction of rotation of the dynamo reverses and is in opposition to the direction of the developed torque. This is characteristic of a brake; the braking characteristic curves are shown dashed in Fig. 9-8.

If an induction dynamo is running normally as a motor and the direction of rotation of the rotating magnetic field is suddenly changed (by reversing the polarity of the three-phase supply), the dynamo will immediately begin to act as a brake, because the slip will now be in the interval $1 \le s \le 2$. This method of braking, called plugging, will bring the rotor rapidly to rest by dissipating the load kinetic energy as $I_2^2 r_2$ loss in the rotor.

In the preceding analysis, it was assumed that the only losses in the motor occurred in the rotor. This assumes that there are no losses in the stator, and parameter E_2 then becomes the controlling constant in Eq. (9-17).

A more accurate analysis which includes the stator resistance and reactance can be made. An equation similar to Eq. (9-17) will be derived in which the voltage V_1 applied to the stator replaces E_2 and the resistance and reactance terms r_2 and X_2 are replaced by equivalent resistance and reactance of rotor plus stator.

9-3 COMPARISON OF CAGE AND WOUND ROTOR MOTORS

The cage rotor is cheap to manufacture and rugged in use. It would always be preferred to the wound rotor motor, if it were not for the fixed nature of its electrical circuit. The fact that the cage rotor is inaccessible electrically means that starting and speed control must be accomplished either by varying the stator supply voltage or by varying the supply frequency. The former method will be inefficient, because supply voltages are fixed and any voltage reduction will involve dissipation in a resistor in the stator circuit. The method of controlling frequency requires costly electronic controls, as will be discussed later.

The wound rotor design makes possible the inclusion of an externally controlled resistor in the rotor electrical circuit. We have already seen that there is value in being able to control the rotor resistance because, although low resistance is desirable in general, it causes the starting torque to be low (perhaps lower than full-load torque).

Some typical torque-speed curves for a wound rotor motor are sketched in Fig. 9-9 for various values of rotor resistance. Notice how motor speed can be controlled, at constant torque, by varying the rotor resistance. This is called slip speed control. Although it is a feasible method of control, it is inefficient, because increasing the motor slip increases energy dissipation.

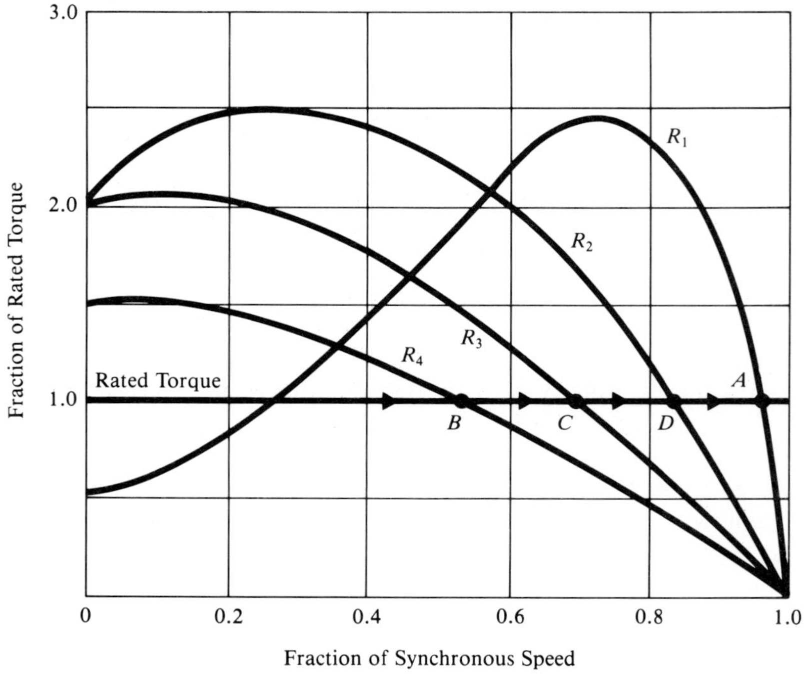

Figure 9-9. Wound rotor motor characteristics.

Figure 9-9 also shows how a wound rotor motor can be started in such a way as to control the starting current and torque. Suppose we want the motor to operate at a slip of 3 percent under full-load. This means the steady-state operating point will be located at point *A* on a motor characteristic curve of relatively low resistance R_1. However, this motor would have a low starting torque and a large starting current. So, to start the motor, a resistance is placed in series with the rotor by means of the slip ring connection. Suppose the total rotor resistance at starting is R_4. The characteristic motor curve for rotor resistance R_4 has a starting torque of about one and one-half times full-load torque, and since the resistance is relatively high, the starting current inrush will not be much greater than full-load current. When the switch is closed with R_4 in the rotor circuit, the motor will accelerate to point *B*. Then the external resistance can be reduced so that the total rotor resistance is R_3, and the rotor will be further accelerated to point *C*. Continuing in this way, the motor operating point will advance to point *D* and then to point *A* (where all the external resistance has been removed).

9-4 THREE-PHASE SQUIRREL CAGE MOTOR REFERENCE DESIGNS

Having noted that the ability to include a variable external resistance in the rotor circuit of a wound rotor motor is a decided advantage in exercising control of torque, speed, and current, we realize there is an incentive to try to design squirrel cage rotors in such a way as to obtain some of these desirable properties. What is really needed is a design that exhibits high resistance at low speeds and low resistance at high speeds. Indeed, this can be achieved by using the concept of a double squirrel cage (see Figure 9-10).

The conductors of the inner cage have lower resistance because they are larger in cross-section than those in the outer cage. On the other hand, the inner cage will have high inductance because it lies deeper inside the iron core of the rotor. So at starting, when the emf frequency in the rotor is high, the reactance of the inner cage is high and the current in it is low; this gives low starting current and good starting torque. As the rotor accelerates and the emf frequency of the

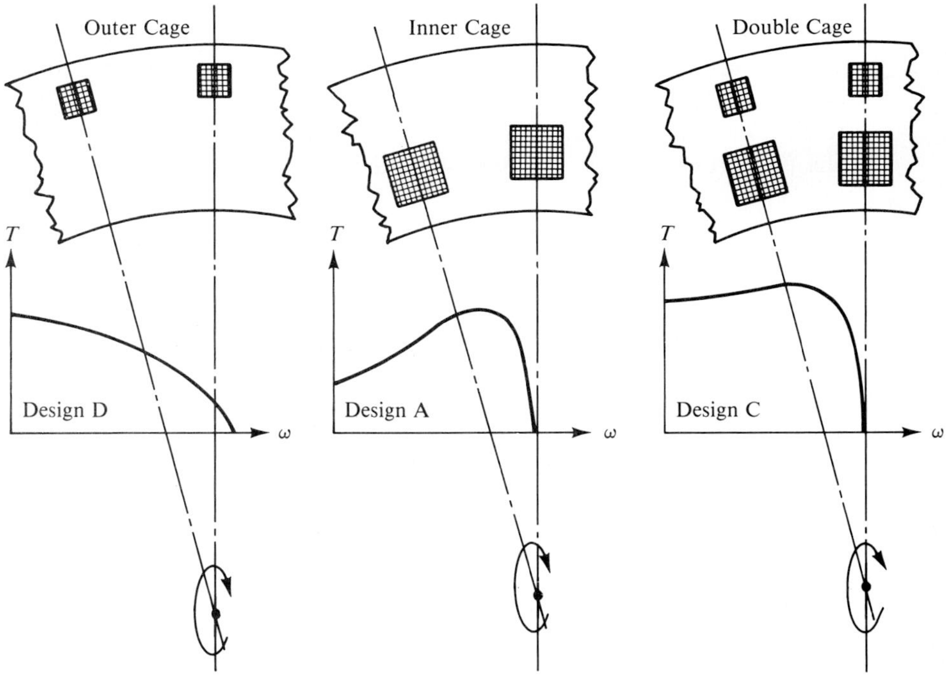

Figure 9-10. Squirrel cage rotor designs.

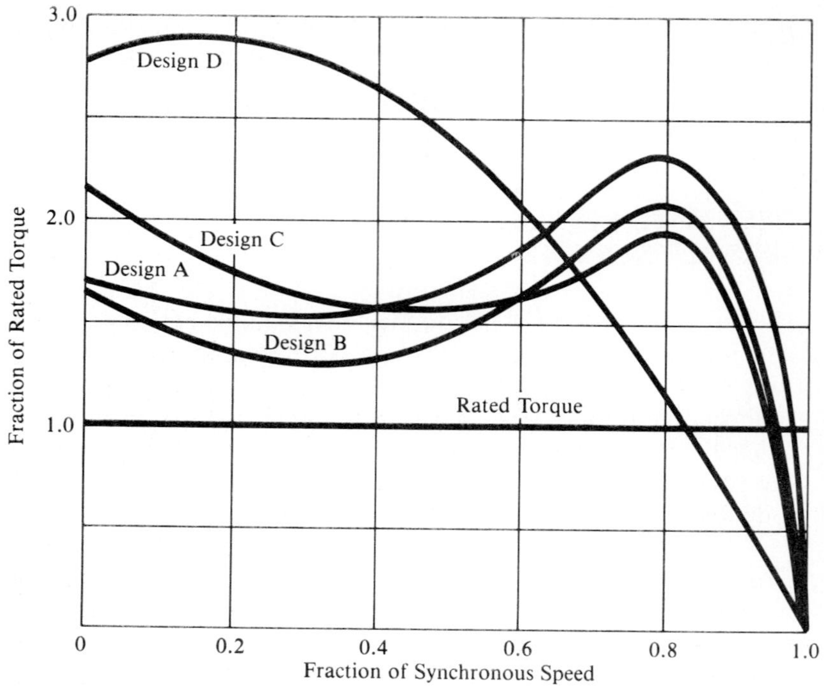

Figure 9-11. NEMA three-phase motor designations.

rotor diminishes, the inner cage carries most of the current, because it now has very low reactance and resistance. The concept of shaping the conductors of the squirrel cage as has been discussed above has led NEMA (the National Electrical Manufacturers Association) to define reference designs called A, B, C, and D. The torque-speed curves characteristic of these designs are sketched in Fig. 9-11. The properties of these designs are tabulated in Table 9-1.

TABLE 9-1. Some Properties of NEMA Squirrel Cage Motors.

Design A	Design B	Design C	Design D
Normal starting torque	Normal starting torque	High starting torque	Very high starting torque
High starting current	Normal starting current	Normal starting current	Low starting current
Low slip	Low slip	Low slip	High slip
High breakdown torque	Normal breakdown torque	Used where break-away torque is higher than full-load torque	Breakdown torque occurs at high slip
Usually designed for a specific purpose	Most common motor, especially in fractional-horsepower sizes	Next most popular motor to Design B	Usually a special-purpose motor

9-5 SINGLE-PHASE INDUCTION MOTORS

The three-phase supply required by a three-phase induction motor consists of three wires (and a ground). The voltage provided between any two wires of the three-wire supply will be a single-phase voltage supply. Of course, the three individual single-phase supplies differ in phase by 120 degrees, and it is this phase difference that makes possible the production of a rotating field in the three-phase motor when the various phases are connected to windings properly spaced around the periphery of the stator.

If one of the wires is disconnected from a three-phase supply provided to an induction motor, the result will be a single-phase pulsating supply in the stator windings rather than a rotating field. Thus, if the rotor is at rest, it will remain at rest because there will be only a pulsating field rather than a rotating field, and this will not cut the conductors of the rotor so as to induce a current in the rotor. We conclude that a single-phase induction motor is not self-starting. Note, though, that if the rotor is rotating, there will be current induced in the rotor even with a single-phase stator field because there will then be a relative motion between the field and the rotor conductors (even though the stator field is not rotating). So, if we can get the motor started, it will continue to operate. In fact, we actually discriminate among the various types of single-phase motors by the various schemes used to start the motors. Most of these single-phase motors will be of the smaller variety (generally, fractional-horsepower motors).

To develop the characteristics of an idealized single-phase motor, we will use the artifice of two contra-rotating fields each of which has half the strength of the single-phase field. One of these rotating fields rotates ccw at synchronous speed ω_s while the other one rotates cw at speed ω_s. The superposition of these two contra-rotating fields produces the equivalent, stationary, pulsating single-phase field.

Assume the rotor of the single-phase induction motor is rotating ccw at rotational speed ω_r. Then the slip between the rotor and the ccw-rotating field is given by the same relationship defined earlier for the three-phase motor:

$$s_f = \frac{\omega_f - \omega_r}{\omega_f} = \frac{\omega_s - \omega_r}{\omega_s} = s \qquad \textbf{(9-21)}$$

The subscript "f" means in the forward (ccw) direction. The torque produced by this ccw-rotating field is in the same direction as the rotor rotation, and hence it corresponds to the motor domain region of Fig. 9-8. Note that

$$0 \le s_f \le 1 \qquad 0 \le s \le 1 \qquad \textbf{(9-22)}$$

The slip between the rotor and the cw-rotating field is given by

$$s_b = \frac{\omega_b - (-\omega_r)}{\omega_b} = \frac{\omega_s + \omega_r}{\omega_s} = 2 - s \qquad \textbf{(9-23)}$$

Here the subscript "b" means in the backward (cw) direction. The torque produced by the cw-rotating field is in opposition to the rotor rotation and corresponds to the brake domain region of Fig. 9-8. Note that

$$1 \le s_b \le 2 \qquad 0 \le s \le 1 \tag{9-24}$$

If we denote the torque in the motor domain as positive and that of the brake domain as negative, then the superposition of these two torques produces the single-phase motor torque sketched in Fig. 9-12. For the curves shown, parameter α was chosen to be about 0.25.

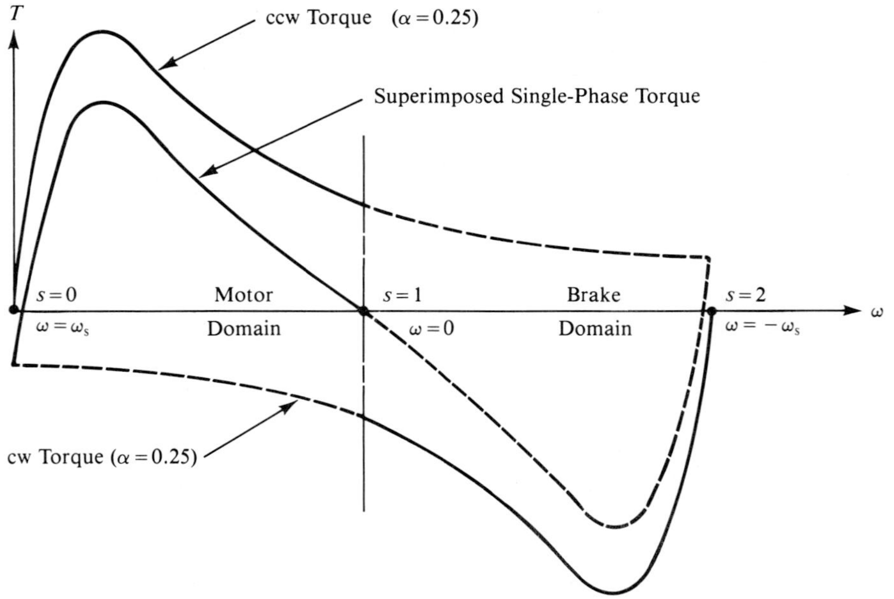

Figure 9-12. Single-phase motor properties.

The single-phase motor torque analytical expression can be obtained from the superposition

$$T_m = T_f + T_b \tag{9-25}$$

Substituting Eq. (9-17) into Eq. (9-25), we obtain

$$T_m = \frac{sE_2^2 r_s}{\omega_s[r_2^2 + s^2 X_2^2]} - \frac{(2-s)E_2^2 r_2}{\omega_s[r_2^2 + (2-s)^2 X_2^2]} \tag{9-26}$$

At a slip of $s = 1$, the single-phase induction motor has no starting torque, as we suggested earlier. Equation (9-26) shows this. Consequently, some scheme must be devised to get this motor started.

9-5-1 Split-Phase Induction Motors

To be able to start a single-phase induction motor, a rotating field must be created, albeit temporarily, that will provide a starting torque. One method used to produce the required rotating field is to provide the stator with an auxiliary (starting) winding situated at 90 degrees with respect to the main winding and connected, electrically, in parallel with the main single-phase winding.

If a phase shift can be created between the currents flowing in the auxiliary and main windings, there will be a phase shift between the magnetic fluxes these two windings produce. The result will be a rotating field suitable for starting the motor but, perhaps, not an optimum one for running conditions.

In the split-phase motor, the auxiliary winding is designed to have high resistance and low reactance. The main winding, on the other hand, has low resistance and high reactance. This combination of two windings can produce a phase shift of some 30 to 40 degrees between the magnetic fields produced. Schematically, the motor configuration is as shown in Fig. 9-13.

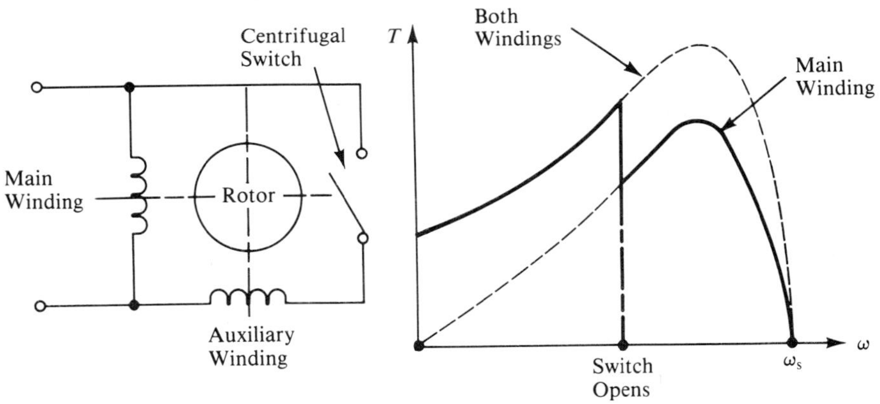

Figure 9-13. Split-phase motor characteristics.

Since the auxiliary (starting) winding is no longer needed once the motor has been started, and since it generally is a hindrance rather than a help during running conditions, most split-phase motors are designed to cut out the auxiliary winding at some designated speed (e.g., using a centrifugally actuated switch). Consequently, the torque characteristic will appear as shown in Fig. 9-13.

9-5-2 Capacitor Induction Motor

The split-phase induction motor is limited basically to the fractional-horsepower range (up to around 1/2 hp). Higher starting torques and horsepowers can be obtained if the phase shift obtained with the split-phase motor is augmented by including a capacitor (in series) in the auxiliary winding. The scheme is shown in Fig. 9-14. The motor indicated in the figure is called a capacitor-start motor

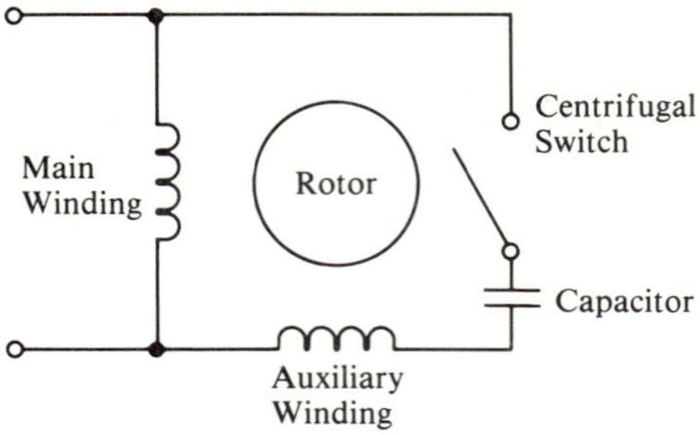

Figure 9-14. Capacitor-start motor.

because the auxiliary winding is disconnected using a switch at some designated speed (same as with the split-phase motor). At operating speed, only the main winding is energized.

Some capacitor induction motors are designed to operate with the auxiliary winding energized at all speeds. They are called capacitor-run motors and are available in powers up to around 10 hp.

9-5-3 Shaded-Pole Motors

In the smaller fractional-horsepower sizes, efficiency of operation is often not a major design consideration. When this is the case, it is possible to design the single-phase induction motor to be self-starting without the necessity of providing a separate, and relatively expensive, auxiliary winding. This is usually done with what is called a salient-pole stator, which means that the stator is actually geometrically configured similar to the design shown in Fig. 9-15.

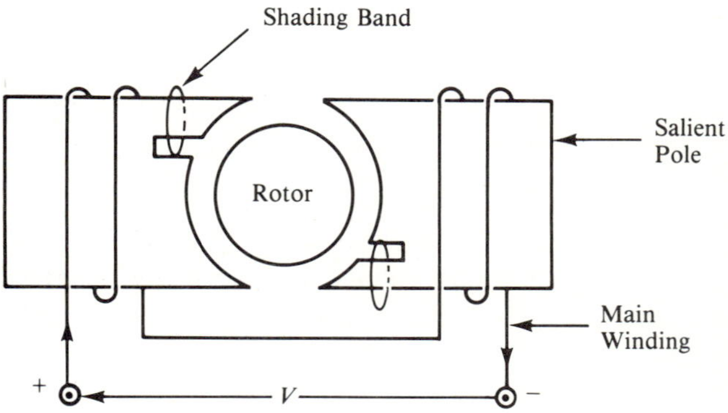

Figure 9-15. Shaded-pole motor.

The salient pole is energized by a main single-phase winding as shown in Fig. 9-15 and it provides the primary magnetic field for the production of torque.

In addition to the main winding, a small portion of the salient pole is circumferenced (shaded) with a copper ring short-circuited upon itself. This small portion has the effect of causing a phase lag of the magnetic flux issuing from the shaded part of the pole as compared to the main magnetic flux. Since this small lagging magnetic flux is also displaced, or shifted, with respect to the location of the main flux, the combination produces a weak rotating magnetic field sufficient to start the motor. The starting torque of these motors will be small and of the same order of magnitude as the running torque. For most applications of the shaded-pole motor, this is not a great disadvantage, because the application will be one of small rated power anyway (of order of magnitude 1/20 hp).

9-6 THE SYNCHRONOUS MOTOR

The AC electrical dynamos discussed up to this point all operate based on the principle that there is an electromagnetic torque associated with the slip between a magnetic field and a current-carrying conductor in the magnetic field. In a sense, these dynamos are electrical clutches analogous to mechanical or hydraulic clutches, which also derive their torques from slip.

The synchronous motor operates differently, based on the principle of relative displacement of two magnetic fields rather than on the rate of this displacement (which would be slip). The mechanical analog of the synchronous motor is not a clutch but, rather, an elastic shaft carrying mechanical torsion.

Imagine, for the moment, an elastic mechanical shaft that transmits a torque by virtue of the elastic torsional deflection of the shaft. The torsional deflection is equivalent to a rotational displacement of one end of the shaft relative to the other end. Both ends of the shaft, however, rotate at the same angular velocity. In other words, the carrying of a torque by the shaft does not require a loss of registration (i.e., an unsynchronization) of one end of the shaft relative to the other end.

Of course, with no torque in the elastic shaft, there will be no relative rotational displacement between the two ends of the shaft. The rotational displacement will increase linearly from zero as the torque is increased from an initial value of zero. So long as the torque does not cause a shear stress in the shaft exceeding the yield point, both ends of the shaft will rotate at the same angular velocity even though they are slightly displaced relative to each other (with the magnitude of the displacement depending on the transmitted torque). If the torque rises to a point such as to cause the shear stress to enter the plastic domain, then there would be a continuous slippage of one end of the shaft relative to the other end, and the synchronization of speed between the two ends would be lost.

The synchronous motor is an electromagnetic analog of the elastic shaft. In a synchronous motor, we have a stator very similar to that of the three-phase induction motor. A three-phase AC supply voltage impressed upon the stator produces a rotating magnetic field having synchronous rotational speed ω_s just as for the three-phase induction motor.

The rotor of the synchronous motor contains windings excited by a DC supply in such a way as to produce an electromagnetic rotor having definite north and south poles. So no slip is required between rotor and stator, as in an induction motor, in order to produce the rotor current (and the resultant electromagnetic torque). In the synchronous motor, torque is produced by the direct interaction between stator rotating field and rotor currents created directly by a DC supply.

Imagine that there is a slip between the electromagnetic rotor and the stator's rotating field. Then there will be no net torque created, because the rotor will experience an uncertainty about the direction in which it should move as the rotating field first passes its north pole and then (before the rotor can respond) passes its south pole. The force effects cancel each other out in a short time interval.

But as the slip diminishes toward zero, the rotor will begin to be able to respond to the rotating field. In effect, the north pole of the rotating field will lock on to the south pole of the rotor, and the rotor will be dragged along with the stator field at zero slip.

If there is no torque to be transmitted, the poles of stator and rotor will line up perfectly. But as a torque is provided to the rotor, there will begin to be a slight electromagnetically elastic displacement between the poles of rotor and stator fields which is analogous to the mechanical elastic shaft displacement we discussed earlier. However, there will be no loss of registration, or synchronization, between the speeds of rotor and rotating field. The rotor angular speed ω_r will be identical to the synchronous speed ω_s of the stator field (hence the name synchronous motor) provided that the torque placed on the rotor remains below the breakdown torque that causes unlocking of the rotor and stator fields.

Since the synchronous motor is not self-starting, it must be brought up to synchronous speed using external means or by designing the rotor to include an induction motor cage in addition to the DC-excited rotor windings. The cage provides enough starting and acceleration torque to bring the rotor to synchronous speed, at which point the cage can no longer be energized because of the disappearance of slip.

Synchronous motors have been built in large sizes of up to 50,000 hp. These extremely expensive machines are used in applications of electric power systems. Synchronous motors have the capability of increasing the power factor of an electrical system and for this reason, as well as for the fact that they do not possess the dissipative feature of slip in the induction motor, they are competitive with induction motors in the larger horsepower sizes. This is especially true for low-speed motors such as might be suitable for direct connection to centrifugal pumps, fans, saws, and metal-rolling equipment.

Synchronous motors, of course, will also be used where precise control of speed is required. Generally, these applications will be typified by light starting loads.

9-6-1 The Reluctance Motor

In the fractional horsepower and very small sizes, there is yet another kind of synchronous motor used for applications requiring exact synchronous speed of a rotor. Timing devices and clock drives make use of this type of motor, which is a single-phase device as opposed to the three-phase synchronous motor discussed above.

The motor is sketched in Fig. 9-16. It has a salient-pole stator excited by a single-phase winding subjected to applied voltage V. The rotor is a salient-pole iron core. Note that the rotor does not require DC excitation as in a conventional synchronous motor.

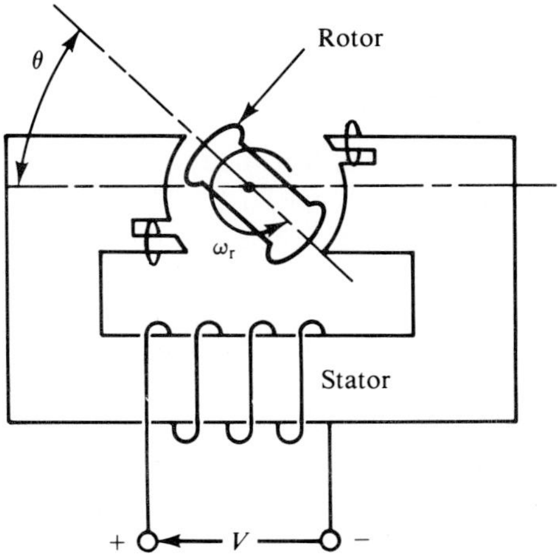

Figure 9-16. Reluctance motor.

The electromagnetic torque exerted on the rotor by the stator electromagnetic field is a function of (a) the magnetic flux provided by the stator field, (b) the angular position of the rotor, and (c) the magnetic reluctance of the magnetic circuit.

The frequency ω_s of the single-phase stator field is the same as that of the supply voltage V. We will designate the frequency of the rotor angular speed as ω_r.

An analysis of the torque developed by this motor requires an understanding of electromagnetic devices, and particularly their reluctance circuits, which is beyond the scope of this book. It is sufficient here to simply state that this torque is a rather unusual function of rotor speed ω_r. It vanishes for all speeds except for a rotor speed equal to the synchronous speed ω_s. The motor is, therefore, a synchronous motor.

It is easy to see that when the rotor frequency ω_r is equal to the supply frequency ω_s, the rotor salient poles will be synchronized with reversals of the magnetic field flux issuing from the salient poles of the stator. In other words, the rotor always reappears at the same location when the alternating magnetic flux assumes, periodically, the previous value. So the motor has reproducible reluctance in this fashion.

9-7 SPECIAL MOTORS

A morphological study of the possible electric motor configurations that can be synthesized from all possible combinations of permanent magnets and electromagnets of the single-phase and polyphase variety, energized by AC, DC, or some other voltage mode, will reveal a very large number of possible types. Thus, we have repulsion, reluctance, hysteresis, induction, selsyn, synchro, servo-, stepper, timing, PM, linear, and many more, types of electric motors.

We have already discussed the more common types of motors about which the mechanical systems designer must be knowledgeable. Now we want to discuss two additional types: the two-phase motor and the stepper motor.

9-7-1 The Two-Phase (Servo-) Motor

The two-phase motor is an induction motor having a squirrel cage type rotor and a stator with two windings placed 90 spatial degrees apart. It differs from the three-phase motor only in having a two-phase supply to the stator instead of a three-phase supply. The difference between the phases, electrically, is 90 degrees rather than the 120 degrees for a three-phase supply.

This electrical two-phase supply and the two stator windings at 90 degrees to each other produce a rotating field whose synchronous speed is determined by the same formula we developed earlier for the three-phase induction motor.

The two-phase induction motor has no intrinsic advantages over the three-phase motor insofar as torque and horsepower are concerned per se. It is as a control motor that the two-phase motor is useful. We call it a control motor in the following sense.

One of the stator windings (the main winding) is excited by a single-phase voltage of constant frequency and amplitude. The other winding (the control winding) is identical to the main winding, but it is excited by a variable single-phase voltage of the same frequency as the main winding but differing in phase by 90 electrical degrees. The voltage amplitude of the control winding can be varied, usually by use of an amplifier, from zero to a maximum value equal to the constant amplitude of the main winding. Usually, the polarity of the control winding can also be controlled, making the rotor's direction of rotation reversible.

With the control voltage equal to zero, there will be no rotating field and the motor will not be self-starting (just as with a single-phase motor). So its stall torque is zero with no voltage applied to the control winding. As the control voltage is increased from zero, the torque capability of the motor will increase to a maximum when the control voltage is equal to the voltage provided by the main winding. The torque-speed curves as a function of control voltage are shown at the left of Fig. 9-17.

The characteristic curves shown correspond to a motor having high rotor resistance and, therefore, are of the type D we discussed for the three-phase induction motor. With high enough resistance, the characteristic curve can be made almost linear as is shown in the figure.

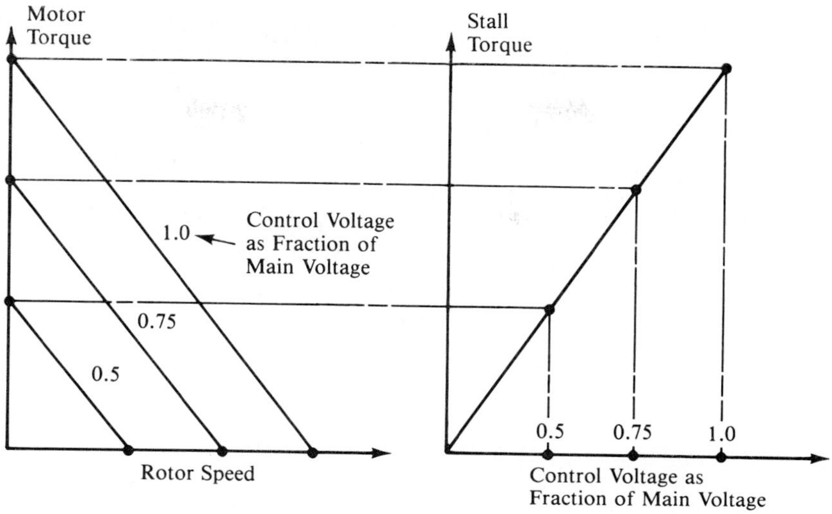

Figure 9-17. Two-phase servo-motor characteristics.

The curve on the right shows the stall torque of the motor as a function of control winding voltage. Clearly, the torque this motor can develop is easily controllable via the control winding voltage. It is also easily reversible simply by changing the control voltage polarity.

The two-phase control motor is well suited for servo applications and is therefore often called the two-phase servo-motor. In this application, the rotor is designed to have very low inertia so that the motor will have fast response to control signals fed back to the control winding amplifier. The torque-speed curve, being linear, is equivalent to linear damping in the servo-system.

Compared to the DC servo-motor, the AC version has the advantage of not requiring a commutator or brushes. It also possesses the rugged qualities of the squirrel cage motor. The disadvantage of the AC servo-motor is that it is generally available only in the smaller capacities (1 kW and less). For larger power requirements, the DC servo-motor is used.

9-7-2 The Stepper Motor

Some mechanical systems require the capability of providing precise control of position or speed of some mechanical component. An example is the carriage of a lathe programmed to turn some specified geometry in a specified time. Robotic devices also require this capability. One type of electrical actuator providing this function is the stepper motor.

Although there are several electrical versions of the stepper motor depending on whether the rotor is an iron core, a permanent magnet, or excited from a DC supply, all are basically synchronous motors of the reluctance type. One way of viewing the stepper motor is to see it as having a rotating field whose speed and total number of revolutions can be controlled electronically. Since the motor is synchronous, the rotor must also have experienced this speed and total number of revolutions (or some fraction thereof).

One version of a stepper motor stator involves two windings placed at 90 spatial degrees to each other. The windings are supplied with separately controlled DC supplies whose polarities can be reversed. These DC supplies are actuated electronically by a series of pulses controlled by a digital electronic circuit (perhaps even from a microprocessor). In this way, the stator coils can be energized in a sequential fashion such as to cause the rotor to rotate at a given speed, and through a controlled number of revolutions, depending on the frequency of the pulse train control and on the total number of pulses.

9-8 APPLICATIONS OF AC MOTORS

AC motors are generally cheaper, more rugged, and require less maintenance than do DC motors. Their major disadvantage has been that they are, basically, less susceptible to speed control than are DC motors. With the advent of solid-state speed control schemes, this situation is changing rapidly, so that AC motors are becoming competitive with DC motors even in applications requiring variable speed.

For constant-speed applications requiring only modest starting torque, AC motors of the squirrel cage type are universally preferred, particularly in environments that require ruggedness and low maintenance.

The choice of a particular type of AC motor for a mechanical system depends on many factors, including:

(a) *Working environment.* Factors to be considered are moisture, dust, grit, temperature, chemicals, volatile vapors, and so on.

(b) *Duty cycle.* The duty cycle is an estimate of the fraction of the time that the motor is off, has a reduced load, runs at full-load, is overloaded, and is started and stopped. Service factors must also be applied, much as was done for gear reducers, to take care of unknown or overload factors. Some information about how to handle these application factors will be available from the motor manufacturer or from trade organizations such as NEMA. Also, the authoritative standards for electric motors is the American National Standard ANSI/NEMA MG1-1978.

(c) *Starting conditions.* It will be necessary to have information about the load sufficient to determine if the motor being considered can provide the starting torque. Under starting conditions, it will generally be necessary to know all of the inertia effects in the power train between the motor and the load. If starting torque conditions are severe, a wound rotor motor may be required. But it should be remembered that such a motor may introduce inefficiencies that could be obviated by choosing a DC motor.

(d) *Running conditions*. The motor must be suitable to meet the torque-speed characteristics of the load over the range of speeds anticipated. Here, matching of the motor characteristic to the load characteristic will determine the steady-state operating point and the nominal horsepower rating of the motor under continuous running conditions. Motor ratings are based on motor life which, in turn, depends on the motor operating temperature. Motor lifetimes are established similar to the manner in which bearing lifetimes are related to bearing capacities using the L_{10} life concept.

(e) *Stopping conditions*. Consideration must be given to how the motor, and the mechanical system attached to it, is to be braked and parked.

In the larger sizes, three-phase AC induction motors have been used in steel mill equipment; in cement-making machines; to drive centrifugal blowers, fans, and pumps; and in industrial machinery of all kinds including machine tools, punch presses, shears, and brakes.

In the smaller sizes, single-phase motors are generally used. Appliances and other products used in the home generally incorporate these motors because of the almost universal use of the 120 V circuit. We find single-phase motors in refrigerators, washers and dryers, furnaces, and similar appliances.

9-9 REFERENCES

9-1. M. G. Say, *Alternating Current Machines*, Wiley, New York, 1976.

9-2. A. B. Carlson and D. G. Gisser, *Electrical Engineering*, Addison-Wesley, Reading, MA, 1981.

9-3. R. H. Engelmann, *Static and Rotating Electromagnetic Devices*, Marcel-Dekker, New York and Basel, 1982.

PROBLEMS

9-1. For the idealized three-phase induction motor, show that

(a) The breakdown torque occurs at a slip of $s = \pm\alpha$

(b) The magnitude of the breakdown torque is $\frac{1}{2}K$

(c) The breakdown torque will be equal to the starting torque for $\alpha = 1$

(d) Maximum starting torque occurs with $\alpha = 1$

9-2. Consider an idealized three-phase induction motor in which $\alpha = 0.2$.

(a) Determine the ratio of breakdown torque to starting torque.

(b) If rated load occurs at a slip of $s = 0.05$, can this motor be started under rated load?

9-3. The power flowing into the rotor of a three-phase induction motor is 25 kW. The power available at the output shaft of the rotor is 24 kW. If the synchronous speed is 3600 rpm, estimate the motor output torque.

9-4. A four-pole, 60 cps, three-phase wound rotor induction motor has a rotor resistance and inductive reactance under locked rotor conditions of 2 Ω and 0.2 Ω respectively.

Estimate the external resistance r_e that will be necessary for the motor to develop the same torque at both $s = 0.05$ and $s = 0.20$.

9-5. Show that the starting torque of an induction motor is approximately inversely proportional to supply frequency for small frequencies, that is,

$$T_m \cong \left(\frac{pE_2^2}{4\pi r_2}\right)\left(\frac{1}{f}\right)$$

where p is the number of poles, r_2 is the rotor resistance, and E_2 is the emf developed in the rotor when it is locked.

9-6. Show that for high frequency, the starting torque of an induction motor is approximately inversely proportional to the cube of supply frequency, that is,

$$T_m \cong \left(\frac{pE_2^2 r_2}{16\pi^3 L_2^2}\right)\left(\frac{1}{f^3}\right)$$

9-7. The torque-speed characteristic of a 60 Hz induction motor is linear between the speeds of 50 rev/sec and 60 rev/sec. The power delivered at 55 rev/sec is 6 kW.

(a) What is the maximum power this motor can deliver in the speed range of 50 to 60 rev/sec?

(b) This induction motor drives a load through a gearbox. The load characteristic is

$$T_\ell = 5 \times 10^{-5} N_\ell^3 \text{ N-m} \qquad (N_\ell \sim \text{rev/sec})$$

What speed ratio is required of the gearbox in order that 6000 W are delivered to the load?

(c) What is the torque-speed demand characteristic of the load as seen by the motor when 6000 W are delivered to the load?

10

The Eddy Current Coupling (Clutch)

The eddy current coupling is the electrical analog of the fluid coupling. In the eddy current coupling, the magnetic field cutting the interface between rotor and drum is analogous to the hydraulic fluid that cuts the interface between impeller and turbine of the fluid coupling.

Eddy current couplings (or drives) have been in use for about 50 years and are available in a range of sizes from the fractional horsepower to very large sizes up to 20,000 hp. Some typical operating characteristics of eddy current couplings are listed below:

- **(a)** They are slip devices having a torque-speed characteristic curve similar to that of the fluid coupling and the Design D type AC induction motor.
- **(b)** They can provide speed regulation of 1 percent (and less).
- **(c)** Speed ranges of order of magnitude 30:1 can be obtained at constant torque.
- **(d)** They are susceptible to both analog and digital feedback speed control techniques.
- **(e)** They can be air cooled in the range of sizes up to about 2000 hp.
- **(f)** In the range 2000 to 20,000 hp, they must be liquid cooled.
- **(g)** They can be purchased integral with drive motors, gearboxes, and brakes.
- **(h)** They are rather widely used in dynamometer devices.

One unique feature of the eddy current drive is that the level of armature excitation required to control the device over its speed range is relatively low. Typically, the excitation level is 1 percent (or less) of the horsepower being transmitted by the coupling. This feature is in contrast to the competing adjustable-speed AC induction motor drive in which all of the transmitted horsepower must pass through the control apparatus.

The eddy current coupling works basically on the same principle as the AC induction motor. However, whereas the rotating field of the induction motor is obtained using AC excitation in an armature (the stator) which remains physically at rest, the rotating field of the eddy current dynamo is obtained by exciting an armature with DC current and then physically rotating the armature. Also, when the eddy current dynamo is used as a clutch, both halves of the coupling will rotate, whereas in the induction motor only the squirrel cage (or wound) rotor will be rotating.

In the eddy current coupling, a rotating field is obtained when an armature, excited by a DC winding, is physically rotated. Generally, this rotating armature is the rotor (inner cylinder) of the coupling (even though it is the analog of the stator of the induction motor). (See Fig. 10-1.)

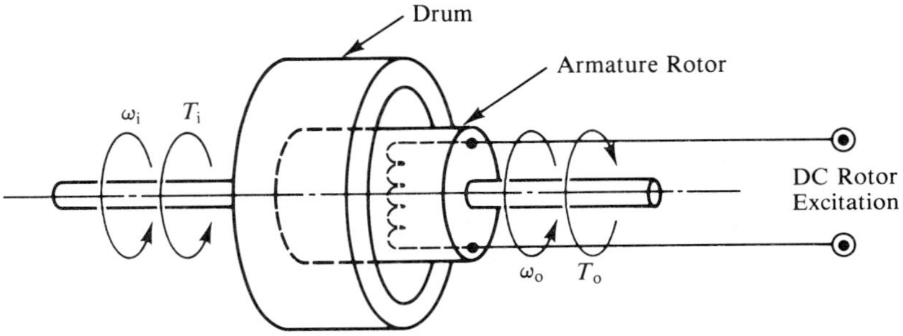

Figure 10-1. Schematic diagram of an eddy current coupling.

In the process of rotating, the magnetic field of the rotor induces eddy currents in the continuous steel, concentric outer cylinder called the drum of the eddy current coupling. This process is analogous to the induction of current in the squirrel cage rotor of the induction motor. It is the electromagnetic interaction between the rotating field and the eddy currents induced in the drum that produces the electromagnetic torque of the eddy current dynamo.

In most eddy current couplings, the drum is the input to the clutch and is driven by the prime mover (which frequently is a squirrel cage induction motor). (See Fig. 10-2.) The driven member (the output) of the clutch is usually the excited rotor. If necessary, the roles of drum and armature can be reversed so that the drum becomes the output (driven) half of the clutch. This is sometimes done to lower the rotating inertia of the output member in order to improve transient response of the total driven system (which includes the output half of the clutch).

When the eddy current clutch is supplied integral with an AC induction motor as a prime mover, we have a device that has essentially a constant-speed input (the AC motor) and a variable-speed output. It is an adjustable-speed drive in which the output speed is controlled by controlling the level of DC excitation of the rotor armature. If a DC supply is available at the proper voltage (standard

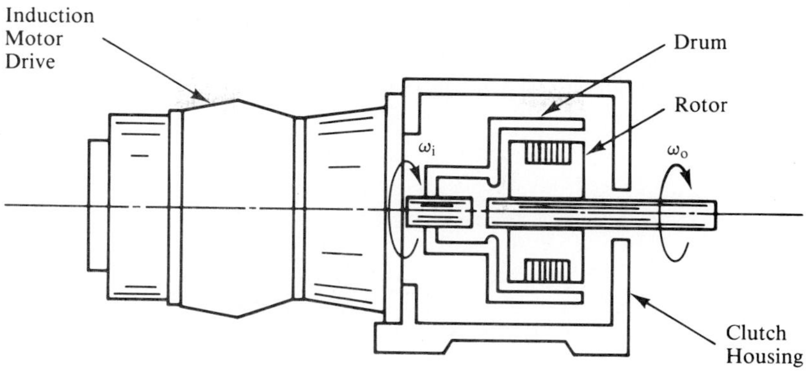

Figure 10-2. Eddy current coupling variable-speed drive.

coil voltages are 45 V DC in one manufacturer's products), it can be used directly to control the eddy current clutch by using a potentiometer or other voltage-modulating device. If only AC power is available, then the eddy current clutch is generally controlled by using a phase-controlled rectifier which is a device not only for rectifying the AC supply to DC but also for providing voltage-level control. This device will be discussed in more detail in Chapter 11.

Typical output characteristic curves at various levels of rotor excitation are shown in Fig. 10-3 for an eddy current clutch driven by a four-pole induction motor ($n_s = 1800$ rpm). Note the drooping torque versus output speed characteristic which is similar to the output characteristic of the fluid coupling.

In Fig. 10-3, it can be seen how the basic level of the torque versus speed curve can be elevated by increasing the excitation of the rotor of the coupling. Note how electrical excitation here is analogous to the amount of fill of the fluid

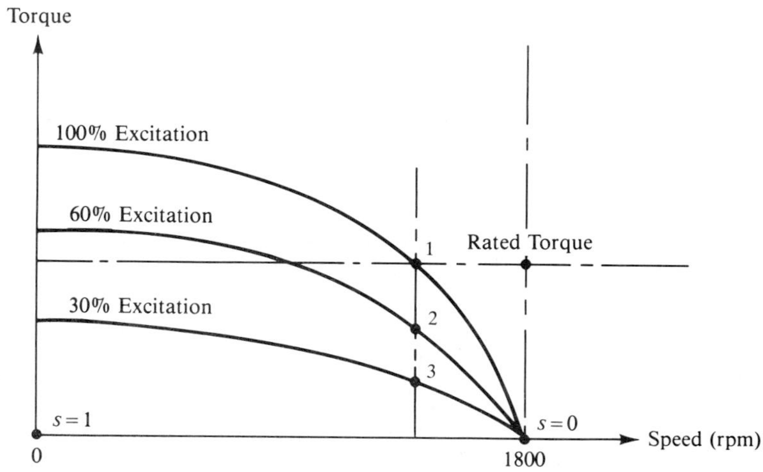

Figure 10-3. Eddy current coupling torque versus excitation curves.

coupling. Therefore, the torque can be controlled by excitation for a given output speed (see points 1 and 2 of Fig. 10-3). This feature makes the eddy current clutch amenable to servo-control, as is indicated in Fig. 10-4.

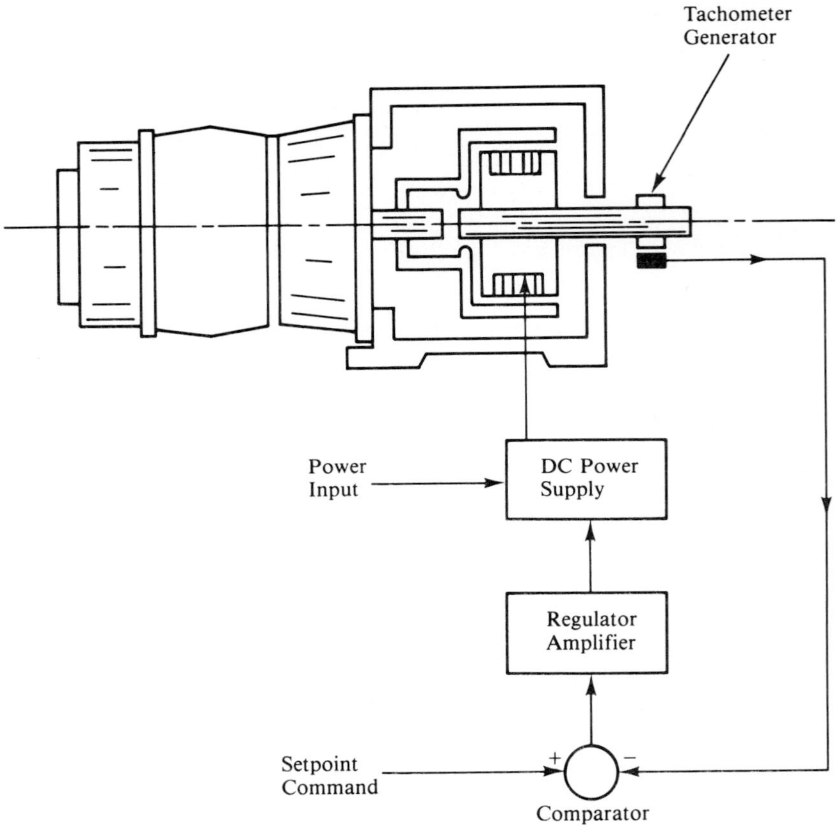

Figure 10-4. Eddy current coupling speed control.

Figure 10-4 shows a closed-loop feedback system used for controlling speed of a mechanical system driven by an AC motor through an eddy current clutch. The desired speed is provided to the comparator as the command signal. A tachometer (of the DC generator type) measures the actual output speed and feeds this information back to the comparator where it is subtracted from the command signal (of course, both command and tachometer signals must be processed in order that they be commensurable). If there is a difference between commanded and measured speeds, an error signal is created by the comparator which is fed to the power amplifier. The amplifier changes the excitation of the eddy current rotor until the speed of the load becomes equal to that of the command signal; then the error signal vanishes. In this fashion, the speed of the load is maintained constant at the commanded value regardless of changes in load torque. The speed regulation and drift obtained with this control scheme can be kept to within 1 percent.

In the larger horsepower sizes, eddy current couplings have been used in applications involving all kinds of rotating devices such as centrifugal pumps, fans, and blowers. In the lower horsepower and fractional horsepower sizes, these devices have been used to drive conveyors, machine tools, web and wire processing machines, and in textile, bagging, and business machines. Some applications require brakes which may be of the friction, spring-set, or eddy current type. We note that the eddy current clutch becomes a brake when we fix either the drum or the rotor.

Finally, we note that the eddy current coupling is not a torque converter (torque multiplier). The same torque exists at both input and output of the clutch (except for small windage losses). Therefore, as was true with the fluid coupling, the efficiency of the eddy current clutch is given by

$$\eta = 1 - s \qquad s = \frac{\omega_i - \omega_o}{\omega_i} \tag{10-1}$$

EXAMPLE 10-1

Figure 10-5 shows the load characteristic and pump characteristics of a typical fluid centrifugal pump application in which a 20 foot static head of load is ascending to a load of 60 feet of head for a flow rate of approximately 35,000 gpm. At the full-load operating point of 60 feet of head, the centrifugal pump operates at 440 rpm in supplying the 35,000 gpm.

It is desired that the discharge be lowered to 25,000 gpm. There are two ways in which this could be accomplished that we want to evaluate. In the first scheme, the pump is directly coupled to a three-phase induction motor with a synchronous speed

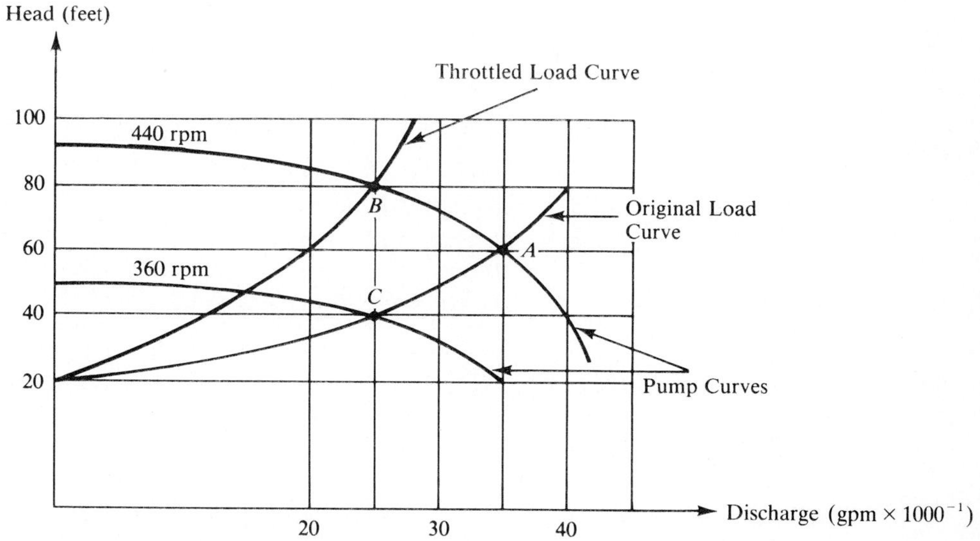

Figure 10-5. Throttling versus clutch horsepower losses.

of 450 rpm. With direct coupling, the decrease in pump output is to be obtained by inserting a throttling valve in the flow loop and throttling the pump until the flow drops to 25,000 gpm. The induction motor can be assumed to be 100 percent efficient and acting essentially as a constant-speed prime mover at 440 rpm.

In the second scheme, the pump is to be coupled to the induction motor using an eddy current clutch. The drop in pump output is to be obtained by lowering the clutch excitation so that the pump speed drops from 440 rpm to 360 rpm (which is the pump speed for delivering 25,000 gpm).

We want to estimate the losses in power associated with these two methods of reducing the flow.

Solution

The load horsepowers associated with points B and C are

$$P_B = \frac{(62.4)(80)\left(\frac{25{,}000}{7.48(60)}\right)}{550} = 506 \text{ hp}$$

$$P_C = \frac{(62.4)(40)\left(\frac{25{,}000}{7.48(60)}\right)}{550} = 253 \text{ hp}$$

Therefore, the loss in horsepower by simply throttling the pump is

$$P_B - P_C = 506 - 253 = 253 \text{ hp}$$

The slip of the eddy current clutch at point C is

$$s_C = \frac{450 - 360}{450} = 0.2$$

The efficiency of the clutch is therefore

$$C = 1 - 0.2 = 0.8$$

Therefore, the input (motor) horsepower to the clutch is

$$P'_C = \frac{253}{0.8} = 316 \text{ hp}$$

The loss in the horsepower in using the clutch is then

$$P'_C - P_C = 316 - 253 = 63 \text{ hp}$$

This shows the virtue of using the clutch rather than the throttling valve. Note that in the above calculation we have assumed, essentially, that the pump efficiency is 100 percent. Even though the efficiency is an approximation, the calculation gives a good estimate of orders of magnitude involved in the savings obtained using the eddy current clutch. ■

10-1 THE EDDY CURRENT DYNAMOMETER

When we hold either the drum or rotor of an eddy current clutch stationary, we have an eddy current brake. If an eddy current brake is used as a load for a prime mover in such a way that it makes possible measurement of torque and power developed by the prime mover (and dissipated in the eddy current brake), then the brake is called an eddy current dynamometer. The dynamometer can be implemented by mounting the eddy current dynamo in a cradle (or trunnion) and having the stator stopped from rotating through the means of a force cell.

Eddy current dynamometers have been used for many years in internal combustion engine testing. Advantages of this type of dynamometer include:

- **(a)** High horsepower capability per pound of dynamometer weight
- **(b)** High torque capability per pound of dynamometer weight
- **(c)** Torque readings that are smooth and simple to instrument
- **(d)** Torque level that can be conveniently controlled electrically
- **(e)** Ability to work with equal facility for either direction of prime mover rotation
- **(f)** Effective cooling and noise insulation

A typical torque curve for an eddy current dynamometer is shown in Fig. 10-6.

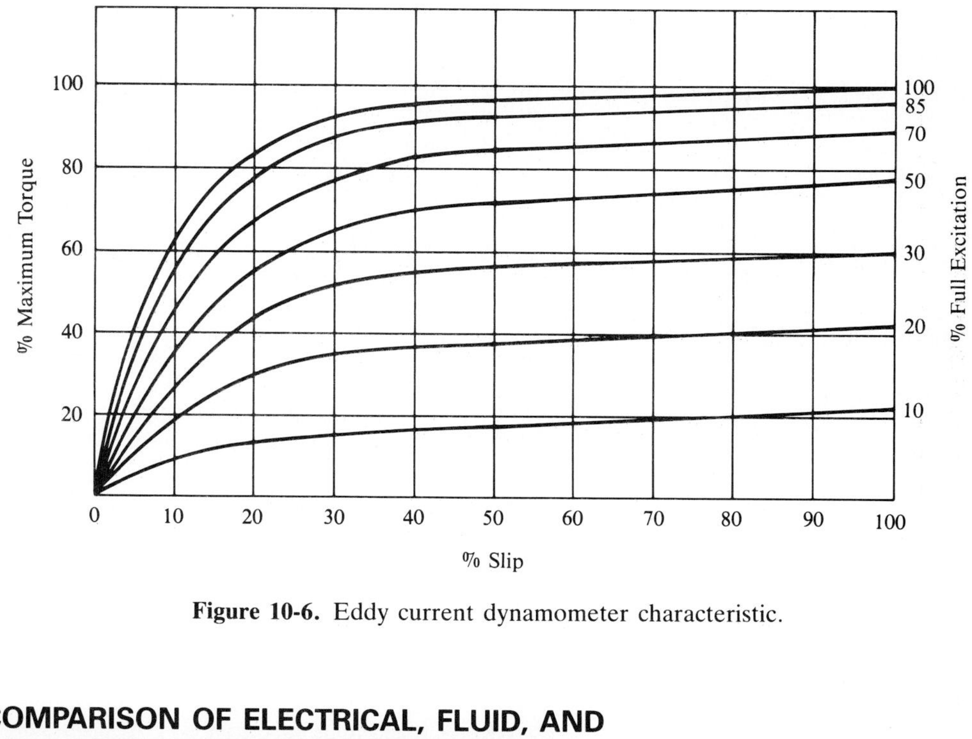

Figure 10-6. Eddy current dynamometer characteristic.

10-2 COMPARISON OF ELECTRICAL, FLUID, AND MECHANICAL VARIABLE-SPEED DRIVES

In the first ten chapters, we have presented a variety of electrical, fluid, and mechanical drives. Nine types of drives have been discussed, each with its own merits and shortcomings. Each also has its own critic and proponent. And these drives vary in their capacities and capabilities. For example, integrated hydrostatic drives are available in a range up to only around 50 hp while induction motors are available in ratings of thousands of horsepower.

For many years, the efficiency of variable-speed drives was not a major factor in discriminating among them. Today, however, the increased cost of energy makes an evaluation of overall system efficiency more important. As an example, a 3 percent difference in efficiency for an installation of 1000 hp running 6 days a week, 50 weeks a year, 16 hours per day, implies a differential in power consumption of about 100,000 kWh or about $5000 to $10,000 per year in power costs depending on local rates. Consequently, the life-cycle costing method is appropriate (which would include capital, maintenance, and operating costs) in selecting a variable-speed drive.

Table 10-1 compares nine different types of potential variable-speed drives (which in some cases is only a relative comparison). The characteristics compared include performance, control, and cost considerations. Keep in mind that the relative importance of the various characteristics will be dictated by the particular application requirements. We hope that, at least in a qualitative way, these comparisons will be of some help in choosing a variable-speed drive.

TABLE 10-1. Comparison of Variable-Speed Drives.

	Electrical				Fluid			Mechanical	
Characteristic	VIM	WRM	DCM	ECD	HSD	HDD	HVD	VPP	TRA
Relative cost	Med	Med–high	Med	Med–high	Low–med	Low–med	Med	Low	Low–med
Size and weight	Low–med	Med	Med	Med–high	Low–med	High	Med	Low	Low
Environmental rating*	Med–high	Low–med	Low	Low–med	High	High	Med	Med	High
Nominal maximum speed	3600	3600	3600	3600	1800	3600	3600	3600	1800
Speed ratio	10:1	10:1	6:1	6:1	40:1	8:1	20:1	10:1	9:1
Speed regulation, open-loop	3%	3%	3%	3%	3%	2%	2%	5%	0.02%
Speed variation, closed-loop	0.1%	0.1%	0.1%	<1%	0.5%	0.5%	0.1%	1.5%	—
Reversible?	Yes	Yes	Yes	No	Yes	No	No	No	Yes
Efficiency**	Med–high	Med–high	Med–high	Med–high	Med–high	Med–high	High	Med	High
Starting torque	150%	150%	100%	150%	250%	0	0	150%	150%

Note: VIM = variable-speed induction motor; WRM = wound rotor motor; DCM = DC shunt motor; ECD = eddy current drive; HSD = hydrostatic drive; HDD = hydrodynamic drive; HVD = hydroviscous drive; VPP = variable-pitch pulley drive; TRA = traction drive

*Resistance to dirty, dusty, abrasive, volatile conditions

**Efficiencies of electrical drives include controllers

Information for Table 10-1 has been taken from a variety of sources including References 10-3 and 10-4. Various catalog data and other technical publications have also been of assistance.

In addition to the tabulated comparison, we note the following general items:

(a) Electrical drives are generally the most responsive and versatile in control. Speed regulation is good in the closed-loop control mode. Electrical drives have ease of reversibility in speed, but high rotor inertia may make rapid reversal difficult. All electrical drives are available with electronic controllers which can perform many functions and can be easily interfaced with sensors. These are the areas in which electrical drives excel.

Electrical drives are limited by torque degradation at both the low- and high-speed ends of the torque-speed curve and suffer generally in having poor starting torque. Induction motor controls can drive down to about 15 percent of synchronous speed. DC motor controls are not recommended for continuous duty below about 40 percent of rated speed. An advantage of electrical drives is their ease in incorporating regenerative or dynamic braking.

AC induction motors are available at reasonable cost to cope with the worst environments. This is their forte. DC motors and eddy current devices, on the other hand, may need expensive and unwieldy ancillaries for use in an abusive environment. The modern electronic controllers for electrical drives are sensitive to electrical noise caused by arc welding, motors starting, and so forth. And these control panels must be kept very clean which may prove to be difficult in an abusive environment.

In regard to maintenance, service, and reliability, electrical drives rate high except for the DC drive and electronic controllers.

(b) Mechanical drives usually incorporate speed reducers as a part of the overall drive system. They are not noted for improved starting torque beyond that of the prime mover which is generally an induction motor (150 percent of rated torque). These drives can provide speeds in excess of 4000 rpm and can also achieve very low speeds. Reversal of a mechanical drive is not possible through the conventional control mechanism. If reversal must be available, it is accomplished through reversal of the prime mover.

Under varying load, the mechanical drive must compensate for both mechanical slip amd prime mover slip (if driven by an electric motor). This, combined with rather slow control response, prevents a speed regulation of less than about 1.5 percent in closed-loop systems. Traction drives, on the other hand, can provide an open-loop slip (creep) as low as 0.02 percent.

Although mechanical drives offer only a few control mechanisms and are not the most responsive in control, they excel in abusive environments. Protection packages are economically available which can readily mitigate the effects of dirty environments. And they can operate readily in applications requiring explosion-proofness.

Mechanical drives require maintenance at scheduled intervals for reliable operation (e.g., oil changes, belt adjustments, traction fluid for traction drives, and so forth). Mitigating these requirements is the simplicity of most mechanical drives, which makes them popular.

(c) Fluid drives comprise hydrostatic, hydrodynamic, and hydroviscous drives. These three drives have quite different characteristics. For example, the hydrostatic drive can operate continuously at high torque in the low-speed region (say, down to 50 rpm). At these lower speeds, the ultimate limitation is torque ripple rather than torque per se. On the other hand, hydrodynamic and hydroviscous drives provide very little torque at low speed and may have to be started under no-load. The starting torque of a hydrostatic drive, in contrast, is typically 250 percent.

Of these three fluid drives, only the hydrostatic drive is capable of providing regenerative or dynamic braking or of coping with an overhauling load. The hydrodynamic and hydroviscous drives, conversely, are excellent for soft-starts.

The fluid coupling is inherently inefficient and must, therefore, be operated at low slips. The same thing would be true of the hydroviscous drive, which cannot be locked up. The typical hydrostatic drive will provide an efficiency of 80 percent in the high-speed range and drop down to around 65 percent at 50 percent of rated speed. Efficiency will drop even more at lower speeds.

Hydrostatic drives are highly controllable and responsive because rotating inertia is low and control movements are small. Closed-loop control can provide a speed regulation of around 0.5 percent. And, of course, the hydrostatic drive is easily reversed by overstroking the pump. Since it can be totally enclosed, this drive can provide immunity from dirty, hostile environments.

The virtue of the hydrodynamic drive is in its ruggedness and compliance as an element in the power train. It does tend to be relatively large with high rotating inertia which does not make it the ideal choice insofar as control and response are concerned. It, too, can operate in the dirtiest environment.

The strengths of the hydroviscous drive (clutch) are controllability in the closed-loop mode (0.1 percent) and its ability to become locked up (to mitigate inefficiencies accruing to slip). It is available in large horsepower sizes and can be integrated readily into an integrated drive incorporating a gear reduction. It does require strict maintenance and careful application in order to be reliable.

In summary, one drive type will not be the best in all applications. In choosing a drive type, the project engineer must consider many criteria, the most important of which is life-cycle costing, which includes the costs of maintenance, unreliability, and service.

10-3 REFERENCES

10-1. J. B. Winther, *Dynamometer Handbook of Basic Theory and Applications*, published by Eaton Corporation, 3122 Fourteenth Avenue, Kenosha, WI, 1975.

10-2. Eddy-Current Drives, various publications by Eaton Corporation, 3122 Fourteenth Avenue, Kenosha, WI.

10-3. G. D. Fillion, "Integrated Hydrostatic Transmission Challenges Electrical and Mechanical Drives," *Machine Design*, October 21, 1982, pp. 88–93.

10-4. J. K. Liu, "Recent Developments in Belt Conveyor Drives," presented at the National Conference on Power Transmission, Chicago, November, 1979; available from Philadelphia Gear Corporation, King Of Prussia, PA.

PROBLEMS

10-1. An eddy current clutch has a torque-speed characteristic as sketched in Fig. 10-6 for various rotor excitations. Suppose the clutch is operating a 70 percent excitation at a slip of 5 percent. Assume the input speed is fixed at 3500 rpm. Suddenly the load torque increases by 100 percent.

- **(a)** Is it possible to maintain the output speed constant using a closed-loop control scheme?
- **(b)** What is the maximum output speed that can be maintained under the circumstances stated?
- **(c)** If no control actions are taken, what will be the new output speed?

11

Electronic Control of Electric Motors

Before the advent of electronics, motors were controlled by using rheostats to control either the current in the field or armature circuits of DC motors or to control the rotor current or stator voltage of AC motors. We have discussed some of these traditional control concepts in Chapters 8 and 9, and they remain viable in many motor applications today in which the control requirements are not highly esoteric.

The early electronic devices (beginning around 1910) were of the vacuum tube type. They made possible an alternative method of control through the electronic processes of rectification and amplification. Then, beginning around 1950, vacuum tubes began to be displaced, as electronic control devices, by solid-state power rectifier and amplification devices such as the p-n junction diode and the p-n-p-n silicon-controlled rectifier (SCR). It seems probable that techniques for controlling electric motors will ultimately be dominated by these solid-state devices and their circuitry because solid-state devices have an inherent compatibility with modern electronic control and instrumentation systems.

The field of electronic control is vast. Entire books have been written on this subject. The detailed quantitative knowledge of the transient electrical conditions existing in motor control circuits, which incorporate diodes or SCRs, is beyond the scope of this book. Such detailed knowledge will include some of the following items:

- **(a)** Transient behavior of *RLC* circuits
- **(b)** Transient performance of diodes, thyristors, and their circuits including inductance (or dv/dt) effects on thyristor gating
- **(c)** Thyristor commutation techniques
- **(d)** Ac line pollution caused by phase-controlled rectifiers or DC pollution caused by the AC supply
- **(e)** Power factor considerations
- **(f)** Effect of over-running loads and regeneration capabilities on diode or thyristor control circuits

Fortunately, we will not have to expose all of these detailed considerations in order to meet our objectives here. We simply want to present a qualitative survey of some of the types of circuits available and to understand how they work in a qualitative way.

We have already shown that DC motors can be readily controlled through field resistance, armature resistance, or applied DC voltage. All of this presupposes, of course, that a DC supply is available. But the industrial power source of this nation is not DC; it is AC with a frequency of 60 cps. So, unless we are going to use batteries as our ultimate power source, some provision must be made, in the DC motor control scheme, for rectifying the AC source into a DC supply. This process is called conversion, and a circuit for accomplishing this rectifying task is called a converter.

In addition to converting the AC supply, we also want to be able to modulate (control) the DC voltage level produced; this voltage modulation will provide the third type of DC motor control mentioned above. One method of doing this is to drive a DC generator (a rectifier) with an AC motor. The AC motor is connected to the AC supply, and the DC generator output voltage is used to drive the DC motor. The DC supply voltage can be controlled simply through the resistance of the generator field winding (see Eqs. [8-5] and [8-6]). This configuration is the Ward-Leonard scheme we discussed in Chapter 8. Although this method has the advantage of being non-dissipative, it is expensive and bulky and is being replaced by more modern alternatives. The modern scheme is to use a solid-state rectifier that incorporates diodes or thyristors.

Actually, we would prefer to use the AC motor rather than the DC motor because of the AC motor's ruggedness and low cost. But the AC motor has more difficult control problems. The problem of speed control with the AC motor is not one of rectification per se despite the fact that we might want to incorporate a rectifier in the AC motor electronic control circuit. The basic problem is that AC motor speed control can be effected only by controlling the stator voltage, the number of poles, or the supply frequency (see Eq. [9-3]). The first method is inefficient, the second cannot provide continuously variable speed, and the third (frequency control) is complex and expensive.

It is not enough just to control the frequency of the AC supply to the AC motor in order to obtain speed control; the supply voltage level must also be controlled, because as the frequency is changed, so is the inductive reactance of the stator. Frequency and voltage control must be integrated together. With the proliferation and improvement in solid-state devices and circuits, this integration can be achieved, and the electronically controlled AC motor may eventually displace the DC motor as the preferred device (from a control viewpoint).

In any event, knowledge of the electronic control of both DC and AC motors is essential to those who intend to design and control mechanical systems. Keep in mind that this control can be either open-loop or closed-loop. Here we are interested basically in the open-loop electronic control schemes, which will require us to discuss some basic solid-state control concepts, including (a) the diode, (b) the thyristor, (c) gating, (d) commutation, (e) the free-wheeling diode,

(f) the phase-controlled rectifier, (g) the chopper, (h) center tap and bridge circuits, and (i) the inversion of DC to AC power (circuits for doing this are called inverters). We shall begin with a study of the solid-state devices themselves: the diode and the thyristor.

11-1 SOLID-STATE POWER DEVICES

The development of power semiconductor devices in the 1950s has made possible the design of many novel electronic circuits for controlling motors. The two devices of interest are the p-n diode and the silicon-controlled rectifier (SCR). In motor control circuits, these solid-state devices act as highly responsive, controllable, one-way electronic switches.

The diode is the electronic analog of the hydraulic check valve. Ideally, it offers zero impedance to current flow in the forward direction and infinite impedance to current flow in the reverse direction.

The SCR is one of a class of solid-state switches called thyristors. We shall use the terms *SCR* and *thyristor* synonymously here. The SCR differs from the diode in that the time at which it begins to conduct in the forward direction can be controlled independently of the forward voltage impressed upon it.

11-1-1 The Power Diode

The power diode is a wafer consisting of the junction of a layer of n-type silicon and a layer of p-type silicon. These layers are semiconductor materials having electrodes soldered to them. A sketch of the p-n junction diode and its electronic symbol is shown in Fig. 11-1.

The characteristics of the diode are also shown in Fig. 11-1. Note that there is very little resistance to current flow in the forward direction. Typically, currents of several hundred amperes can flow in the forward direction with a voltage drop of order of magnitude of only 1 V. On the other hand, resistance to current flow in the reverse direction is very high. Typically, the current leakage in the reverse direction will be only a few milliamperes when the reverse voltage is as high as 1 kV.

Thus, the diode acts as a one-way switch permitting current to flow in one direction but not in the other direction. Once the diode begins to conduct in the forward direction, the solid-state electronics within the p-n junction are such that the diode will continue to conduct until the forward current is reduced to a small value called the holding current.

Incorporated into a simple circuit excited by an AC source, the diode provides half-wave rectification as shown in Fig. 11-2. The load in the circuit shown at the left is a pure resistance, so the current will be in phase with the applied voltage. Since the diode permits current flow only in the forward direction, this circuit accomplishes half-wave rectification; there is no current flow in the load when the applied voltage negatively biases the diode. The current will have fallen to a value below the holding current at the end of the positive half-period, because the current is in phase with the applied voltage. This fall in

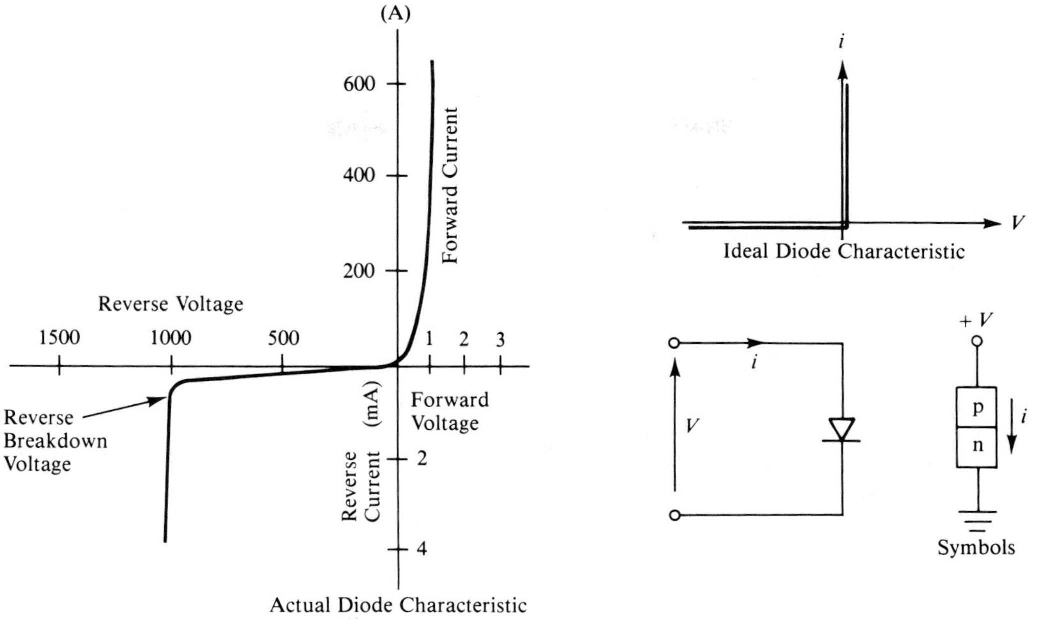

Figure 11-1. Semiconductor diode characteristics.

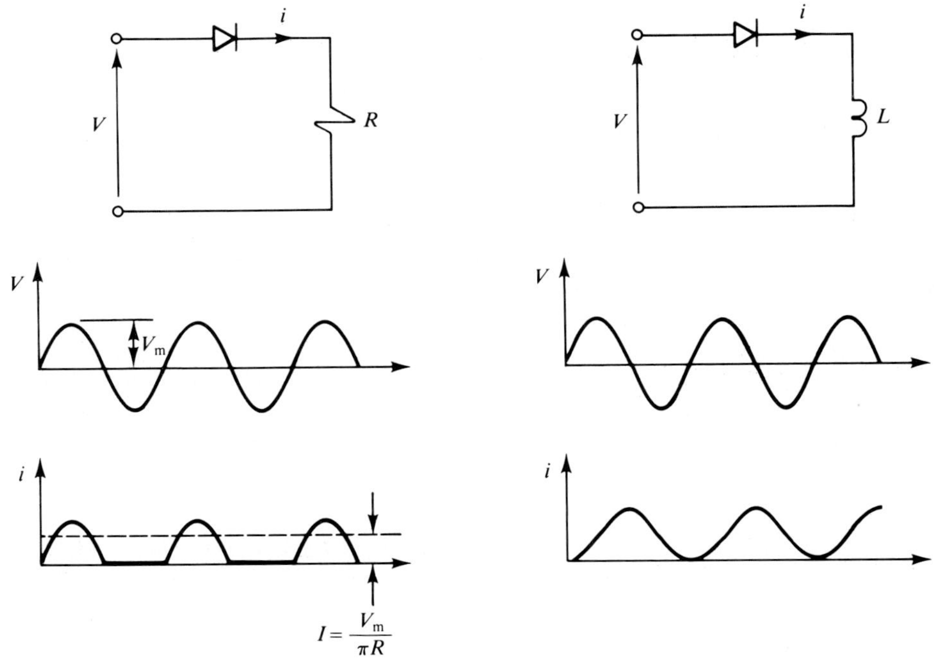

Figure 11-2. Resistance and inductance rectification characteristics.

current permits the diode to be turned off (commutated) at the end of the positive half-period. The average DC current in this pure resistance load can easily be shown to be

$$I = \frac{V_m}{\pi R} \tag{11-1}$$

The average DC load voltage is then

$$v_a = IR = \frac{V_m}{\pi} \tag{11-2}$$

If the load is a pure inductance, as shown at the right of Fig. 11-2, we get a quite different current flow. Now the current will be out of phase with the supply voltage. An analysis (rather complicated) of the load current will show it to be as sketched at the right of Fig. 11-2.

Recall that the relationship between current and applied voltage for a pure inductance is

$$\frac{di}{dt} = \frac{V(t)}{L} \tag{11-3}$$

Consequently, the current rises to a maximum during the half-period the applied voltage is forward biasing the diode (remember, the diode has negligible resistance to current flow under forward biasing).

During the half-period when the applied voltage is negatively biasing the diode, the diode nevertheless continues to conduct in the forward direction because the current is not zero (as it was for the pure resistance) at the end of the previous half-period. The diode will not turn off (or block) even in the face of the reversed voltage bias until the current has fallen below the holding current. Whereas the diode blocks current flow for the pure resistance, we now have a falling current as the magnetic energy previously stored in the inductance discharges. Thus, electrical energy is regeneratively fed back into the power supply. The net voltage produced across the inductance will be zero, and so will be the net power consumption.

Electric motors have appreciable (even large) inductance, and we generally do not want our motor control circuit to act in a regenerative fashion because we want to pump power into the load rather than back into the AC supply (although this is not necessarily true if we desire regenerative braking). This regenerative effect caused by the presence of the inductance can be avoided by incorporating a shunt branch that incorporates another diode (d_2 in Fig. 11-3).

Diode d_2 is called a free-wheeling diode. It permits the inductance to retain its magnetic energy rather than pumping it back into the AC supply during the negative half-period when diode d_1 is voltage biased in the reverse direction. Diode d_2 does this by short-circuiting the load during the negative half-period when diode d_2 is turned on by the negative voltage bias (this is a positive voltage bias insofar as d_2 is concerned). Turning on diode d_2 robs diode d_1 of current, and diode d_1 turns off. When d_1 turns off, the current free-wheels around the shunted circuit containing the inductance and diode d_2.

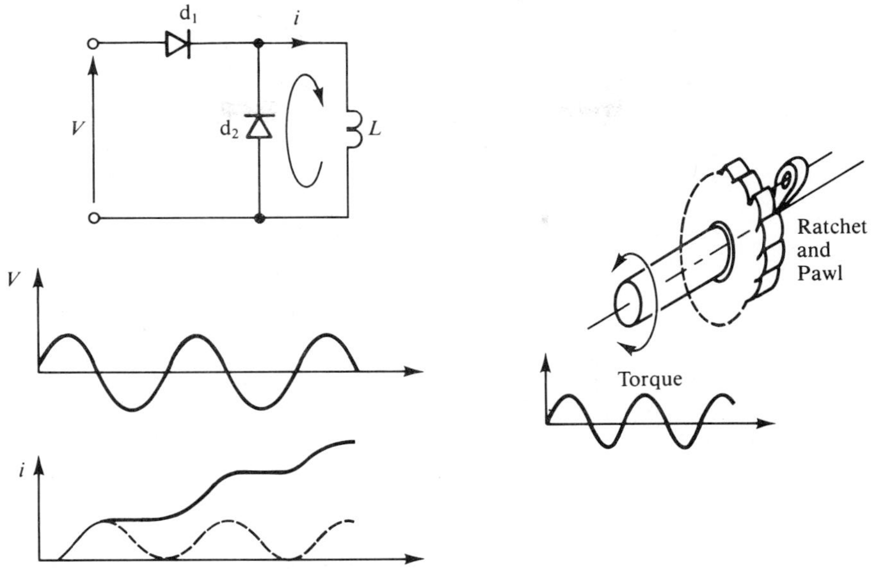

Figure 11-3. Free-wheeling diode in an inductive circuit.

The mechanical analog of the free-wheeling diode is the over-running clutch (as in a ratchet and pawl mechanism); this is no doubt where diode d_2 gets its name.

With no resistance in the circuit of Fig. 11-3, the current would continue to rise with each full cycle of applied voltage. The inductance would simply continue to store more and more energy without ever using it to produce power (see the current trace in Fig. 11-3). To produce power, a passive circuit such as that shown in Fig. 11-3 must contain resistance.

Of course, in a real motor circuit we have a counter-emf as well as resistance and inductance. And the motor is driving a torque load. The real motor makes the electronic control circuit even more complicated, as indicated in Fig. 11-4.

Now diode d_1 will turn on only when the difference $V - E$ positively biases it. Positive bias will occur only at time t_1, corresponding to angle θ_1, as shown (remember, $\theta = \omega t$). At time t_1, the current begins to rise. It will continue to rise until time t_2, corresponding to θ_2, at which time the current will begin to fall but continue to flow through diode d_1 until the applied voltage V falls to zero (at $\theta = \pi$). Now diode d_2 will be turned on by the negative voltage bias (note that this is a positive voltage bias on d_2) and diode d_1 will turn off. The current will then free-wheel through diode d_2 but will decay because of the power consumption either by the motor load or by the resistance R (or both).

The current trace will be as shown in Fig. 11-4. During the time diode d_2 is conducting, the rate at which current i falls depends on the magnitude of inductance L. If the inductance is large enough, the current may be continuous. As shown, it is discontinuous.

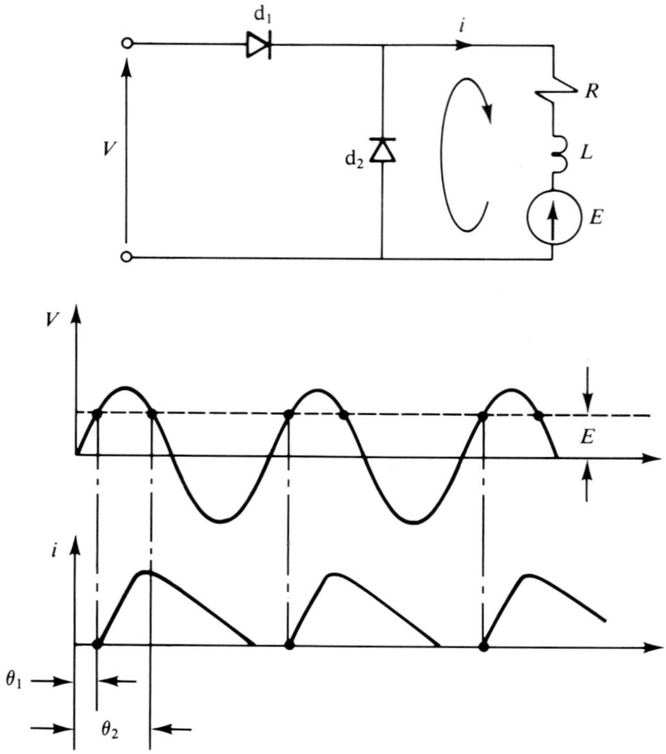

Figure 11-4. Diode rectification in a motor circuit.

None of the circuits discussed above allow independent control of the time (or phase angle θ) at which diode d_1 is to be turned on in order to conduct in the forward direction. Diode d_1 will simply turn on when the voltage across it is positively biased. This diode action is a serious limitation of the power diode when used as a rectifier; it does not make possible the independent control of the level of the DC output produced by the half-wave rectifiers discussed above.

11-1-2 The Silicon-Controlled Rectifier (SCR)

The SCR, or thyristor, is a slightly more complex combination of n-type and p-type semiconductor layers which has properties similar to the power diode but with an additional electrode called the trigger. A sketch of this type of thyristor and its electronic symbol is presented in Fig. 11-5.

The trigger electrode makes it possible to control the turn-on of the device independently of the forwardly biasing voltage. The characteristics are shown in Fig. 11-5.

The current-voltage characteristic of the thyristor under reverse voltage bias is very similar to that of the power diode. Resistance to current flow in the reverse direction is very high, so the device, in effect, blocks reverse current flow.

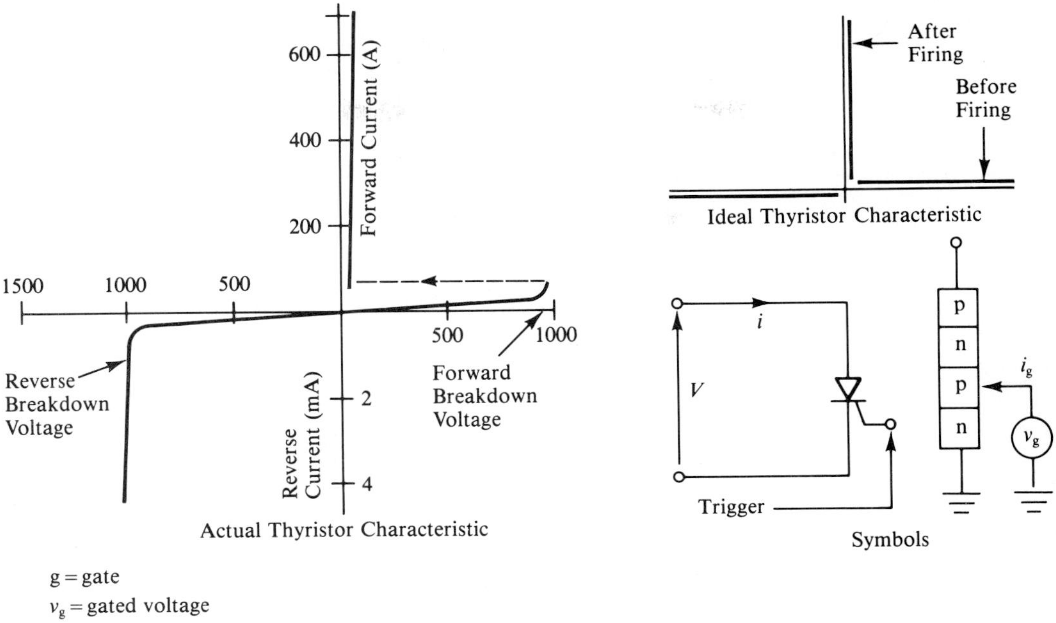

Figure 11-5. Semiconductor thyristor characteristics.

The current-voltage characteristic of the thyristor under forward voltage bias is quite different from that of the diode and depends on the current gated through the trigger electrode. With zero gate (trigger) current, the forward characteristic is similar to the reverse blocking characteristic up to the turn-on (or breakdown) voltage, at which the thyristor suddenly reverts to the diode type of characteristic of large forward current flow under negligible resistance.

As the current gated through the trigger increases in magnitude, the breakdown voltage, at which the thyristor is turned on for conduction in the forward direction, will diminish (see Fig. 11-5). Once the thyristor is in the conducting mode, it will remain in this state until the forward current falls below the holding current, at which point forward current will once again be blocked. This means that the trigger need only be pulsed to turn on the thyristor. Once turned on, the thyristor is immune to what happens further at the trigger.

Thus, we see that the forward applied voltage at which the thyristor will break down to act like a diode can be controlled by a current pulse admitted by the trigger. In a diode, this breakdown voltage cannot be controlled; it is for practical purposes zero.

Being able to control the level of applied voltage at which the thyristor is turned on to conduct in the forward direction is tantamount to being able to control the phase angle at which the thyristor will begin to conduct. This control is explained in terms of Fig. 11-6 for a DC motor as the circuit load.

The only difference between Figs. 11-4 and 11-6 is that thyristor th_1 has replaced diode d_1. Whereas in the diode circuit of Fig. 11-4, diode d_1 was turned

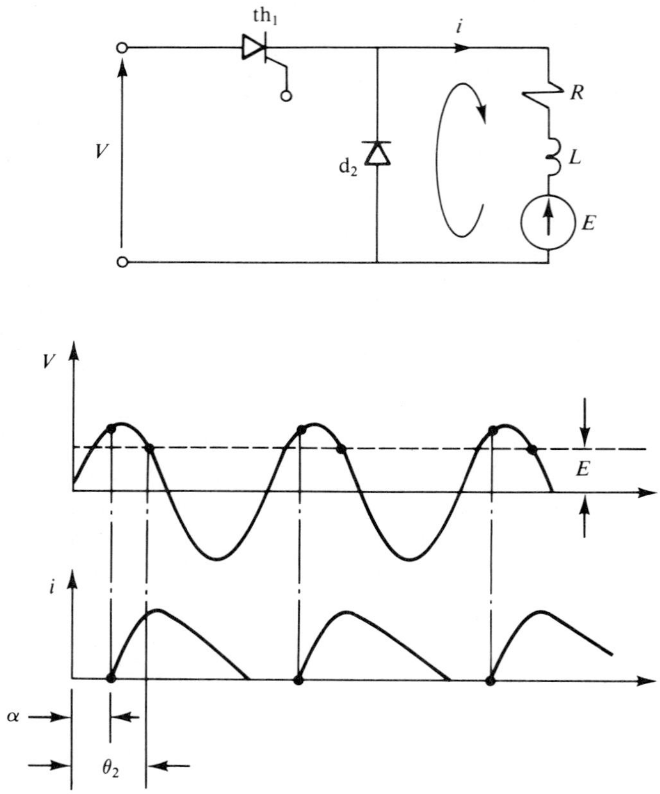

Figure 11-6. Thyristor rectification in a motor circuit.

on when the difference $V - E$ became positive, now thyristor th_1 will be turned on only when the difference $V - E - V_b$ becomes positive, where V_b is the thyristor breakdown voltage. In other words, a higher forward bias voltage is required than before, which will have the effect of turning on the thyristor (called firing) at some phase angle α.

Once the thyristor has fired, the voltage across it drops to $V - E$ as in the case of the diode. At phase angle θ_2, the current in the thyristor will be continued by the inductance in the circuit until phase angle $\theta = \pi$ is reached. At this point, the applied voltage biases diode d_2 to conduct, which short-circuits the load. Thus the thyristor will be caused to turn off, and the free-wheeling current through diode d_2 will decay because of the load.

The process by which the thyristor is turned off is called commutation. In the circuit just discussed, the commutation process is called natural (or line) commutation because it is initiated by the applied voltage source itself at $\theta = \pi$. At $\theta = \pi$, the applied voltage turns on diode d_2 which, in turn, robs th_1 of current. This process turns off, or commutates, th_1.

11-2 PHASE-CONTROLLED RECTIFICATION

Speed control of a DC motor by controlling the level of DC voltage applied to it can be accomplished in several ways. The traditional way is to place a rheostat in series with the load in order to reduce the voltage applied to the motor. The rheostat may have to be capable of handling an appreciable power dissipation.

Another way of controlling the DC voltage level is not to change the magnitude of the level applied to the load but, rather, to use a switch to disconnect the voltage source from the load periodically. The effective voltage level then applied to the motor will depend on the fraction of the time that the fixed level is applied to the load. For example, if the circuit is disconnected 50

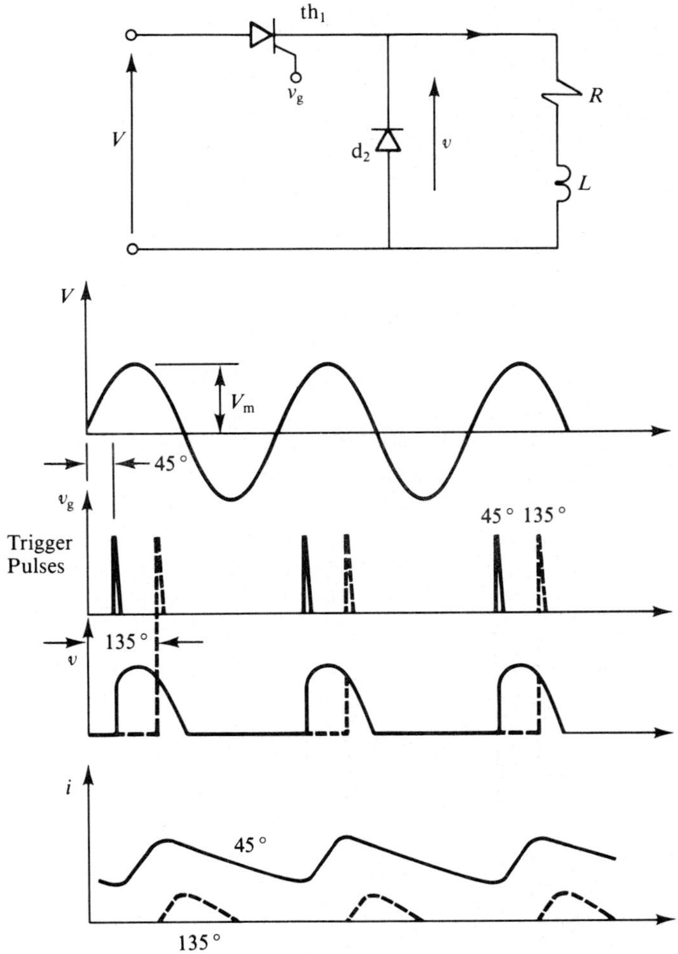

Figure 11-7. Thyristor rectification for various firing angles.

percent of the time, a voltage level of 100 V will produce an average applied voltage of only 50 V. This process is called chopping. With a rheostat to accomplish this voltage reduction, 50 percent of the power will be dissipated, which is wasteful. Chopping can eliminate most of this loss.

The thyristor circuit of Fig. 11-6 not only rectifies the AC voltage source into a DC supply to the motor (albeit a crude DC supply) but it also has the capability of producing the chopping effect discussed above. Chopping is done by controlling the phase angle α at which the thyristor is fired. To do this requires only that there be a suitable method of controlling the thyristor trigger. The entire process is called phase-controlled rectification.

Figure 11-7 is a sketch of AC supply voltage, trigger pulses, voltage across the load, and current in the load for a simple *RL* load. The solid curves result when the trigger is pulsed to gate current into the thyristor at a phase angle $\alpha = 45$ degrees. The dashed curves represent the same outputs when the trigger is pulsed at a phase angle of 135 degrees. It is apparent how the current in the load responds to a change in the phase angle α at which the thyristor is fired. The level diminishes as firing angle α increases; it is a maximum for $\alpha = 0$ degrees and zero for $\alpha = 180$ degrees. The level of DC current (and hence load voltage) is controlled by controlling phase angle α; hence the name phase-controlled rectifier.

11-3 FULL-WAVE CONVERTERS

All of the converter circuits discussed thus far have been half-wave rectifiers. Extension of these circuits to produce full-wave rectification is straightforward and involves simply using additional diodes or thyristors to invert and rectify the half-period when the applied voltage is negative in bias.

A circuit for accomplishing full-wave rectification using only two diodes is sketched in Fig. 11-8. The scheme shown is called a single-phase full-wave rectifier with center tap transformer. The purpose of the transformer is to electrically isolate and divide the AC supply circuit as a source for the two branches of the rectifier circuit. The current waveform produced by this circuit for a pure resistance load is shown in Fig. 11-8. Note that the mean value of the DC current is now twice that of the half-wave rectifier:

$$I = \frac{2V_m}{\pi R} \qquad \textbf{(11-4)}$$

During the half-cycle $(0 \leq \theta \leq \pi)$, diode d_1 conducts while diode d_2 is off. Then during the succeeding half-cycle, $(\pi \leq \theta \leq 2\pi)$, when the voltage bias is negative, diode d_2 conducts while diode d_1 is off.

An alternative circuit for accomplishing full-wave rectification is shown in Fig. 11-9. It is a bridge circuit that requires four diodes. The advantage of the bridge circuit is that the peak voltage applied to a nonconducting diode is just half of that in the center tap circuit.

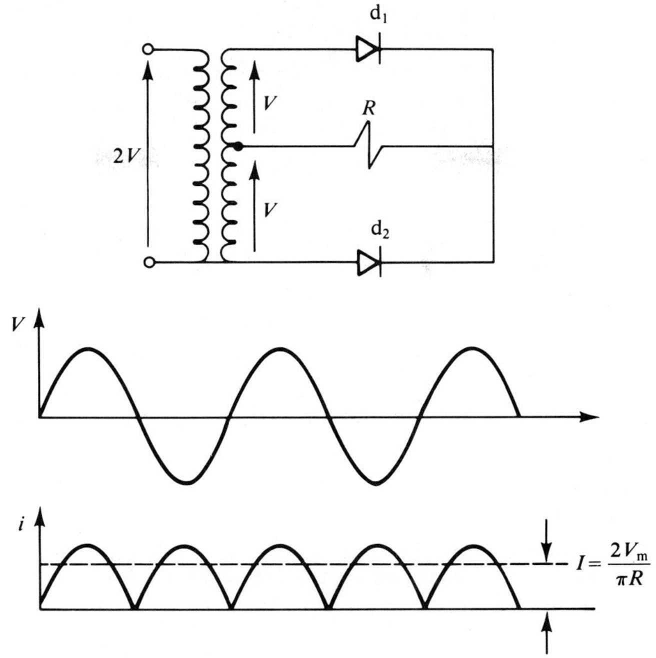

Figure 11-8. Full-wave center tap converter.

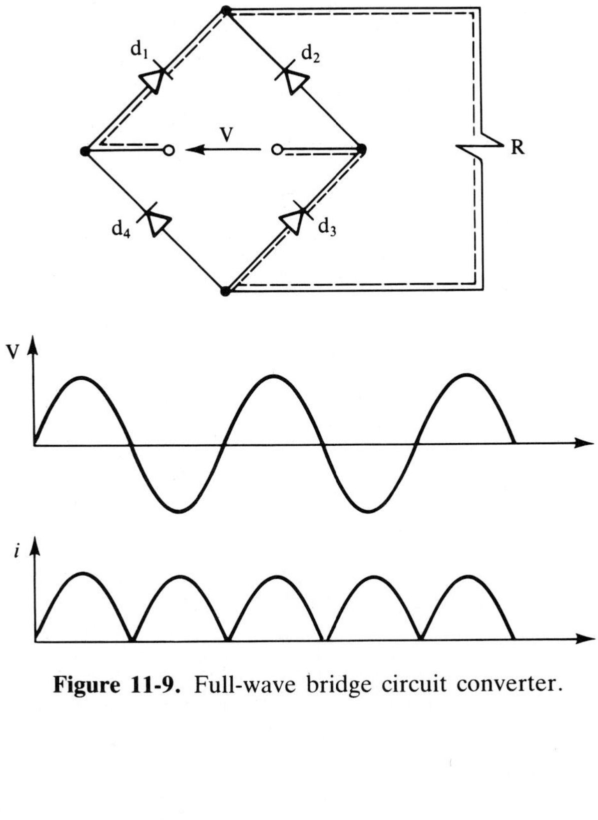

Figure 11-9. Full-wave bridge circuit converter.

Note that during the half-cycle ($0 \leq \theta \leq \pi$), diodes d_1 and d_3 conduct and diodes d_2 and d_4 are off, while during the half-cycle ($\pi \leq \theta \leq 2\pi$), just the opposite occurs. The current path during the positive half-cycle ($0 \leq \theta \leq \pi$) is shown by the dashed line in Fig. 11-9.

The circuits of Figs. 11-8 and 11-9 are not phase-controlled rectifiers because they use diodes exclusively. The element of phase control (and hence voltage level control) can be introduced in these circuits by using thyristors to replace some, or all, of the diodes. Many combinations could be considered, including those that do, and do not, incorporate free-wheeling diodes. Also, it must be remembered that actual motor circuits include inductance and back-emf in addition to just pure resistance. These various combinations make the circuit analyses rather complex.

11-4 DC MOTOR ELECTRONIC CONTROL

The ultimate source of power for a DC motor could be either a three-phase or a single-phase AC supply, or it could be a DC supply (e.g., batteries). When the supply is AC the DC motor controller is generally a phase-controlled rectifier (converter), and when the supply is DC the DC motor controller is an electronic circuit called a chopper. In selecting a DC motor controller, some factors to be considered in addition to the ultimate power source are:

(a) Braking or regeneration requirements
(b) The nature of the equipment and the load the DC motor is driving (in particular, the necessity of handling over-running loads)
(c) Open-loop versus closed-loop control

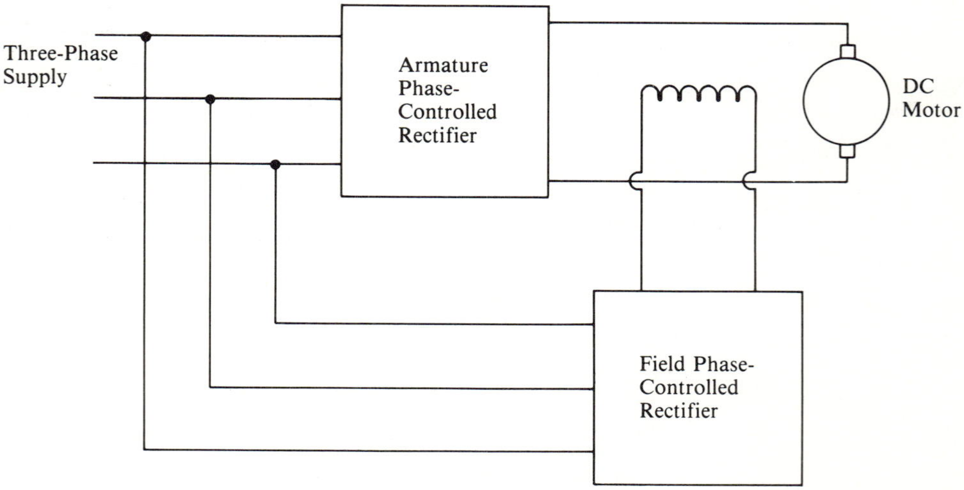

Figure 11-10. General three-phase DC motor control.

(d) The need to interact with microprocessors or otheı computational equipment
(e) Capital cost of the motor, electronic controller, transformers, and so on
(f) Operating and electrical costs
(g) AC line pollution caused by the installation

When the power source is a three-phase AC supply, the most general control scheme using phase-controlled rectifiers is as shown in Fig. 11-10. Note that the basic mechanism of control is the voltage control achieved by phase-controlled rectification as we discussed in Section 11-2. In general, both the armature and field voltages could be controlled by separate phase-controlled rectifiers, as shown in Fig. 11-10. These devices will result in speed control characteristics like that shown in Fig. 11-11.

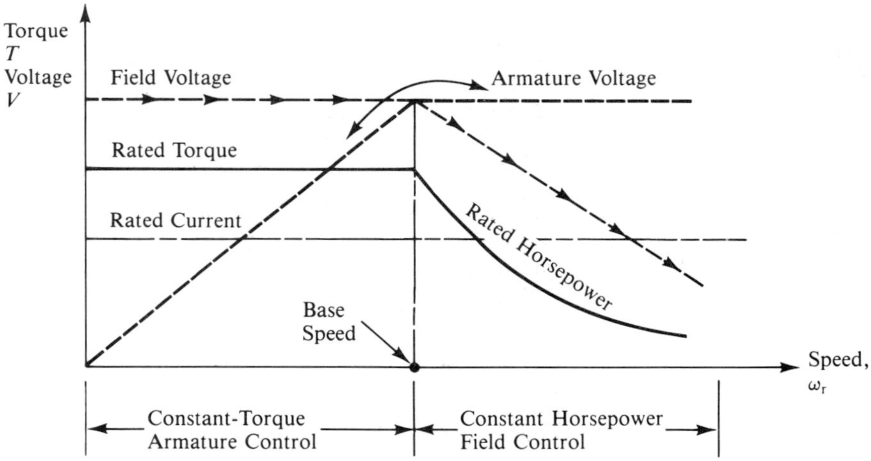

Figure 11-11. Speed control characteristics of the DC motor.

11-5 DC MOTOR SPEED CHARACTERISTICS

The basic equilibrium equations for the DC motor were developed in Chapter 8, and are repeated here for easy reference:

$$T = K_3 I_a I_f \tag{8-3}$$

$$E = K_5 I_f \omega_r \tag{8-5}$$

$$V = E + I_a R \tag{8-6}$$

If we combine Eqs. (8-5) and (8-6), the result is

$$\omega_r = \frac{V - I_a R}{K_5 I_f} \tag{11-5}$$

Recall that the base speed of a DC motor is its speed under full field excitation and full (rated) voltage applied to the armature. When the motor operates at base speed conditions, the armature current will correspond to rated torque in accordance with Eq. (8-3). It will be the maximum current that the motor can carry continuously in accordance with its nameplate rating.

Equation (11-5) implies two fundamental modes of DC motor speed control. In the first mode, we assume the field excitation constant at its full value (maximum I_f). The torque will then vary directly as the armature current, and vice versa. With the assumption that the motor operates at its maximum allowable armature current with full field excitation, Eq. (8-3) shows this will correspond to constant rated torque operation. Speed control in this mode will then be possible in the range of zero to base speed by varying the armature voltage from a low value to rated armature voltage. Since the term $I_a R$ of Eq. (11-5) will generally be around 10 percent (or less) of the applied voltage, Eq. (11-5) becomes approximately

$$\omega_r \cong kV \tag{11-6}$$

where k is a constant. This mode of control, called armature voltage control, is sketched at the left (below base speed) of Fig. 11-11.

So, with armature voltage control and full field excitation, we find that the speed ω_r increases essentially linearly with armature voltage. This type of control is readily achieved with phase-controlled rectification of AC voltage applied to the armature. Several possible electronic configurations for armature control will be discussed below.

Once the armature voltage has been increased to its full rated value, we will not be able to increase motor speed further by this method, because maintaining constant rated torque would require the power to exceed the motor rating. Consequently, speed control in the range beyond base speed requires that the back-emf E be maintained constant because V, I_a, and R are now fixed in Eq. (8-6). The constant back-emf requires that the field excitation I_f be diminished in inverse proportion to speed ω_r. Equation (11-5) shows that diminution of I_f will cause the speed to increase above base speed. Also, the result will be to operate at constant horsepower in this speed range, as shown at the right of Fig. 11-11.

11-6 DC ARMATURE VOLTAGE CONTROL

In most applications, only the armature voltage will be controlled because field control makes possible the control of speed only in excess of base speed. Generally, we require control in the range from zero to base speed, and this means varying the armature voltage. As indicated earlier, we can use either a center tap or a bridge circuit for phase controlling and rectifying an AC supply. Here we shall discuss only the bridge circuit, which requires four solid-state semiconductor devices for a single-phase supply.

11-6-1 Single-Phase AC Supply

The various electronic circuits used for phase-controlled rectification of a single-phase AC supply are basically similar. They differ only in choice of thyristor versus diode for the various switches or in incorporating a free-wheeling diode. Although there is also the choice between half-wave and full-wave rectifiers, we will consider only full-wave rectifiers here. Although half-wave rectifiers are cheaper in requiring fewer semiconductor devices, they produce greater ripple with attendant I^2R losses which are smaller in full-wave circuits. So unless the motor being controlled is of low horsepower, the full-wave circuit will be preferred.

Figure 11-12 shows a single-phase, half-controlled rectifier control circuit that utilizes two thyristors and two diodes in the bridge circuit plus a free-wheeling diode. The purpose of the free-wheeling diode is to smooth out current conduction in the circuit.

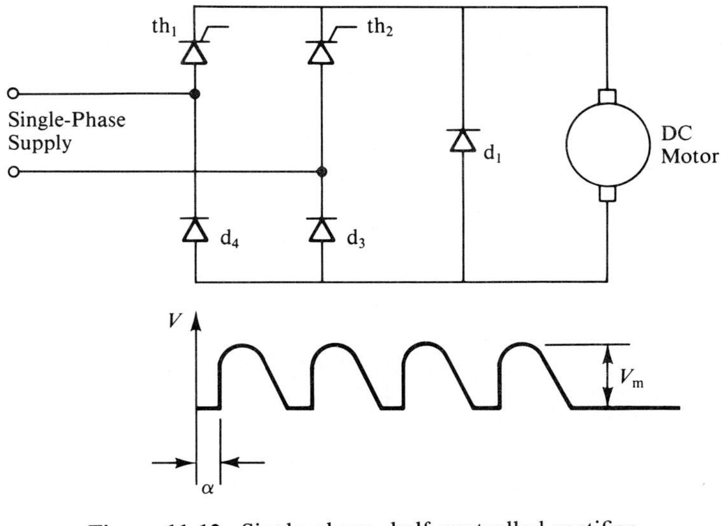

Figure 11-12. Single-phase, half-controlled rectifier.

If we assume that both thyristors are fired at the same angle α, the voltage output from the phase-controlled rectifier (which is also the voltage applied to the motor armature) will be as shown in Fig. 11-12. The average (DC) value of this voltage output can be readily shown to be

$$v_a = \frac{V_m}{\pi}(1 + \cos\alpha) \tag{11-7}$$

where V_m is the amplitude of the single-phase AC supply.

The torque-speed relationship for the DC motor then becomes, from Eqs. (11-5) and (8-3),

$$\omega_r = \frac{V_m(1 + \cos\alpha)}{\pi K_5 I_f} - \frac{RT}{K_3 K_5 I_f^2} \tag{11-8}$$

This equation shows how speed is controlled by the firing angle α for a given torque.

The control scheme of Fig. 11-12 is called a half-controlled bridge because only half of the devices in the bridge are phase-controllable. This circuit is not capable of providing regeneration, that is, the feeding of power back into the supply. It is a limitation in mechanical systems such as cranes or hoists where regenerative braking may be desirable.

A fully controlled bridge circuit capable of providing the regenerating mode of control is sketched in Fig. 11-13. Since a free-wheeling diode would prevent the regeneration mode, it is absent from this circuit. All of the semiconductor devices are thyristors.

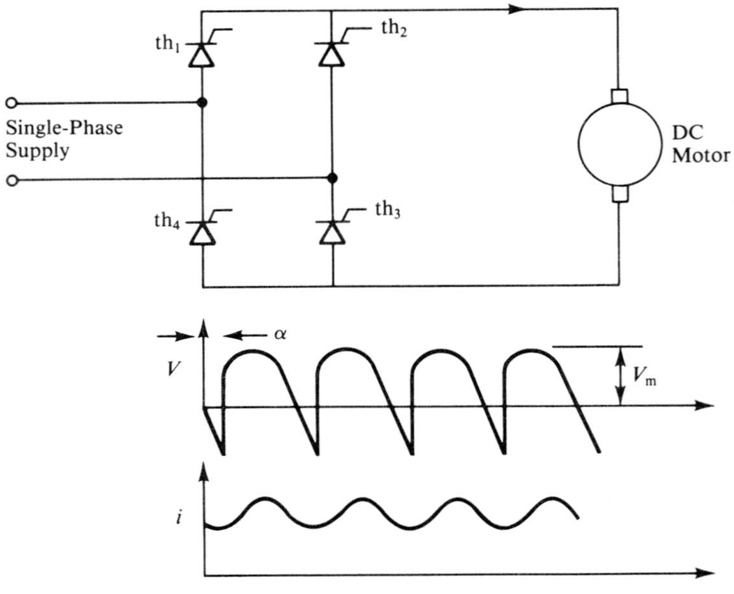

Figure 11-13. Single-phase, fully controlled rectifier.

In the fully controlled bridge circuit, all thyristors are fired at the same angle α. During the positive half-cycle of the AC supply voltage, thyristors th_1 and th_3 are fired. Then, during the succeeding negative half-cycle, thyristors th_2 and th_4 are fired. The firing sequence produces full-wave rectification with an average (DC) output voltage of

$$v_a = \frac{2V_m}{\pi}(\cos\alpha) \tag{11-9}$$

Load voltage and current relationships are sketched in Fig. 11-13. The regenerative mode is obtained by reversing the polarity of the field excitation.

EXAMPLE 11-1

A 220 V separately excited motor has a rated speed of 1000 rpm at rated voltage, a rated armature current of 40 A, and a field current of 2 A. The armature resistance is 0.2 Ω. Assume the motor is controlled by a fully controlled rectifier excited by a 220 V single-phase AC supply. Determine

(a) The motor speed at rated armature current if the firing angle is $\alpha = 55$ degrees
(b) The firing angle required in the regenerative mode to keep the motor current at the rated value of 40 A
(c) The power fed back to the power supply during the regenerative mode

Solution

Under rated conditions, the back-emf will be given by

$$E = V - I_a R = 220 - 40(0.2) = 212 \text{ V}$$

Then the magnetizing constant K_5 is given by

$$E = K_5 I_f \omega_r$$

$$K_5 = \frac{212}{1000(2)} = 0.106 \frac{\text{V}}{\text{amp-rpm}}$$

With a firing angle of $\alpha = 55$ degrees, the applied voltage is

$$v_a = \frac{2V_m}{\pi} \cos\alpha = \frac{2}{\pi}(\sqrt{2})(220)\cos(55) = 113.6 \text{ V}$$

Then the back-emf will be

$$E = v_a - I_a R = 113.6 - 40(0.2) = 105.6 \text{ V}$$

The motor speed will then be

$$\omega_r = \frac{E}{K_5 I_f} = \frac{105.6}{0.106(2)} = 498 \text{ rpm}$$

To operate in the regenerative mode, the field polarity is reversed so at the instant of reversal the motor will be acting as a generator with a developed emf of $E = 105.6$ V. The voltage then developed at the terminals of the generator will be given by

$$v_a = E - I_a R = 105.6 - 40(0.2) = 97.6 \text{ V}$$

The firing α will then by given by

$$-v_a = \frac{2V_m}{\pi}\cos\alpha$$

$$-97.6 = \frac{2}{\pi}(\sqrt{2})(220)\cos\alpha$$

$$\alpha = 119.5 \text{ degrees}$$

The minus sign is required because polarity of the motor terminals is reversed when the motor acts as a generator.

The power fed back to the supply will be

$$P = 97.6(40) = 3904 \text{ W}$$ ■

11-6-2 Three-Phase AC Supply

Extension of the single-phase rectifier circuit to the three-phase case is straightforward. Again, several combinations of diodes and thyristors are possible. Here we will discuss only the fully controlled case which uses all thyristors and no diodes (see Fig. 11-14).

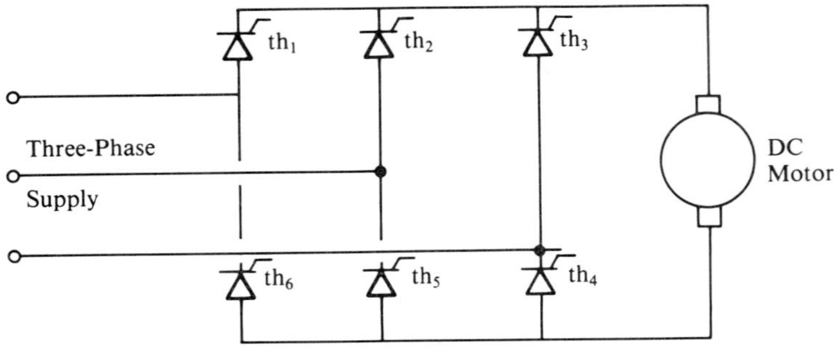

Figure 11-14. Three-phase, fully controlled rectifier.

The average (DC) output voltage provided by this three-phase, fully controlled bridge circuit is

$$v_a = \frac{3\sqrt{3}}{2\pi} V_m \cos\alpha \tag{11-10}$$

EXAMPLE 11-2

Assume the motor of Example 11-1 is controlled by a three-phase, fully controlled rectifier excited by a three-phase AC supply. Determine the range of firing angle α required to control the speed of this motor in the range 500 to 1000 rpm under constant torque. Assume the firing angle is zero for the 1000 rpm speed. Also, what must be the rms voltage of the supply?

Solution

For 1000 rpm, we require the full 220 V to be applied to the armature. Therefore,

$$220 = \frac{3\sqrt{3}}{2\pi} V_m \cos(0)$$

or

$$\frac{3\sqrt{3}}{2\pi} V_m = 220$$

This will require a three-phase supply with an amplitude of $V_m = 266$ V or an rms supply voltage of $266/\sqrt{2} = 188$ V/phase.

For 1000 rpm, the armature *IR* drop is 8 V, so the counter-emf will be 212 V. At 500 rpm, this counter-emf will be halved, or 106 V. So for 500 rpm, the output from the phase-controlled rectifier must be

$$v_a = 106 + 8 = 114 \text{ V}$$

Then, from Eq. (11-10), we find that

$$114 = 220 \cos \alpha$$

or

$$\alpha = 58.7 \text{ degrees}$$

Thus, the firing angle is confined to the range

$$0 \text{ degrees} \leq \alpha \leq 58.7 \text{ degrees}$$ ■

11-6-3 DC Supply (Chopper Control)

Our basic principle of electronic control for DC motors is to chop the supply voltage in order to provide a variable output voltage that can be applied to the DC motor armature. With an AC source, this principle is implemented using the principle of phase-controlled rectification in which thyristors are commutated naturally by the alternating nature of the ultimate supply voltage. Natural commutation is called line commutation.

This basic principle can also be used when the ultimate source of power is a DC supply. Now, however, there must be some special means for providing thyristor commutation (turn-off), because the supply is a constant voltage rather than an alternating one and cannot be depended upon as a means of commutation. In other words, we must have forced commutation in motor controls energized from DC sources. Forced commutation generally involves additional circuits which incorporate capacitors or inductors and additional switching devices (which could be thyristors). This mode of control is called chopper control rather than phase control.

A circuit, called the Jones chopper, that provides the requisite mechanism of forced commutation for a DC supply (in this case, a battery) is shown in Fig. 11-15. Analysis of this electrical circuit is beyond the scope of this book. It will be

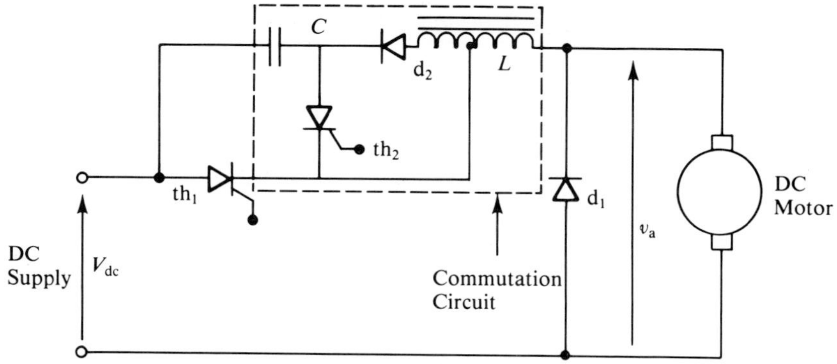

Figure 11-15. Jones chopper circuit.

sufficient for our purposes simply to note that thyristor th_1 is the device that chops the supply voltage to achieve voltage control of the motor. Thyristor th_2, diode d_2, capacitor C, and the magnetically coupled inductance L are all part of the electrical circuit required to provide commutation of th_1.

Ideally, the output of the Jones chopper would be a supply voltage such as is sketched in Fig. 11-16. If t_1 represents the on-time of the chopper circuit and t_2 represents the off-time, then voltage v_a available to be applied to the motor is

$$v_a = \beta V_{dc} \tag{11-11}$$

where

$$\beta = \frac{t_1}{t_1 + t_2} \tag{11-12}$$

Voltage V_{dc} is the battery voltage. The current trace through the motor is also sketched in Fig. 11-16. It is continued during the off-time by the motor inductance and free-wheeling diode d_1.

The chopper voltage control scheme discussed above is used with forklifts, golf carts, and other battery-powered vehicles. Circuits can also be designed to provide regenerative braking.

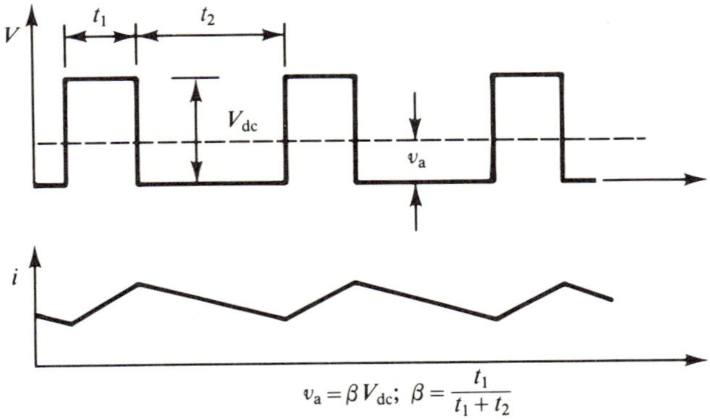

Figure 11-16. Chopper voltage and current relationships.

11-7 AC MOTOR ELECTRONIC CONTROL

We have already seen that one mode of control for induction motors is slip control, achieved by varying the resistance of a wound rotor. Although this method is simple and inexpensive, it provides a limited means of speed control and it is energy inefficient; as slip increases, so does power dissipation.

An alternative method of control is to vary the synchronous speed of the induction motor. In this fashion, we could obtain control of the rotor speed while keeping the slip, and hence efficiency, constant. Recall that the expression for synchronous speed is

$$n_s = \left(\frac{120}{p}\right) f \text{ rpm} \tag{9-3}$$

where p is the number of poles and f is the frequency of the supply that excites the stator windings.

By changing the number of poles of the motor, only a limited number of discrete changes in motor speed can be obtained. Pole changing may be satisfactory in a few applications. However, the more common speed control requirement will involve continuous speed control perhaps from zero speed up to a maximum value dictated by the torque, current, and voltage limitations of the dynamo in question. Continuous control will be possible only through the control of frequency f, as Eq. (9-3) shows.

11-8 FREQUENCY CONTROL OF INDUCTION MOTORS

The constraints of current, torque, and voltage for which a particular AC induction dynamo is rated will make it impossible to control its speed through control of supply frequency alone. As we shall see, the voltage applied to the motor will have to be controlled concomitantly with frequency. For a given dynamo, the resulting speed control characteristics will be analogous to the traction speed control characteristic we discussed for a particular DC dynamo in Section 11-5.

By assuming an idealized stator, we derived expressions in Section 9-2 for the current and torque developed in the rotor of an induction motor, namely,

$$I_2 = \frac{E_2}{\sqrt{\left(\frac{r_2}{s}\right)^2 + X_2^2}} \tag{9-10}$$

$$T_m = \frac{sE_2^2 r_2}{\omega_s[r_2^2 + (sX_2)^2]} \tag{9-17}$$

The locked rotor electromotive force E_2 is directly proportional to the magnetic flux developed by the rotating field of the stator. This, in turn, is proportional to the voltage V applied to the stator. The locked rotor inductive reactance X_2 is proportional to the frequency at which E_2 is applied to the rotor conductors. This, in turn, is proportional to synchronous speed ω_s which, from Eq. (9-3), we see is proportional to the stator frequency f.

Equation (9-10) shows, therefore, that if the voltage V supplied to the stator is held fixed, current I_2 induced in the induction motor rotor would increase as synchronous speed is reduced. This means that in the speed range below rated speed of the motor, the applied voltage must be reduced in constant proportion to the supply frequency to insure that the rotor current does not exceed its rated value as the speed of the motor is reduced.

We will define the rated speed to be the synchronous speed of the motor that produces an inductive reactance X_2 in the motor such that rated current will flow in the rotor when rated voltage is applied to the stator. We see then that when the applied voltage is reduced as the supply (control) frequency is reduced, the rotor current I_2 will be maintained constant at its rated value. This constant current provides constant magnetic flux Φ (which is proportional to E_2/ω) and constant torque in the speed control domain from zero to rated speed.

Rated speed will occur when rated voltage is applied to the motor and rated current flows in the rotor. It will not be possible to increase the motor voltage above its rated value. Consequently, speed control in the range above rated speed will necessarily be accompanied by diminishing rotor currents caused by the increasing inductive reactance as supply frequency is increased and E_2 is held fixed. The motor characteristic will then be one of constant horsepower as the torque falls off with increasing speed in the range beyond rated speed.

The resulting overall speed characteristics will be similar to the DC motor traction characteristic we discussed earlier, as illustrated in Fig. 11-17. One difference between the DC and AC motor characteristics is that in Fig. 11-17 the traction characteristic merely forms the envelope for the various torque-speed characteristic curves of the induction motor for the various synchronous speeds. The envelope shown was chosen to correspond to the breakdown torque. This is for convenience only. A similar envelope will correspond to any other unique torque of these characteristic curves, because the curves remain similar in shape to each other as synchronous speed is changed.

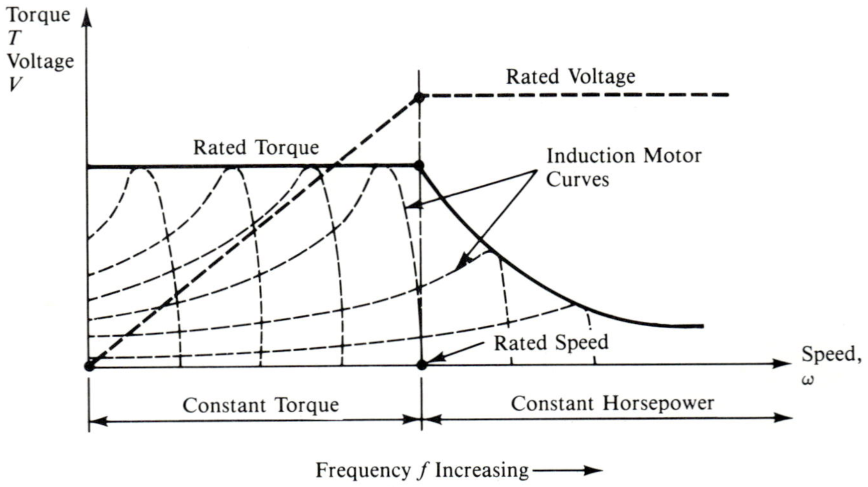

Figure 11-17. Speed control characteristics of the induction motor.

11-9 ELECTRONIC CIRCUITS FOR INDUCTION MOTOR CONTROL

The electronic circuits for the control of induction motors may involve several combinations of diode rectifiers, phase-controlled rectifiers, choppers, inverters, and triggering circuits. An inverter is simply the opposite of a converter (rectifier). Its function is to take a DC input and invert it into an AC output. We have already seen, in Section 11-6-3, that when the input is a DC source, we have no means of naturally commutating the thyristors involved. Thus, inverters require triggering circuitry for providing the requisite commutation.

A simplified diagram of a thyristor inverter is sketched in Fig. 11-18. An actual circuit would contain additional diodes and capacitors for current control and commutation purposes. The output of an inverter such as that shown in Fig. 11-18 depends on how the thyristors are triggered. Either a simple square wave or a stepped square wave can be produced (as is shown).

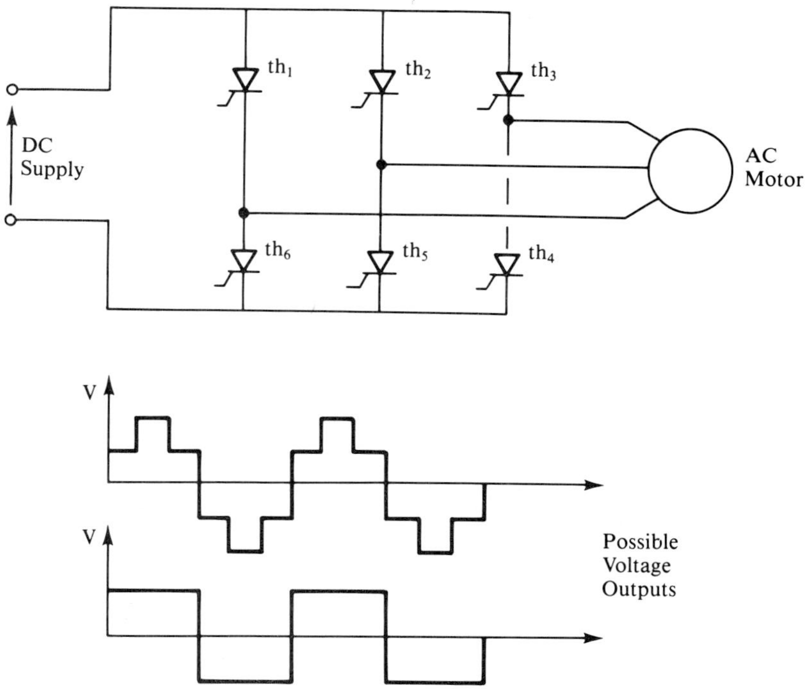

Figure 11-18. Three-phase, fully controlled inverter.

Of the various electronic circuits that have been developed for the control of induction motors, we will discuss only three: the variable input voltage and variable output voltage DC link controllers and the cycloconverter.

11-9-1 Variable Voltage Input Controller (DC Link)

One common electronic scheme for controlling an AC induction motor is to first convert the original AC supply (which is of fixed frequency) to DC. This DC power is used as the input to an inverter which changes the DC back to an AC output of controllable frequency. Then this AC output is applied to the induction motor. Since the DC power occurs at the interface between the converter and inverter, it can be viewed as linking these two electronic devices. Consequently, the controller is called a DC link controller. One such controller is sketched in Fig. 11-19.

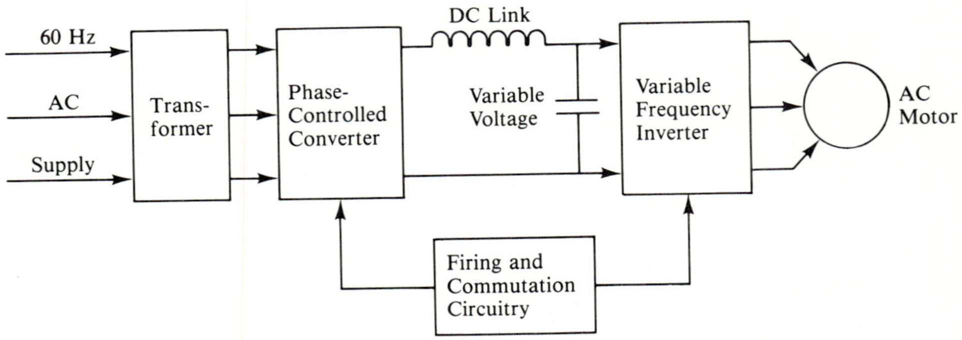

Figure 11-19. Variable voltage input DC link controller.

The DC link controller of Fig. 11-19 uses a phase-controlled rectifier as the converter. This control scheme produces a DC link of variable (controllable) voltage. Thus, the input to the inverter is a variable DC voltage, and this is where this controller gets its name. This electronic circuit makes possible the simultaneous control of AC voltage and AC frequency which will be applied to the induction motor. The actual control system is implemented by means of a complex firing and commutation circuit in addition to the converter and inverter. This control circuitry consists of an adjustable-frequency oscillator along with the logic circuitry required to distribute firing pulses and to provide forced commutation for the inverter.

11-9-2 Variable Voltage Output Controller (DC Link)

An alternative scheme for AC induction motor speed control, which eliminates the need for a phase-controlled converter, is shown in Fig. 11-20. In this DC link controller, a simple diode rectifier produces a DC link of constant voltage. The

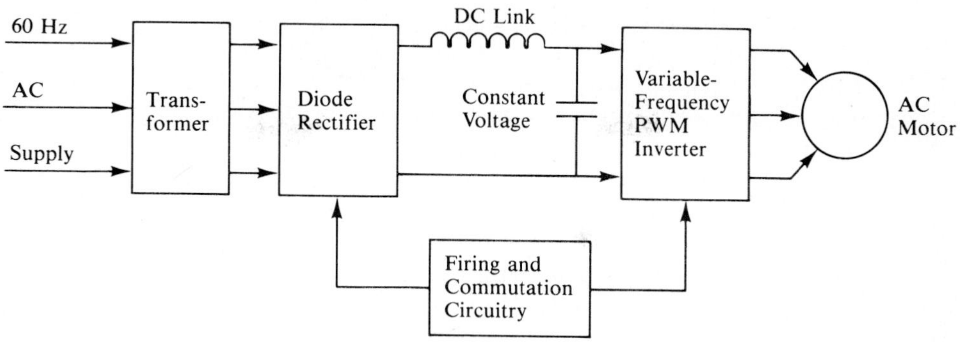

Figure 11-20. Variable voltage output DC link controller.

requisite voltage control applied to the AC motor is then obtained with the inverter rather than with the converter as in the previous controller. This is done by means of pulse width modulation (PWM). Pulse width modulation is simply a chopping process by means of which a DC source of constant voltage can be changed into a variable-voltage source in a manner similar to that discussed in Section 11-6-3. Figure 11-21 indicates how pulse width modulation can be used to generate an approximate sine wave AC voltage. Since voltage is controlled by the inverter rather than by a phase-controlled converter, this controller is called a variable voltage output controller.

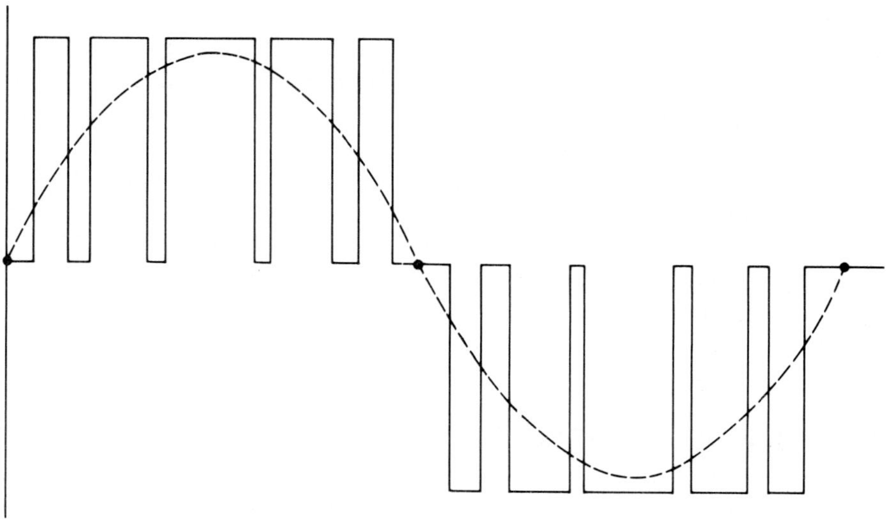

Figure 11-21. Pulse width modulation (PWM) approximation of a sine wave.

11-9-3 Cycloconverter

The cycloconverter is an electronic controller for AC induction motors in which the DC link is eliminated. Conversion of the fixed-frequency AC supply into a variable-frequency AC to be applied to the motor is implemented directly. The cycloconverter implements voltage control along with frequency control.

The mechanism of simultaneous voltage and frequency control provided by the cycloconverter is by means of thyristor circuits which operate on the same principles we elaborated previously. Basically, what the cycloconverter does is provide positive full-wave rectification of a fixed number of input frequency pulses followed by negative full-wave rectification of an equal number of input frequency pulses. At the same time, the pulses are undergoing voltage amplitude modulation in a manner analogous to pulse width modulation (see Fig. 11-22).

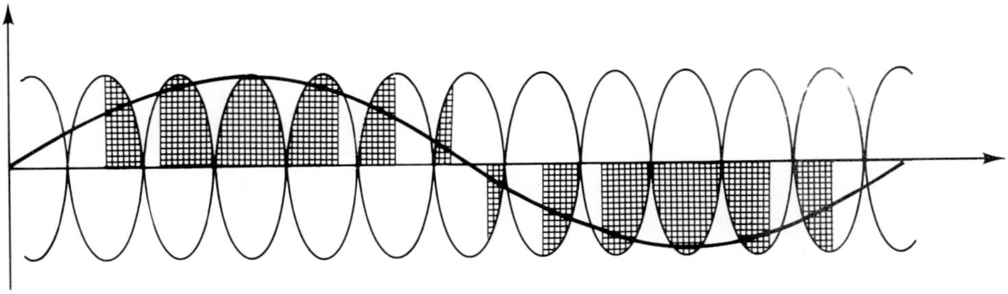

Figure 11-22. Synthesis of a low-frequency waveform from a high-frequency waveform.

The result of this process is a rather crude AC output of a frequency that is necessarily less than that of the supply frequency. To minimize distortion of the AC output produced, the output frequency can be at most only one-fourth to one-third of the supply frequency. This means that speed control will be available only in the range from zero to one-fourth or one-third of the motor rated synchronous speed. Consequently, the cycloconverter is generally limited to lower-speed motors such as might be used in steel rolling mills or in traction applications. Another disadvantage of the cycloconverter is that it requires a large number of thyristors (18) for three-phase power frequency conversion.

11-10 CLOSED-LOOP MOTOR CONTROL

All of the electronic motor control concepts discussed up to this point have been in the context of open-loop control. All that is needed to extend these controllers to operate in the closed-loop mode is a feedback link from the motor output speed back to a system controller that has the capability of modifying the firing or commutation circuitry in response to the error signal created by the feedback controller. Figure 11-23 delineates the concept in simplified block diagram form.

Of course, the blocks of Fig. 11-23 can be quite complicated individually, and their makeup will be dictated by the type of motor (AC or DC). For example, if the electronic motor controller is a cycloconverter, then the feedback controller will contain a controllable-frequency oscillator and a phase-delayed pulse generator.

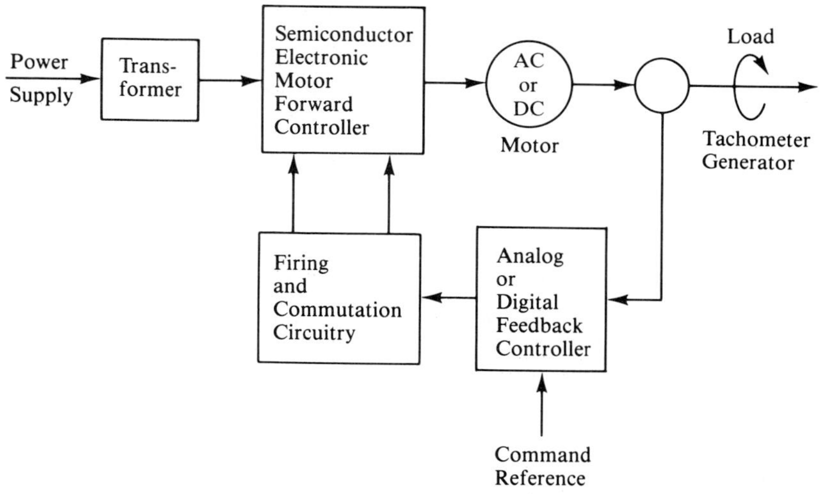

Figure 11-23. Closed-loop motor control system.

If the control is required to be highly responsive to complex speed and load programs, the feedback controller could be a microprocessor with the capability of being interfaced with other computers or instrumentation. The control problem then requires software development. Although software development can be a difficult task, the microprocessor-controlled closed-loop motor controller can eliminate many of the nagging problems of analog feedback schemes such as tuning and drift.

11-11 EDDY CURRENT COUPLING CONTROL

The phase-controlled rectifier makes a simple and convenient control method for the eddy current coupling. A schematic diagram showing how an eddy current coupling can be controlled in the closed-loop mode appears in Fig. 11-24.

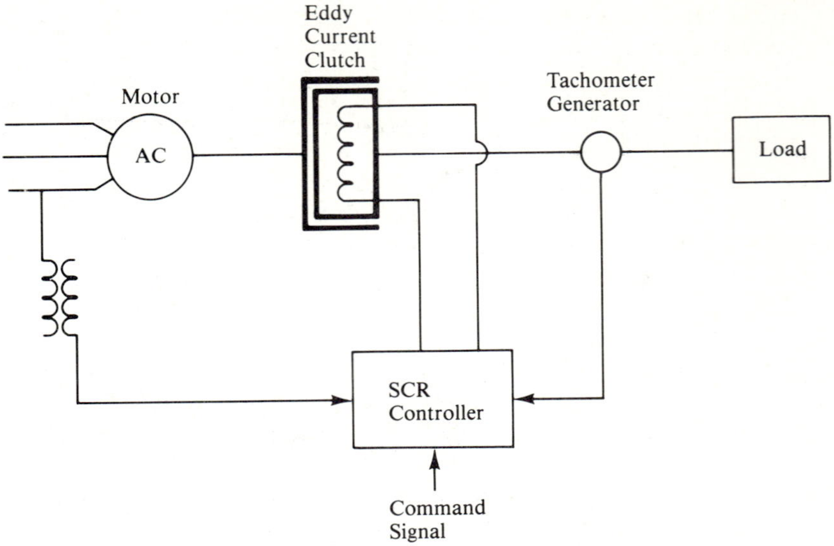

Figure 11-24. SCR closed-loop control of an eddy current clutch.

11-12 SUMMARY

The mechanical engineer concerned with incorporating an electric motor and its controller in a mechanical system will have to make decisions about (a) using a DC or an AC motor, (b) whether to use an open-loop or a closed-loop control system, and (c) whether to use a conventional controller or one of the more modern electronic controllers.

Two of the most important factors involved in making these decisions are cost and safety. The functional requirements can often be satisfied by any one of several electrical drives. With regard to cost, it must be remembered that total cost consists of capital costs plus operating costs. One incentive in using an electronic motor controller is to reduce operating costs. At the same time, this choice generally will provide a more compact drive than other alternatives.

If an AC motor is to be used as a prime mover, the decision must be made about whether to use a squirrel cage motor, a wound rotor motor, or a synchronous motor. The AC motor is generally preferred over the DC motor when:

- **(a)** An explosion-proof or otherwise totally enclosed motor is required
- **(b)** Space requirements preclude the use of the heavier and bulkier DC motor
- **(c)** An adverse environment exists that will require inordinate maintenance procedures for the DC motor
- **(d)** Very high speed is required (the cage induction motor is better able to resist centrifugal forces than is the DC motor)

TABLE 11-1. Characteristics of Electric Motors and Their Controllers.

Motor Type	Relative Motor Cost	Electronic Controller	Relative Controller Cost	Comments	Applications and Relative Power Range
DC shunt	1.0	Phase-controlled rectifier, natural commutation	1.0	High motor cost but low controller cost. Good overall controllability, efficiency, reliability, cost. Major disadvantages are maintenance costs for commutators and brushes. Most common package today.	0.25 kW–5 MW traction applications. All kinds of machinery. Pumps, blowers, fans, rolling mills, other industrial drives.
		Chopper control, forced commutation	1.2	Power transistors can be used to replace thyristors.	0.25 kW–100 kW traction applications. Battery-powered vehicles such as golf carts, lift trucks.
AC squirrel cage	0.40	DC link forced-commutation controller	3.0	Low motor cost but high controller cost. Advantage is in ruggedness, low motor maintenance, explosion-proofness. Overall cost exceeds cost of DC shunt motor package. Will become more economical in future.	1 kW–10 MW machine tools, pumps, blowers, fans, conveyors. Industrial drives in hazardous environments. Textile machinery.
		Cycloconverter	2.5	Disadvantage is that speeds up to only about 1/3 of synchronous speeds can be obtained.	Low-speed industrial and traction drives.
AC wound rotor	0.80	Naturally commutated diode rectifier	0.75	Limited speed range. Overall cost is low, but slip rings and brushes must be maintained.	50 kW–15 MW baler feed pumps, rolling mills, compressors, wind tunnels. Heavy equipment.
AC synchronous	0.60	DC link self-commutated controller	1.4	Starters for gas turbines, hydroelectric generators, wind tunnels, synchronous condensors. Large compressors, blowers, fans.	100 kW–75 MW. Good controllability, reliability. Flexible system, medium cost for large kW applications.

(e) Machinery synchronization is required (in this event, a synchronous motor may avoid the necessity for precise closed-loop control)

Additional characteristics of AC motors are tabulated in Table 11-1.

The robustness, ruggedness, and high horsepower-to-weight ratio of the induction motor will continue to produce electronic control advances for this motor, which will ultimately bring the total cost of this type of drive down to where it will compete with DC motor drives. Today, however, DC electric motor drives dominate the variable-speed drive market, mainly because they provide the following benefits.

(a) Closed-loop control is easier to implement and provides a high degree of control over speed and torque.
(b) The electronic controllers for DC motors are less complex than for AC motors and are naturally commutated.
(c) Less line pollution is caused by the DC motor controller.
(d) The overall cost of motor and controller is less than for the AC drive system.

Open-loop control systems will be adequate for stand-alone systems or for systems not otherwise mechanically coupled to other drives or by the machinery process itself (such as would be the case for machines producing a web that must be closely tension controlled). Some machine tools, crane and winch drives, feeders, presses, pumps, blowers, and fans are amenable to open-loop control.

When the mechanical system involves drives that are coupled, or where precise speed control (or position control) is required at several machine locations, or where synchronization of various parts of the machine is required, closed-loop control will be necessary. The classic example is the machine producing a web of material (which could be film, nylon, paper, rubber, or steel) in which tension must be closely controlled to avoid distortion or rupture of the web. In such a machine, the DC closed-loop-controlled drive will often be used. The most sophisticated example is the microprocessor-controlled DC drive sketched in Fig. 11-25.

In summary, the squirrel cage induction motor is the preferred electric motor because of its low cost, simplicity of design, ruggedness, high-speed potential, safety, and maintainability. However, it is encumbered with a costly, complex electronic controller to convert it into a variable-speed drive. The DC motor is inherently a variable-speed drive and can be controlled electronically rather simply. However, the design of the DC motor is complex, which limits its speed potential and requires costly maintenance. In this classic conflict, it is the DC system that prevails today. This situation will change as advances in microelectronic technology continue.

A selection table indicating the type of drive suitable for various mechanical systems is presented as Table 11-2.

The catalog from which this table was extracted also contains a procedure by which a motor controller can be selected (see Reference 11-11).

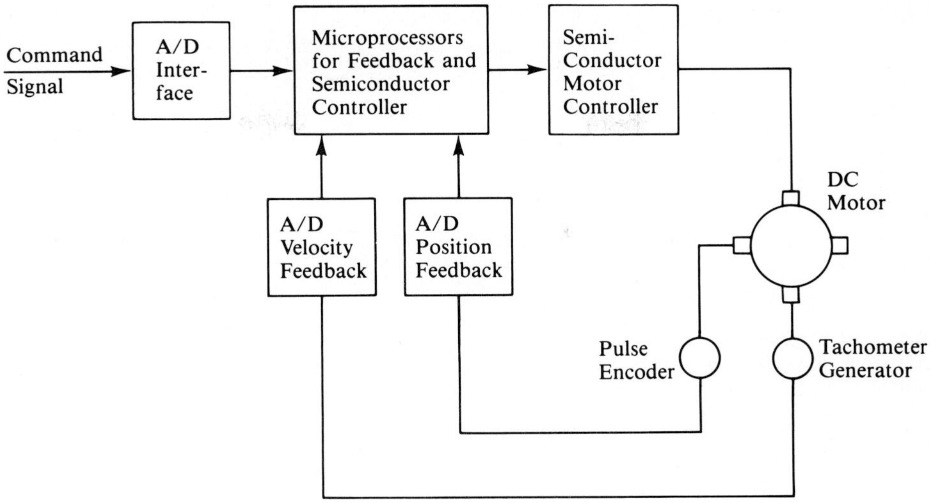

Figure 11-25. Microprocessor-controlled DC drive.

TABLE 11-2. Motor Application Selection Table. (Reprinted by permission of Reliance Electric Co.)

		Motor Speed Range*			Load Torque as Percent of Full-Load Drive Torque		
		2:1	4:1	10:1			
Name of Application	Drive Best Suited	Motor/Controller Selection Table Number			Break-away	Accel-erating	Peak Running
Actuators:							
Screw-down (rolling mills)	ES	—	—	—	200	150	125
Positioning	ES	—	—	—	150	110	100
Agitators:							
Liquid	AC	2	3	4	100	100	100
Slurry	AC	5	6	7	150	100	100
Blowers, centrifugal:							
Valve closed	AC	2	3	4	30	50	40
Valve open	AC	5	6	7	40	110	100
Blowers, positive-displacement, rotary, bypassed	AC	2	3	4	40	40	100
Calenders, textile or paper	ES	—	—	—	75	110	100
Card machines, textile	AC/DC	5	6	7	100	110	100
Centrifuges (extractors)	AC	5	6	7	40	60	125
Chippers, wood, starting empty	AC/DC	8	9	10	50	40	200
Compressors, axial-vane, loaded	AC	2	3	4	40	100	100
Compressors, reciprocating, start unloaded	AC/DC	2	3	4	100	50	100
Conveyors, belt (loaded)	AC/DC	5	6	7	150	130	100
Conveyors, drag (or apron)	AC/DC	8	9	10	175	150	100
Conveyors, screw (loaded)	AD/DC	8	9	10	200	100	100
Conveyors, Shaker-type (vibrating)	AC	5	6	7	150	150	75
Coolers, hot solids, rotary (loaded)	DC	—	—	—	175	140	100

TABLE 11-2 (continued)

Name of Application	Drive Best Suited	Motor Speed Range* 2:1	4:1	10:1	Load Torque as Percent of Full-Load Drive Torque: Break-away	Accel-erating	Peak Running
		Motor/Controller Selection Table Number					
Cranes, traveling:							
Bridge motion	ES	—	—	—	100	300	100
Trolley motion	ES	—	—	—	100	200	100
Hoist motion	ES	—	—	—	50	200	190
Draw presses (flywheel)	AC/DC	8	9	10	50	50	200
Drill presses	AC/DC	5	6	7	25	50	150
Edgers (starting unloaded)	ES	—	—	—	40	30	200
Elevators, bucket (starting loaded)	DC	—	—	—	150	175	150
Elevators, freight (loaded)	DC	—	—	—	100	125	100
Elevators, man lift	DC	—	—	—	50	125	100
Elevators, personnel (loaded)	DC	—	—	—	110	150	100
Escalators, stairways (starting unloaded)	AC	2	3	4	50	75	100
Extruders (rubber or plastic)	DC	—	—	—	150	150	100
Fans, centrifugal, ambient:							
Valve closed	AC	2	3	4	25	60	50
Valve open	AC	5	6	7	25	110	100
Fans, centrifugal, hot gases:							
Valve closed	AC	2	3	4	25	60	100
Valve open	AC	8	9	10	25	200	175
Fans, propeller, axial-flow	AC	5	6	7	40	110	100
Feeders, belt (loaded)	AC/DC	5	6	7	100	120	100
Feeders, distributing, oscillating drive	AC/DC	5	6	7	150	150	100
Feeders, screw, compacting rolls	AC/DC	5	6	7	150	100	100
Feeders, screw, filter-cake	AC/DC	5	6	7	150	100	100
Feeders, screw, dry	AC/DC	8	9	10	175	100	100
Feeders, vibrating, motor-driven	AC/DC	5	6	7	150	150	100
Frames, spinning, textile	AC	5	6	7	50	125	100
Grinders: metal	AC	2	3	4	25	50	100
Hoists, skip	ES	—	—	—	100	150	100
Indexers	DC	—	—	—	150	200	150
Ironers, laundry (mangles)	AC/DC	5	6	7	50	50	125
Jointers, woodworking	AC/DC	5	6	7	50	125	125
Kilns, rotary (loaded)	DC	—	—	—	250	125	125
Looms, textile, without clutch	DC	—	—	—	125	125	150
Machines, boring (loaded)	DC	—	—	—	150	150	100
Machines, bottling	AC/DC	5	6	7	150	50	100
Machines, buffing, automatic	AC	2	3	4	50	75	100
Machines, cinder-block, vibrating	AC	5	6	7	150	150	70
Machines, keyseating	AC	2	3	4	25	50	100
Machines, polishing	AC	2	3	4	50	75	100
Mills, flour, grinding	AC	2	3	4	50	75	100
Mills, rolling metal:							
Billet, skelp and sheet, bar	ES	—	—	—	50	30	200
Brass and copper finishing	ES	—	—	—	120	100	200
Brass and copper roughing	ES	—	—	—	40	30	200
Merchant mill trains	ES	—	—	—	50	30	200
Plate	ES	—	—	—	40	30	250
Reels, wire or strip	ES	—	—	—	100	100	100
Rod	ES	—	—	—	90	50	200
Sheet and tin (cold rolling)	ES	—	—	—	150	110	200

TABLE 11-2 (continued)

Name of Application	Drive Best Suited	Motor Speed Range* 2:1	4:1	10:1	Load Torque as Percent of Full-Load Drive Torque: Break-away	Accel-erating	Peak Running
		Motor/Controller Selection Table Number					
Strip, hot	ES	—	—	—	40	30	200
Structural and rail finishing	ES	—	—	—	40	30	200
Structural and rail roughing	ES	—	—	—	40	30	250
Tube	ES	—	—	—	50	30	200
Tube piercing and expanding	ES	—	—	—	50	30	250
Tube reeling	ES	—	—	—	50	30	200
Mills, rubber	DC	—	—	—	100	100	200
Mills, saw, band	AC/DC	8	9	10	50	75	200
Mixers, chemical	AC	8	9	10	175	75	100
Mixers, concrete	AC	2	3	4	40	50	100
Mixers, dough	AC	8	9	10	175	125	100
Mixers, liquid	AC	2	3	4	100	100	100
Mixers, sand, centrifugal	AC	2	3	4	50	100	100
Mixers, sand, screw	AC	8	9	10	175	100	100
Mixers, slurry	AC	5	6	7	150	125	100
Mixers, solids	AC	8	9	10	175	125	175
Planers, metalworking	DC	5	6	7	50	150	150
Planers, woodworking	AC/DC	5	6	7	50	125	150
Plows, conveyor, belt (ore)	DC	—	—	—	150	150	200
Positioners, indexing (machine tool)	DC	—	—	—	50	200	100
Presses, pellet (flywheel)	AC/DC	5	6	7	150	75	150
Presses, printing, production-type	DC	—	—	—	100	150	150
Presses, punch (flywheel)	AC/DC	5	6	7	150	75	100
Puller, car	ES	—	—	—	150	110	100
Pumps, adjustable-blade, vertical	AC	5	6	7	50	40	125
Pumps, centrifugal, discharge open	AC	2	3	4	40	100	100
Pumps, oil-field, flywheel	AC	8	9	10	150	200	200
Pumps, oil, lubricating	AC	5	6	7	40	150	150
Pumps, oil fuel	AC	5	6	7	40	150	150
Pumps, propeller	AC	2	3	4	40	100	100
Pumps, reciprocating, positive-displacement	AC/DC	8	9	10	175	30	175
Pumps, screw-type, started dry	AC/DC	2	3	4	75	30	100
Pumps, screw-type, primed, discharge open	AC/DC	5	6	7	150	100	100
Pumps, slurry-handling, discharge open	AC/DC	5	6	7	150	100	100
Pumps, turbine, centrifugal, deep-well	AC	2	3	4	50	100	100
Pumps, vacuum (paper-mill service)	AC	5	6	7	60	100	150
Pumps, vacuum (other applications)	AC	2	3	4	40	60	100
Pumps, vacuum, reciprocating	AC	5	6	7	150	60	150
Pumps, vane-type, positive-displacement	AC/DC	8	9	10	150	150	175
Rolls, bending	DC	—	—	—	150	150	100
Rolls, crushing (sugarcane)	AC	5	6	7	50	110	125
Rolls, flaking	AC	2	3	4	30	50	100
Sanders, woodworking, disk or belt	AC/DC	2	3	4	30	50	100
Saws, band, metalworking	AC/DC	2	3	4	30	50	100
Saws, circular, metal, cutoff	AC/DC	5	6	7	25	50	150
Saws, circular, wood, production	AC/DC	5	6	7	50	30	150
Saws, edger (see Edgers)							

TABLE 11-2 (continued)

Name of Application	Drive Best Suited	Motor Speed Range* 2:1	4:1	10:1	Load Torque as Percent of Full-Load Drive Torque: Break-away	Accel-erating	Peak Running
		Motor/Controller Selection Table Number					
Saws, gang	AC/DC	5	6	7	60	30	150
Screens, centrifugal (centrifuges)	AC	5	6	7	40	60	125
Screens, vibrating	AC	5	6	7	50	150	70
Separators, air (fan-type)	AC	2	3	4	40	100	100
Shakers, foundry or car	DC	—	—	—	50	150	70
Shears, flywheel-type	AC/DC	5	6	7	50	50	120
Shovels, dragline, hoisting motion	ES	—	—	—	50	150	100
Shovels, dragline, platform motion	ES	—	—	—	50	100	100
Shovels, large, digging motion	ES	—	—	—	50	200	200
Shovels, large, platform motion	ES	—	—	—	50	100	100
Tension-maintaining drives	DC	—	—	—	100	100	100
Textile machinery	AC/DC	5	6	7	150	100	90
Tools, machine	DC	—	—	—	100	150	100
Tools, machine, broaching, automatic	DC	—	—	—	50	150	150
Tools, machine, lathe, metal production	DC	—	—	—	50	200	200
Tools, machine, mill, boring production metal	DC	—	—	—	100	125	100
Tools, machine, milling, production	DC	—	—	—	100	100	100
Tools, machine, planer, production, metal (see Planers)							
Tools, machine, shaper, metal, automatic	DC	—	—	—	50	75	150
Vehicles, freight	DC	—	—	—	200	200	200
Vehicles, passenger	DC	—	—	—	100	400	200
Walkways, mechanized	AC	2	3	4	50	50	100
Washers, laundry	AC	2	3	4	25	75	100
Winches	DC	—	—	—	125	150	100

Note: AC = a good a-c drive application; AC/DC = an application suited for either an a-c or d-c drive; DC = normally a d-c drive application; ES = normally requires an engineered drive system

*Motor Speed Range of 2:1 is 1/2 to Full Nameplate Speed, of 4:1 is 1/4 to Full Nameplate Speed, and of 10:1 is 1/10 to Full Nameplate Speed

11-13 REFERENCES

11-1. M. G. Say and E. O. Taylor, *Direct Current Machines*, Wiley, New York, 1980.

11-2. M. G. Say, *Alternating Current Machines*, Wiley, New York, 1976.

11-3. S. A. Nasar and L. E. Unnewehr, *Electromechanics and Electric Machines*, Wiley, New York, 1983.

11-4. A. B. Carlson and D. G. Gisser, *Electrical Engineering*, Addison-Wesley, Reading, MA, 1981.

11-5. R. L. Jaeschke, *Controlling Power Transmission Systems*, Penton/IPC, Cleveland, 1978.

11-6. B. K. Bose, *A Review of AC Drives Technology*, Report No. 81CRD127, published by General Electric Co., Schenectady, NY, June, 1981.

11-7. "Electric versus Hydraulic Drives," *Proceedings of the Institution of Mechanical Engineers*, London, October, 1983; available from Society of Automotive Engineers (SAE), 400 Commonweath Drive, Warrendale, PA.

11-8. "Developments in Variable Speed Drives for Fluid Machinery," *Proceedings of the Institution of Mechanical Engineers*, London, June, 1981; available from Society of Automotive Engineers (SAE), 400 Commonweath Drive, Warrendale, PA.

11-9. *1983–1984 Power Transmission Design Handbook*, Penton/IPC, Cleveland, 1984.

11-10. M. Knebusch, "Micro-Controlled DC Drives," *Machine Design*, June 7, 1984, pp. 103–107.

11-11. *Selection and Application of AC Drives*, Application Manual D-9084-1, published by Reliance Electric Co., 24703 Euclid Avenue, Cleveland, OH, 1984.

PROBLEMS

11-1. A center tap, single-phase, full-wave rectifier for a pure resistive load is shown in Fig. P11-1. The thyristor firing angle for th_1 is $\alpha_1 = 45$ degrees, and the firing angle for th_2 is $\alpha_2 = 90$ degrees.

(a) Draw a sketch of the current in the load resistance (compared with the applied voltage) in the interval $0 \leq \theta \leq 4\pi$.

(b) Derive (or write) a relationship for the average DC voltage across the load as a function of α_1, α_2, and V_m, where V_m is the amplitude of the AC supply.

(c) What is the role of the free-wheeling diode in this circuit?

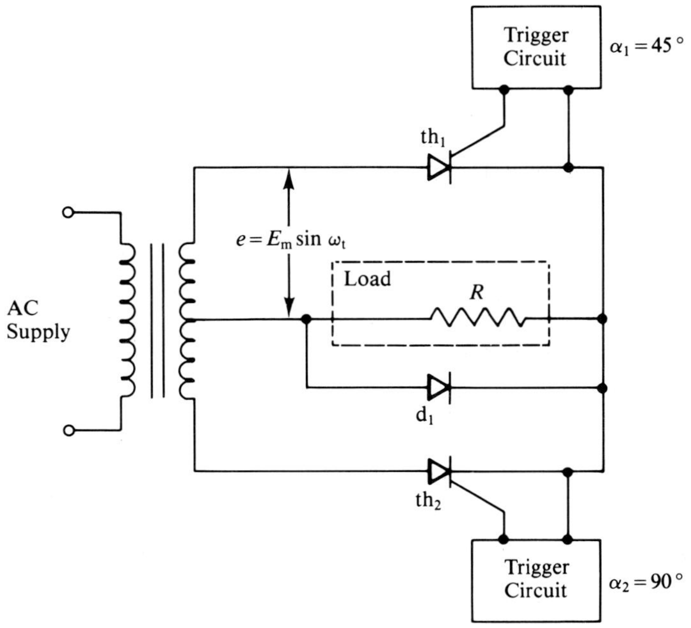

Figure P11-1. Figure for Problem 11-1.

11-2. For the circuit sketched in Fig. 11-7, define the sequence of events that occur successively from the angle $\theta = 0$. During what interval does th_1 conduct, and during what interval does d_2 conduct? Show that the average (DC) value of voltage applied to the load is

$$v_a = \frac{V_m}{2\pi}(1 + \cos\alpha)$$

11-3. Answer the following questions with either a short sentence or a numerical answer.

(a) The output from a chopper is shown in Fig. P11-3. If the applied battery voltage is 24 V, what is the chopper output voltage?

(b) What is the fundamental mode of DC electronic motor control?

(c) Why is it necessary to control voltage simultaneously with frequency in the electronic control of the AC induction motor?

(d) What is the essential difference, in terms of output, between a diode rectifier and a phase-controlled rectifier?

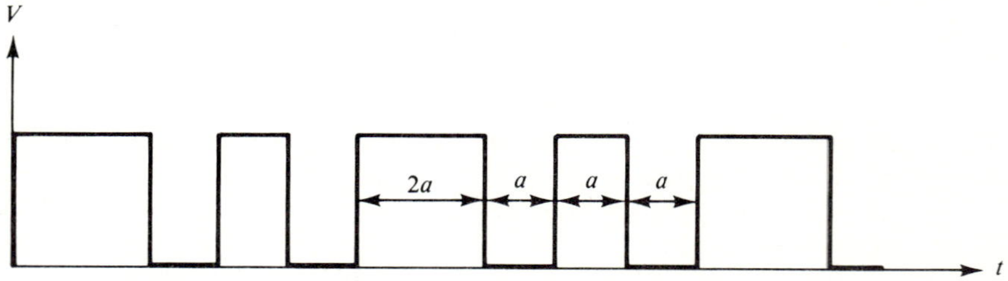

Figure P11-3. Figure for Problem 11-2.

Part D

MECHANICAL SYSTEMS CONTROL

Prior to 1950, there were practically no formal courses in mechanical engineering dealing with the control of mechanical systems. As a result, the mechanical designer was faced with some rather confusing and unrelated principles and techniques regarding control systems.

Beginning in the mid-1950s, control theory courses began to appear in mechanical engineering curricula in which the basic concepts of transfer function, feedback, and stability of closed-loop systems were developed. Today the graduating mechanical engineer is well equipped to deal with both open- and closed-loop systems of an analog nature.

Unfortunately, there remain areas of machine control that are not covered in formal control theory courses but which are, nevertheless, very important. One such area is machine sequence control. In this type of control, certain components and actuators of a mechanical system are caused to be energized in definite sequences and combinations. A good example is the injection molding machine discussed in Chapter 4.

Actually, the sequence (or on-off) type of control of machine components is a rather old art compared to the esoteric control concepts of continuous feedback systems. Nevertheless, sequence control concepts remain viable in a wide range of mechanical systems including those that incorporate electrical, pneumatic, hydraulic, and mechanical components. We have already seen, in Part B of this book, that various electromechanical devices are used in conjunction with fluid power circuits. For example, solenoid-actuated valves are often used to dictate the on-off motion of fluid actuators, and limit switches may be used, in turn, to dictate to the solenoids.

In a sequence control system, we are generally not concerned with system transients, stability, and feedback per se. Instead, focus is on causing discrete events to occur in the proper order and to be of proper duration. The fundamental engineering construct required to accomplish this is the electrical relay ladder diagram. The ladder diagram represents the circuitry and logic devices required to implement relaying, timing, counting, simultaneity, and the other functions of sequence control.

Prior to 1970, electrical ladders were hardwired and implemented using a range of electromechanical hardware devices including relays, timers, counters, limit switches, solenoids, and so forth. Recently, there has been a steady evolution toward the use of solid-state components and microprocessors in industrial control systems. Despite this, there will continue to be many applications in which traditional electromechanical devices will be used to accomplish the electrical-hydraulic-mechanical interfacing. This type of control equipment is widely known in industry (and particularly to maintenance personnel).

Even more important, for our educational purposes here, the ladder circuit concept provides a simple type of control scheme by means of which we can present a unified introduction to two of the most important developments in the history of mechanical systems control. We are referring to the microprocessor and the programmable controller.

There is little doubt that the microprocessor is the wave of the future in mechanical systems control either in a dedicated mode or as a programmable type of controller. These new and fascinating logic devices are finding increased application in all kinds of equipment and will definitely have a pronounced effect on the mission of the systems designer.

The microprocessor and programmable controller can be implemented in a vast array of both analog and digital control schemes. It will be impossible here for us to hope to expose these devices in the full breadth of their potential applications. Yet it is incumbent on those who intend to design mechanical systems to have at least a rudimentary understanding of how these controllers are implemented. We will accomplish this by using the ladder type of control circuit as a unifying theme.

12

Electrical Control Circuits

The complete specification of an electrical control system for a machine may involve the following items:

(a) Delineation of functions, events, and sequences required for the machine to operate as intended
(b) Block diagram indicating the control system logic which will include sequence, duration, simultaneity, and connectivity of events
(c) Elementary or schematic circuit diagram that implements the logic of the block diagram electrically
(d) Wiring diagram for use by manufacturing and maintenance personnel
(e) Panel layout(s) of components and devices by means of which control is to be initiated, maintained, and terminated
(f) Parts list of control system components and devices
(g) Identification of and compliance with standards for safety, drawings, and other industrial practices

Although the mechanical engineer may have responsibility for items (a) and (b), the overall responsibility for implementing the items will generally rest with an electrical engineer. It is essential that the mechanical engineer be able to understand and interpret item (c) in its particular form as a ladder diagram; this is the focus of the present chapter.

12-1 ELECTRICAL RELAYS

The key element in an electrical ladder diagram is the electrical relay. A relay is a remotely controlled and electrically operated switch containing one or more pairs of contacts that are available to effect the operation of other devices in the same, or another, electrical circuit. Use of relays in an electrical circuit makes possible the multiplication of switches, sequencing, timing, on-off control, amplification, and many other functions required to implement control system logic.

The recent development of microelectronic logic gates has permitted the implementation of the relay function by the powerful process of programming (i.e., software). We will discuss this in more detail in Chapter 14. For our present purposes, we will limit our study to relays of the electromechanical type (i.e., hardware).

An electromechanical relay is illustrated in Fig. 12-1. It consists of a spring-loaded, pivoted lever that is the armature of an electromagnet. The electromagnet is energized when voltage V_s is applied to the solenoid coil of the electromagnet. When the coil is energized, the armature is pulled toward the coil.

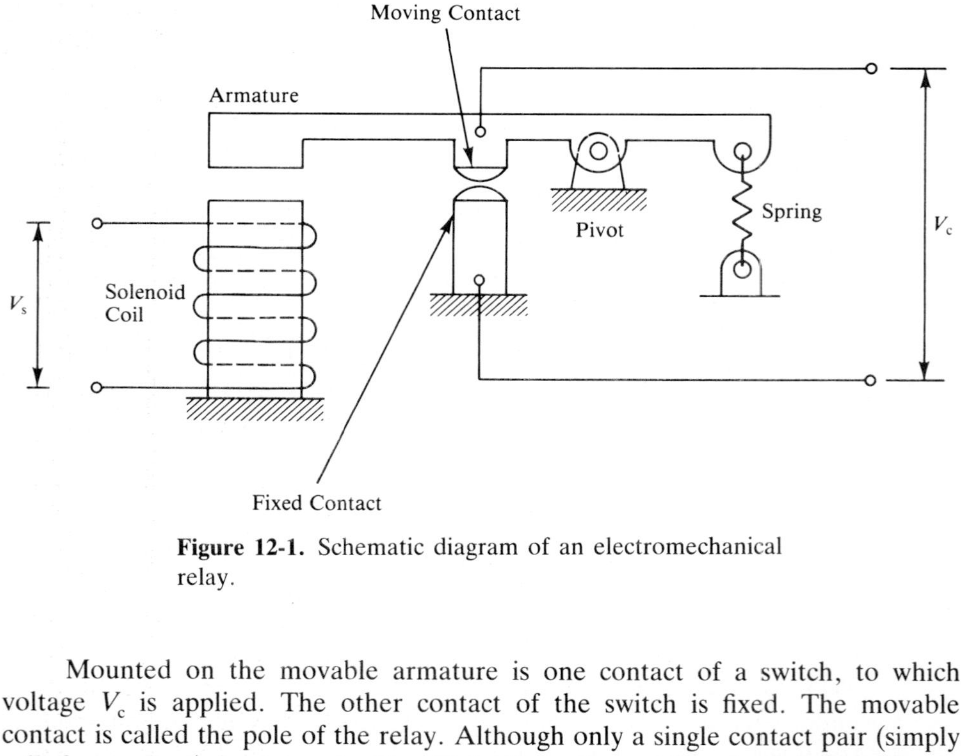

Figure 12-1. Schematic diagram of an electromechanical relay.

Mounted on the movable armature is one contact of a switch, to which voltage V_c is applied. The other contact of the switch is fixed. The movable contact is called the pole of the relay. Although only a single contact pair (simply called a contact) is shown in Fig. 12-1, the action of the single solenoid may actuate a number of contacts. Thus, the action of a single switch (used to energize the solenoid) can be multiplied by using multiple contacts on the armature.

Voltage V_s is the signal sent to the solenoid of the relay by some external device (perhaps a limit switch). This voltage is usually momentary (say, as might be provided by a spring-return pushbutton that is pressed and then released) and relatively low in magnitude (say, 110 V). Thus, some method of latching the contacts in the closed position (if they are normally open) may be necessary. Latching might be done mechanically or electrically. If it is done mechanically, the solenoid can then be de-energized.

Latching of the relay contacts can also be done electrically by dedicating one of the contacts of a multiple-contact relay to the maintenance of current in the solenoid. All that is required is to place the contact in a circuit branch in parallel with the device that originally sent signal V_s. This idea will be discussed later in more detail.

There are two major kinds of relays: power relays and control relays. In a power relay, voltage V_c might be very large, in which case the purpose of the relay is to isolate personnel from the task of closing the high-voltage contacts. To actuate the relay, personnel (or other control circuits) can be located remotely and need to be exposed only to low voltages (say, 110 V). Only the relay itself is exposed to high voltage. In this example, the relay is an amplifier.

Our interest here is not in power relays but, rather, in control relays. In a control relay, our purpose is to actuate some control operation in some remote circuit by sending a signal to the relay. As an example, if the cam on a piston actuates a limit switch, the limit switch will send a signal to a relay which will cause a contact to close in some other circuit which might be used, say, for actuating a solenoid-operated valve for reversing the direction of the piston.

12-1-1 Relay Types and Symbols

Relays are manufactured in a diverse variety of types. The number of contacts and their arrangement will be determined by the application. A treatise on the types of relays, their terminology and classes of service, and on principles and applications, has been prepared by the National Association of Relay Manufacturers (Reference 12-1). This treatise delineates twenty different kinds of relay configurations. Four of the more important forms are shown in Fig. 12-2.

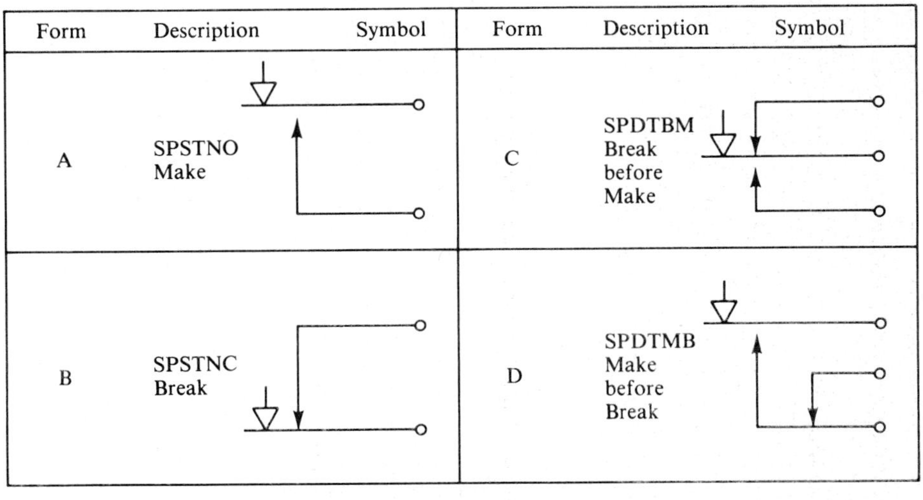

Figure 12-2. Four common forms of relays.

Form A is a single-pole, single-throw, normally open (SPSTNO) relay. For this relay, the signal voltage V_s will result in the closing of a single, normally open, contact. Form C is a single-pole, double-throw, break-make (SPDTBM) relay. Here, the signal voltage V_s opens one normally closed contact and closes one normally open contact. The moving contact is called the pole of the relay.

In selecting a relay, such things as coil current, operate—transfer–bounce–and release times, may be important practical considerations. Here, however, we are interested only in the logical function provided by the relay as a component in an electrical ladder diagram. For this purpose, the symbolism sketched in Fig. 12-3 will be used.

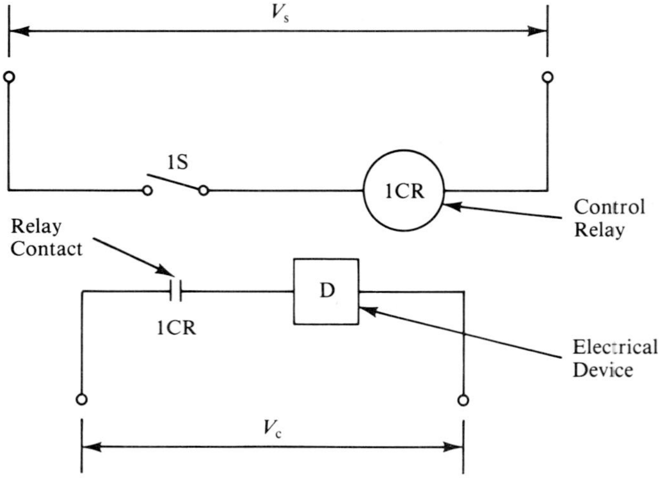

Figure 12-3. Symbolic representation of a relay.

The control relay solenoid actuator is symbolized by 1CR within a circle. It is energized by voltage V_s via mechanical switch 1S. The relay contact ⊣⊢ is symbolized as 1CR which means that it is a contact belonging to the relay 1CR. This contact (shown as normally open) is simply another switch which closes a circuit having supply voltage V_c when 1CR is energized. Although the signal and contact voltages are shown as separate sources in Fig. 12-3, they usually have the same source, in which case we would call the circuit a ladder circuit. An example is shown in Fig. 12-4 where switch 1S has been replaced by a contact from some other relay.

Figure 12-4 shows why the relay gets its name. By energizing the relay (not shown), which has a normally open contact 1CR (shown) in ladder branch 1, we ultimately relay this signal to ladder branch 2 via control relay 2CR. When it is energized, ladder branch 2 becomes a closed circuit through the device labeled D. Device D could be a signal lamp, a solenoid, a timer, another control relay, or a similar electrical component.

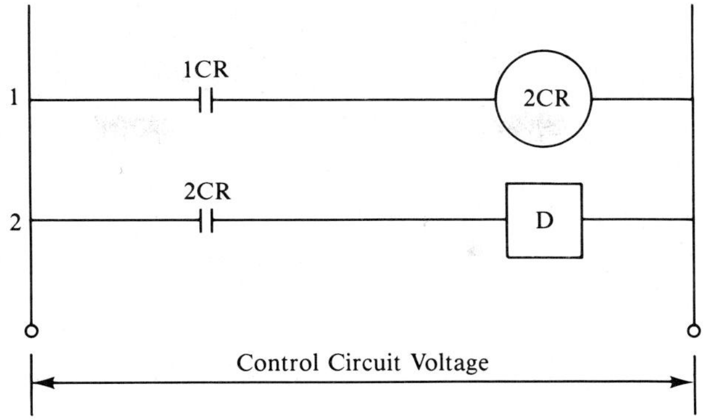

Figure 12-4. Relaying a signal.

12-1-2 The Motor Starter

An excellent illustration of the use of control relays and their schematics is provided by the pushbutton motor control circuit. Keep in mind that although it may be acceptable to simply close a mechanical switch in order to start a low-voltage, fractional horsepower motor, this will not be done for larger motors of the higher voltages. The voltage available for control circuitry will usually be single-phase, 110/120 V, while most industrial motors will be three-phase, 220/440 V, or higher. Relays will be used to start these larger motors.

An elementary ladder diagram illustrating how a motor starter works is presented in Fig. 12-5. Here we show a three-phase induction motor of relatively high voltage and a single-phase control circuit obtained by tapping one of the motor supply phases through a stepdown transformer.

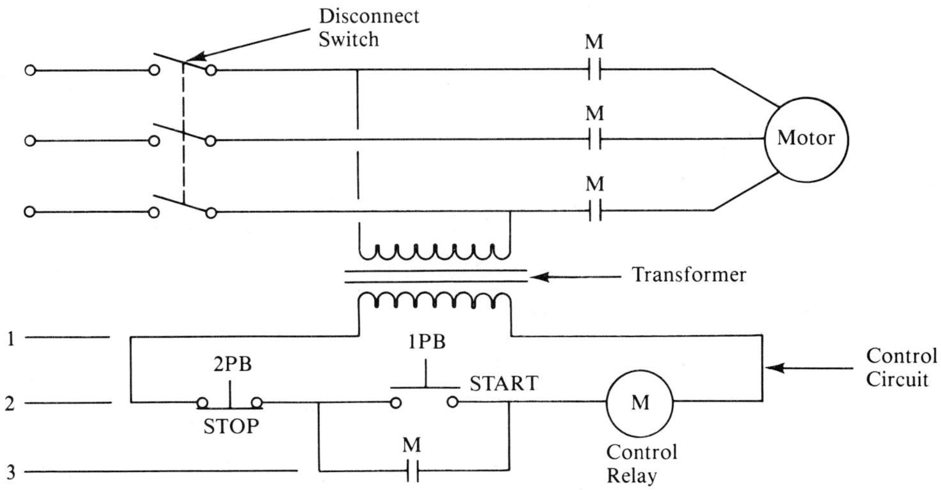

Figure 12-5. Motor starter.

The motor control relay is designated as Ⓜ in the second branch of the control circuit. It has four normally open contacts designated with the symbol ⊣⊢. Three of these contacts are in the three lines of the three-phase power circuit for the motor. The other contact is placed in parallel with pushbutton 1PB.

Also shown are mechanical knife switches which serve as the main disconnect for the power circuit. The intention is that these knife switches will never be operated unless the circuit is open (which will be true when the relay contacts are open). Thus, the operator of these mechanical switches will never be subjected to electrical arcing at the switches.

We assume that all circuits are de-energized. To start the motor, the main disconnect knife switches are closed with the power circuit open. This action will provide power to the control circuit via the transformer but not to the motor, because all contacts are open. We assume the STOP pushbutton is closed and pushbutton 1PB is open.

Now pushbutton 1PB is pressed, which momentarily energizes control relay Ⓜ, causing all four contacts ⊣⊢ to close. The contact in parallel with 1PB closes and latches the control relay so that the relay will remain energized even after pushbutton 1PB is released. This contact serves as the electrical latch we discussed earlier. At the same time, all three of the contacts in the three power lines going to the motor are closed, which energizes the motor, which now accelerates to its steady-state rpm.

To stop the motor, the normally closed STOP pushbutton is opened. This action momentarily de-energizes control relay Ⓜ so that all of its contacts open. The contacts in the power lines open and the motor is disconnected from its power supply. The contact in parallel with 1PB opens so that the control circuit is unlatched and will remain open even if the STOP button is released. Now the knife switches can be opened to isolate the whole system from the power supply.

A real motor starter includes additional circuit elements including fuses, thermal overload switches, and an indicator lamp. All of these devices are incorporated into a single enclosure chassis called a motor starter. They can be purchased as off-the-shelf items.

12-1-3 Time Delay Relays

In mechanical systems that are sequence controlled, it is often necessary to have a controllable time delay between consecutive machine events. In the injection molding machine, for example, the process of curing the molded part may require a certain number of seconds. The time delay must be designed into the control system for closing and opening the mold halves.

There are two ways of implementing time delays in an electrical ladder circuit. The simplest way is to use a time delay relay such as that indicated schematically in Fig. 12-6. Time delay relay 2TR is located in branch 1; it is energized by contact 1CR of a control relay (not shown). Relay 2TR has two types of contacts: 2TR-1 and 2TR-2, shown in branches 2 and 3 respectively. Contact 2TR-1 is just like an ordinary control relay contact; it closes instantaneously after relay 2TR is energized. Contact 2TR-2 is a time delay contact that is

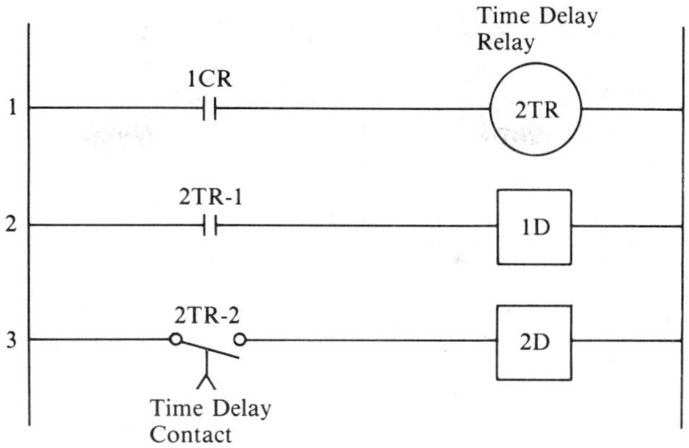

Figure 12-6. Time delay relay and its contacts.

normally open but that closes only after a time delay once 2TR is energized. Consequently, when contact 1CR closes, there will be a prescribed time delay between the times at which devices 1D and 2D are energized. The time delay relay may have a variety of contacts assigned to it as are illustrated subsequently in Table 12-1.

EXAMPLE 12-1

A system having two motors is to be controlled in such a way that the second motor is started when the first one is stopped. The second motor is to run for only a fixed period of time because it operates a cooling fan. Design an electrical ladder diagram to accomplish these requirements using a time delay relay 1TR that has a normally closed instantaneous contact and a normally open, timed-open (upon de-energization) contact. The motor control relays are designated as 1M and 2M.

The circuit is shown in Fig. 12-7. In this circuit, notice that when STOP botton 2PB is pushed to stop motor 1M, the instantaneous contact 1TR-1 will close. The

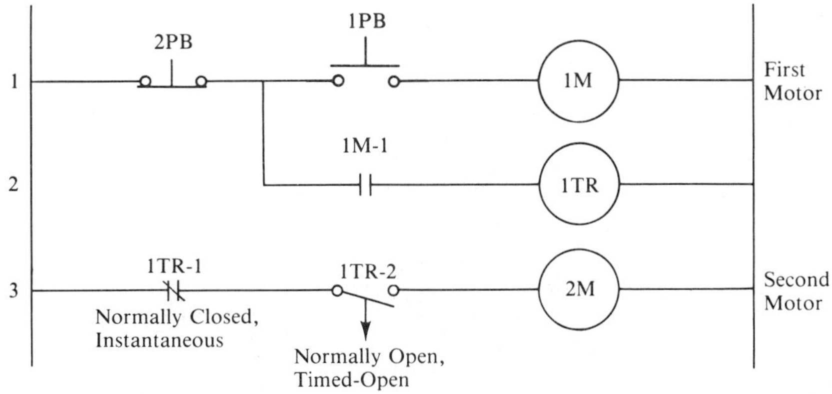

Figure 12-7. Circuit diagram for Example 12-1.

normally open contact 1TR-2 will have been closed prior to 2PB being pushed; therefore motor 2M will start when 2PB is pushed. After a designated time delay, contact 1TR-2 will open and motor 2M will stop.

After motor 2M stops, the circuit has been restored to its initial condition so that the cycle can be repeated. ■

12-1-4 Timers and Counters

An alternative method of implementing a timing function is to use a timer. Whereas a time delay relay is a solenoid-armature type of electromechanical device, similar to Fig. 12-1, a timer is a device driven by a small synchronous motor by means of a clutch. Timers of the reset type generally have a set of three contacts. Each contact is affected by three conditions: (a) reset, (b) timing, and (c) timed-out. These conditions are represented symbolically by three small circles above each contact as shown in Fig. 12-8.

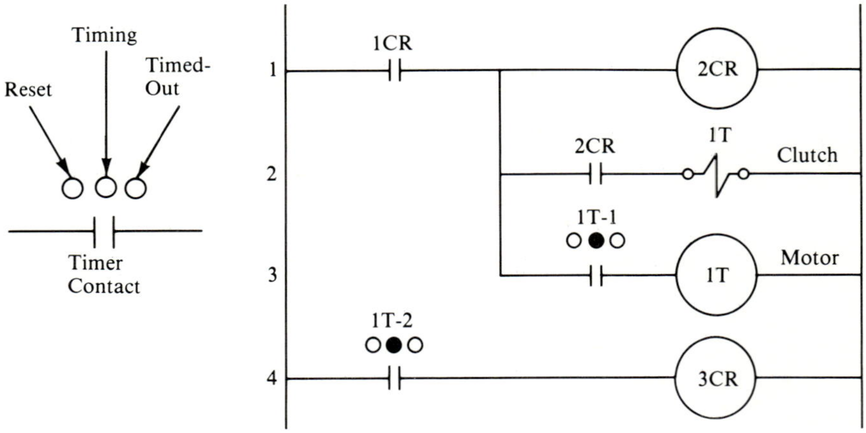

Figure 12-8. Timer and its contacts.

An open circle indicates that the contact is open when subject to the condition represented by the circle. A solid circle means the contact is closed. In branch 4 of Fig. 12-8, for example, timer contact 1T-2 is indicated as being open in the reset condition, closed in the timing condition, and open in the timed-out condition.

To understand how a timer and its contacts work, it is important to remember that:

(a) The contact changes from its reset condition to the timing condition when the clutch is energized

(b) The contact changes from the timing condition to the timed-out condition after the motor has driven the timer for the pre-set (set by hand) amount of time

(c) The contact changes back to the reset condition when the clutch is de-energized

The timer motor and clutch must be indicated on the circuit diagram as shown in Fig. 12-8. In this circuit, control relay 2CR is energized by contact 1CR. In turn, the timer clutch is energized by contact 2CR. Timer contact 1T-1 changes to the timing condition after the timer clutch is energized which, since the contact is closed (see the solid circle above 1T-1), energizes the timer motor. Control relay 3CR is also energized by contact 1T-2. After the timer motor runs for the pre-set time, both contacts 1T-1 and 1T-2 will open when they change from the timing to the timed-out condition. This will de-energize the timer motor and control relay 3CR. Finally, if contact 1CR is opened, the timer clutch will be de-energized which will cause both contacts 1T-1 and 1T-2 to change to their reset conditions. Note that these contacts are open under the reset condition.

Counters also have a motor and a clutch to drive them. They are also pre-set by hand. Symbolically, a counter is represented in a manner similar to a timer. The three circles above a counter contact represent (a) reset, (b) counting, and (c) counted-out.

12-2 ELECTRICAL LADDER DIAGRAMS

Consider the simple circuit shown in Fig. 12-9. This circuit contains five devices: two pushbuttons, two control relays, and one solenoid. Three contacts are shown, all of which belong to control relay 1CR. Two of these contacts are normally open while the third is normally closed. Contacts belonging to control relay 2CR are not shown. Presumably, they reside in some other circuit not shown.

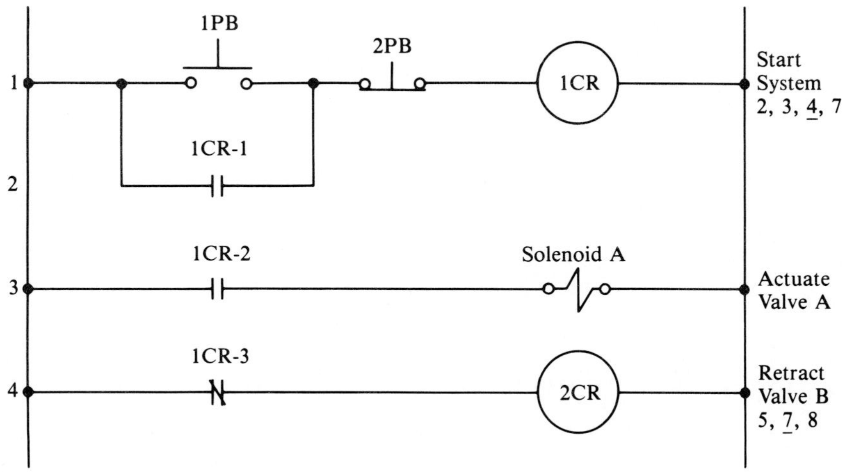

Figure 12-9. Simple electrical ladder diagram.

Figure 12-9 is called a ladder diagram because the same voltage source is used for the three main branches of the circuit. The diagram therefore resembles a ladder with the voltage supply lines being the ladder rails and the branches being the ladder rungs.

Electrical ladder diagrams are always drawn showing switches and relay contacts in their normal positions when the circuit is de-energized. Contacts controlled by a relay are always assumed to be in their normal positions when the relay is de-energized. One exception to these rules occurs when a switch, though normally open, is being held closed when the circuit is de-energized. We will show an example of this later. In Fig. 12-9, for example, relay 1CR is assumed to be de-energized because pushbutton 1PB is normally open and is not being held closed. This means that contacts 1CR in branches 2 and 3 are open (because they are normally open) and contact 1CR in branch 4 is closed (because it is normally closed). Note that relay 2CR is energized when relay 1CR is de-energized.

When pushbutton 1PB is pushed, relay 1CR is energized and the normally open contacts 1CR in branches 2 and 3 will close and the normally closed contact 1CR in branch 4 will open. So the result of pressing 1PB is:

(a) To close contact 1CR-1 in branch 2, which will keep relay 1CR energized even after 1PB is released (an electrical latch)
(b) To close contact 1CR-2 in branch 3, which energizes solenoid A
(c) To open contact 1CR-3 in branch 4, which de-energizes relay 2CR

Thus, the net result of pressing 1PB is to energize solenoid A and de-energize relay 2CR.

Next, suppose normally closed pushbutton 2PB is pressed. This will open circuit branch 1, which de-energizes relay 1CR. The result will now be:

(a) To open contact 1CR-1 in branch 2, which de-energizes (unlatches) relay 1CR
(b) To open contact 1CR-2 in branch 3, which de-energizes solenoid A
(c) To close contact 1CR-3 in branch 4, which energizes relay 2CR

Note that each branch of a ladder diagram is assigned a numerical value on the left side of the diagram (including sub-branches) for reference purposes. Notes appear at the right side of the diagram that give an indication of what the output device in the branch will accomplish. These notes help the reader to interpret the overall circuit logic. The numbers at the right of branch 1 indicate the branches in which contacts belonging to the relays are located. The underline denotes a normally closed contact.

12-2-1 Ladder Circuit Devices and Symbols

We can classify the various devices incorporated into electrical ladder circuits into three broad classes: input devices, output devices, and intermediate devices. Input devices usually are switches of a variety of types. As a general rule, they are located on the left side of the ladder diagram. Relay contacts can be considered as input devices in branch circuits even though they are, in another sense, intermediate devices. Symbols of the more common input devices used in mechanical system controls are given in Table 12-1.

TABLE 12-1. Input Devices.

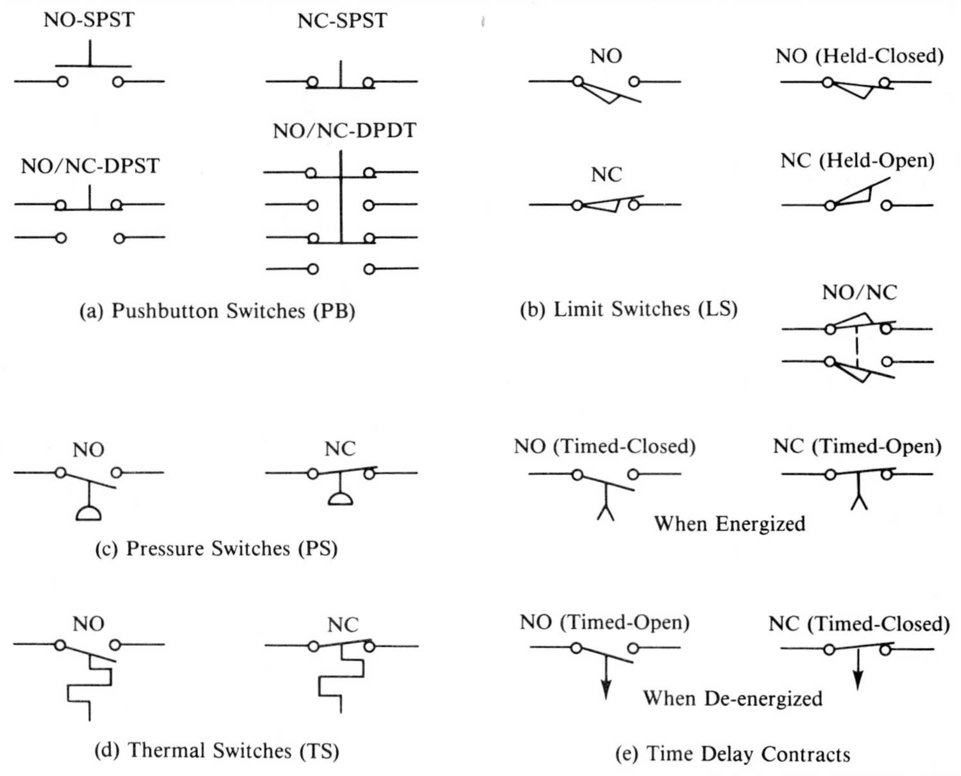

Output devices are generally actuators of a variety of types. They should always be shown at the right side of the ladder diagram. Common actuators are solenoids, motors, and control relays. Symbols used to designate output devices are presented in Table 12-2.

12-2-2 Ladder Circuit Design

There are pitfalls to be avoided by the student or neophyte circuit designer in constructing ladder circuits. Although we cannot hope to exhaust all of these potential errors, several are discussed below.

As a general rule, it is good practice to include only one output device in a major ladder branch. In particular, output devices should not be placed in series

TABLE 12-2. Output Devices.

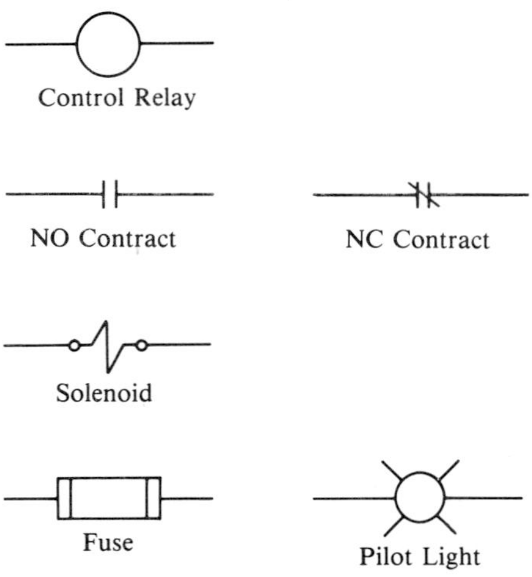

in a branch! The reason is that failure of one of the devices will compromise the branch. In Fig. 12-10, three different methods of energizing two control relays with a single switch are shown. In some cases, two output devices in parallel, as shown in Fig. 12-10(b), will be acceptable, particularly if one of them is an indicator lamp.

Another potential problem in ladder circuits is called relay racing. An illustration is shown in Fig. 12-11. In this circuit, an attempt is being made to latch control relay 2CR in an energized condition by pressing pushbutton 1PB.

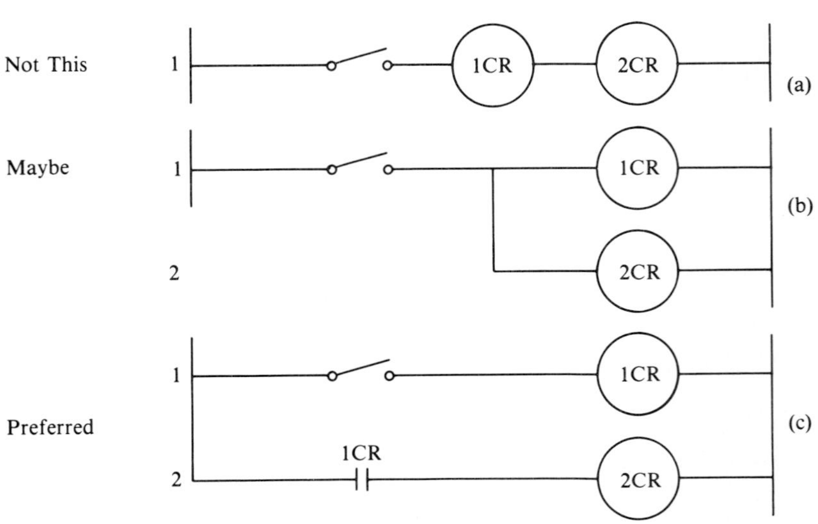

Figure 12-10. Ways to energize control relays.

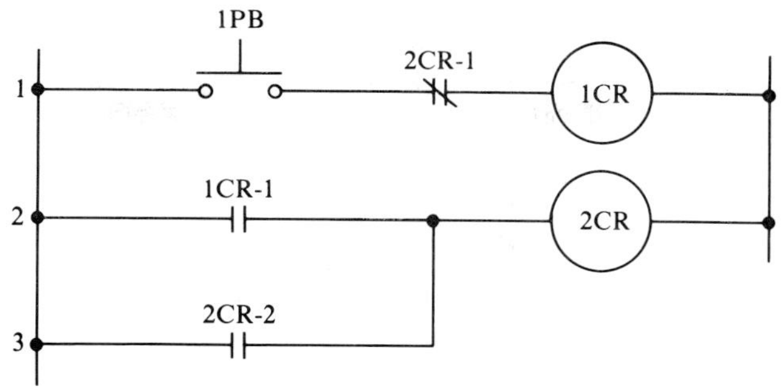

Figure 12-11. Problem of relay racing.

However, a little reflection about this circuit will show that unless contact 2CR-2 can close faster than contact 2CR-1 can open, the latching process will fail. The circuit can be made to work reliably only if contact 2CR-1 can be made slower than contact 2CR-2. This could be accomplished by replacing relay 2CR with a time delay relay.

Electrical devices should not be placed in the vertical legs of the ladder diagram. One exception to this rule occurs when a contact of a master relay is used to isolate one whole section of a ladder diagram as shown in Fig. 12-12.

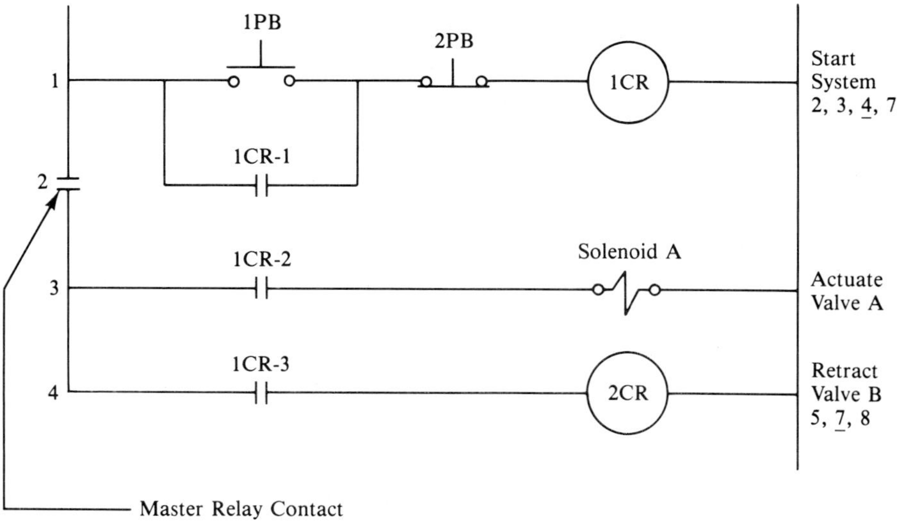

Figure 12-12. Contact of a master relay.

12-2-3 Analysis of a Ladder Diagram

In the introductory section of Part B of this book, we discussed the fluid power and electrical ladder circuits required to actuate two hydraulic cylinders sequen-

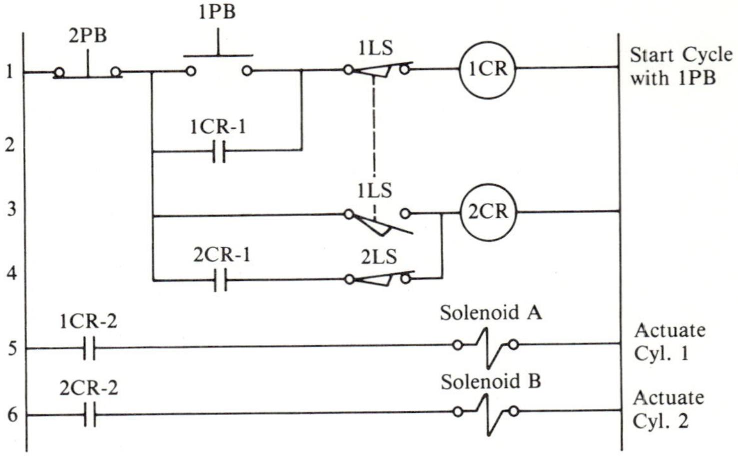

Figure 12-13. Ladder diagram for control of two cylinders.

tially (see Fig. 4-1). The details about how the ladder circuit works were deferred at that point but will now be discussed.

The ladder diagram is reproduced as Fig. 12-13. The circuit works as follows:

(a) Pressing the START button 1PB energizes control relay 1CR. This causes the normally open contacts 1CR-1 and 1CR-2 to close in ladder branches 2 and 5 respectively. Closing contact 1CR-1 in branch 2 latches relay 1CR so pushbutton 1PB can be released. Closing contact 1CR-2 in branch 5 causes solenoid A to be energized. This causes the four-way directional control valve (of Fig. 4-1) to port oil to the blank-end of cylinder 1, which moves its piston to the right.

(b) Eventually, the cam on cylinder 1 (Fig. 4-1) will actuate the double-pole (NO/NC) limit switch 1LS. Note how the two contacts of this limit switch, located in branches 1 and 3, are indicated as tied together by the dashed line. The dash indicates that the normally open contact (in branch 3) will close at the same time the normally closed contact (in branch 1) will open. Thus, actuation of 1LS does two things. First, control relay 1CR is de-energized which causes the directional control valve for cylinder 1 to reverse the direction of its piston (i.e., solenoid A is de-energized). Cylinder 1 now has its piston moving leftward. At the same time, the normally open contact of 1LS closes and energizes control relay 2CR. This closes the normally open contacts 2CR-1 and 2CR-2 in branches 4 and 6 respectively. Closing 2CR-1 simply latches relay 2CR in place, while closing 2CR-2 energizes solenoid B. The directional control valve for cylinder 2 now ports oil into the blank-end of cylinder 2, moving its piston to the left.

(c) Eventually, the piston of cylinder 2 will actuate normally closed limit switch 2LS and open it. This de-energizes relay 2CR which, in turn, releases solenoid B of the directional control valve for cylinder 2. The piston of cylinder 2 will now reverse and move to the right. In the meantime, the leftward-moving piston of cylinder 1 will have reached the top of its stroke and will be stalled against its cylinder head.

(d) The rightward-moving piston of cylinder 2 will eventually reach the top of its stroke, where it too will stall against its cylinder head. At this time, all of the pump outlet will be bypassed through the relief valve. At this point, either the pump can be turned off or the cycle can be repeated by again pressing pushbutton 1PB. The function of normally closed pushbutton 2PB is to de-activate the circuit for safety purposes.

EXAMPLE 12-2

A hydraulically operated milling machine is to be used to cut hexagon heads on some large bolts. The part to be machined is mounted in a chuck which is mounted on a vertical slide reciprocated by a hydraulic cylinder. As the part is reciprocated vertically, it feeds past a revolving end mill which does the machining (see Fig. 12-14).

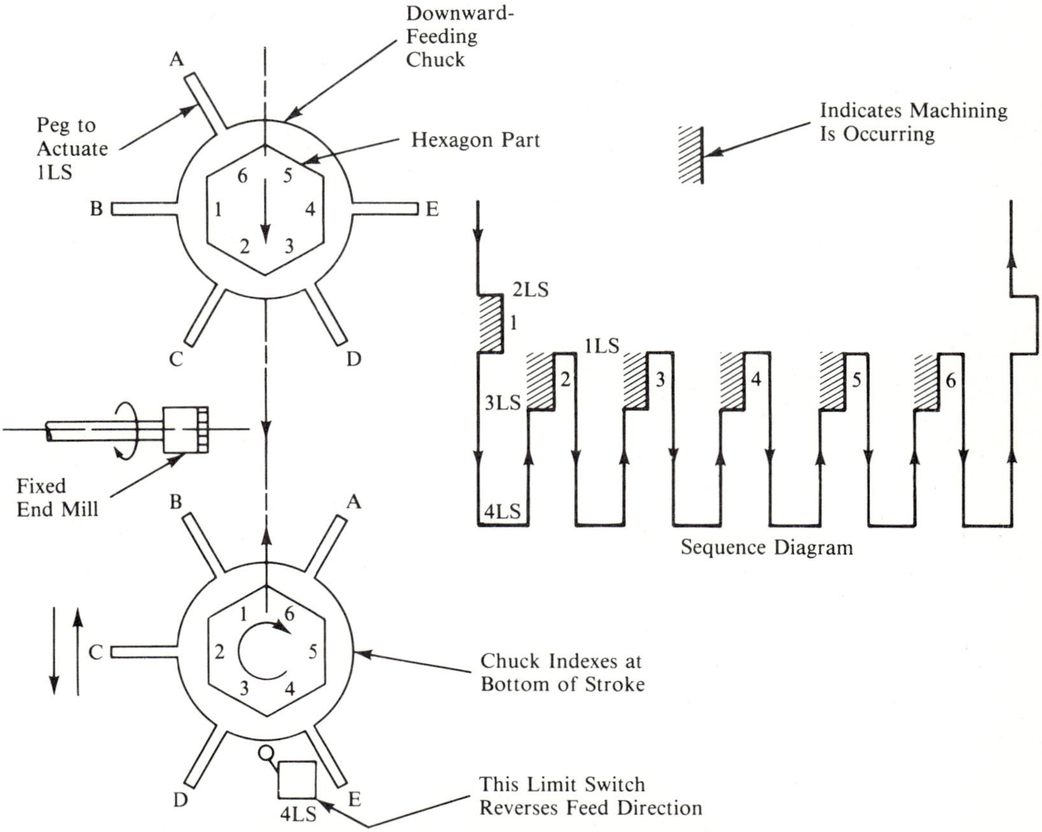

Figure 12-14. Schematic diagram for Example 12-2.

As the machining cycle is initiated, the slide-mounted chuck will be caused to feed downward, and side 1 of the hexagon will be machined on the downstroke. After passing the end mill on the downstroke, the chuck will eventually trip a limit switch 4LS which will reverse the direction of the chuck so that it will start to feed upward. At the same time, the chuck is caused to index mechanically by 60 degrees (cw), so now side 2 will be exposed to the end mill on the upstroke.

At the top of the stroke, a peg (labeled A) will actuate a limit switch 1LS which will stop the upward feed and cause the piston to reverse direction. But, this time, no machining will occur on the downstroke. The slide will go all the way to the bottom where it will index mechanically so that side 3 will be exposed to the end mill on the upstroke. The process will continue until all sides are machined and will eventually stop with the slide stopped at the top of the stroke because peg F has been omitted from the chuck and cannot, therefore, actuate limit switch 1LS. A diagram of the cutting strokes is shown in Fig. 12-14, and the hydraulic circuit is shown in Fig. 12-15.

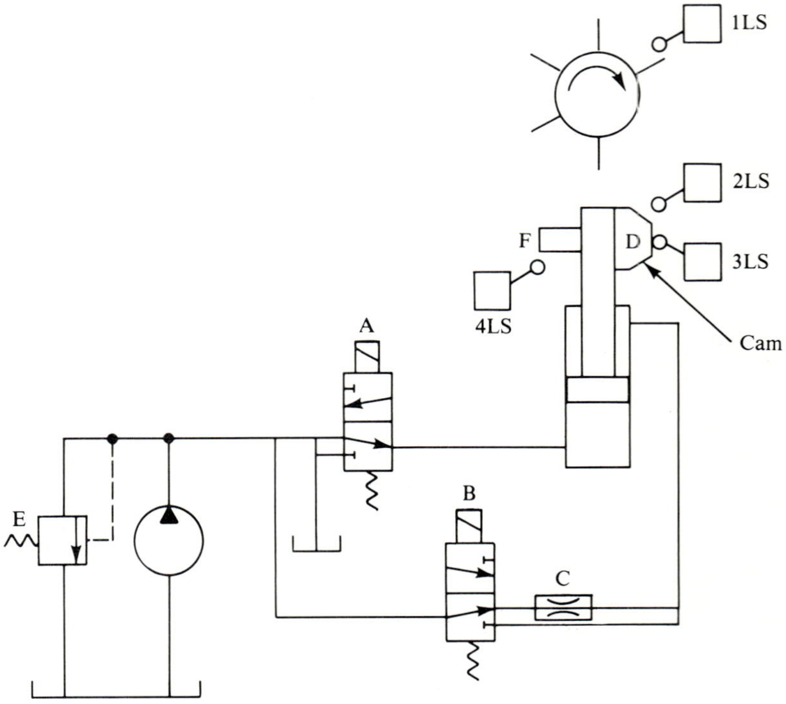

Figure 12-15. Hydraulic circuit.

The hydraulic circuit is regenerative, so the feed rates of the piston on the upstroke will be the same as on the downstroke. Note that a metering valve C is placed in the circuit, or bypassed, depending on the position of three-way valve B. The relief valve E has been set such that whenever metering valve C is in the circuit, some of the pump output will be bypassed over relief valve E.

When solenoid B is not actuated, valve B is in its spring-return position which places metering valve C in the circuit. If solenoid B is actuated, then the metering valve is bypassed.

Three-way valve A is the directional control valve that controls the direction of the piston. In the spring-return position, the pump output is directed into the blank-end of the cylinder and the piston moves upward. When solenoid A is energized, the valve will port the blank-end to the reservoir and the piston will be caused to descend.

The system incorporates four limit switches actuated by various cams and levers. The length of cam D is such that the cam will engage limit switch 2LS or 3LS for a length of time suitable for machining one side of the hexagon. Limit switches 1LS and 4LS are for reversing the piston direction.

The electrical circuit is shown in Fig. 12-16. This circuit incorporates nine main components: pushbutton 1PB, control relays 1CR and 2CR, solenoids A and B, normally closed limit switches 2LS, 3LS, and 4LS, and normally open limit switch 1LS. All of the relay contacts are normally open except the normally closed contact in branch 7 of the ladder diagram.

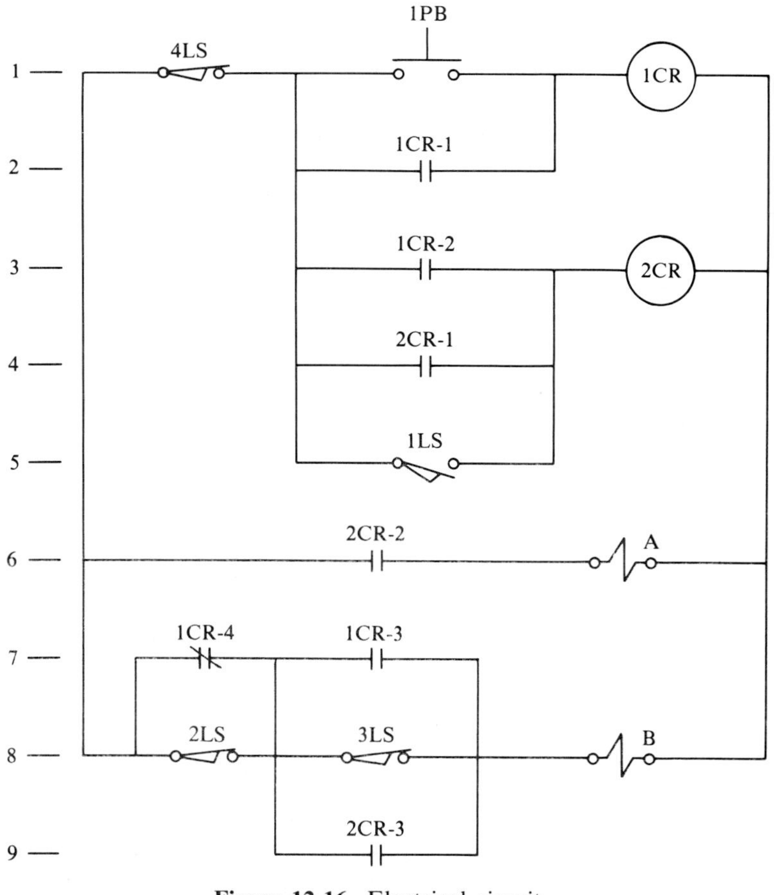

Figure 12-16. Electrical circuit.

The electrohydraulic circuit works as follows:

(a) Pressing pushbotton 1PB energizes control relay 1CR. This causes normally open contacts to close in branches 2, 3, and 7, and the normally closed contact in branch 7 opens. The contact in branch 2 latches 1CR, and the contact in branch 3 energizes control relay 2CR. Energizing 2CR causes normally open contacts to close in branches 4, 6, and 9. The contact in branch 4 latches 2CR in place unless limit switch 4LS opens. The contact in branch 6 energizes solenoid A which actuates the three-way valve labeled A. The contact in branch 9 energizes solenoid B which actuates the three-way valve labeled B. Note that the normally open contact of 1CR in branch 7 is also closed, which would actuate solenoid B even if the contact in branch 9 was open.

(b) As the piston feeds downward at maximum speed, cam D actuates normally closed limit switch 2LS and opens it. This de-energizes solenoid B (remember, the normally closed contact in branch 7 is open), which shifts three-way valve B to its spring-return position. This causes the flow into the rod-end of the cylinder to be metered through metering valve C, which slows the descent of the piston, while cam D remains in contact with 2LS. Thus, machining of the first side of the hexagon occurs during downfeed at the metered rate.

(c) After machining of the first side of the hexagon is completed, limit switch 2LS loses contact with cam D and closes, which again actuates solenoid B. This will cause the downward feed rate to be resumed at its maximum value. Even though cam D actuates limit switch 3LS during this time, the actuation cannot open branch 8 of the circuit because the contacts in branches 7 and 9 are closed which shorts 3LS.

(d) At the bottom of the stroke, cam F actuates normally closed limit switch 4LS and the chuck holding the hexagon is mechanically indexed through 60 degrees. Switch 4LS de-energizes both control relays, and all contacts assume their normal positions. The result is to de-energize solenoid A but not solenoid B (because both limit switches 2LS and 3LS are closed at this time). The net result is to shift three-way valve A to its spring-return position which ports the pump outlet to the blank-end of the cylinder which reverses its piston direction. The piston now ascends at maximum speed because the hydraulic circuit is regenerative. Note that 4LS again closes after the piston has reversed direction and moves upward.

(e) As the piston ascends at maximum rate, normally closed limit switch 3LS is opened by cam D and solenoid B is de-energized (contacts in branches 7 and 9 are open at this time). Three-way valve B is shifted to its spring-return position, and the return flow from the rod-end of the cylinder is now metered through valve C (speed diminishes because relief valve E will open). Machining now occurs on the second side of the hexagon during the upstroke.

(f) After machining of the second side of the hexagon is completed, cam D loses contact with limit switch 3LS and solenoid B is again energized which causes the upfeed to be resumed at maximum speed because metering valve C is now bypassed. As cam D actuates 2LS on its upstroke, there will be no effect because 2LS is shorted by the normally closed contact in branch 7.

(g) As the piston (and attached slide) reach the top of the stroke, a peg A extending from the chuck will contact normally open limit switch 1LS. This will energize control relay 2CR, and 2CR's normally open contacts will close to latch 2CR and energize solenoid A. The piston will now be caused to reverse direction to feed downward at maximum rate. But this time, no machining will be done on the downstroke because both 2LS and 3LS are shorted by the normally closed contact in branch 7 and the closed (normally open) contact in branch 9. So side 2 machined on the previous upstroke simply passes the end mill at maximum speed.

(h) At the bottom of the stroke, limit switch 4LS is actuated, and steps (d), (e), and (f) are repeated for machining sides 3, 4, 5, and 6 of the hexagon.

(i) After the sixth side of the hexagon is machined, another cycle will occur in which no further machining will take place but side 1 of the hexagon will be indexed back to its original position. After this occurs, the upward-feeding piston will not contact a peg on the chuck because the last peg on the chuck has been deleted and limit switch 1LS cannot be actuated. The piston will simply remain in its top position with all of the pump outlet being bypassed through relief valve E. The machined part will be unloaded and a new part mounted in the chuck. Then a new cycle is initiated by pressing pushbutton 1PB. ■

EXAMPLE 12-3

A jug-filling operation consists of three belt conveyors located in series as sketched in Fig. 12-17. Each conveyor section is powered by an individual motor, which we identify as 1M, 2M, and 3M. Limit switches 1LS, 2LS, and 3LS are located as shown. At the start of the cycle, all three belts are turned on by a START pushbutton which energizes a control relay 1CR having seven contacts.

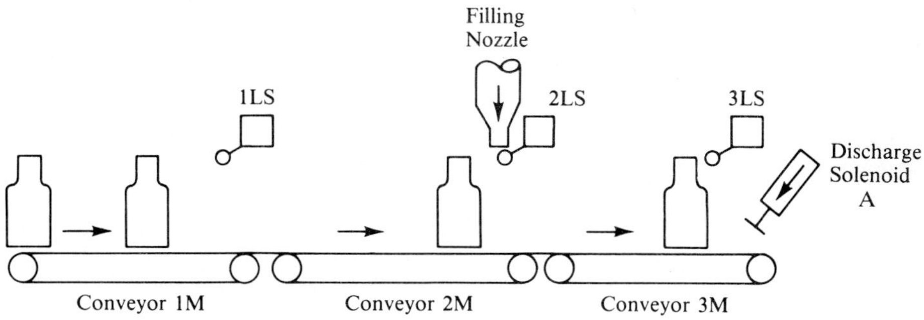

Figure 12-17. Jug-filling operation for Example 12-3.

Conveyor 1 carries empty jugs. In transferring from conveyor 1 to conveyor 2, the leading jug trips limit switch 1LS. This stops conveyor 1, but the jug continues to move on conveyor 2. At some point on conveyor 2, the jug trips limit switch 2LS which stops conveyor 2 for 6 seconds with the jug located beneath the filling station. At the end of 6 seconds, the jug is filled and conveyors 1 and 2 are restarted which transfers the filled jug to conveyor 3 while at the same time new unfilled jugs are delivered from conveyor 1 to conveyor 2.

When the filled jug trips limit switch 3LS, conveyor 3 stops and solenoid A is actuated to sweep the filled jug into a carton holding four filled jugs.

A portion of the electrical ladder diagram required to implement this jug-filling operation is shown in Fig. 12-18. This example is a good illustration of the use of various kinds of relays and their contacts including time delay contacts. It is well worthwhile for the student to study this ladder diagram in order to learn how

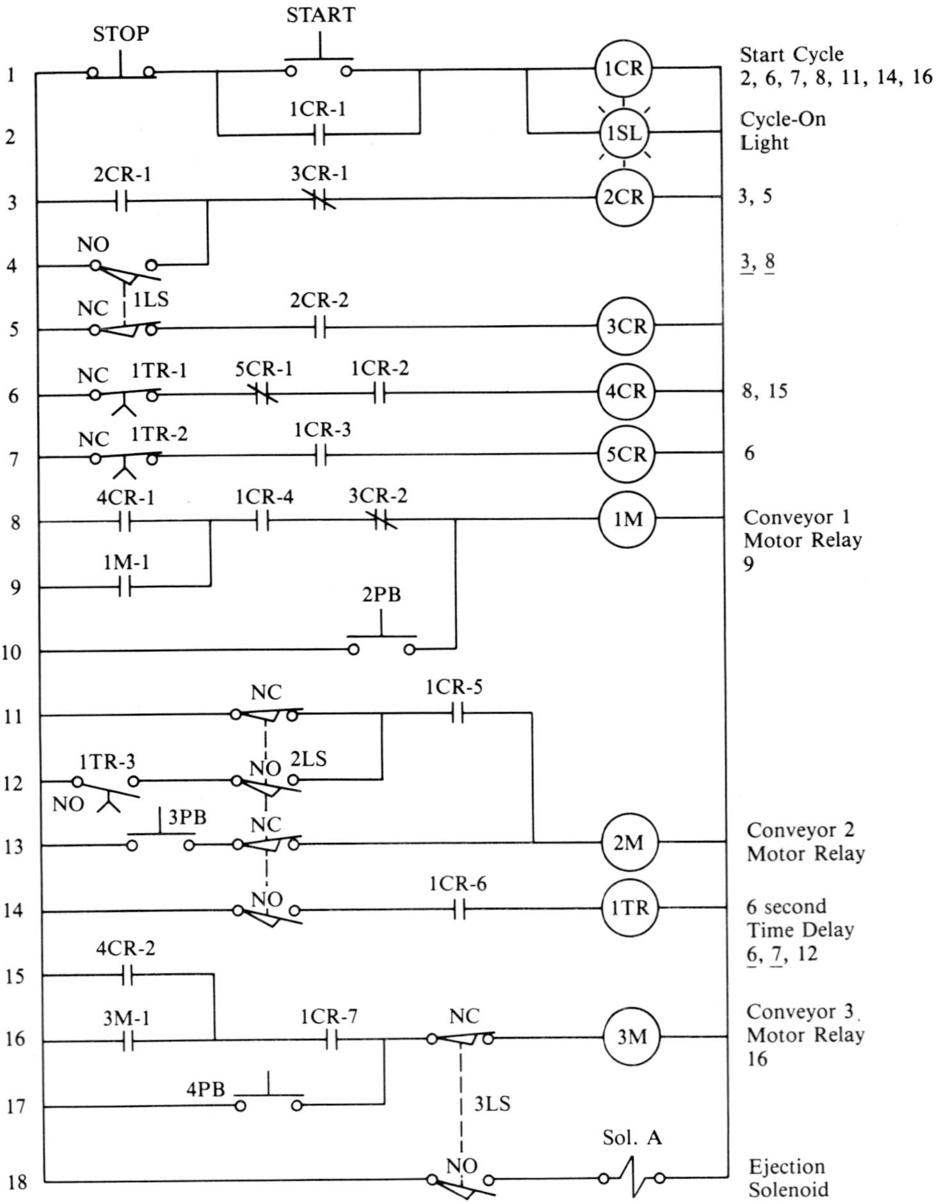

Figure 12-18. Partial ladder diagram for Example 12-3.

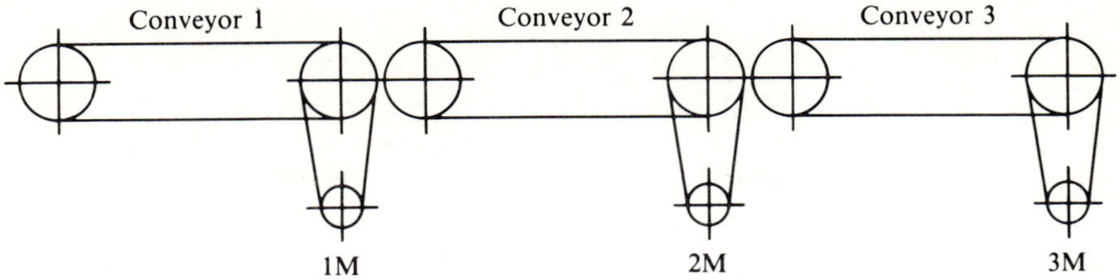

Figure P12-3. Figure for Problem 12-3.

12-3. Three belt conveyors are placed in tandem as sketched in Fig. P12-3. They are to operate in the following way:

(a) Conveyor 1 can be started alone.

(b) Conveyor 2 can be started only if conveyor 1 is running.

(c) Conveyor 3 can be started only if both conveyors 1 and 2 are already running.

(d) The conveyors may be stopped individually in reverse order, or all three may be stopped simultaneously.

Identify the motor control relays for the three conveyors as 1M, 2M, and 3M, and design a ladder circuit to accomplish the machine sequence using only:

(a) The three control relays, 1M, 2M, and 3M

(b) Three SPSTNO START pushbuttons, 1PB, 2PB, and 3PB

(c) Three SPSTNC STOP pushbuttons, 4PB, 5PB, and 6PB

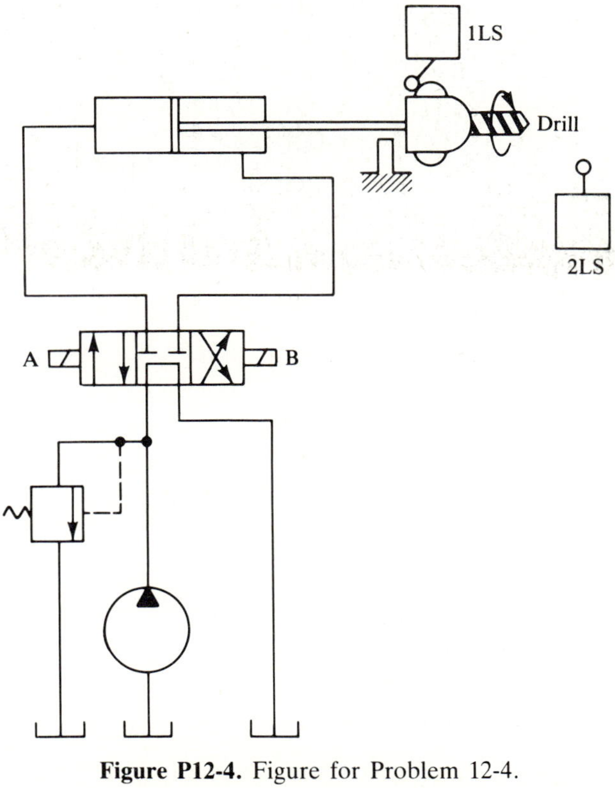

Figure P12-4. Figure for Problem 12-4.

12-4. Figure P12-4 shows an electric drill actuated by a hydraulic cylinder controlled by a three-position, solenoid-actuated, spring-centered, directional control valve. Do not concern yourself with how to actuate the electric motor that turns the drill bit. Instead, design an electrical ladder circuit that will cause the drill body to translate in the following way:

(a) Pressing pushbutton 1PB will cause the drill to move to the right until it contacts normally open limit switch 2LS.

(b) Contact with 2LS causes the drill to reverse its direction and move to the left until it contacts normally closed limit switch 1LS.

(c) Contact with 1LS restores the circuit to its original condition prior to 1PB being pushed.

Your ladder diagram will require no more than six branches. Limit switch 1LS is to be a held-open, normally closed switch in the rest position. Show it this way on the ladder diagram. Only two control relays will be required.

12-5. Figure P12-5 shows two hydraulic cylinders controlled by solenoid-actuated, spring-return, directional control valves A and B. The system is shown in its rest position. Both limit switches 1LS and 2LS are normally closed, but 2LS is being held open in the rest position. The system is to work in the following way:

(a) When pushbutton 1PB is pressed, both cylinders begin to extend (though, perhaps, not precisely at the same time).

(b) When 1LS is contacted, cylinder 1 will reverse direction and begin to retract.

(c) When 2LS is contacted, cylinder 2 will reverse direction and begin to retract.

(d) After cylinder 2 is stalled to the left, the ladder circuit must be in its original condition prior to 1PB being pressed.

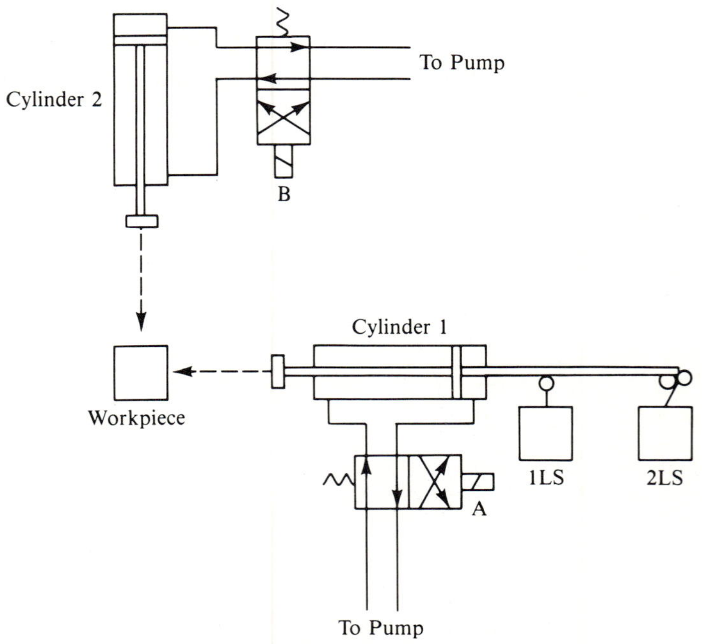

Figure P12-5. Figure for Problem 12-5.

(e) It must be possible to stop the whole system at any time by pressing a pushbutton 2PB.

Design a ladder diagram to accomplish the above logic that uses only four branches and one control relay, 1CR.

12-6. Traveling cutoff knives and saws are used in a variety of machines designed to produce square cuts and segmentation of cylinders being manufactured continuously. Cutting bars of soap from a continuously extruded soap cylinder is an example.

Figure P12-6 illustrates a traveling circular saw used to segment a continuously extruded aluminum tubing into equal lengths. The saw is articulated such that its rotating blade can be traversed perpendicular to the tubing axis at the same time it is caused to translate parallel to the tubing axis at the same time the tubing is extruded. This action produces a square cut. After the tubes are cut, they roll down a ramp, where they are bundled.

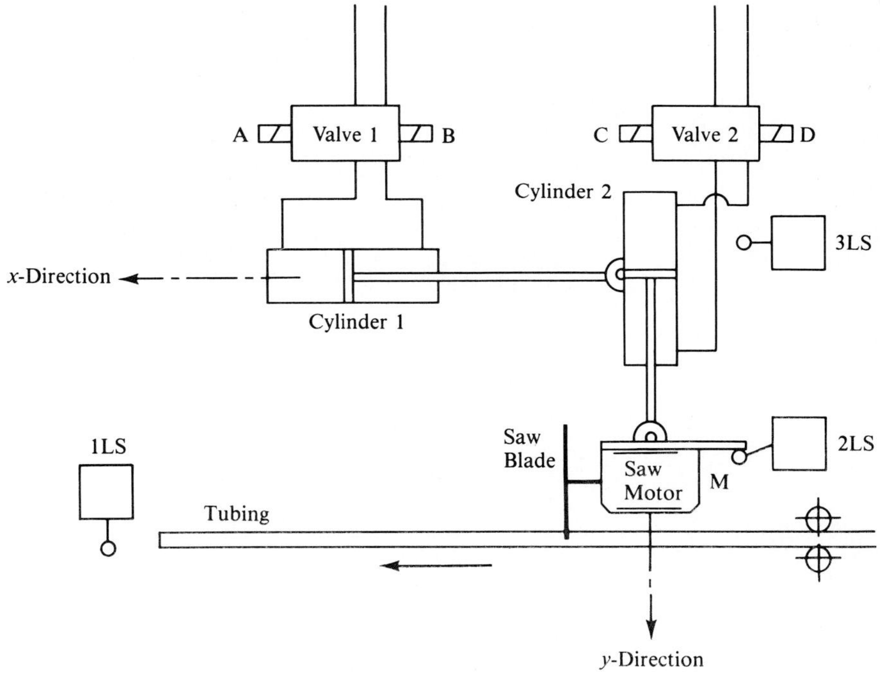

Figure P12-6. Figure for Problem 12-6.

Two cylinders and their directional control valves are used to accomplish the requisite motions of the saw spindle. Limit switches are provided to reverse the directions of traverse. Design a ladder circuit to accomplish the following sequence of operations:

(a) Pressing the START pushbutton 1PB turns on the saw (this will occur prior to the start-up of the extrusion process). This also energizes the valve solenoids and limit switches.

(b) When the tubing reaches limit switch 1LS, the saw will begin to traverse along the x-direction (propelled by cylinder 1) and the y-direction (propelled by cylinder 2) simultaneously (from its rest position).

(c) When the transverse cut is completed, the saw head will trip limit switch 2LS which reverses the piston of cylinder 2 and, hence, the direction of traverse along the y axis.
(d) When the saw head is fully retracted, it trips limit switch 3LS which reverses the direction of cylinder 1.
(e) Of course, operation (d) must be designed to be completed before event (b) reoccurs in order for this cycle to continue repetitively. This is simply a matter of suitably controlling the cylinder speeds.
(f) Pressing the STOP pushbutton 2PB turns off the saw and de-energizes the valve solenoids.

Assume the saw head is articulated such that it will rest somewhere between limit switches 2LS and 3LS when the system is off. Design a ladder circuit to accomplish the required machine sequence. Show the motor starter and designate its relay as 1M.

12-7. A drilling machine is to be designed to drill a line of equally spaced holes in a horizontal surface. The drill head is traversed horizontally by a chain drive powered by a hydraulic motor (see Fig. P12-7[a]). The hydraulic motor is actuated by a four-way, three-position, solenoid-actuated, closed-center, directional control valve. Solenoid A is actuated each time a hole has been drilled in order to produce the spacing between holes. Solenoid B is actuated after all holes have been drilled in order to return the drill head to its initial position.

The vertical traverse of the drill head (to accomplish the actual drilling) is achieved by mounting the drill head on the piston of a vertically oriented hydraulic cylinder (see Fig. P12-7[b]). The piston of the cylinder is actuated by a four-way, two-position, solenoid-actuated, spring-return, directional control valve. Limit switches 1LS and 2LS are actuated by the drill head at each end of its vertical stroke in order to reverse the downward motion and to stop the upward motion. Switch 1LS is normally open while switch 2LS is normally closed.

The spacing between holes is controlled by gearing that connects the horizontally traversing chain to a mechanical counter. The gearing ratio is such that one revolution of the counter corresponds to 1/100 of an inch of horizontal traverse. The

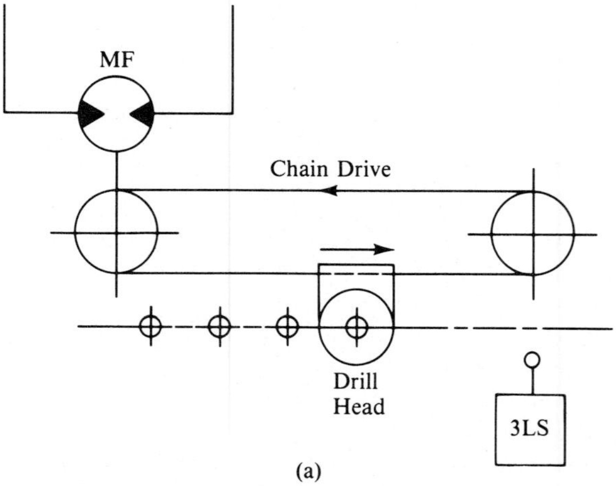

(a)

Figure P12-7. Figure for Problem 12-7.

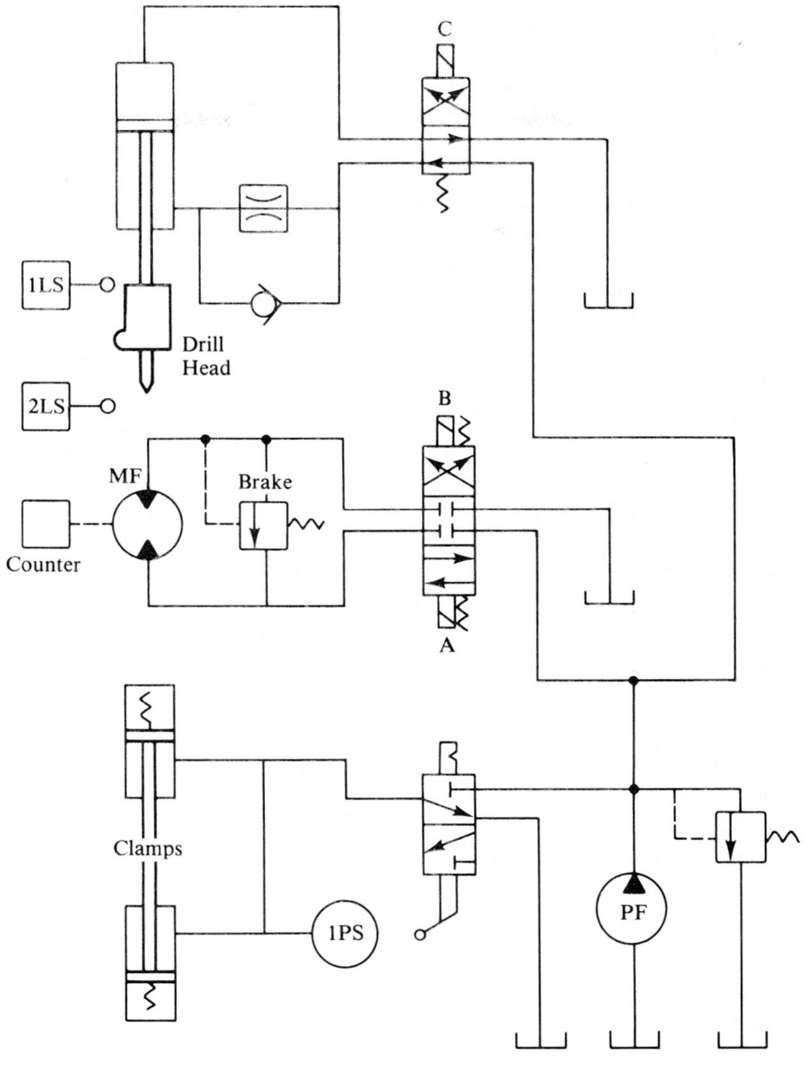

(b)

Figure P12-7 (continued)

counter has two wheels. One wheel is set by hand to determine the actual spacing desired between holes. If, for example, the spacing is to be 1/2 inch, then this wheel is set to 50. The other wheel is the actual counter; when it has counted to the number set on the first wheel (50 in this example), it closes a normally open counter contact. The counter is reset mechanically by a solenoid once each cycle.

Develop a sequence of operations to carry out this drilling operation and design an electrical ladder circuit to implement it.

13

Digital Control Concepts

The fluid power and electrical control circuits discussed previously constitute what we might call the open-loop, on-off control of mechanical systems. A sequence of machine operations is obtained by simply having motors, solenoids, valves, limit switches, pushbuttons, relays, contacts, and timers either ON or OFF in some kind of programmed fashion. Many mechanical systems (though not all) in the design domain of the mechanical engineer are susceptible to this simple kind of control.

The controls we have discussed up to this point are all said to be hardwired, and they have been implemented using hydromechanical or electromechanical hardware (valves, limit switches, relays, timers, etc.) defined by the hydraulic circuit and electrical relay ladder diagrams. Furthermore, design of the electrical ladder circuits has not been based on any formal logical procedures; instead, it has been a trial-and-error or intuitive synthesis process.

Since the ON and OFF states of the fluid and electrical circuit components mentioned above can be represented by the digital logic levels 1 and 0, respectively, it would seem plausible that all of the power of Boolean logic might be brought to bear on the open-loop, sequential control of mechanical systems. Indeed, this can be done. This fact has tremendous implications. It means, first of all, that the techniques of Boolean algebra will provide a more formal and systematic approach to the design of the fluid power circuits and their associated relay ladder diagrams. This formal approach is thoroughly discussed in Reference 13-1, and since it is not essential to our objectives here, we will not elaborate these methods in any detail. For our simple illustrative examples, we will give an inkling of one possible formal approach but, in general, will continue to muddle through the synthesis of the ladder diagram by intuition and experience.

More important, the possibility of representing the on-off nature of these particular circuit components by the binary digits (BITS) 1 and 0 implies that all of the power of microelectronic control will be available. This means that there will be alternatives to the implementation of control by means of electromechanical relay hardware. Three of these alternatives are:

(a) Control using dedicated electronic logic gate circuits
(b) Control using microprocessors
(c) Control using programmable controllers

The latter two methods are most important because they make possible implementation of much of the control of mechanical systems through either firmware or software as opposed to hardware. Firmware is what we might call dedicated software; generally, it is stored in a ROM (Read Only Memory).

Of course, the application of digital control systems, the microprocessor, and the programmable controller extends far beyond the mere sequence control of machinery. The applications are richly diverse and include both analog and digital control schemes and the incorporation of computers in the control system. But the basic elements of digital control are common to all microelectronic control systems whether they be dedicated logic gate circuits, microprocessors, programmable controllers, or any synthesis of these three types of logic devices. Here we will develop these concepts in the context of the machine sequence control system and its associated ladder circuits.

13-1 THE TRANSISTOR AS A RELAY

Two of the most important technological inventions in the history of humankind are the transistor and the integrated circuit. These devices have not only brought about a revolution in electronics but they also lie at the heart of the modern digital computer.

In the context of digital control systems, the most important application of the transistor is as a relay type of switch. It is the transistor that embodies the physical implementation of TRUE or FALSE, ON or OFF, and BIT 1 or BIT 0. Furthermore, the transistor can do this rapidly, unambiguously, economically, and microscopically. Logic gates are built from transistors; registers, decoders, memories, and control units are built from logic gates; and computers are built from registers, decoders, memories, and control units. So we see that the transistor switch is the ultimate building block of both computers and digital control systems. In an integrated circuit, it is the technological electronic atom.

The study of the solid-state electronics of the transistor is a comprehensive subject by itself. Fortunately, we do not need to cover this topic here. Of the various kinds of transistors available, we will select the common-emitter n-p-n junction transistor in order to illustrate the use of the transistor as a relay and switch.

The symbol for, and the elements of, the n-p-n transistor are sketched in Fig. 13-1. In the sketch at the right, the emitter ground voltage has been made common to both the base and the collector. This mode of connecting the three-terminal device is called the common-emitter mode. In this mode, the input to the device has terminals B and E while the output has terminals C and E.

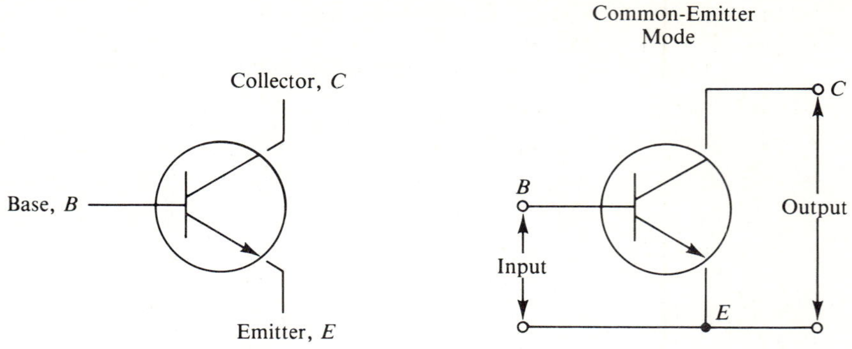

Figure 13-1. n-p-n junction transistor.

The way in which the transistor operates in the common-emitter mode depends, in part, on the voltage applied at the input. Here we will assume that the voltage applied to the input is high enough that when the transistor is ON, the voltage between collector and emitter will be very small (of order 0.2 V). This device is called a saturated transistor.

The way in which the transistor acts as a switch is illustrated in Fig. 13-2. With an input voltage of $V_{BE} = 0$, the transistor will be OFF, which means that there is no current flow from collector to emitter. This will result in an output voltage $V_{CE} \neq 0$, because there is a voltage applied at terminal C. So, when the input is OFF (digital 0), the output is ON (digital 1).

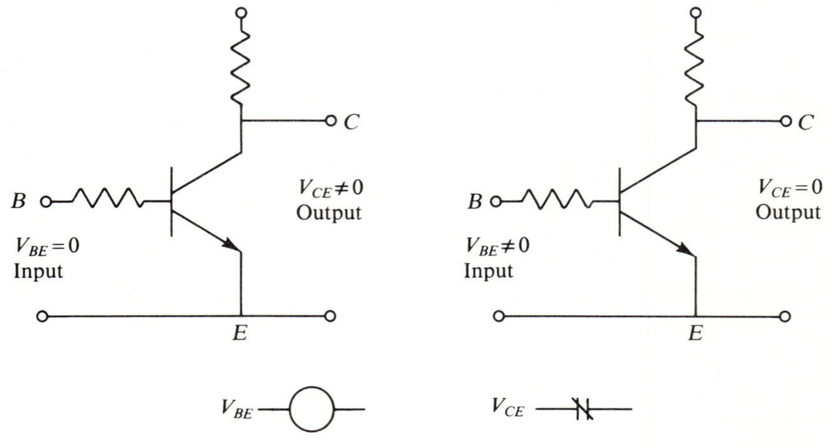

Figure 13-2. Transistor as a switch.

When the input voltage V_{BE} becomes high enough, the transistor will be turned ON. In other words, the non-zero input voltage produces current flow from base to emitter that is high enough to precipitate a current flow from collector to emitter across an essentially zero impedance. This process produces a voltage drop which is essentially $V_{CE} = 0$. So, when the input is ON (digital 1), the output is OFF (digital 0).

The transistor operates electrically in a manner analogous to a relay and its contact. When a relay is energized, its normally open contact will close to permit current flow. Similarly, when a transistor is energized, its output permits a current flow. However, digitally, the transistor is analogous to a relay having a normally closed contact in the sense that when the relay is energized (ON), its contact is open (OFF). The analogy is sketched in Fig. 13-2.

The n-p-n transistor in the common-emitter mode operates in such a way that the output state is always opposite to the input state. This device is called an inverter. The symbols and truth table for the inverter as a logical device are shown in Fig. 13-3. When the inverter is drawn by itself, the symbol at the left is usually used. But when it is combined with other logical devices, which we shall discuss later, the triangle is usually omitted and only the small circle is used to indicate the inversion process.

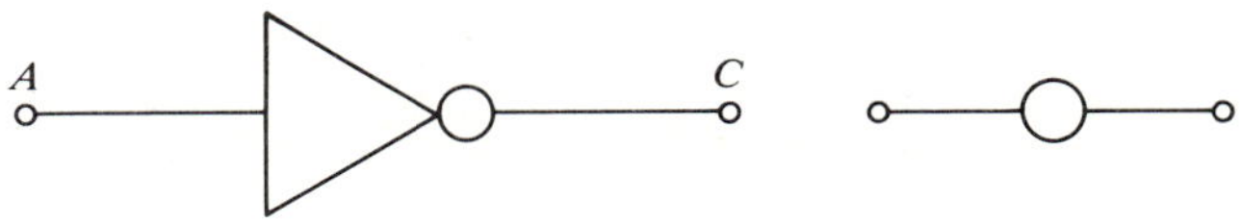

A	*C*
0	1
1	0

Figure 13-3. Symbols and truth table for the inverter.

13-2 THE BINARY NUMBER SYSTEM

A coordinate intellectual achievement that preceded the invention of the transistor by almost a century, but which is certainly on a par with it, was the discovery that human logical reasoning could be represented numerically. One of the great names associated with this discovery is George Boole, an Irishman, who developed an algebra for implementing logical reasoning.

Perhaps the simplest mode of logical reasoning is in terms of some proposition being either TRUE or FALSE. The technical counterpart of this is a circuit being closed or open, that is, ON or OFF, and the numerical counterpart involves the binary digits 1 and 0. In any case, this mode of analysis involves only two unambiguously distinguishable states which certainly provides for simplicity.

The numerical system intrinsic to the two-state concepts delineated above is the binary number system. In this system, we will represent TRUE with the binary digit 1 and FALSE with the binary digit 0.

If binary digit 1 (or TRUE) is equated to the ON circuit condition, with digit 0 (FALSE) equated to the OFF condition, we call this a positive logic system. The negative logic system is just the opposite.

In terms of electrical circuitry for representing these states, the ground potential will represent digital 0 and a nominal voltage of 5 V will represent digital 1 in the positive logic system.

In our familiar decimal number system, a number—say, 4532—is written as "four thousand, five hundred, and thirty-two." Arithmetically, we mean that this number has a base of 10 (decimal) and is represented by

$$\begin{aligned} 4532 &= 4 \times 10^3 + 5 \times 10^2 + 3 \times 10^1 + 2 \times 10^0 \\ &= 4000 + 500 + 30 + 2 \end{aligned}$$

In this system, the coefficients of the powers of ten are digits having ten distinguishable states (0 through 9).

The binary number system has a base of 2 (binary). Numbers are represented in this system in a manner analogous to that for the decimal system. Now, however, the coefficients will be digits having only two distinguishable states (0 and 1), and they will be coefficients of powers of two. As an example, the number 4532 is written in binary as

$$\begin{aligned} 4532 = &\ 1 \times 2^{12} + 0 \times 2^{11} + 0 \times 2^{10} + 0 \times 2^9 + 1 \times 2^8 + 1 \times 2^7 + 0 \times 2^6 \\ &+ 1 \times 2^5 + 1 \times 2^4 + 0 \times 2^3 + 1 \times 2^2 + 0 \times 2^1 + 0 \times 2^0 \\ = &\ 4096 + 256 + 128 + 32 + 16 + 4 \end{aligned}$$

The shorthand notation for this binary number is

$$4532 = 1000110110100$$

We see that 13 BITS are required to represent this number in binary. This is in contrast to the 4 DITS (decimal digits) required to represent the number 4532. So the trade-off in going to the binary form is to accept the larger number of digits required for the smaller number of states each digit may assume.

The value, or weight, of each bit in the binary system can be shown as

2^{11}	2^{10}	2^9	2^8	2^7	2^6	2^5	2^4	2^3	2^2	2^1	2^0
2048	1024	512	256	128	64	32	16	8	4	2	1

(nibble: 2^3 to 2^0; byte: 2^7 to 2^0)

We have shown the weights up to only 12 bits. Each 4-bit sequence is called a nibble, while each 8-bit sequence is called a byte. A 12-bit number consists of 3 nibbles or $1\frac{1}{2}$ bytes.

For a 4-bit binary number, the universe of possible combinations is shown in Table 13-1. Note that 4 bits can represent $2^4 = 16$ distinct numbers. It follows that 8 bits can represent 256 numbers and that 12 bits can represent 4096 numbers. The so-called 64K memory consists of an array of 16-bit addresses which, in reality, is equivalent to 65,536 (2^{16}) distinct addresses.

TABLE 13.1. Universe of 4-Bit Combinations.

Binary	Sum	Decimal
0000	0+0+0+0	0
0001	0+0+0+0	1
0010	0+0+2+0	2
0011	0+0+2+1	3
0100	0+4+0+0	4
0101	0+4+0+1	5
0110	0+4+2+0	6
0111	0+4+2+1	7
1000	8+0+0+0	8
1001	8+0+0+1	9
1010	8+0+2+0	10
1011	8+0+2+1	11
1100	8+4+0+0	12
1101	8+4+0+1	13
1110	8+4+2+0	14
1111	8+4+2+1	15

EXAMPLE 13-1

A technique for converting a decimal number to a binary number is shown below using the number 52.

$$\begin{array}{rll} & \downarrow \text{remainder} & \\ 52/2 = 26 + 0 & \text{(LSB)} & \\ 26/2 = 13 + 0 & & \\ 13/2 = \ \ 6 + 1 & & 110100 = 52 \\ 6/2 = \ \ 3 + 0 & & \\ 3/2 = \ \ 1 + 1 & & \\ 1/2 = \ \ 0 + 1 & \text{(MSB)} & \end{array}$$

Simply divide the number successively by 2, keeping track of where there is a remainder. Assigning digits 1 and 0 to where there is (1), and is not (0), we find that a remainder will produce the requisite binary number. Note that the most significant bit (MSB) will be the last remainder and the least significant bit (LSB) will be the first remainder. ■

13-2-1 Binary Arithmetic

The binary number system has application in digital control in several different contexts. Three important uses are in (a) logical operations, (b) coding, and (c) arithmetic operations. In this section, we will discuss some simple binary arithmetic operations including addition, subtraction, and multiplication. Keep in mind that we are not interested in arithmetic operations per se; the logical operations are more important. However, a knowledge of the simple arithmetic operations will be required in order to understand why certain logic gate circuits have been invented to carry out certain functions. One such circuit is the binary adder used in microprocessors.

Addition. As in decimal addition, in binary addition we must be concerned with the problem of carry-forward. The rules for the addition of bits are illustrated by the following universe of 2-bit sums:

1	1	0	0
1	0	1	0
10	01	01	00

(carry) MSB ↑↑ LSB (sum)

The least significant bit of these 2-bit sums represents the actual sum while the most significant bit represents the carry-forward (called the carry).

The arithmetic operation of addition is shown in tabular form in Table 13-2. This table is called the truth table for binary addition. The significance of truth tables will become readily apparent as we proceed.

TABLE 13-2. Truth Table for Binary Addition.

Bit A	Bit B	Carry C	Sum S
0	0	0	0
0	1	0	1
1	0	0	1
1	1	1	0

$$\begin{array}{r} A \\ \underline{B} \\ CS \end{array}$$

The results of adding three bits are shown in Table 13-3. Again, the carry bit is at the left (MSB) while the sum is the bit at the right (LSB).

TABLE 13-3. Rules for Adding 3 Bits.

1	1	1	1	0	0	0	0
1	1	0	0	1	1	0	0
1	0	1	0	1	0	1	0
11	10	10	01	10	01	01	00

MSB ↑↑ LSB

EXAMPLE 13-2

Add the numbers 52 and 26 as shown.

First number	0110100 = 52
Second number	0011010 = 26
Carry	0110000
Sum	01001110 = 78

The bits are simply added in order from right to left. Note the non-zero carries in the sixth and seventh columns from the right. ■

Subtraction. In Example 13-2, our numbers 52 and 26 were implicitly positive. For reasons we shall now discuss, we placed a digit 0 at the left of the most significant non-zero bit to denote that the numbers 52 and 26 were positive numbers.

Several schemes can be used to formalize the subtraction operation in binary arithmetic. These various methods are characterized by the manner in which the signs (+ and −) are to be represented in binary. As we have already indicated, the positive sign will be assigned in binary by placing the digit 0 to the left of the most significant non-zero digit.

Of the methods of subtraction available, the two's complement method will be the one of interest to us. In this method, the negative sign will be indicated by a digit 1 to the left of the most significant bit of the number. We will illustrate how this works by means of examples.

A negative number in binary is obtained by taking the two's complement of the positive value. The two's complement is obtained by complementing (inverting) each bit of the positive number (called the one's complement) and then adding the digit 1. Suppose we want to convert +52 to −52. Then we have

$$\begin{array}{rrl} +52 = & \underline{0\,1\,1\,0\,1\,0\,0} & \\ & 1\,0\,0\,1\,0\,1\,1 & \text{(one's complement)} \\ & \underline{\,1} & \text{(add digit 1)} \\ -52 = & 1\,0\,0\,1\,1\,0\,0 & \text{(two's complement)} \end{array}$$

Note that by placing a digit 0 just to the left of the most significant digit 1 of the positive number, we will always obtain digit 1 at the leftmost place of the two's complement. This is the digit that indicates this is a negative number.

We see that the digits to the right of the leftmost digit 1 do not properly represent the magnitude of the number −52. A number in two's complement form (as will be indicated by a digit 1 at the leftmost place) can have its magnitude regained by recomplementing and again adding digit 1. This process is illustrated in Example 13-3.

EXAMPLE 13-3

Determine the magnitude of the two's complement 1001100 = −52.

$$\begin{array}{rrl} -52 = & 1\,0\,0\,1\,1\,0\,0 & \\ & 0\,1\,1\,0\,0\,1\,1 & \text{(one's recomplement)} \\ & \underline{\,1} & \text{(add digit 1)} \\ +52 = & 0\,1\,1\,0\,1\,0\,0 & \text{(+52 regained)} \end{array}$$

■

Having defined how to manipulate negative numbers by means of the two's complement method, we can now define the binary operation of subtraction. Simply convert the negative numbers to two's complement form and add them to the positive numbers! Several illustrations are given in Example 13-4.

EXAMPLE 13-4

Add the numbers −52 and +26.

$$
\begin{array}{rrl}
-52 = & 1001100 & \text{(two's complement)} \\
+\underline{26} = & \underline{0011010} & \\
-26 = & 1100110 & \text{(two's complement)} \\
 & 0011001 & \text{(one's recomplement)} \\
 & \underline{\qquad\quad 1} & \text{(add digit 1)} \\
+26 = & 0011010 & \text{(magnitude of } -26\text{)}
\end{array}
$$

Add the numbers −52 and −26.

$$
\begin{array}{rrl}
-52 = & 1001100 & \text{(two's complement)} \\
-\underline{26} = & \underline{1100110} & \text{(two's complement)} \\
-78 = & 10110010 & \text{(two's complement)} \\
 & 01001101 & \text{(one's complement)} \\
 & \underline{\qquad\quad 1} & \text{(add digit 1)} \\
+78 = & 01001110 & \text{(magnitude of } -78\text{)}
\end{array}
$$

Add the numbers +52 and −26.

$$
\begin{array}{rrl}
+52 = & 0110100 & \\
-\underline{26} = & \underline{1100110} & \text{(two's complement)} \\
+26 = & 10011010 &
\end{array}
$$

In this last illustration, we see a potential problem. The carry of digit 1 in the leftmost place of the answer implies that the answer is negative when in fact it is positive. In this instance, we must discard the last carry to get the correct answer. This ambiguity can be avoided by always adding an extra digit 0 to positive numbers and an extra digit 1 to negative numbers, and always discarding the carry. To illustrate, we repeat the last two examples:

$$
\begin{array}{rrl}
-52 = & 11001100 & \\
-26 = & \underline{11100110} & \\
-78 = & 110110010 & \text{(two's complement)}
\end{array}
$$

↑ discard the carry

$$
\begin{array}{rrl}
+52 = & 00110100 & \\
-26 = & \underline{11100110} & \\
+26 = & 100011010 &
\end{array}
$$

↑ discard the carry

By discarding the carry, we now get the correct answers; the method works! ■

Multiplication. The process of binary multiplication is similar to ordinary decimal multiplication and perhaps is best explained in terms of an illustration.

EXAMPLE 13-5

Multiply the numbers +52 and +26.

```
              +52 = 0110100
              +26 = 0011010
                    0000000
                   0110100
                  0000000
                 0110100
                0110100----
    (carries)   1111100000
    (product)   10101001000 = 1352
```

Note that each partial product is either zero or the same as the first number. This means that the multiplication operation is equivalent to a shifting and adding operation which is admirably suited to the microprocessor capability. ■

Division can similarly be decomposed into shifting and addition, but since we have no interest in division, we will not discuss this binary arithmetic operation.

13-2-2 Binary Codes

We have seen that arithmetic can be carried out in binary just as readily as it can be done in decimal. Furthermore, it is the binary number system that is intrinsic to implementation of logic controllers used in digital control schemes for mechanical systems.

Despite the advantages the binary number system embodies with respect to the internal processes occurring in microelectronic devices, we still live in a world that is decimally oriented. For example, even though the hand calculator is a computer based on binary logic and arithmetic, its output is displayed in decimals. This means that there has been a conversion process from binary to decimal in producing the output. And when numbers are input to the hand calculator, there must be a conversion process from decimal to binary.

Binary Coded Decimal (BCD). We know there is a one-to-one correspondence between any decimal number and its binary equivalent (e.g., 26 = 011010). However, the conversion process would be rather unwieldy when the numbers get large.

The idea behind the binary coded decimal (BCD) is to develop a code for converting decimal to binary that encodes only the decimal digits 0 through 9. Since there are ten states to be represented, we will need only 4 bits to do the job, as illustrated in Table 13-4.

TABLE 13-4. Binary Coded Decimal (BCD).

Decimal	Binary Code
0	0000
1	0001
2	0010
3	0011
4	0100
5	0101
6	0110
7	0111
8	1000
9	1001

Using the code defined in Table 13-4, we find that the decimal number 26 is coded in two 4-bit nibbles as

$$\begin{array}{ccc} 26 = & 0010 & 0110 \\ & 2 & 6 \end{array}$$

The general rule for converting decimal to binary is to translate each decimal digit into its 4-bit equivalent and then to write down the 4-bit equivalents in the same sequence as the decimal digits. For example:

$$526 = 0101 \quad 0010 \quad 0110$$

We see that in BCD, we can represent the decimals 0 through 99 with one byte (8 bits).

Hexadecimal Coded Binary (HCB). There is an even greater incentive to have a method of coding in which binary numbers can be represented by fewer digits. Consider the binary number 01000110110100 (= 4532). It will be almost impossible to work with strings of 1s and 0s such as this without making errors. A shorthand notation is needed for coding and naming such unwieldy representations. The BCD coding system cannot be used in reverse to accomplish the desired shorthand because it does not encode the 4-bit binary numbers from 10 through 15.

A method for encoding all sixteen 4-bit binary numbers is displayed in Table 13-5. This method is called the hexadecimal representation. The hexadecimal code is widely used to represent the operation codes, instructions, and addresses used in microprocessor programming. It is often found displayed on the input keyboard to a microprocessor.

In hexadecimal, the number 4532 would be represented as

$$\begin{array}{ccccc} 4532 = & 0001 & 0001 & 1011 & 0100 \\ & 1 & 1 & B & 4 \end{array}$$
$$= 11B4 \text{ (hex)}$$

TABLE 13-5. Hexadecimal Coded Binary (HCB).

Binary	Hexadecimal Code
0 0 0 0	0
0 0 0 1	1
0 0 1 0	2
0 0 1 1	3
0 1 0 0	4
0 1 0 1	5
0 1 1 0	6
0 1 1 1	7
1 0 0 0	8
1 0 0 1	9
1 0 1 0	A
1 0 1 1	B
1 1 0 0	C
1 1 0 1	D
1 1 1 0	E
1 1 1 1	F

TABLE 13-6. Some 7-Bit ASCII Codes.

Character	Code	Character	Code
Space	010 0000	F	100 0110
*	010 1010	G	100 0111
+	010 1011	H	100 1000
−	010 1101	I	100 1001
/	010 1111	J	100 1010
0	011 0000	K	100 1011
1	011 0001	L	100 1100
2	011 0010	M	100 1101
3	011 0011	N	100 1110
4	011 0100	O	100 1111
5	011 0101	P	101 0000
6	011 0110	Q	101 0001
7	011 0111	R	101 0010
8	011 1000	S	101 0011
9	011 1001	T	101 0100
=	011 1101	U	101 0101
A	100 0001	V	101 0110
B	100 0010	W	101 0111
C	100 0011	X	101 1000
D	100 0100	Y	101 1001
E	100 0101	Z	101 1010
	Carriage Control Codes		
Line Feed	000 1010	Carriage Return	000 1101

In contrast to the BCD system, the hexadecimal system is actually a number system having a base of 16. Whereas the decimal number system has ten distinct states for each digit and the binary system has two, the hexadecimal system has sixteen states for each digit. While it would be possible to contemplate arithmetic operations in hexadecimal, our use will be confined to coding binary numbers.

The ASCII Code. In communicating with microprocessors or programmable controllers, it will be necessary to use symbols other than just numbers. In addition to letters of the alphabet, it will be necessary to include such things as asterisks, punctuation marks, inequality symbols, and so forth. In order that there be a standard way of coding such symbols, the American Standard Code for Information Interchange (ASCII) has adopted a code that uses seven bits. Table 13-6 presents some of the more important symbols and their coded counterparts in binary.

13-3 BINARY LOGIC

Our next objective is to develop the concepts that make possible a transition from the physical circuit viewpoint of machine sequence control to the more abstract approach used in implementing the control microelectronically. The transition is one from relays to transistors to logic gates to functional logic packages to microprocessors.

A microelectronic digital control system is made up of thousands of constituent, small subsystems each of which performs a logical process. These subsystems are called logic gates; they are built from transistors usually integrated onto a small chip. The elemental logic gates are those of NOT, AND, and OR. These devices implement logical processes by means of the binary number system.

Of course, the basic logic concepts we want to expose are independent of the hardware used to implement them. The hardware could be mechanical, pneumatic, hydraulic, electromechanical, or electronic (or implementation could be in software). To illustrate, we shall use manually operated mechanical switches in introducing the subject.

13-3-1 Digital Switching Logic

Consider the simple electrical circuit shown in Fig. 13-4. We have a battery, two manually operated mechanical switches, and a resistive load all connected in series.

It is clear from Fig. 13-4 that there will be a voltage across resistor *C* if, and only if, both switches *A* AND *B* are closed. We will define the states of the switches to be ON when they are closed and OFF when they are open. Resistor *C* is defined as ON when there is a voltage across it and OFF when it has no voltage drop.

In logical terms, we could make the circuit of Fig. 13-4 implement a logical operation by identifying TRUE with ON and FALSE with OFF. In this case, we

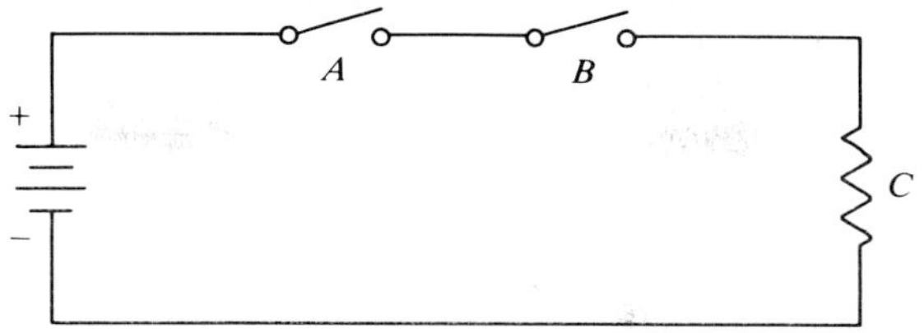

Figure 13-4. Mechanical digital switching circuit.

would identify these states with some specific kinds of statements whose truth or falsity is to be demonstrated.

A more pertinent identification for us here is in terms of the binary number system. We shall identify the ON state with binary digit 1 and the OFF state with binary digit 0. Then, the possible combinations of switch and resistor states can be represented as shown in Table 13-7.

TABLE 13-7. Truth Table for AND Operation.

A	*B*	*C*
0	0	0
0	1	0
1	0	0
1	1	1

Table 13-7 is called the truth table for the switching circuit of Fig. 13-4. It is called the truth table for the AND operation because output *C* is binary 1 if, and only if, both switches *A* AND *B* are binary 1. Otherwise, the output is binary 0. Note how we have been able to represent the states of a switching circuit numerically using the binary number system. Conversely, we could look at this situation in reverse, which is to view the circuit of Fig. 13-4 as implementing the truth table 13-7 in hardware. The implications of this seemingly simple correlation are of immense significance.

The manual switches of Fig. 13-4 could just as well have been the normally open contacts of control relays used to energize some device in a branch of a ladder circuit (see Fig. 13-5). If we let BIT 0 represent the normal (OFF) condition of a normally open contact and BIT 1 represent the closed (ON)

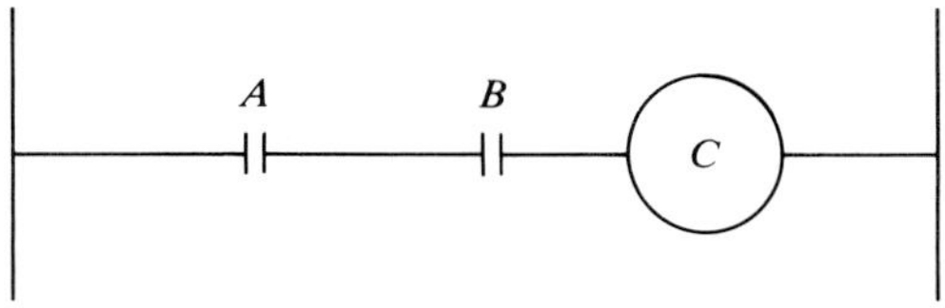

Figure 13-5. Implementing AND with relay contacts.

condition, then output C will again be energized (bit 1), if, and only if, both contacts A AND B have logic levels 1. The same truth table as before applies, and we will have implemented the AND operation by two (normally open) relay contacts A and B in series with output C.

Truth table 13-7 can be represented algebraically by

$$C = A \cdot B \tag{13-1}$$

We find the somewhat peculiar result that multiplication represents the logical AND operation. If inputs A and B are 1 or 0, we have ordinary multiplication. However, as we shall see, the more general rules for logical multiplication are quite different from those of ordinary algebra.

The logical AND operation can be provided in any number of ways using transistors. Using resistor-transistor logic (RTL), a circuit utilizing saturated transistors is shown in Fig. 13-6 that will implement truth table 13-7. Transistors A

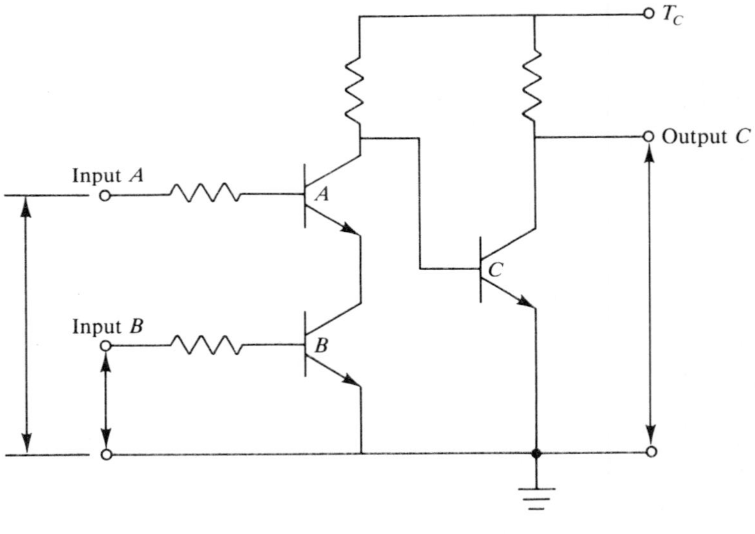

Figure 13-6. Implementing AND with transistors.

and B control a third transistor, C, as shown. Unless both transistors A and B are ON (in which case current from terminal T_C is shunted through transistors A and B), there will be a base to emitter current flowing in transistor C (from terminal T_C), which turns transistor C ON. This state produces negligible voltage signal at the output of transistor C (logic level 0) because of the collector to emitter current flow through negligible impedance. With both transistors A and B ON, current flows through these transistors from T_C to ground rather than through transistor C, which is then OFF (logic level 1 at the output of C).

Suppose we have two manual switches or normally open contacts in parallel rather than in series (see Fig. 13-7). The corresponding truth table is shown in

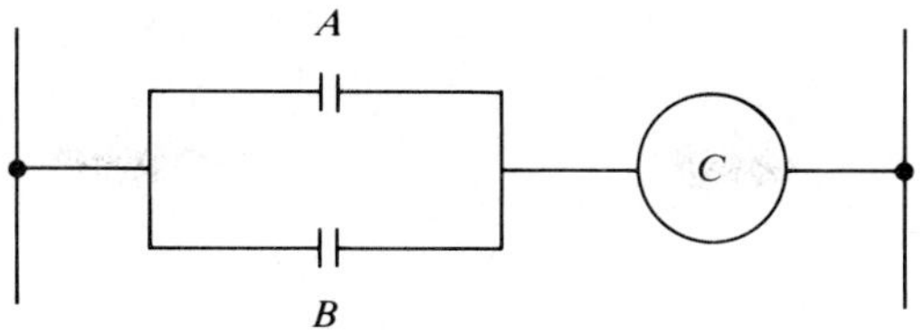

Figure 13-7. Implementing OR with relay contacts.

Table 13-8. This is the truth table for the OR operation, because now output C will be binary 1 if either contact A OR contact B (OR both) are binary 1 (ON).

TABLE 13-8. Truth Table for OR Operation.

A	B	C
0	0	0
0	1	1
1	0	1
1	1	1

Algebraically, the logical OR operation of Table 13-8 is represented by

$$C = A + B \tag{13-2}$$

Equation 13-2 requires that $1 = 1 + 1$, which is an identity peculiar to this form of algebra. This identity will be elaborated in more detail later.

13-3-2 Digital Logic Gates

The discussion of the previous section on digital switching shows how some simple electrical circuits incorporating manual switches, relays, or transistors can be interpreted as binary logic circuits. These same concepts could have also been developed using simple pneumatic or hydraulic circuits and components. A little reflection about this will inevitably lead to the concept of a logical device as an entity independent of any particular physical components. This type of device is called a logic gate. Instead of processing electrical, pneumatic, or hydraulic signals, the logic gate is conceived simply to process binary information in the form of bits 1 or 0.

The symbol for the AND logic gate is shown in Fig. 13-8. The truth table is the same as Table 13-7. The AND logic gate functions in such a way that logic level 1 will exist at output C if, and only if, logic levels 1 exist at input A AND input B. Otherwise, the output has logic level 0.

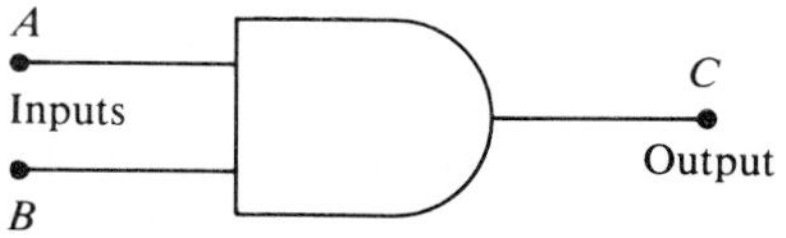

Figure 13-8. Symbol for the AND gate.

The term *logic gate* is applied to devices such as that shown in Fig. 13-8 because these devices can be viewed as having the ability to selectively pass input signals. To illustrate, suppose we interpret the AND symbol and its truth table to be as shown in Fig. 13-9. We imagine a signal to be present at input B. It could be either 0 or 1. The signal will be transmitted to the output only if control input A has logic level 1. This situation is shown by the last two rows of the truth table. If control input A has logic level 0, then the signal at terminal B does not matter; the output will always be logic level 0, as shown in the top two rows of the truth table. In other words, control A makes the device function as a gate, letting signal B through when A has logic level 1 and preventing signal B from being transmitted when A has logic level 0.

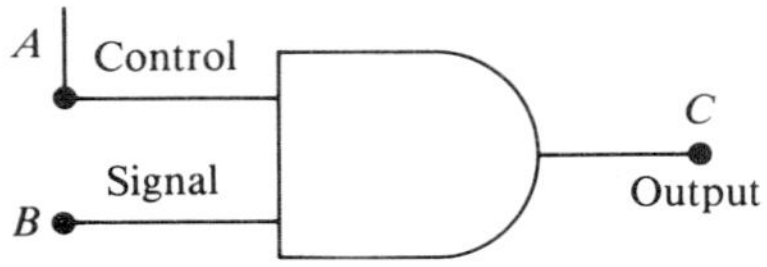

Control	Signal	Output
A	B	C
0	0	0
0	1	0
1	0	0
1	1	1

Figure 13-9. AND operation as a gate.

An application of the AND gate is shown in Fig. 13-10. We assume an infinite pulse train is applied at terminal B as shown at the right of Fig. 13-10. Terminal A is to be used as a control to permit only a finite number of pulses to be transmitted to output C. So it is set at logic level 0 except for a specific time t when it is set at logic level 1. The output will then be as shown in Fig. 13-10. Since terminal A either inhibits or enables the passage of pulses applied at B, the process is called gating or enabling. The frequency of the pulse train could be determined by counting the pulses appearing at C and then dividing by t.

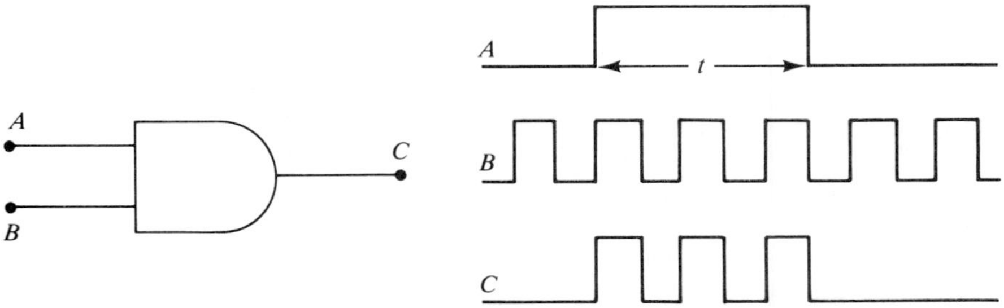

Figure 13-10. Input A enabling an AND gate.

The logic gate symbol for the OR operation is shown in Fig. 13-11. This gate functions such that logic level 1 appears at the output if, and only if, input A OR input B OR both have logic level 1. Otherwise, the output is logic level 0. Also

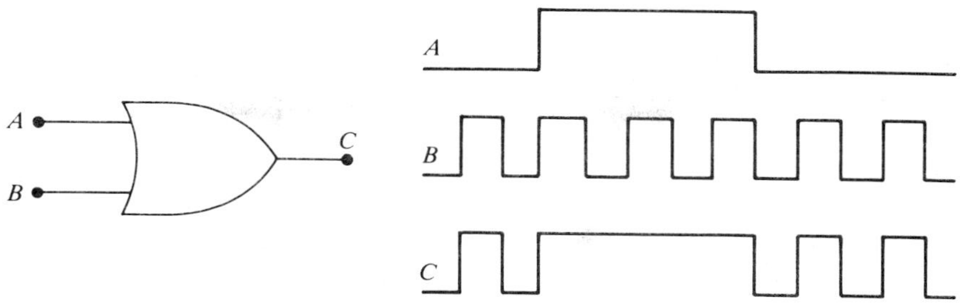

Figure 13-11. Gating properties of the OR gate.

shown in Fig. 13-11 is an illustration of how the OR gate transmits the inputs used in Fig. 13-10 for the AND gate.

The third of the three fundamental logic gates is the NOT gate. The NOT gate is nothing more than the inverter described earlier and illustrated in Fig. 13-3. The word NOT is used to indicate that the output is NOT (and never is) equal to the input; it is always opposite (or complementary) to the input. Complementarity is indicated algebraically by placing a bar above the input to denote the output. Logic level 1 at the input produces logic level 0 at the output, and vice versa. If the input is A, then the output is $\bar{A}$ (NOT A).

By combining the NOT gate with the AND and OR gates, we obtain two additional elemental logic gates, called the NAND (meaning NOT AND) and NOR (meaning NOT OR) gates. The truth tables for the NAND and NOR logical operations are simply the inverted outputs of the AND and OR operations. The NAND and NOR gates also have their relay equivalents and algebraic representations (note that if AND is represented by $A \cdot B$, then NAND is represented by $\overline{A \cdot B}$; similarly, NOR is represented by $\overline{A + B}$). All of these elemental logic gates and their various representations are summarized in Table 13-9.

Note that an additional gate, called the exclusive OR (or XOR) gate, has been included in Table 13-9. This gate produces the same output as the OR gate with the exception of the case $A = 1$, $B = 1$, where we see that C has been inverted to $C = 0$. In other words, this type of OR is being excluded in producing logic level 1 at the output. The relay implementation of XOR and its algebraic expression are also indicated in Table 13-9. We see that the truth table for the XOR operation is identical to the truth table for binary addition (Table 13-2) if the carry column is ignored. The XOR logic is also the same as that required to implement the familiar unstairs-downstairs light switch circuit.

EXAMPLE 13-6

In implementing relay ladder logic with a microprocessor, it frequently will be necessary to check the contents of a register to determine the logic level of selected bits. This can be done by a process called masking which uses the AND operation to set the nonselected bits to logic level 0. The process is sketched in Fig. 13-12.

TABLE 13-9. Elementary Logic Gate Equivalents.

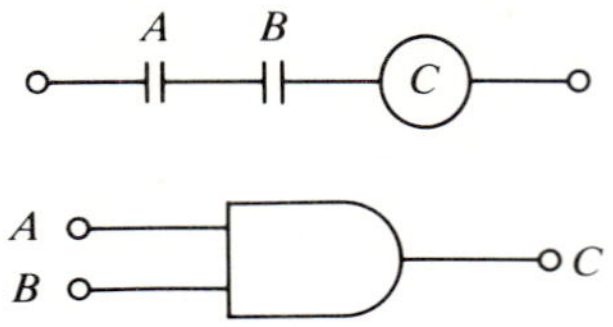

A	B	C
0	0	0
0	1	0
1	0	0
1	1	1

AND

$C = A \cdot B$

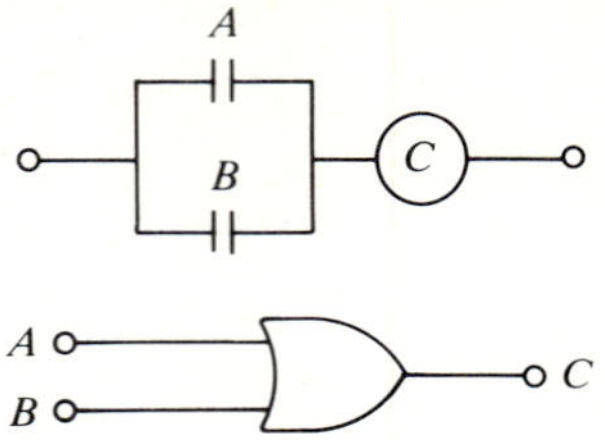

A	B	C
0	0	0
0	1	1
1	0	1
1	1	1

OR

$C = A + B$

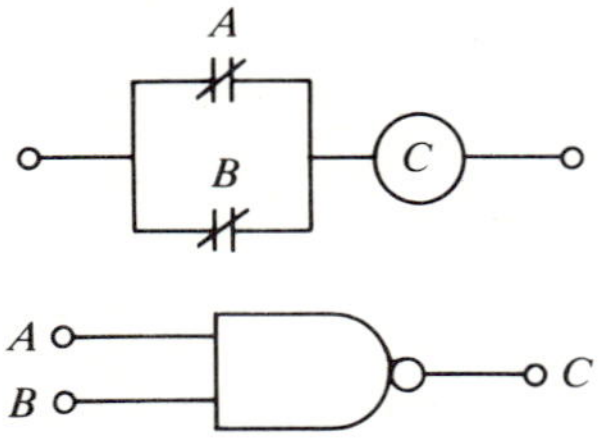

A	B	C
0	0	1
0	1	1
1	0	1
1	1	0

NAND

$C = \overline{A \cdot B}$

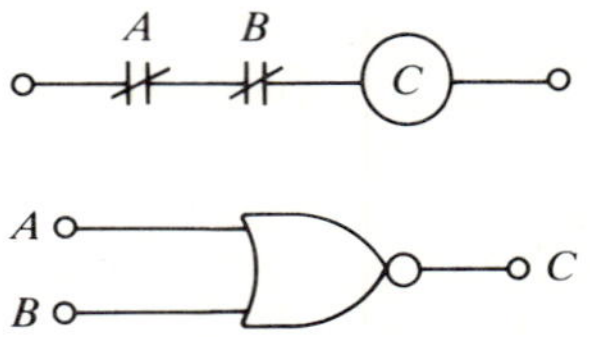

A	B	C
0	0	1
0	1	0
1	0	0
1	1	0

NOR

$C = \overline{A + B}$

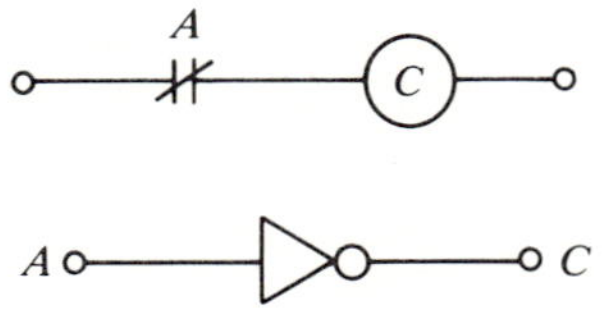

A	C
0	1
1	0

NOT

$C = \overline{A}$

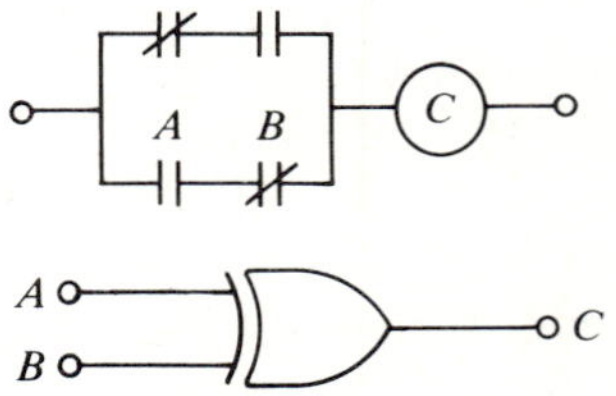

A	B	C
0	0	0
0	1	1
1	0	1
1	1	0

XOR

$C = A \oplus B = \overline{A} \cdot B + A \cdot \overline{B}$

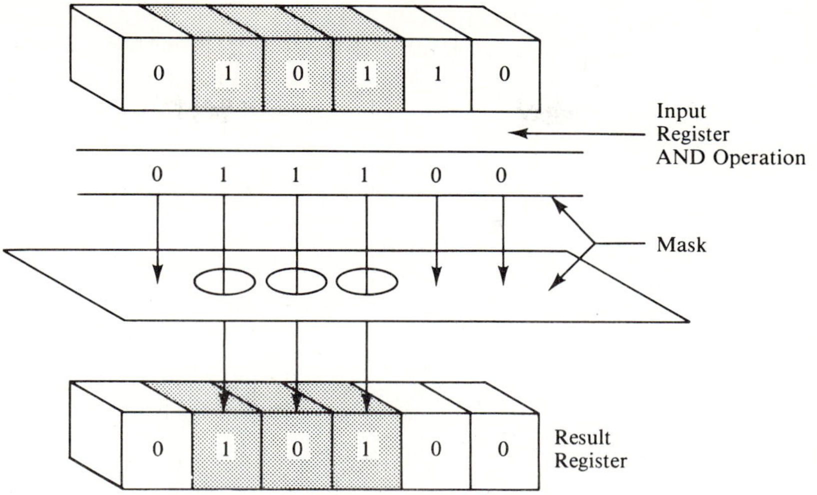

Figure 13-12. Selecting bits by masking.

The number in the register is 010110, but only the third, fourth, and fifth bits from the right are of interest. These bits can be selected by "ANDing" the register contents with the mask 011100. Here the mask is the control of an AND gate. Each bit in the register is ANDed, in turn, with the corresponding bit of the mask, and the result is stored in a result register. Only the bits in the third, fourth, and fifth places from the right will be preserved by this selective filtering process, because logic level 1 occurs in these positions of the mask (and logic 0 elsewhere). Note that there is no carry because this is not an addition operation.

If the mask were to contain all 0s, then the masking operation could be used to clear a register by storing the result back into the original register. Or if it is desired to set a register to all 1s, this could be done by using an OR mask containing all 1s. ■

EXAMPLE 13-7

We wish to analyze the circuit consisting of the logic gates shown in Fig. 13-13 and show that it produces the truth table for the XOR operation defined in Table 13-9.

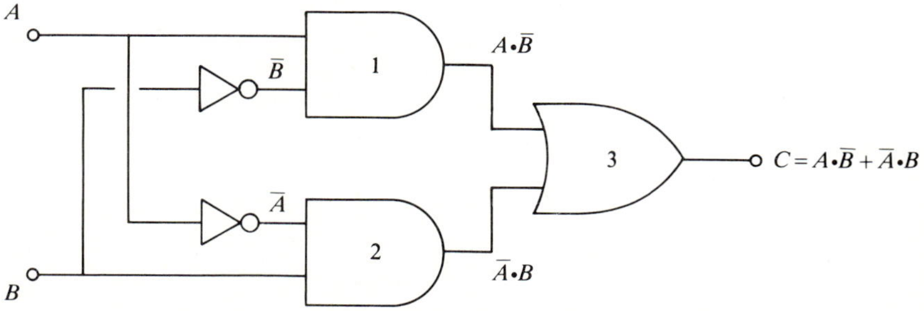

Figure 13-13. Gate circuit for the exclusive OR (XOR) gate.

Input A proceeds directly into AND gate 1 but is inverted to $\bar{A}$ at the input to AND gate 2. Input B proceeds directly into AND gate 2 but is inverted to $\bar{B}$ at the input to AND gate 1. The OR gate takes the outputs from these AND gates and produces output C.

A truth table showing inputs, output, and intermediate values appears as Table 13-10. We see that output C has the properties of the XOR operation. This circuit, therefore, is a way of implementing an XOR gate from NOT, AND, and OR gates. ■

TABLE 13-10. Truth Table for Fig. 13-13.

Inputs		Intermediate Values				Output
A	B	$\bar{A}$	$\bar{B}$	$A \cdot \bar{B}$	$\bar{A} \cdot B$	C
0	0	1	1	0	0	0
0	1	1	0	0	1	1
1	0	0	1	1	0	1
1	1	0	0	0	0	0

13-4 BOOLEAN ALGEBRA

The concept of a logic gate is an abstraction under which we can generalize particular pneumatic, hydraulic, and electronic modes of implementing machine control logic. However, we have defined the logic gate only in terms of a symbol and truth table. Ideally, we would like to have an algebra for manipulating the inputs and outputs of logic gates in a purely formal manner. This algebra will provide a power of analysis and synthesis that goes far beyond an intuitive or visual approach.

Such an algebra was invented by George Boole (1815–1864), an Irishman, who wanted to develop a calculus for dealing with the "laws of thought." This Boolean algebra differs from ordinary algebra in that it relates variables each of which can assume only the states 1 or 0. Although this two-state algebra may seem strange and unfamiliar, it is actually simpler than ordinary algebra.

The commutative, associative, and distributive laws of Boolean algebra are similar to those of ordinary algebra. There are, however, additional powerful theorems which are quite different from ordinary algebra. All of these rules and theorems have their switching circuit and logic gate circuit equivalents. And the content of these algebraic theorems is consistent with the objective of representing a calculus for logical operations.

The most important Boolean theorems are listed in Table 13-11. Note that there is a certain symmetry between the AND type and OR type of identities. This symmetrical property is known as the principle of duality. Perhaps this dual nature of the theorems is best seen in the two theorems of DeMorgan, which are very useful in simplifying Boolean algebraic expressions. These theorems can be easily proved using the truth table method.

TABLE 13-11. Boolean Algebra Identities.

	Theorems in One Variable AND	OR
$\bar{\bar{A}} = A$	$A \cdot 0 = 0$	$A + 0 = A$
$\bar{0} = 1$	$A \cdot 1 = A$	$A + 1 = 1$
$\bar{1} = 0$	$A \cdot A = A$	$A + A = A$
	$A \cdot \bar{A} = 0$	$A + \bar{A} = 1$

	Theorems in More Than One Variable AND	OR
Commutative	$A \cdot B = B \cdot A$	$A + B = B + A$
Associative	$(A \cdot B) \cdot C = A \cdot (B \cdot C)$	$(A + B) + C = A + (B + C)$
Distributive	$A \cdot (B + C) = A \cdot B + A \cdot C$	$A + B \cdot C = (A + B) \cdot (A + C)$
Absorption	$A \cdot (A + B) = A$	$A + A \cdot B = A$
DeMorgan's	$\overline{A \cdot B} = \bar{A} + \bar{B}$	$\overline{A + B} = \bar{A} \cdot \bar{B}$

EXAMPLE 13-8

Listing the states that variables A and B can assume, prove DeMorgan's theorems.

A	B	$\bar{A}$	$\bar{B}$	$A \cdot B$	$A + B$	$\overline{A \cdot B}$	$\overline{A + B}$	$\bar{A} + \bar{B}$	$\bar{A} \cdot \bar{B}$
0	0	1	1	0	0	1	1	1	1
0	1	1	0	0	1	1	0	1	0
1	0	0	1	0	1	1	0	1	0
1	1	0	0	1	1	0	0	0	0

Since the seventh column is identical to the ninth column and the eighth column is identical to the tenth column, DeMorgan's theorems have been proved because every possible combination of states has been tested for variables A and B. This is a good illustration of the use of a truth table. ■

EXAMPLE 13-9

Using the Boolean theorems, simplify the expression

$$(A + C) \cdot (A + \bar{C}) \cdot (\bar{B} + C)$$

From the distributive law, we see that

$$(A + C) \cdot (A + \bar{C}) = A + C\bar{C}$$

But $C\bar{C} = 0$ from the theorems in one variable. Therefore,

$$(A + C) \cdot (A + \bar{C}) \cdot (\bar{B} + C) = A(\bar{B} + C)$$ ■

EXAMPLE 13-10

Simplify the expression

$$AB + BC + \bar{A}C$$

The steps are listed below.

$$
\begin{aligned}
AB + BC + \bar{A}C &= AB + BC \cdot 1 + \bar{A}C \\
&= AB + BC(A + \bar{A}) + \bar{A}C \\
&= AB + BCA + BC\bar{A} + \bar{A}C \\
&= AB + ABC + \bar{A}CB + \bar{A}C \\
&= AB + ABC + \bar{A}C + \bar{A}CB \\
&= AB + \bar{A}C
\end{aligned}
$$

The last step made use of the absorption theorem. ■

EXAMPLE 13-11

Numerous corollary theorems can be proved using the theorems from Table 13-11. As an illustration utilizing DeMorgan's theorem, consider the expression

$$(\bar{A}\bar{B} + C)\bar{A}D$$

Complementing the whole expression, we obtain, from DeMorgan's theorems,

$$
\begin{aligned}
\overline{(\bar{A}\bar{B} + C)\bar{A}D} &= \overline{\bar{A}\bar{B} + C} + \bar{\bar{A}} + \bar{D} \\
&= \overline{\bar{A}\bar{B} + C} + A + \bar{D} \\
&= \overline{\bar{A}\bar{B}} \cdot \bar{C} + A + \bar{D} \\
&= (\bar{\bar{A}} + \bar{\bar{B}}) \cdot \bar{C} + A + \bar{D} \\
&= (A + B)\bar{C} + A + \bar{D}
\end{aligned}
$$

We see that the complement of a rather involved expression can be decomposed into constituent terms in this fashion. ■

13-4-1 Realization of a Boolean Expression in a Logic Gate Array

The correspondence between the Boolean + and · signs and the logical OR and AND operations implies that a Boolean expression can be realized from a synthesis of elementary logic gates. Indeed, this can be done.

Consider the final expression obtained in Example 13-11, that is,

$$(A + B)\bar{C} + A + \bar{D}$$

Remembering that + signifies OR, that · signifies AND, and that ¯ signifies NOT, the above expression is realized at the output of the final OR gate in the circuit shown in Fig. 13-14. Note that an OR gate having three inputs is shown in the figure. The truth table for this OR gate is simply an extension of that for two inputs; there will be a logic 1 output provided at least one of the inputs has logic level 1.

There is an incentive to construct logic gate circuits using only a single type of gate because of the way logic gate circuit packages are marketed commercially.

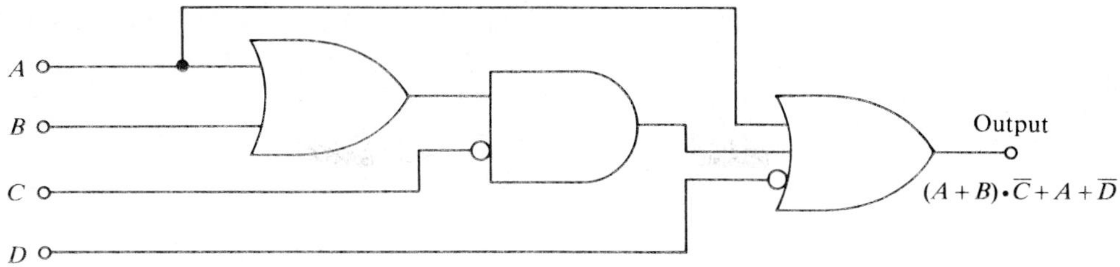

Figure 13-14. Implementation of $(A + B) \cdot \bar{C} + A + \bar{D}$.

And, of these logic gate packages, the NAND gate package is a popular choice because it is capable of easily implementing the NOT, AND, and OR functions.

As an illustration of the universal nature of the NAND gate, the XOR function we considered previously in Fig. 13-13 is implemented using only NAND gates in Fig. 13-15. It can be seen in Fig. 13-15 that the NOT function can be implemented by tying the two inputs of a NAND gate together. This is easily proved using DeMorgan's theorem as follows:

$$\overline{A \cdot A} = \bar{A} + \bar{A} = \bar{A} \tag{13-3}$$

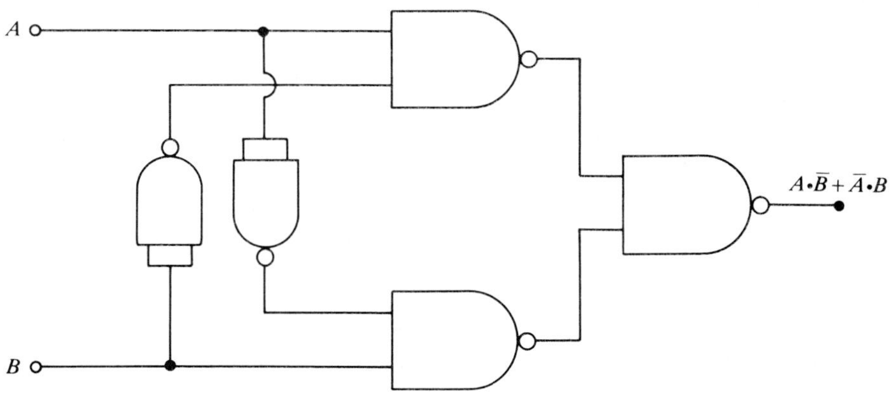

Figure 13-15. Implementation of XOR using NAND gates.

The fact that the arrangement of Fig. 13-15 implements XOR is also demonstrated from DeMorgan's theorems, that is,

$$\overline{\overline{(A \cdot \bar{B})} \cdot \overline{(B \cdot \bar{A})}} = A \cdot \bar{B} + \bar{A} \cdot B \tag{13-4}$$

and

$$A \cdot \bar{B} + \bar{A} \cdot B = \overline{\overline{A \cdot \bar{B} + B \cdot \bar{A}}} \tag{13-5}$$

The Half-Adder. As a final illustration of how Boolean expressions can be realized using logic gates, we will synthesize a gate structure for implementing the truth table for binary addition shown in Table 13-2. Note that the carry column C requires simply an AND gate. The sum S is simply the output of an XOR gate which could be implemented using either Fig. 13-13 or Fig. 13-15. Here, however, we will use the Boolean theorems to develop a logic gate circuit using fewer gates than found in Figs. 13-13 and 13-15.

The Boolean expression for the sum is

$$S = A \cdot \bar{B} + \bar{A} \cdot B \qquad \text{(XOR function)} \tag{13-6}$$

Using the Boolean identities ($A \cdot \bar{A} = 0$), we obtain

$$S = \bar{B} \cdot (A + B) + \bar{A} \cdot (A + B) \tag{13-7}$$

or

$$S = (\bar{B} + \bar{A}) \cdot (A + B) \tag{13-8}$$

Using DeMorgan's theorem for $\bar{B} + \bar{A}$ yields

$$S = (A + B) \cdot (\overline{A \cdot B}) \tag{13-9}$$

This sum S is implemented in Fig. 13-16 using AND, OR, and NOT gates. We see that 2-bit binary addition can be implemented using only three gates and an inverter. This circuit is called a half-adder because it cannot incorporate previous carries into its input.

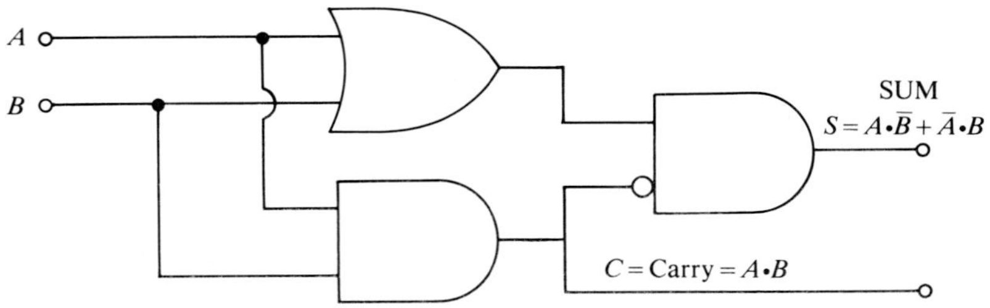

Figure 13-16. Half-adder logic gate circuit.

13-4-2 Synthesis of a Truth Table Using Minterms

It should be apparent at this point that there is an equivalence between relay ladder circuits, logic gate circuits, truth tables, and Boolean algebraic expressions (as exemplified by Table 13-9). And we have just seen how to synthesize a logic gate circuit for representing a Boolean expression.

In many instances, the most primitive information we may have about the control requirements for a mechanical system will be in truth table form. For example, let's assume that our mechanical system is one whose functional requirements can be satisfied with a sequence control scheme of the on-off type. An injection molding machine or a hydraulic backhoe digger are good examples.

One of the first things the designer must do is to analyze the machine cycle and define, in some methodical way, the sequence (or order) in which certain actions will be initiated, sustained, and terminated. This study will result in some kind of flow diagram or sequence diagram involving the actuation of motors, pumps, cylinders, and valves. An example was provided in Fig. 4-1 where two hydraulic cylinders were required to act sequentially in a prescribed fashion.

The result of the study may be the development of a truth table for representing the conditions for energizing an actuator (or several actuators). Consequently, a method for synthesizing the Boolean algebraic equivalence of a truth table would be a powerful tool in designing the control system. Of the several methods available, the most useful is the method of minterms.

As an illustration of the minterm method, consider the arbitrary truth table for three inputs illustrated as Table 13-12. Consider the first row for which the output variable D is to be logic level 1. This is the second row, where the inputs are $A = 0$, $B = 0$, and $C = 1$. This row of the truth table can be represented by the Boolean ANDing of $\bar{A}$, $\bar{B}$, and C, namely, by the expression $\bar{A} \cdot \bar{B} \cdot C$. For every other row of the truth table, this expression produces the value 0.

TABLE 13-12. Arbitrary Truth Table for Three Inputs.

Inputs			Output
A	B	C	D
0	0	0	0
0	0	1	1
0	1	0	1
0	1	1	0
1	0	0	0
1	0	1	0
1	1	0	1
1	1	1	1

Similarly, for the third row of the truth table, the expression $\bar{A} \cdot B \cdot \bar{C}$ produces an output of logic level 1 (and an output of logic level 0 for every other row of the truth table). Proceeding in this fashion for every row of the truth table for which the output is to be logic level 1, we see that the superposition of all of these constituent ANDed expressions will produce a Boolean expression that represents the truth table. In other words,

$$D = \bar{A} \cdot \bar{B} \cdot C + \bar{A} \cdot B \cdot \bar{C} + A \cdot B \cdot \bar{C} + A \cdot B \cdot C \qquad \textbf{(13-10)}$$

Each product term of Eq. (13-10) is called a minterm, and the representation of the truth table by Eq. (13-10) is called its minterm canonical form. This form is also sometimes called the standard sum-of-products form.

So the method of minterms involves representing each row of a truth table having an output of logic level 1 by a minterm in which the inputs having logic level 0 are complemented while those having logic level 1 are uncomplemented. Then these minterms are added together (ORed) to provide the minterm canonical form for the truth table. Note that the rows having an output of logic level 0 are not even used, so they might as well be deleted from the truth table insofar as the standard sum-of-products representation is concerned.

EXAMPLE 13-12

Truth table 13-13 represents the states of a control system in which the output is a control relay 4CR in its energized state (logic level 1). The inputs consist of two pushbuttons, 1PB and 2PB, and three relay contacts, 1CR, 2CR, and 3CR. It must be remembered that the total universe of bit patterns with five inputs is $2^5 = 32$. Here, however, only seven of these bit patterns will result in actuating 4CR, so Table 13-13 is only a part of the total truth table, but it is the only portion we need to be concerned with.

TABLE 13-13. States of a Control System.

Inputs					Output
1PB	2CR	1CR	2PB	3CR	4CR
0	0	0	1	0	1
0	0	1	0	0	1
0	0	1	1	0	1
0	1	0	0	0	1
0	1	0	1	0	1
0	1	1	0	0	1
0	1	1	1	0	1

The minterm for the first row of the truth table is

$$\overline{1PB} \cdot \overline{2CR} \cdot \overline{1CR} \cdot 2PB \cdot \overline{3CR}$$

This minterm was constructed by complementing the inputs having logic level 0 and using the input uncomplemented when it has logic level 1. Proceeding as before, we find that the minterm canonical form of the truth table is

$$\begin{aligned} CR4 = {} & \overline{1PB} \cdot \overline{2CR} \cdot \overline{1CR} \cdot 2PB \cdot \overline{3CR} + \overline{1PB} \cdot \overline{2CR} \cdot 1CR \cdot \overline{2PB} \cdot \overline{3CR} \\ & + \overline{1PB} \cdot \overline{2CR} \cdot 1CR \cdot 2PB \cdot \overline{3CR} + \overline{1PB} \cdot 2CR \cdot \overline{1CR} \cdot \overline{2PB} \cdot \overline{3CR} \\ & + \overline{1PB} \cdot 2CR \cdot \overline{1CR} \cdot 2PB \cdot \overline{3CR} + \overline{1PB} \cdot 2CR \cdot 1CR \cdot \overline{2PB} \cdot \overline{3CR} \\ & + \overline{1PB} \cdot 2CR \cdot 1CR \cdot 2PB \cdot \overline{3CR} \end{aligned} \tag{a}$$

By factoring, we can simplify this to

$$CR4 = \overline{1PB} \cdot \overline{3CR} \cdot [\overline{2CR} \cdot (\overline{1CR} \cdot 2PB + 1CR) + 2CR] \tag{b}$$

In obtaining Eq. (13-12), we used the Boolean identity $\bar{A} + A = 1$.

Using yet another Boolean identity, namely, $A + \bar{A} \cdot B = A + B$, we can further reduce Eq. (13-12) to

$$4CR = \overline{1PB} \cdot (2CR + 1CR + 2PB) \cdot \overline{3CR} \tag{c}$$

It is not possible to further simplify the above Boolean expression. ■

Inspecting Eq. (c) carefully, we begin to see the usefulness of being able to represent a truth table by its equivalent Boolean algebraic equation. Recalling that the · is equivalent to AND while + is equivalent to OR, it is now a simple matter to construct a logic gate array that represents Eq. (c) and, therefore, truth table 13-13. And, recalling that the AND operation is equivalent to relay ladder devices in series while the OR operation implies devices in parallel, we can synthesize a relay ladder type of circuit for implementing this logic. Note that the complement of a variable is equivalent to a normally closed device while an uncomplemented variable is represented by a normally open device. These circuits are shown in Fig. 13-17.

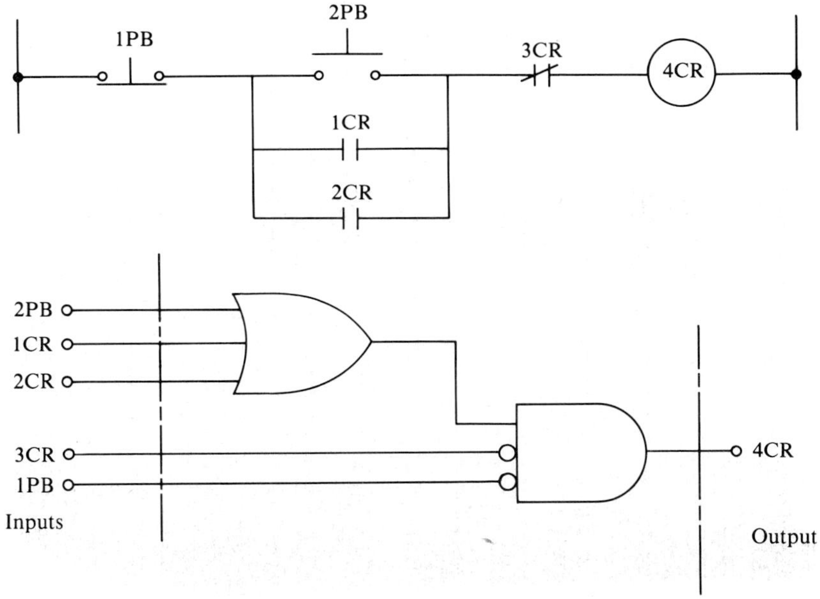

Figure 13-17. Equivalent relay ladder and logic gate circuits for truth table 13-13.

The circuits of Fig. 13-17 are methods by which the logic of truth table 13-13 can be implemented in hardware. They are called hardwired, or dedicated, hardware circuits, which means they are not programmable methods of implementing this logic. Changes can be made only by physically rewiring the circuits.

Finally, note that inputs are placed at the left and outputs are placed at the right in these diagrams.

13-5 FUNCTIONAL LOGIC CIRCUITS

We have just seen how some relatively simple logic circuits can be synthesized from the elementary logic gates summarized in Table 13-9. The possible combinations of these gates for producing useful circuits are virtually limitless. However, certain circuits recur frequently. For example, a microprocessor is a complex logic system made up of certain functional logic circuits synthesized from the elementary AND, OR, NOT, NAND, and NOR gates. Examples are adders, decoders, registers, flip-flops, and memories.

Our purpose here is not to delve into the internal workings of a microprocessor in any great detail. This would require a complete book in itself. We just want to discuss some simple functional logic circuits that will provide insight into how a microprocessor rewires itself with each instruction, how an addressing system works, and how a register works.

13-5-1 Decoders

The typical memory used in an 8-bit microprocessor will have the capability of storing 8-bit bytes in 65,536 memory locations (a 64K memory). These 65,536 locations can be addressed using only 16 bits (the addresses will range from 0000

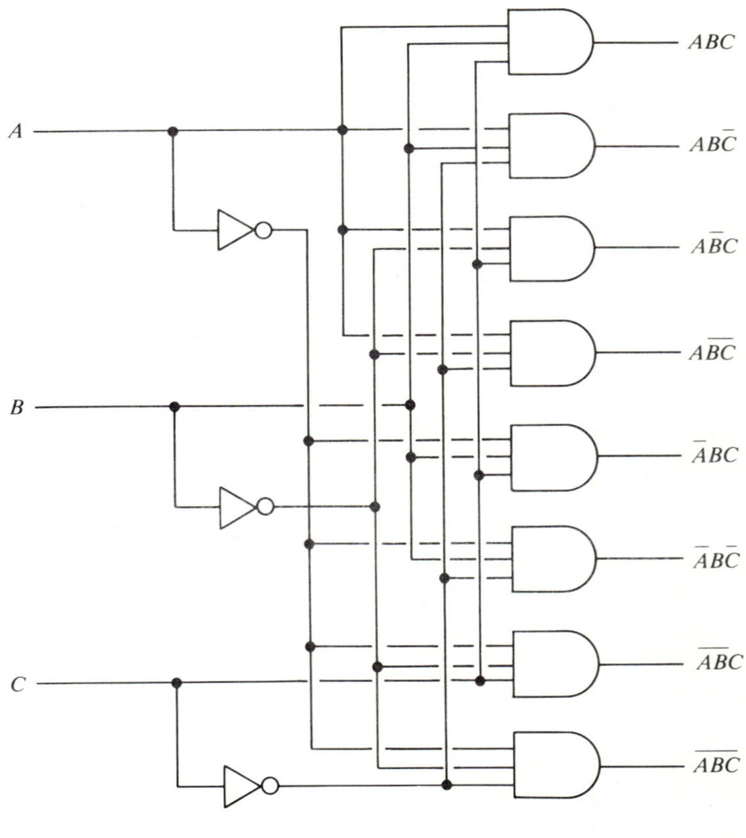

Figure 13-18. One-out-of-eight decoder.

0000 0000 0000 to 1111 1111 1111 1111). The process of selecting a memory location from among the 65,536 available using only the 16-bit inputs is called decoding, and the logic gate circuit for implementing this process is called a decoder. We will illustrate the concept of a decoder using only eight memory locations which can be addressed by using only 3 bits (000 to 111).

Consider the logic gate circuit sketched in Fig. 13-18. This circuit is called a three-to-eight decoder (or in terms of only the outputs, a one-out-of-eight decoder). The eight distinct outputs of the decoder are outputs of AND gates such that for any combination of the 3 input bits only one of the outputs will be at logic level 1; all others will be at logic level 0. Note that each output is a distinct minterm of the A, B, and C bits. The truth table for this decoder is displayed as Table 13-14.

TABLE 13-14. Truth Table for Three-to-Eight Decoder (Fig. 13-18).

Inputs (Address)			Active Output (Location)	
A	B	C	Gate	Minterm
0	0	0	1	ABC
0	0	1	2	$AB\bar{C}$
0	1	0	3	$A\bar{B}C$
0	1	1	4	$A\overline{BC}$
1	0	0	5	$\bar{A}BC$
1	0	1	6	$\bar{A}B\bar{C}$
1	1	0	7	$\overline{AB}C$
1	1	1	8	$\overline{ABC}$

Decoders such as that shown in Fig. 13-18 are supplied as part of digital logic families marketed by various electronics manufacturers. One such family is the transistor-transistor logic (TTL) family described in Reference 13-2. In this family of devices, the three-to-eight decoder package is given the number 7442 and is referred to as "the 7442."

In a more general sense, the process of decoding can be considered to be implemented by a logic gate circuit that takes an input bit pattern and decodes it in such a way as to activate (place logic level 1) a specific set of outputs. Decoding is the key step whereby software (instructions) activates a unique set of gates that send out only the signals necessary to perform an instruction. This process will be discussed again later.

EXAMPLE 13-13

Show that the logic gate circuit sketched in Fig. 13-19 has the capability of addressing two, and only two, of the 256 eight-bit addresses in the range 0000 0000 to 1111 1111.

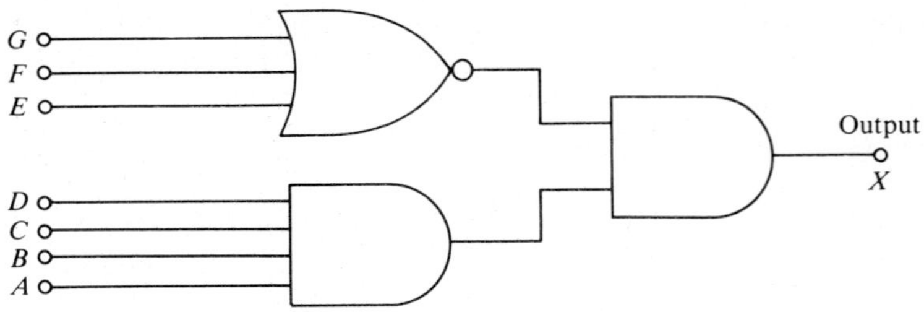

Figure 13-19. Decoding circuit for Example 13-13.

Solution

The output of the four-input AND gate is $A \cdot B \cdot C \cdot D$ while the output of the three-input NOR gate is $\overline{E + F + G}$. Consequently, the final output X is given by

$$X = (A \cdot B \cdot C \cdot D) \cdot (\overline{E + F + G}) \qquad \textbf{(1)}$$

Using DeMorgan's theorem, this is the same as

$$X = A \cdot B \cdot C \cdot D \cdot \bar{E} \cdot \bar{F} \cdot \bar{G} \qquad \textbf{(2)}$$

If we now identify A as the most significant bit of an 8-bit address and G as the bit next to the least significant bit, this logic gate circuit will address locations 1111 0000 and 1111 0001 because the least significant bit can be either at logic level 1 or 0. ■

13-5-2 The S-R Flip-Flop

Up to this point, all of the logic gates and circuits discussed have been of the combinational type. That is, the output(s) are completely defined by the current state of the inputs. In particular, the order in which the inputs have been set is of no consequence in dictating the state of the output of a combinational circuit.

Although they are essential in digital control schemes, combinational logic gate circuits do not possess one of the prime requisites of a total control system. This is the necessity to have circuits that can remember previous signals and that can discriminate the order in which input signals have been set. Such logic gate circuits are called sequential circuits because the output does depend on the order (or sequence) in which the input signals have been set or reset (S-R). Sequential circuits are used in constructing microprocessor components such as registers and memories.

The basic element of sequential logic is the set-reset (S-R) flip-flop. It is a simple memory circuit using two gates in such a way that a cross-coupled feedback is created between the outputs and inputs of the gates. Two different versions for implementing an S-R flip-flop are shown in Fig. 13-20. The truth tables for these flip-flops are also shown.

Consider the S-R flip-flop that uses the NAND gates as in Fig. 13-20(a). Remember that with a NAND gate the output will be logic level 1 if any input is logic level 0. Consequently, if both S and R inputs are at logic level 0, then both

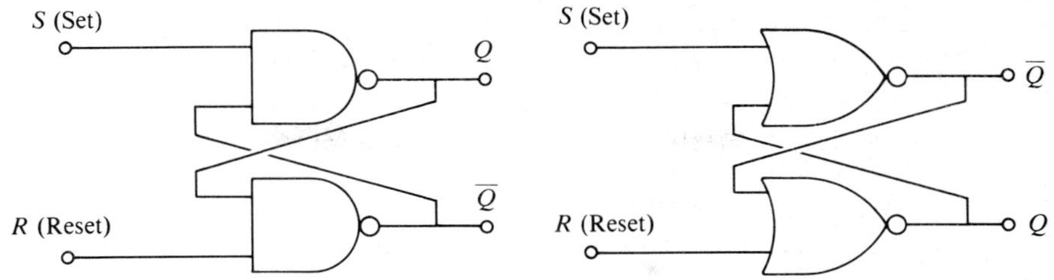

Inputs		Outputs	
S	R	Q	$\bar{Q}$
1	0	0	1
0	1	1	0
0	0	Undefined	
1	0→1	0	1
0→1	1	1	0

(a) Using NAND Gates

Inputs		Outputs	
S	R	Q	$\bar{Q}$
1	0	1	0
0	1	0	1
1	1	Undefined	
1→0	0	1	0
0	1→0	0	1

(b) Using NOR Gates

Figure 13-20. Set-reset (S-R) flip-flop.

outputs Q and $\bar{Q}$ will be at logic level 1, which is a contradiction to the use of the symbols Q and $\bar{Q}$. Therefore, this state will be avoided in the use of the flip-flop constructed of NAND gates.

For the defined states in the truth table of Fig. 13-20(a), outputs Q and $\bar{Q}$ will always have opposite logic levels. To see how this device works, suppose the set (S) input is held at logic level 1. If the reset (R) input is then placed at logic level 0, the output $\bar{Q}$ will definitely be logic level 1 since this is a NAND gate. Feeding output $\bar{Q}$ back to the input of the other NAND gate produces an output of $Q = 0$, because now both inputs are at logic level 1 (because S was held at 1). But the value $Q = 0$ (feedback) guarantees that output $\bar{Q}$ will be at logic level 1 because it comes from a NAND gate. Therefore, the act of setting $R = 0$ latches the values $Q = 0$, $\bar{Q} = 1$ irrespective of future values of R provided S is held at logic level 1. This is why the S-R flip-flop is also called a latch. This latching property is indicated by the fourth row of the truth table.

Similarly, if the input reset (R) is held at logic level 1, then the act of setting the input set (S) at logic level 0 will latch the outputs to $Q = 1$, $\bar{Q} = 0$ irrespective of future values of S.

In use, input terminals S and R will normally both be at logic level 1. The state of the outputs will then depend on which of the S or R inputs was last set at logic level 0. Suppose S was at logic level 0 more recently than R. Then the output state will be $Q = 1$, $\bar{Q} = 0$. If now input R is set at a momentary logic level of 0, the device will flip-flop to the output state $Q = 0$, $\bar{Q} = 1$. Again, the state of the inputs will return to $S = 1$, $R = 1$ after the momentary null of input R is restored to logic level 1, and the flip-flop will be latched at the state $Q = 0$, $\bar{Q} = 1$.

A similar analysis can be made for the NOR gate S-R flip-flop. Here, however, the inputs are both normally at logic level 0 and the device is flip-flopped by placing a momentary logic level of 1 at one of the inputs. For this implementation, the state $S = 1$, $R = 1$ is undefined.

The discussion of flip-flops could be continued to include many additional types such as the *JK*, *D*, and *T* flip-flops. These devices are used in constructing the basic building blocks of microprocessors such as counters and registers. Except for the *D* flip-flop, we refer the reader to one of the many books available on microprocessors or digital circuits for further information (e.g., see Reference 13-3).

The *D* Flip-Flop. Consider the S-R flip-flop constructed from NOR gates as shown in Fig. 13-20(b). Since there is no virtue in having both inputs take up the value of logic level 1 simultaneously, one way of preventing this is to tie *S* and *R* together in such a way that they are always inverted (i.e., $S = \bar{R}$). But doing this will also prevent the desirable input state $S = 0$, $R = 0$ from being retained. A way to achieve both of these desirable changes is to use the flip-flop sketched in Fig. 13-21.

C	D	Q	$\bar{Q}$
1	1	1	0
1	0	0	1
0	1 Last	1	0
0	0 Last	0	1

Figure 13-21. *D* flip-flop (or data latch).

The NOR gate version of the S-R flip-flop is preceded by two AND gates and an inverter. The terminal labeled *C* is a control, or enabling, input (also called a clocked or gated input). In other words, if input *C* is at logic level 0, then both *S* and *R* will be at logic level 0 irrespective of the value of input *D*. But if control input *C* is at logic level 1, then the value $D = 1$ is able to transmit as $S = 1$, $R = 0$, while $D = 0$ is able to transmit as $S = 0$, $R = 1$.

It is clear that this device functions essentially the same as the S-R flip-flop but with the undefined state eliminated. The truth table for this *D* flip-flop is displayed in Fig. 13-21.

The control input enables the flip-flop to capture the state of input D at the time of the pulse of logic 1 at control C. In a microprocessor, the timing of these enabling pulses at C is controlled by a clock. This clock is the means by which many simultaneous operations of this type are synchronized.

EXAMPLE 13-14

One of the fundamental building blocks of microprocessors and programmable controllers is the register. A register is a memory element that can be used to store a byte of binary data in an 8-bit system. Individual registers have unique locations and generally are named. An example in the microprocessor is the accumulator, which is simply a register that is used in every arithmetic and logical operation. Other registers are the instruction, index, and flag registers.

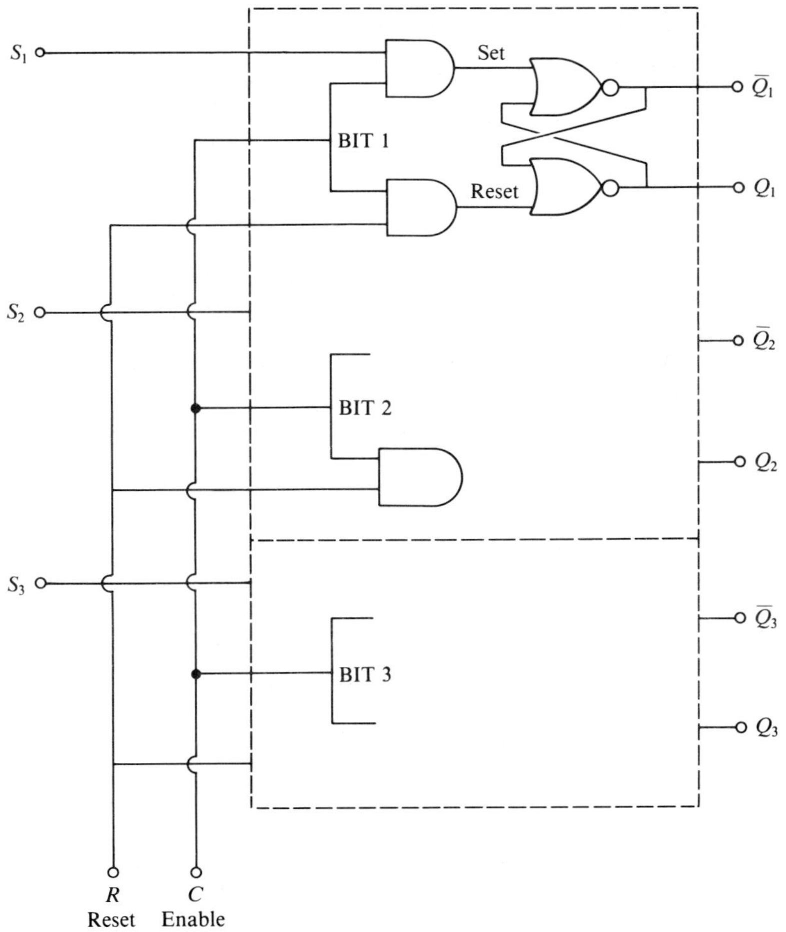

Figure 13-22. Three-bit register comprised of three S-R flip-flops.

A memory can be considered to be merely a number of registers that are consecutively addressable. Logically, a register is simply a collection of flip-flops that are synchronized when they are actuated. A 3-bit register consisting of S-R flip-flops is illustrated in Fig. 13-22.

Note how the enabling inputs of the three flip-flops are synchronized by tying them together to a common signal line. This enable line is connected to a timing and control section of the logic unit of which this register is a component. When the enabling input C is at logic level 1, the register will accept a waiting binary signal $S_1S_2S_3$ which it stores in the register as $Q_1Q_2Q_3$. The register can be reset to all 0s if the reset line R is set to logic level 1 while the register is enabled ($C = 1$). ■

13-6 SUMMARY

In this chapter, we have tried to systematically summarize the basic concepts of binary numbers, binary logic, switching circuits, logic gates, Boolean algebra, and functional logic circuits necessary to an understanding of how microprocessors and programmable controllers operate. We have also shown the equivalences between relay circuits, logic gate circuits, truth tables, and Boolean formulas for some simple control structures. Even though the subject of digital control is broad and comprehensive, neither sophisticated mathematics nor a large number of prerequisite courses are required in order to comprehend its bare bones as we have outlined.

The mechanical systems designer should master these essential concepts of digital control because they are the wave of the future. At the very least, the systems designer must be able to comprehend how control systems incorporate microprocessors or programmable controllers as black boxes and the capability of these black boxes to make machine actuators responsive to machine sensors and transducers. These devices will be discussed in the next two chapters.

13-7 REFERENCES

13-1. R. W. Henke, *Introduction to Fluid Power Circuits and Systems*, Addison-Wesley, Reading, MA, 1970.

13-2. D. Lancaster, *TTL Cookbook*, published by Howard W. Sams and Co., 4300 West Sixty-Second Street, Indianapolis, IN.

13-3. D. D. Givone and R. P. Roesser, *Microprocessors/Microcomputers: An Introduction*, McGraw-Hill, New York, 1980.

PROBLEMS

13-1. Convert the following binary numbers into their decimal equivalents:

(a) 01101110 **(b)** 11110000
(c) 01000101 **(d)** 00110110
(e) 11111111 **(f)** 1001

13-2. Convert the following decimal numbers into their binary equivalents:
(a) 467 (b) 63
(c) 54 (d) 102
(e) 99 (f) 705

13-3. Perform the following binary arithmetic operations:
(a) 110110 + 11010 (b) 10110 − 1011
(c) 1011 · 1100 (d) 1101 + 1100

13-4. Represent the following decimal numbers as binary coded decimals:
(a) 139 (b) 6875
(c) 705 (d) 12
(e) 99 (f) 1022

13-5. Convert the following binary numbers into their hexadecimal equivalents:
(a) 01101010 (b) 11111111
(c) 000111110000 (d) 1001
(e) 00001011 (f) 11101100

13-6. Show that the hexadecimal range 40 to F0 is capable of addressing 176 locations.

13-7. Write down the ASCII code for the message HOT STUFF.

13-8. An input A is inverted and applied to an OR gate. An input B is applied directly to the OR gate. Construct the truth table for this logic gate circuit.

13-9. Show how to construct an AND gate using inverters and an OR gate.

13-10. Show how a three-input AND gate can be synthesized from two simple AND gates having two inputs each.

13-11. Develop the truth table for the logic gate circuit shown in Fig. P13-11.

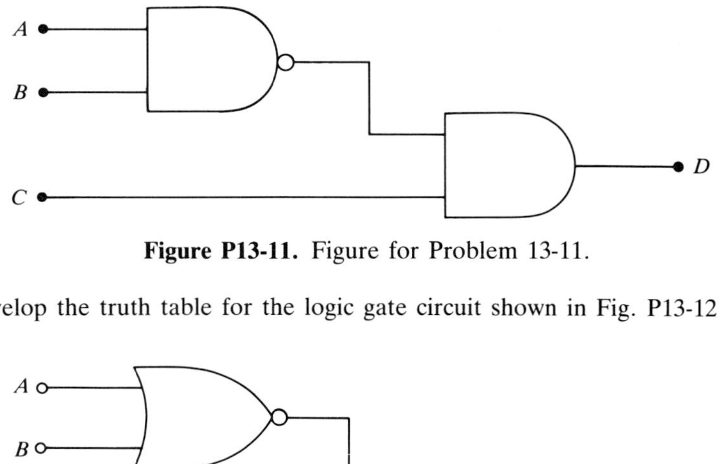

Figure P13-11. Figure for Problem 13-11.

13-12. Develop the truth table for the logic gate circuit shown in Fig. P13-12.

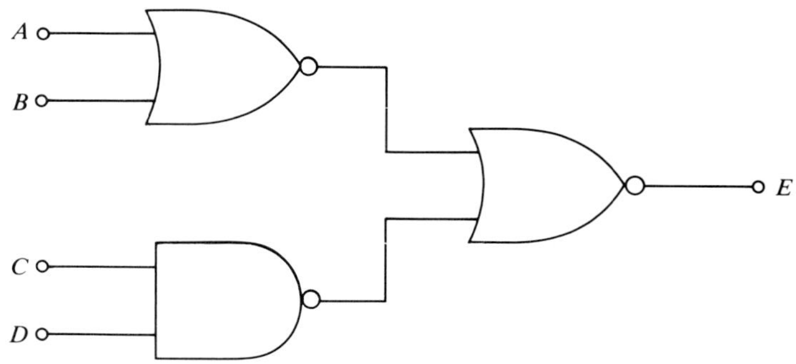

Figure P13-12. Figure for Problem 13-12.

13-13. Satisfy yourself that the relay equivalent of the XOR truth table shown in Table 13-9 is correct.

13-14. Using a truth table, prove that

$$A + \bar{A} \cdot B = A + B$$

13-15. Using a truth table, prove that

$$A \cdot B + \bar{A} \cdot B = B$$

13-16. Using a truth table, prove that

$$A \cdot (B + C) = A \cdot B + A \cdot C$$

13-17. Using two OR gates, an AND gate, and two inverters, implement the function

$$(A + B) \cdot (\bar{A} + \bar{B}) = C$$

13-18. Implement the function of Problem 13-17 using two AND gates, an OR gate, and two inverters.

13-19. Show how an OR function can be synthesized using only NAND gates. (Hint: First develop a Boolean expression.)

13-20. Using minterms, synthesize a Boolean expression for representing the truth table of Fig. P13-20.

A	*B*	*C*	*D*
0	0	0	1
0	0	1	0
0	1	0	0
0	1	1	1
1	0	0	0
1	0	1	0
1	1	0	1
1	1	1	1

Figure P13-20. Figure for Problem 13-20.

13-21. Using minterms, synthesize a Boolean expression for representing the truth table of Fig. P13-21.

A	*B*	*C*	*D*
0	0	0	0
0	0	1	1
0	1	0	1
0	1	1	1
1	0	0	0
1	0	1	1
1	1	0	0
1	1	1	1

Figure P13-21. Figure for Problem 13-21.

13-22. Using the Boolean algebra theorems of Table 13-11, prove that

$$A \cdot \bar{B} + \bar{A} \cdot \bar{B} + A \cdot B \cdot \bar{C} + \bar{A} \cdot B \cdot \bar{C} = \bar{B} + \bar{C}$$

13-23. Using the Boolean algebra theorems of Table 13-11, prove that

$$(A + B) \cdot (\bar{A} + C) = A \cdot C + \bar{A} \cdot B$$

13-24. Using only NOR gates and an inverter, implement the function

$$E = A \cdot (B + C \cdot D)$$

You will need to use a three-input NOR gate.

13-25. Construct a two-to-four (or one-out-of-four) decoder using four NOR gates and two inverters (NOT gates).

13-26. Reason out the truth table for the S-R flip-flop of Fig. 13-20(b) (which uses NOR gates).

13-27. Consider a control system for which A, B, C, D, and E are inputs and X, Y, and Z are outputs. The system is to work in the following way:

(a) If input A, or B, or C is ON, then X is ON while Y and Z are OFF.

(b) If input A, and B, and C are OFF, then X is OFF and Y and Z are determined by inputs D and E as follows:

- If D is ON and E is ON, then Y and Z are ON.
- If D is ON and E is OFF, then Y and Z are ON.
- If D is OFF and E is ON, then Y is OFF and Z is ON.
- If D is OFF and E is OFF, then Y and Z are OFF.

Determine Boolean expressions for the outputs using the theory of minterms and sketch a logic gate circuit for implementing this control logic.

14

Microelectronic Control

Advantages of implementing on-off control of mechanical systems using hardwired, electromechanical relay hardware include the following:

(a) It is an accepted technology with which many older engineers, technicians, electricians, and maintenance personnel are familiar. They will not resist its use or be afraid of it.
(b) It is a dependable technology with an enviable history of reliability and ruggedness.
(c) It is a well-developed technology that is not in a state of evolution, diversity, or disorder.

But electromechanical relay ladder implementation using hardware is necessarily a dedicated method of control. If changes in the circuitry hardware are required, maintenance personnel must actually physically change the hardwired circuits. It is also becoming apparent that it is a more costly means of control than some emerging alternatives.

Implementation of relay ladder logic using dedicated microelectronic logic gate circuitry, as we discussed in Example 13-12, may overcome some of the cost problems, but it also represents a dedicated method of machine control. But of all the on-off control alternatives, the logic gate circuitry approach may have the greatest advantage in speed of operation (which could be critical in real-time control applications).

Development of the microprocessor as a low-cost control element has created the possibility of inexpensive, programmable control of on-off mechanical systems (and of many other kinds of analog control schemes as well). The versatility, adaptability, and logical superiority of the microprocessor makes it, figuratively, the wave of the future in mechanical systems control either directly or as the heart of the programmable controller.

The disadvantage of microprocessor-based control is that few technical people are thoroughly trained in its direct use as a binary digit manipulator. And the software development can be quite expensive. Note the implication, now, that if changes in logic circuitry are required in our control system at some point in time, we won't need maintenance people to get out their pliers and sidecutters; we will need programmers to reprogram the microprocessor to provide the new circuit required.

This chapter serves as a brief introduction to microprocessor-based control. We will first provide a glimpse of the concepts involved with direct microprocessor application. This discussion will be followed by a short discussion of the programmable controller. Since the range of applications of both of these devices is vast, we will restrict our illustrations to the relay ladder type.

14-1 MICROPROCESSOR CONTROL

In Chapter 13, it was shown that control logic for mechanical systems can be implemented using dedicated logic gate circuits constructed from transistor circuits. A microprocessor is also a logic gate controller constructed from transistor circuits but is far more versatile and complex. The microprocessor is programmable rather than being a dedicated logic controller. In fact, a microprocessor can be viewed as a logic gate circuit that rewires itself with each instruction. Decoding is the key step in this process that does the rewiring. The bit pattern of each instruction activates a unique set of gates in the decoder that send out only the signals necessary to perform that instruction. The microprocessor is also characterized by large-scale integration (LSI) of transistors within a single silicon chip. It contains thousands of transistors integrated within a very small space.

The microprocessor (MPU) is simply the central processor unit (CPU) of a programmable logic controller (PLC). By itself, the microprocessor cannot implement mechanical systems control. Not only must it be programmed but it is no substitute for devices such as limit switches, transducers, valves, solenoids, and similar hardware required to obtain signals from the mechanical system or to actuate mechanical actuators. In other words, the microprocessor will still require inputs to it (which it will manipulate), and the mechanical system will require outputs from it. We call this interfacing the microprocessor with input and output (via the input and output ports).

Additional devices are necessary before a microprocessor can perform a specified control task. A memory (or memories) is required to store both data and instructions. The complete set of components required to implement storage, transfer, arithmetic, logic, control, and input/output functions is called a microprocessor system (or microcomputer). The CPU is only one component in this system, as shown in Fig. 14-1.

We see that, conceptually, this scheme is no different from the many control systems we confront every day. A soft drink machine, for example, produces an output (a bottle or can) in response to our inputs (a coin and pushing a pushbutton). The machine does not replace the coin or soft drink. The interfaces

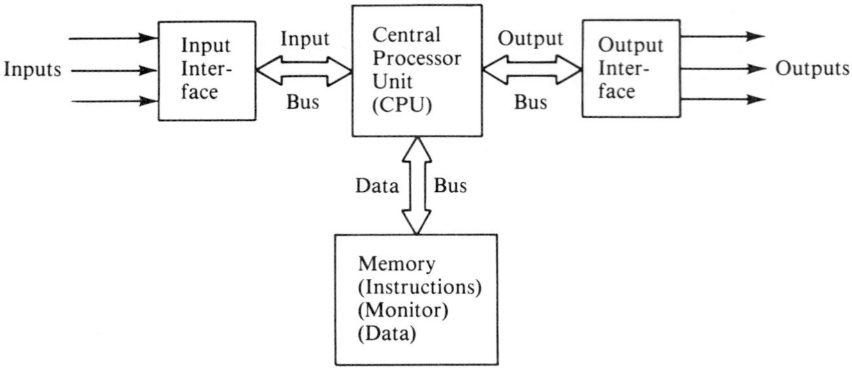

Figure 14-1. Microprocessor system.

shown in Fig. 14-1 are necessary because the CPU is a digital bit manipulator whereas the inputs and outputs from, and back to, the mechanical system may be either digital or analog physical signals.

The capability and cost of a microprocessor depend on the bit width(s) the microprocessor utilizes. The simplest microprocessors are 4-bit manipulators, while the latest machines can manipulate up to 32 bits. All of our discussion and examples in this book presume the microprocessor has 8-bit words for data, instructions, and arithmetic and logical operations. The addressing capability, however, will be 16 bits wide, permitting the addressing of a 64K memory (65,536 memory locations). The 16 bits required for the addressing can be manipulated by the microprocessor in the form of two 8-bit bytes.

Note that an 8-bit microprocessor can have up to 256 (2^8) distinct instruction codes. Not all of these codes will be used for instructions. For example, the Intel 8085 microprocessor is an 8-bit manipulator having some 100 instructions.

In addition to the requirement that the microprocessor be interfaced to input and output devices for our mechanical system, we also must execute the necessary control functions and algorithms by programming the microprocessor. A program is simply a sequence of instructions that the microprocessor must recognize as strings of binary digits. These instruction bytes must be decoded by the microprocessor control unit and implemented in terms of operations involving busses, registers, and arithmetic or logical operations.

The instructions that the microprocessor recognizes directly must be in binary code. At this most primitive level of programming, the language is called machine language (or machine code). Although an experienced microprocessor programmer can work directly in machine language, this technique is tedious and extremely vulnerable to errors. Consequently, a mnemonic equivalent has been developed for each primitive machine instruction. Programming using these mnemonic instructions is called assembly language programming, and will be discussed in some detail later.

In addition to the low-level machine and assembly languages, microprocessors can be programmed in higher-level languages such as BASIC and

PASCAL. However, this process will require the appropriate compiler or interpreter in order to convert the high-level program into the primitive code that the microprocessor requires to implement control operations. As with all computers, doing this will exact a heavy cost in terms of machine language overhead. The additional machine instructions required to implement a high-level language program may well make a real-time control problem impossible to implement in a high-level language so that assembly language may be required instead.

Finally, although microprocessor systems have the same characteristics as digital computers, they are unique due to their small size and very low cost. These incredible devices are finding an increasing spectrum of applications in manufacturing systems, automotive controls, appliances, instrumentation and data logging, and in the control of mechanical systems.

14-1-1 General Architecture of a Microprocessor System

A more detailed architecture of some of the more important functional modules of a microprocessor system is shown in Fig. 14-2. The central processor unit (CPU) is the microprocessor itself, which is enclosed within the dashed rectangle. This unit is connected to the input and output units and to the ROM and RAM memories by the data bus. The data bus is an 8-bit bus in an 8-bit microprocessor system, which means that it is simply a bundle of eight parallel wires for transmitting data between (and among) the various modules in the form of logic levels 1 and 0. The term ROM means Read Only Memory, and RAM means Random Access Memory. The ROM is where the program for the system monitor or operating system resides. This program cannot generally be changed or written into. The ROM can produce only output (or read only). The RAM is the location of the current program to be executed by this microprocessor system. Data are also stored in this memory. The RAM can be read both into and out of.

In addition to the data bus, there are two additional busses. The address bus is a 16-bit bus for selecting specific locations in either the ROM or RAM memories or for selecting input or output addresses as dictated by the control unit within the CPU. The control bus consists of lines going into the control unit and outgoing lines that control timing and the various system components. Timing and control functions are quite complicated and will be explained in more detail below.

14-1-2 Central Processing Unit (CPU)

As noted, the central processing unit (or microprocessor) is shown within the dashed rectangle of Fig. 14-2. It consists of an arithmetic logic unit (the ALU), a control unit, various registers, and busses for connecting the various units.

All arithmetic and logical operations are carried out within the ALU. The ALU gets its data from two registers via busses. One of the registers, the accumulator, is the most important register in the CPU. The other register that

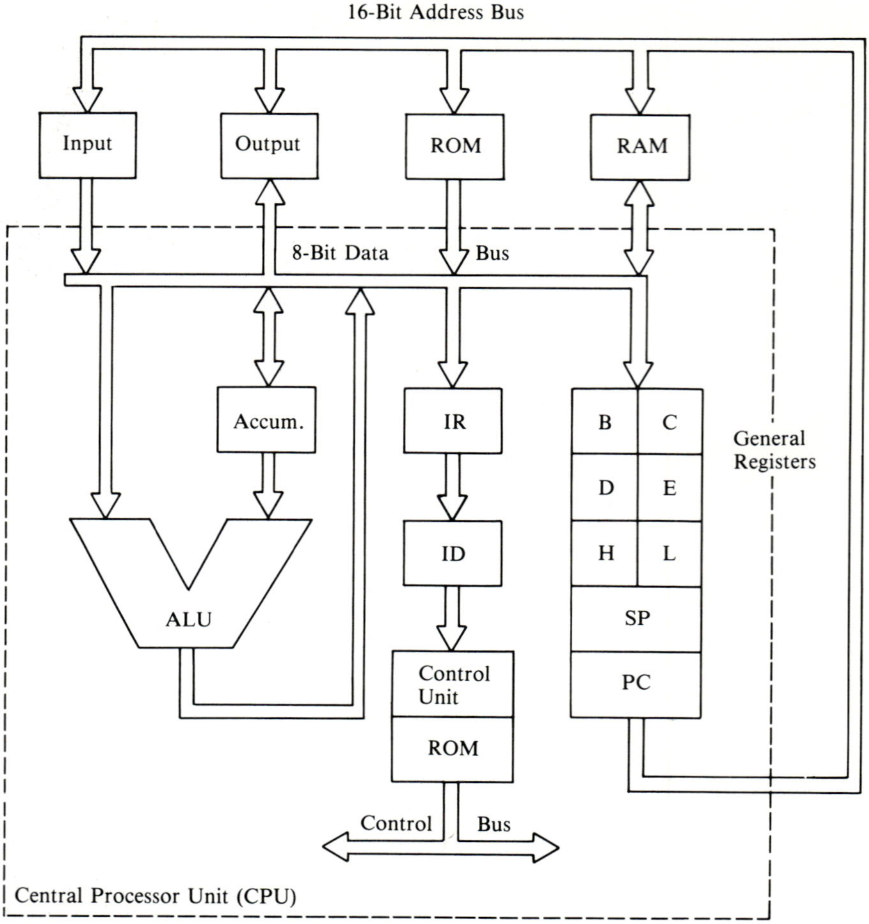

Figure 14-2. Illustrative 8-bit microprocessor system architecture.

feeds the ALU can be selected from among the several available, namely, the B, C, D, E, H, and L registers. After the ALU operates on the contents of the two registers that feed it, the result is stored back in the accumulator register via the ALU output bus which feeds back to the 8-bit data bus. Of course, all of these operations must be sequenced and timed by the control unit.

A typical arithmetic operation carried out by the ALU involves adding the contents of the accumulator to the contents of another register, say C, and storing the result in the accumulator. The assembly language instruction for this operation is ADD C. The ALU must communicate the carry from this arithmetic operation. It does this through a flip-flop whose state is contained in a register (not shown) called the flag register.

The heart of the microprocessor is the control unit which governs sequencing and timing of operations. The control unit fetches, decodes, and executes instructions. An instruction is first fetched from memory and then placed in a register, called the instruction register (IR). Then the instruction is decoded by an instruction decoder (ID). Finally, the instruction is executed by the control unit, resulting in the appropriate data transfers and operations. An illustration of this decoding and execution process is provided in Example 14-1. Note that the control unit contains its own ROM (which is inaccessible to the user). The control unit ROM stores the microprogram that defines the instruction set of the microprocessor and how the instructions are to be executed.

The boxes labeled B, C, D, E, H, and L are general-purpose registers which make it possible for the microprocessor to manipulate data at high speed. The program counter (PC) is a 16-bit register containing the address of the next instruction to be executed. It is automatically incremented after each instruction has been executed. The box labeled SP is called the stack pointer. Discussion of the stack and stack pointer is beyond the scope of this book.

EXAMPLE 14-1

A microprocessor works by shuttling digital words (bytes) within a network of logic gate circuits, each having a specific capability such as data storage (memory), arithmetic and logical operations (ALU), decoding and sequencing (control unit), and interfacing with input/output (I/O) devices. These units are connected by electronic switches and busses. To perform a complex control function requiring sequenced combinations of the various units, the switches open and close to wire and rewire the building blocks together as dictated by instructions contained in the RAM. Thus, a microprocessor program tells the microprocessor how to rewire itself from one machine language instruction to the next.

The simplified diagrams shown in Fig. 14-3 illustrate this process. A simple program, in machine code, is used to transfer two input values into the microprocessor, add them together, and then store the result in the memory. As we mentioned earlier, the program (called a microprogram) that governs this internal control process of decoding and rewiring the various circuits, in order to execute the user's instructions, resides in the ROM of the control unit. This microprogram can be visualized as a book of recipes, each page of which is the control recipe for executing an instruction. ■

14-1-3 Microprocessor Programming

Programming is the formal methodical process by means of which the engineer communicates to the microprocessor hardware how to implement a control task. Preparation of a program usually involves the following steps:

(a) Defining the control problem to be solved
(b) Identifying serial, parallel, sequential, and timing operations
(c) Constructing a flowchart for conceptually structuring a control scheme
(d) Selecting a programming language
(e) Implementing the flowchart using instructions (i.e., writing a program)
(f) Assembling, compiling, or interpreting the program
(g) Debugging and checking the program

How an MPU Rewires Itself

A microprocessor works by shuttling digital words (bytes) within a network of electronic circuits, each having a specific capability, such as data storage, arithmetic logic, control logic, and interfacing to external devices. These circuits are permanently interconnected with one another through busses and electronic switches. The function that the MPU performs at any given instant depends on which switches are opened and closed. This switching action, and the resulting data-transfer paths, is determined by the machine-language program contained in memory. The program steps are selected from the MPU instruction set. These simplified diagrams show how data is transferred during three machine cycles.

Control unit contains circuitry that synchronizes all switching. Sequencing circuits decode directives from memory and activate necessary electronic switches to reroute data paths.

Memory Location	Machine Code	Instruction
010	0110 1110	I/O register A content to accumulator
011	0001 0100	Store accumulator content in Location 100
012	1101 1100	I/O register B content to accumulator
013	0100 0110	Add accumulator to Location 100, store result in accumulator
014	0011 0001	Store accumulator content in Location 120
015	0000 0001	Halt

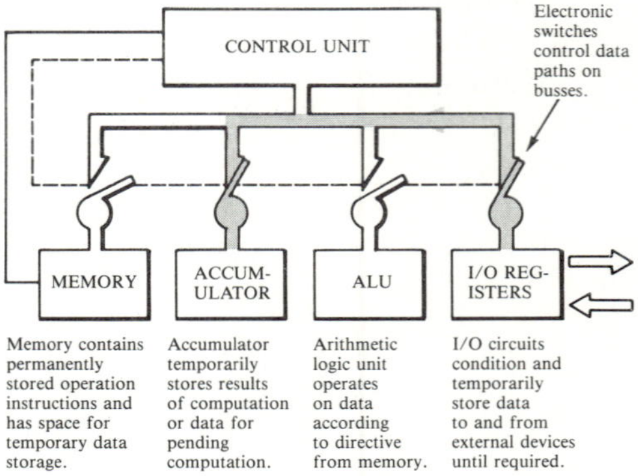

Memory Location	Machine Code	Instruction
010	0110 1110	I/O register A content to accumulator
011	0001 0100	Store accumulator content in Location 100
012	1101 1100	I/O register B content to accumulator
013	0100 0110	Add accumulator to Location 100, store result in accumulator
014	0011 0001	Store accumulator content in Location 120
015	0000 0001	Halt

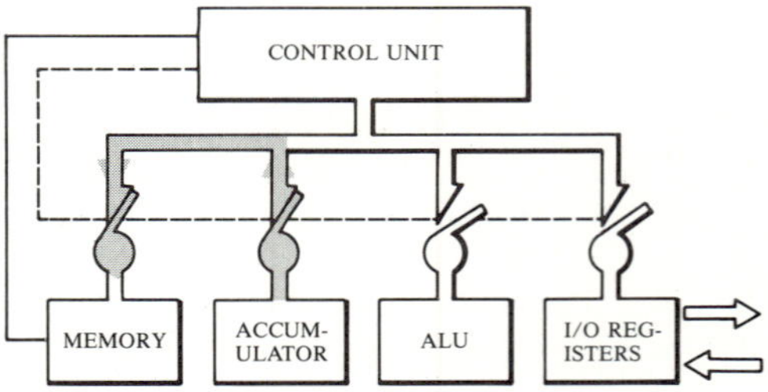

Memory Location	Machine Code	Instruction
010	0110 1110	I/O register A content to accumulator
011	0001 0100	Store accumulator content in Location 100
012	1101 1100	I/O register B content to accumulator
013	0100 0110	Add accumulator to Location 100, store result in accumulator
014	0011 0001	Store accumulator content in Location 120
015	0000 0001	Halt

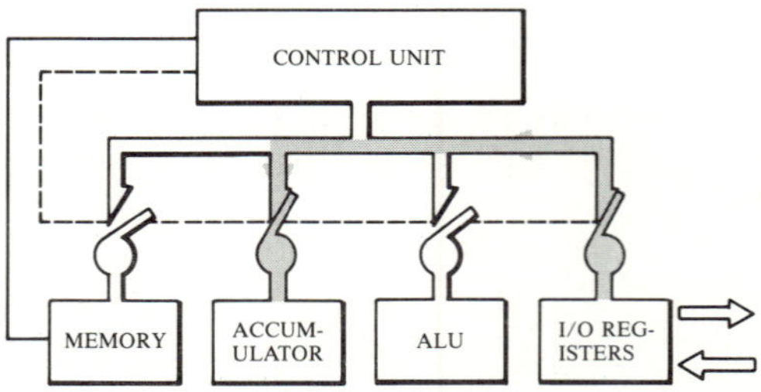

Most engineers and engineering students are familiar with flowcharts which are simply formal block diagrams for structuring a control concept. The flowchart organizes the logical and algorithmic events needed to implement the control desired.

Three types of symbols are used in flowcharts. Computational, logical, or algorithmic tasks are represented by rectangles. The diamond shape is used for decision points. Start and stop points are represented by ovals. Comments are placed inside the symbols in an effort to clarify the flowchart.

A flowchart that represents the relay ladder logic of Example 13-12 is illustrated in Fig. 14-4. In this control scheme, the status of inputs 1CR, 2CR,

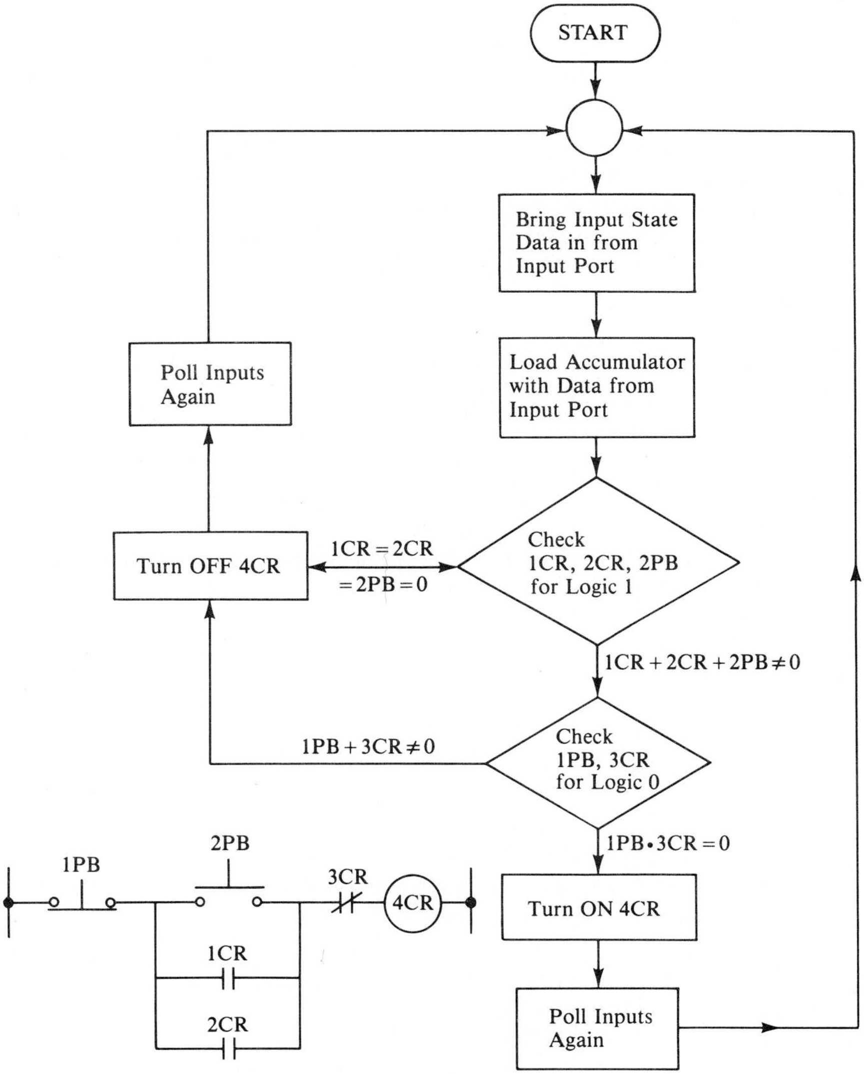

Figure 14-4. Flowchart for Example 13-12.

3CR, 1PB, and 2PB is being checked periodically at a frequency that depends on the time it takes the microprocessor to complete the loops shown in the figure. Since the frequency is very high for the simple application shown, we can regard this checking process to be practically continuous. This process is called polling; the control system consists of tight loops in which the inputs are being continuously asked what their status is. The single output, 4CR, is either ON or OFF depending on the polled status of the five inputs.

We stated previously that assembly language is a dictionary of mnemonic codes that represents microprocessor instructions in binary form (machine language). Each commercial microprocessor has its unique assembly language instruction set although all instruction sets are similar in their construction.

A mnemonic is an alphabetical abbreviation for a binary expression designed to help the user remember what that particular binary code was designed to do. Because mnemonics are more easily recognized and understood than the 1s and 0s of machine code, they are most often used in defining the assembly language instruction set. An instruction set is simply the repertoire of operations a microprocessor can perform.

Once a program has been written in assembly language, it must be translated into machine code. This is done by a software assembler program (called an assembler). Although a program can be written directly in machine code, the reading, writing, and correcting process of the program is tedious and prone to error. And when the program is defined in basic machine code, more than likely it will be in the form of its hexadecimal equivalent. This hexadecimal equivalent has the virtue of reducing eight digits to only two.

The assembly language instruction set for a microprocessor will typically contain some 100 instructions distributed among five basic groups: (a) data transfer, (b) arithmetic, (c) logical, (d) branching, and (e) control (i.e., input and output). Some typical instructions are shown in Table 14-1. The instructions in the table do not necessarily correspond with any particular microprocessor; they are presented for illustrative purposes only.

In reviewing Table 14-1, we see that there are different types of instructions characterized by their addressing modes; some instructions are 2-byte instructions while others are 3-byte instructions. In every case, though, the first byte is the OP code of the instruction (generally in hexadecimal code). An OP code is simply the 8-bit binary code decoded by the instruction decoder and implemented by the control unit of the microprocessor.

In an 8-bit microprocessor, a 3-byte instruction will often be used when the memory address of a data item is the operand of the instruction (remember that since the address bus is 16-bit in an 8-bit microprocessor, 2 bytes are needed for an address). This addressing mode is called a direct mode.

Sometimes the operand of an instruction will be the data themselves rather than an address designating where the data are located. In this case, we have what is called the immediate addressing mode. This mode could be either a 2-byte instruction or a 3-byte instruction depending on whether the data are 8-bit or 16-bit. The immediate mode has the advantage of involving a smaller number of machine cycles than the direct mode (and it therefore is speedier).

TABLE 14-1. Hypothetical Instruction Set.

Hex OP Code	Mnemonic Code	Meaning	Instruction Type
3E	LDA I	Load the accumulator with the immediate byte 2 of the instruction.	Data transfer
78	LAT	Load the accumulator with the content of the temporary register.	Data transfer
47	LDT	Load the temporary register with the content of the accumulator.	Data transfer
2A	LDA X	Load the accumulator with the content of the memory location being addressed by the index register.	Data transfer
2B	LDX ONE	Load the index register with the content of the input port. The content of the index register is issued as an address to access the memory.	Data transfer
E6	AND I	AND the content of the accumulator with the content of the immediate byte 2 of the instruction. The result is placed in the accumulator.	Logical
C3	JMP XXXX	Jump to the instruction labeled XXXX.	Branching (unconditional)
C4	SNA	Check the accumulator and skip the next instruction if the accumulator is not 0.	Branching (conditional)
C5	SZA	Check the accumulator and skip the next instruction if the accumulator is 0.	Branching (conditional)
DB	IN ONE	Load the accumulator with the content of the input port.	Control
D2	OUT TWO	Makes port THREE an output port when 01 (hex) is in the accumulator.	Control
D3	OUT THREE	Place the content of the accumulator at the output port.	Control

EXAMPLE 14-2

Assume that the status of the five inputs 1PB, 2CR, 1CR, 2PB, and 3CR of Example 13-12 and Fig. 14-4 represents the five least significant bits of an 8-bit word that has 0 in the three most significant bits. Assume, further, that this 8-bit word is provided to an input port of a microprocessor system. Write an assembly language program that will implement the flowchart of Fig. 14-4. Use the illustrative instruction set of Table 14-1. To actuate output 4CR, provide logic level 1 (in the form of the byte 00000001) to the appropriate output port of the microprocessor. Relay 4CR will be turned OFF by logic level 0 (i.e., byte 00000000).

The required assembly language program is shown in Table 14-2. ■

TABLE 14-2. Assembly Language Program for Brute Force Method of Implementing Fig. 14-4.

Label	Instruction	Remarks
START	IN ONE	The input word is read into the accumulator as an 8-bit byte.
	LDT	This operation places the input word into a temporary register as well as retaining it in the accumulator.
	AND I 00001110	This is a masking operation (see Example 13-6) designed to filter out bits 1PB and 3CR. Thus, the accumulator contains only the status of bits 1CR, 2CR, and 2PB.
	SNA	If there is a logic level 1 in any bit position in the accumulator, the Boolean expression $1CR + 2CR + 2PB = 1$ is TRUE and control is passed to the instruction following the next instruction.
	JMP STOP	This step is executed only if the expression $1CR + 2CR + 2PB = 1$ is FALSE. Jumping to STOP will turn OFF output 4CR.
	LAT	The accumulator is now reloaded with the original status byte.
	AND I 00010001	This is a masking operation designed to filter out bits 1CR, 2CR, and 2PB. The accumulator will contain only the status of bits 1PB and 3CR.
	SZA	If there is a logic level 1 in any bit position in the accumulator, the Boolean expression $1PB \cdot 3CR = 0$ is FALSE and control is passed to the next instruction. If the expression is TRUE, control is passed to the instruction following the next instruction.
	JMP STOP	This step is executed only if the expression $1PB \cdot 3CR = 0$ is FALSE. Jumping to STOP will turn OFF output 4CR.
	LDA I 00000001	The output signal for turning ON output 4CR is loaded in the accumulator.
	OUT THREE	The signal 00000001 is placed at the microprocessor output port and turns ON output 4CR.
	JMP START	Now the whole process is repeated. This is called polling.
STOP	LDA I 00000000	The output signal for turning OFF output 4CR is loaded into the accumulator.
	OUT THREE	The signal 00000000 is placed at the microprocessor output port and turns OFF output 4CR.
	JMP START	The inputs are polled again, and the whole process is repeated.

14-1-4 The Look-Up Table Method

The programming process illustrated in Example 14-2 and Table 14-2 is called brute force programming because the logic represented in the flowchart of Fig. 14-4 was implemented in a direct, methodical, step-by-step fashion. Note that this software routine involves some ten instructions to turn ON relay 4CR. These instructions will require valuable execution time, which in this particularly simple illustration may be of no consequence but which, in a more complex control problem, might well compromise the real-time response requirements of the control system.

A simple alternative to brute force programming is the look-up table method. This approach can often eliminate execution steps and, consequently, save time.

In the look-up table method, the input states are used as memory address pointers and the contents of the memory locations are the required output states. In Example 14-2, there are five inputs and consequently 32 (2^5) possible input bit patterns which can be used to address 32 consecutive memory locations. Of the 32 bit patterns, 7 should result in turning ON relay 4CR (see Table 13-13), so these 7 memory locations will contain logic level 1 (i.e., 00000001). The remaining 25 memory locations will contain logic level 0 (i.e., 00000000), since they correspond to bit patterns (addresses) that are to result in 4CR's being turned OFF.

EXAMPLE 14-3

Table 14-3 represents the truth table for a sequence control system having four inputs and three outputs. The inputs could be the conditions of limit switches, pushbuttons, thermostats, pressure switches, liquid levels, or similar kinds of devices characterized by on-off status. The outputs could be control relays, solenoid

TABLE 14-3. Truth Table for Sequence Control System.

State	Input State				Output State		
	I_1	I_2	I_3	I_4	O_1	O_2	O_3
1	0	0	0	0	1	1	0
2	0	0	0	1	0	0	0
3	0	0	1	0	0	1	0
4	0	0	1	1	Not possible		
5	0	1	0	0	1	1	0
6	0	1	0	1	1	0	0
7	0	1	1	0	1	1	0
8	0	1	1	1	Not possible		
9	1	0	0	0	1	1	0
10	1	0	0	1	1	0	0
11	1	0	1	0	0	1	1
12	1	0	1	1	1	1	0
13	1	1	0	0	1	0	0
14	1	1	0	1	0	1	1
15	1	1	1	0	0	0	0
16	1	1	1	1	0	0	1

actuators, control valves, or similar kinds of output devices characterized by on-off status.

Note that two of the input states cannot occur. This may well be because the combination of ON and OFF states represented by the input bit patterns are mutually exclusive (as, for example, when two limit switches cannot both be ON at the same time while the remaining inputs cannot both be OFF at the same time). For these two input states, the corresponding memory locations will never be accessed (because their addresses will never occur).

A flowchart and sketch are presented in Fig. 14-5 which indicate how the look-up table method of implementing this truth table is structured. The assembly language program corresponding to this control scheme is presented in Table 14-4. ■

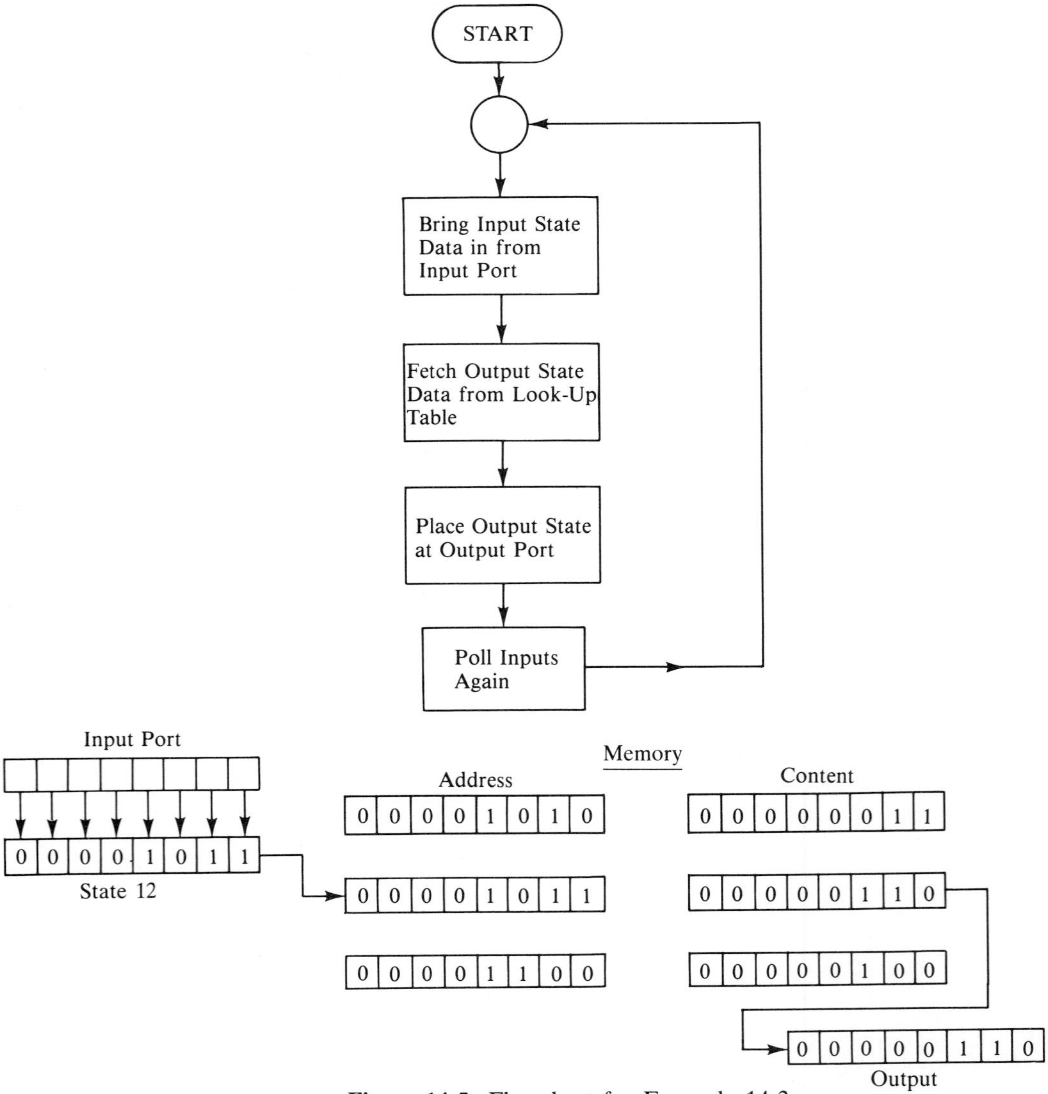

Figure 14-5. Flowchart for Example 14-3.

TABLE 14-4. Assembly Language Program for Look-Up Table Method of Implementing Fig. 14-5.

Label	Instruction	Remarks
START	LDX ONE	The memory is accessed at the memory location whose address is the input bit pattern.
	LDA X	The content of the memory location is placed in the accumulator. This is the output bit pattern.
	OUT THREE	The output state is placed at the microprocessor output port.
	JMP START	The inputs are polled again, and the look-up table process is repeated.

14-1-5 Turning On Selected Actuators

Conceptually, a very simple assembly language program is sufficient to turn on a selected set of mechanical system actuators (electric motors, hydraulic motors, cylinders, etc.). We use logic level 1 to turn on an actuator by placing this bit at the microprocessor output port (of course, there must be a suitable interface between the output port and the actual actuator; this could be a solid-state relay, optical device, or transistor).

Suppose that we have eight actuators and that the output port of the microprocessor is an 8-bit port. Then we can assign each bit location in the output byte to one of the actuators. For example, byte 00000111 will turn on actuators A_0, A_1, and A_2, while actuators A_3, A_4, A_5, A_6, and A_7 remain off.

The assembly language program for turning on a selected set of actuators (according to the hexadecimal byte XY) is written as shown in Table 14-5. The

TABLE 14-5. Assembly Language Program for Turning On a Set of Actuators.

Hex Instruction	Mnemonic Code	Remarks
3E 01	LDA I	This instruction places 01 in the accumulator.
D2 03	OUT TWO	This makes port THREE an output port.
3E XY	LDA I	The hexadecimal XY is placed in the accumulator.
D3 03	OUT THREE	The hexadecimal byte XY is placed at output port THREE, which will turn on the selected actuators.
76	HALT	This instruction stops the microprocessor.

final instruction, HALT, turns off the microprocessor. A RESET operation can be arranged to turn off all eight actuators.

14-1-6 Interfacing the Microprocessor to a Mechanical System

If a microprocessor system is to be incorporated into the control scheme for a mechanical system, it must be interfaced to it. The interfacing will involve both hardware and software. Sometimes a given interface function can be implemented using either hardware or software. The use of hardware generally simplifies the design and increases the system's time response but may increase the system's total cost. Although software can reduce hardware costs, the price paid is slower real-time response of the system. Also, software can be surprisingly expensive.

There are several kinds of external hardware devices to which the microprocessor system may be interfaced:

- **(a)** Mechanical system sensors which provide input to the microprocessor system. They might be digital or analog sensors. On-off sensors such as limit switches are examples of digital sensors; a thermocouple, which provides a voltage output, is an example of an analog sensor.
- **(b)** Mechanical system actuators which require output from the microprocessor system. They also may be either digital or analog. A solenoid actuator is an example of a digital actuator.
- **(c)** A keyboard which will provide a human-machine interactive capability and by means of which data or programs can be input or changed.
- **(d)** Displays for indicating the status of the mechanical system or of the control system. These could include CRTs, light-emitting diodes (LEDs), indicator lamps, or printers.
- **(e)** Another computer or memory unit(s).

The interfacing of these external devices will, in turn, require the use of the following hardware components, some of which may be purchased off-the-shelf as integrated circuits on chips:

- **(a)** Analog-to-digital converters (ADCs)
- **(b)** Digital-to-analog converters (DACs)
- **(c)** Signal conditioners (amplifiers, SCRs, etc.)
- **(d)** General-purpose busses and multiplexers
- **(e)** Scanners, decoders, and encoders
- **(f)** Buffers

An illustration of some of these interface requirements is shown in Fig. 14-6.

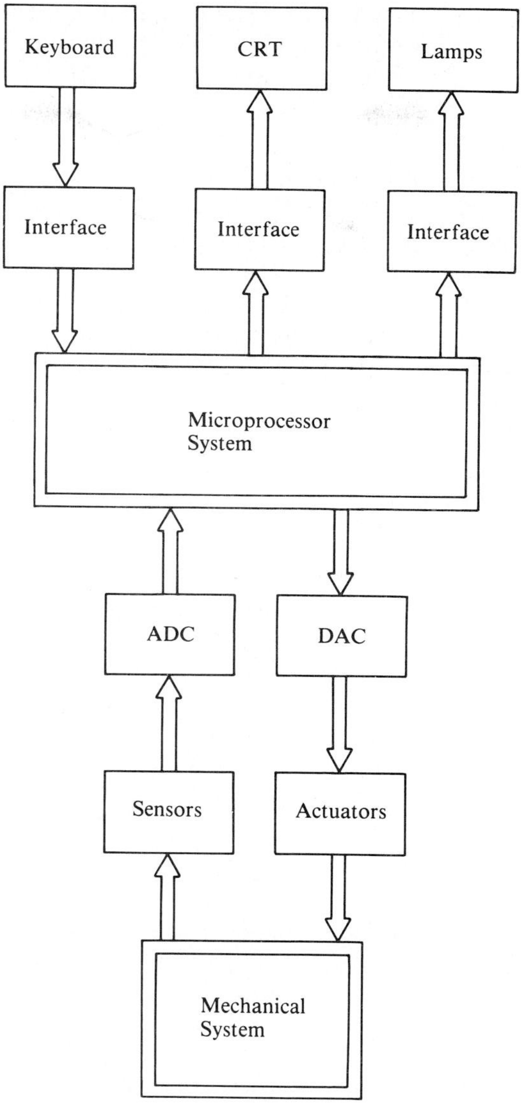

Figure 14-6. Microprocessor peripheral interfaces.

Software techniques by means of which these peripheral devices are made to communicate with the microprocessor include:

(a) Polling
(b) Interrupted input/output (interrupts)
(c) Direct memory access (DMA)
(d) Change-of-state tables

Communications or data transfer between the microprocessor and its peripherals can occur in parallel or serially. Components are available, off-the-shelf, for implementing these functions. Since this topic is beyond the scope of this book, detailed information must be obtained from manufacturers.

Analog-to-Digital Conversion (ADC). It will be sufficient for our purposes here to regard the analog-to-digital converter as a black box that converts an analog input into an 8-bit output (see Fig. 14-7[a]). These 8 bits will provide a resolution of approximately 0.4 percent. The analog-to-digital mechanism inside the black box implements either a successive approximation scheme or the dual-ramp approximation. Here, we will discuss only the method of successive approximations.

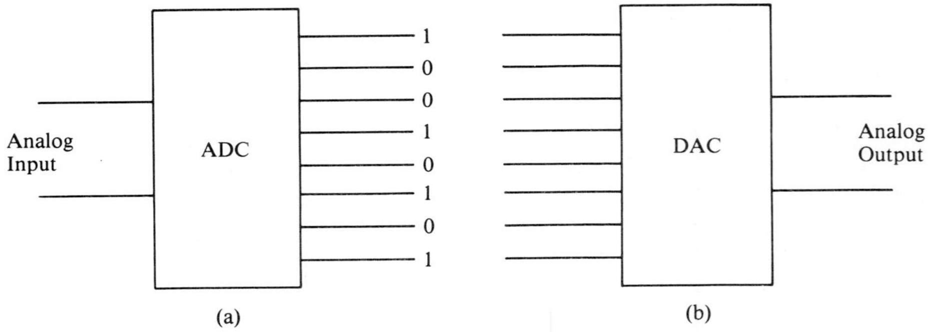

Figure 14-7. Black box representations of ADC and DAC.

The method of successive approximations, in making an analog-to-digital conversion, is to compare, successively, the magnitude of the analog signal with a reference signal. At each successive comparison, the reference is modified to more closely confine the analog value. And at each successive comparison it is possible to assign a bit value 1 or 0 which depends on whether the analog value is greater than or less than the reference value. The method is structured in Fig. 14-8, where we see that the analog input has been approximated to be the digital value 10010101 (hex 95, dec 149).

Analog-to-digital converters are available on a single chip at a low cost. In an actual machine application, the ADC will generally be connected to the external analog signals by a multiplexer and sample-and-hold circuits.

Digital-to-Analog Conversion (DAC). As in the case of the analog-to-digital converter, the digital-to-analog converter can be visualized simply as a black box (see Fig. 14-7[b]). The mechanism inside the box consists generally of a weighted resistor network and a precise reference voltage. Switches are closed or open in response to the digital code input. Conceptually, this process is the inverse of the ADC concept.

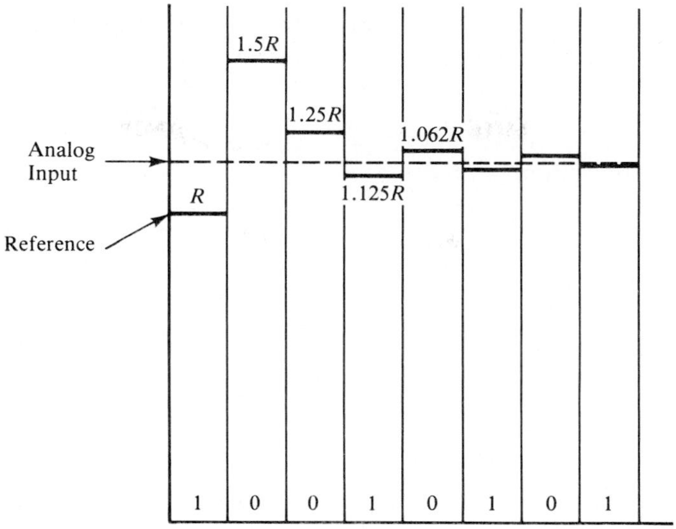

Figure 14-8. ADC by successive approximations.

In selecting ADC or DAC devices, particular attention must be paid to:

(a) Resolution requirements
(b) Time response
(c) Linearity, repeatability, and accuracy
(d) Signal ranges
(e) Temperature and other environmental effects

Polling. In addition to the interface hardwire devices such as ADCs, DACs, multiplexers, and signal conditioners, interfacing will also require appropriate software for communications between the microprocessor system and the mechanical system.

There are two general methods of programming a microprocessor to send or receive information from external peripheral devices. The first is called polling in which the microprocessor is programmed to periodically check the devices connected to its data bus to see if they are ready to be serviced. This procedure is implemented by creating a polling loop in the flowchart for the control system. We discussed this process earlier in connection with Fig. 14-4. This continuous process of periodically checking peripheral devices and asking a question of the device, and receiving an answer, is called handshaking.

Advantages of the polling technique are that it is conceptually simple to program, requires minimal hardware, and is synchronous with program execution. No events can occur that might disrupt the scheduled polling procedure. Also, the predictability of the order of the polling process is a major advantage.

The main disadvantage of the polling technique is its inherent inefficiency in that the devices being polled will often not require service. This process causes a loss of control system response capability, which means real-time requirements of the mechanical system may not be met.

Interrupts. When a polling technique cannot provide a fast enough response to mechanical system control requirements, or when the polling process begins to interfere with other microprocessor requirements, the method of using interrupts can be employed. This is known as an asynchronous technique because peripherals will be serviced only when they are ready to transfer data. The technique is analogous to the situation in which a passenger on a bus wishes to get off. The passenger causes a bell to be rung which will interrupt the bus service.

When peripheral devices are ready to transfer data, they send interrupt signals to the microprocessor, which will then stop what it was doing in order to service the peripheral device(s). This process consumes less computer time than the non-interrupt process because there is no wasted time in polling devices that are not ready to be serviced. Thus, the major advantage of the interrupt technique of communicating is an increase in control system response time. The major disadvantage is the asynchronous nature of the process. Also, debugging of an interrupt program can be difficult because an interrupt has no specific location in the main program. An interruption might occur between any two consecutive instructions.

Direct Memory Access (DMA). If the interrupt technique still results in unacceptable real-time response requirements, there is a third alternative. This method is called direct memory access because the techniques involved bypass the CPU entirely. A DMA controller takes over and transfers one or more words directly from the memory to the peripheral device. This process is complex and beyond the scope of this book. For more detail, consult Reference 14-1.

A complete understanding of the process of interfacing a microprocessor system to its peripheral devices requires discussion of additional concepts such as buffers, multiplexers, demultiplexers, sample-and-hold circuits, and others. Again, we refer the student to the excellent books now available on microprocessors, such as Reference 14-1.

14-1-7 Examples of Microprocessor Control of Mechanical Systems

The focus of our previous discussion of microprocessor control was on mechanical systems susceptible to on-off sequential control. A good example is the plastic injection molding machine in which the microprocessor might simply implement the relay ladder control logic for the machine.

Implementation of relay ladder logic, while it provides a convenient basis for introducing microprocessor control concepts, does not fully utilize the tremendous power of the microprocessor as an element in a control system. In fact, as we shall see, there is a more convenient way of implementing relay ladder control by using the programmable controller which is a microprocessor-based controller tailored to handle sequential control.

In this section, we will provide a glimpse of how the microprocessor can be used in more complex analog control schemes including conventional closed-loop control of both electrical and hydraulic actuators. These applications can be classified into several categories, as follows.

The Microprocessor as a Command Signal Generator. Perhaps the simplest application of the microprocessor in a closed-loop control system is as a command signal generator. Such a scheme is illustrated in Fig. 14-9. In this application, the microprocessor can not only be programmed to contain any number of command signal profiles but it can also serve as an interface for a host computer, for a keyboard (for operator input), or for a variety of input devices such as pushbuttons, limit switches, and transducers. Often, this arrangement is also capable of implementing minor changes in the system performance via software rather than by hardware changes. Note that in an analog system it will be necessary to provide a digital-to-analog converter in order to interface the microprocessor to the control system.

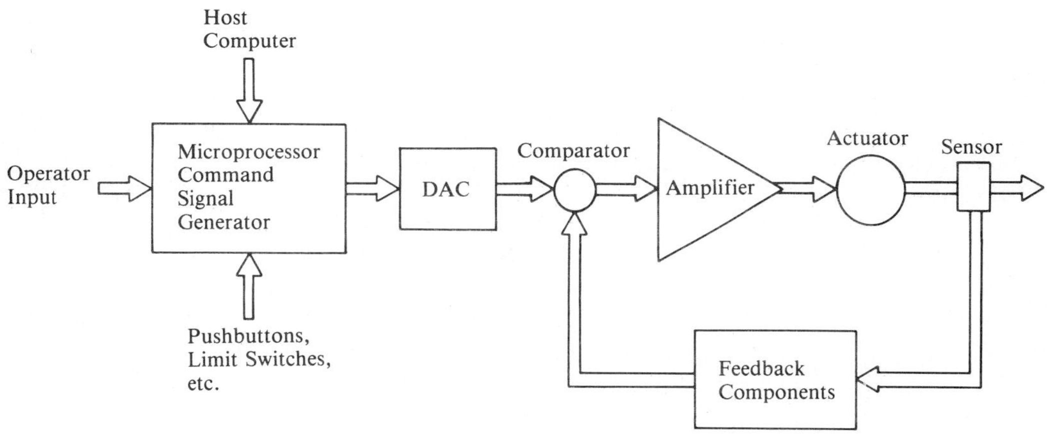

Figure 14-9. Microprocessor as a command signal generator.

The Microprocessor as a Software Comparator. In some mechanical systems, precise control of both speed and position of an actuator (electrical or hydraulic) will be required. The conventional method is to use the classical closed-loop servo-control system sketched in Fig. 14-10. This control scheme is used in robot positioning systems, in numerically controlled machine tools, and in the control of hydraulic and electric motors. The system shown is basically a positioning system; the inner velocity control loop is used to improve stability and damping of the overall system.

The complexity of the mechanical and electronic hardware required to provide precision control of position, velocity, and acceleration with a system such as that shown in Fig. 14-10 makes it attractive to contemplate eliminating

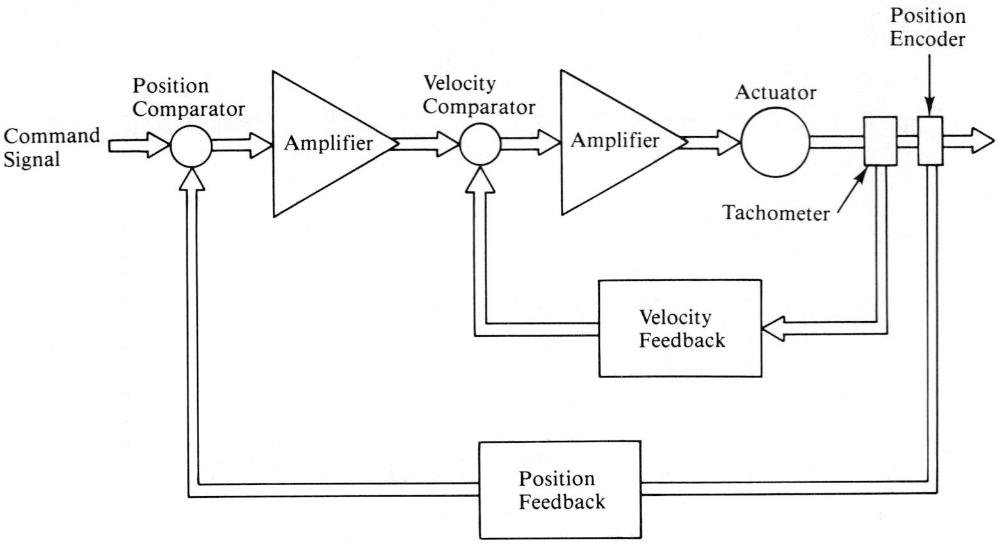

Figure 14-10. Classical closed-loop control system.

some (or most) of the hardware in favor of software. This can be done using a microprocessor to perform either position loop calculations, or both position and velocity loop calculations, in lieu of using the conventional comparator(s) and error conditioners.

In most applications, it is convenient to maintain implementation of the velocity loop by using hardware, and only the position comparator (and associated electronics) is replaced with the microprocessor. Such a scheme is sketched in Fig. 14-11. Note that the position command generator is integrated into this type of microprocessor control scheme by incorporating it into the microprocessor associated memory. Also, a digital-to-analog converter is needed to interface the microprocessor to the analog velocity feedback loop.

Advantages of the microprocessor control are that the microprocessor also can perform checks to determine if the system is working properly and can detect hardware failures. Some hardware design problems can also be circumvented.

The disadvantage of the microprocessor control is the potential loss in speed at which feedback signals are processed. Consequently, the software must be well written. To serve effectively, the microprocessor must process information at a rate consistent with the bandpass of the conventional servo-system to which it purports to be superior.

An alternative scheme in which the microprocessor replaces both position and velocity comparators is shown in Fig. 14-12. In this arrangement, the position encoder provides the dual function of position and velocity feedback. The rate at which the encoder turns gives the velocity, while the number of pulses it produces gives the position. One manufacturer of this arrangement claims that better accuracy and speed (which are generally somewhat mutually exclusive) can be

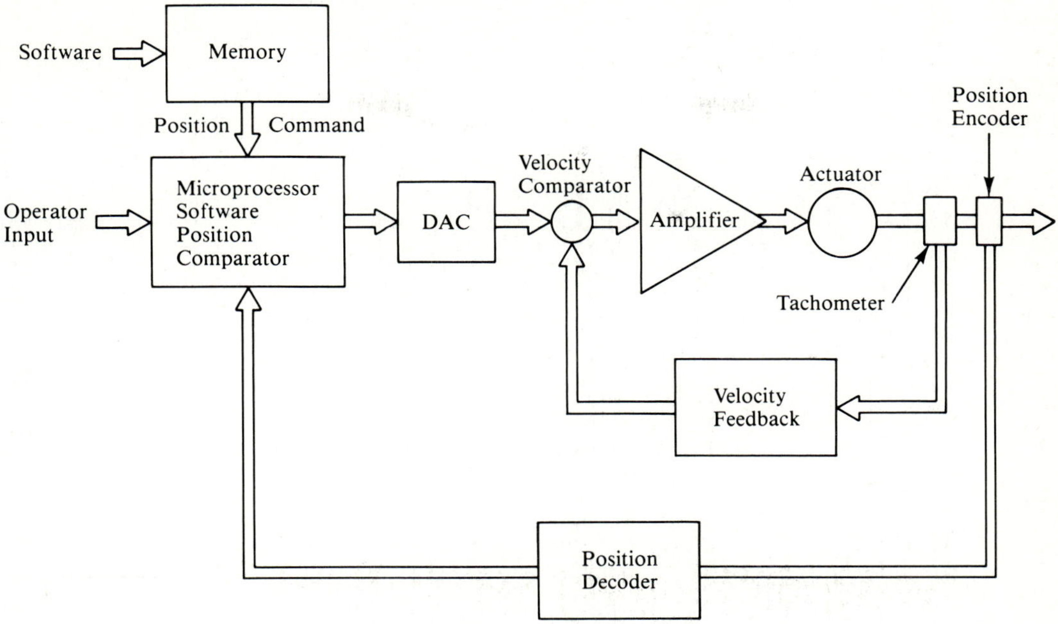

Figure 14-11. Microprocessor as a software comparator.

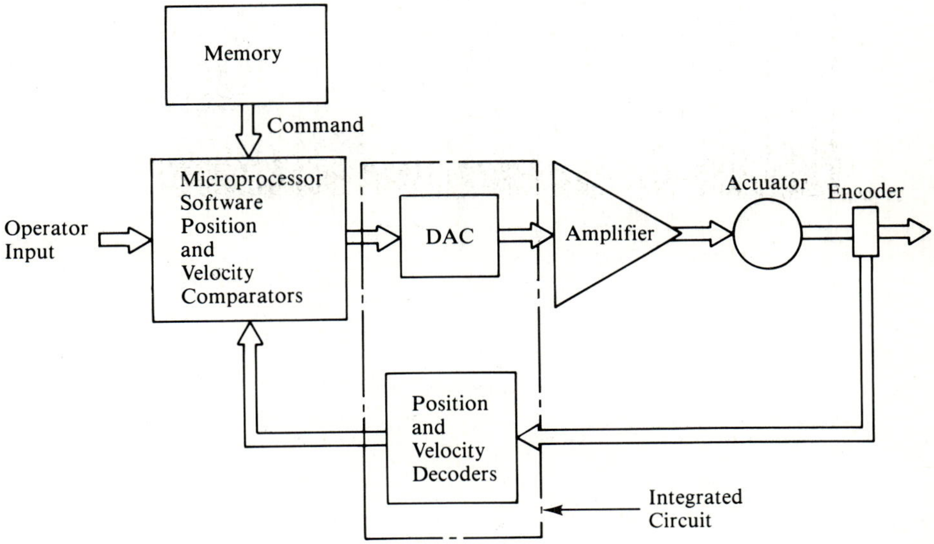

Figure 14-12. Totally microprocessor-controlled closed-loop system.

obtained compared to the conventional analog servo-system. Note that the DAC and position decoder can be implemented with a single integrated circuit.

Microprocessor Control of Stepping Motors. Recall that in Chapter 9 a stepping motor was defined as a reluctance motor whose rotor moves in discrete angular steps in response to the application of pulses to its stator windings. A single pulse advances the shaft position by one angular step, which typically is 1.8 degrees. A train of pulses produces a rotation in rapid steps. For example, a train of 500 pulses (with 1.8 degrees per pulse) will produce a total rotation of $2\frac{1}{2}$ revolutions. This positioning can be done, when properly controlled, with great precision, so closed-loop control is not necessary.

With loads of significant inertia, acceleration and deceleration must be carefully controlled to prevent the stepping motor from overshooting its desired destination. The conventional hardware for producing such control involves several electronic components including a counter, an acceleration/deceleration controller, a ramping oscillator, and logic gates. One drawback of such a hardwired system is that it is often difficult to synchronize the components.

An alternative control scheme which can reduce the cost and complexity of the conventional system is to use software implemented by a microprocessor. An example is shown in Fig. 14-13. In this method, the microprocessor controls the stepping motor acceleration and deceleration by generating a variable motor pulse rate through software. This process controls the time between pulses using pre-calculated constants stored in the microprocessor system memory.

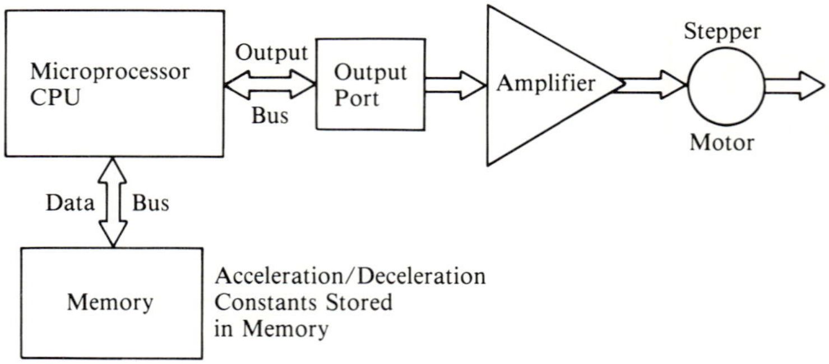

Figure 14-13. Microprocessor-controlled stepping motor.

14-1-8 Summary

A microprocessor is a programmable logic gate circuit array consisting of thousands of miniature transistors integrated on a single silicon chip. It performs the same function as the central processing unit of the traditional computer but does so at very low cost (a few dollars) and in a small space.

A microprocessor system (a microcomputer) consists of a microprocessor plus memory and input/output capability. These devices are connected together by the data bus and address bus. The microprocessor contains a third bus, the control bus, which synchronizes and controls the microprocessor system when it is executing a program.

The microprocessor memories contain several programs. A microprogram is permanently burned into the control unit ROM which controls the execution of the repertoire of instructions of the microprocessor instruction set. A monitor program, or operating system, is stored in the main ROM memory for facilitating operator interactive capability. And, finally, the program to be executed in order to implement a particular mechanical system control is stored in the RAM memory.

In this section, we have provided an introduction to microprocessor operation and implementation. To simplify this discussion, we restricted applications to those of the simple relay ladder type characteristic of mechanical systems that are sequentially, on-off controlled. A few more general closed-loop analog control system applications were also discussed qualitatively to illustrate how the microprocessor can be integrated, conceptually, into a general control scheme for an actuator or a system.

In the next section, we will again analyze a microprocessor-based control system, but it will be done in the particular architecture of the programmable controller. This device has been tailored specifically to simplify the application of microprocessor-based control when it involves relay ladder logic.

14-2 PROGRAMMABLE CONTROLLERs (PLCs)

To be able to implement a variety of on-off discrete control schemes using the microprocessor, the engineer must devote many hours of study to the details of how a microprocessor works. This study will include learning how data are transferred among the registers, accumulator, and memory; how data are converted and transferred at the input and output ports; how the stack is used; what the instructions are; and many similar operations. Learning how a microprocessor works means learning how this fascinating device actually manipulates bits, bytes, and words. Although the knowledge gained will be essential to the engineer's ability to implement all kinds of control systems including closed-loop feedback control, process control, and machine sequence (discrete) control, much of this knowledge will be superfluous insofar as machine sequence control is concerned.

For relay ladder logical control schemes, there is a much simpler alternative to using the microprocessor which involves higher-level programming languages designed specifically to implement relay ladder control. This alternative is the microprocessor-based programmable logic controller (PLC).

The programmable controller was originally developed as a replacement for hardwired industrial sequence control systems. It is a logical processor that has been internally configured to perform sequential, discrete-state operations on external control hardware. Recent advances in programmable controller engineering have made the PLC much more versatile and useful in applications that go beyond the relay ladder circuits for which it was originally designed. The PLC can also be integrated into analog control configurations requiring proportional, integral, and derivative (PID) control. Additionally, the microprocessor-based programmable controller incorporates many valuable characteristics such as good operator interface, easy repair and maintenance, conceptual simplicity, and resistance to hostile environments.

A major advantage of the PLC is that the interfacing problem with inputs and outputs of the relay ladder type has already been solved by implementation directly in the device. The PLC, for example, can be interfaced directly to 115 V AC circuits. Additional features of programmable controllers include the following:

(a) ROM and RAM memories are incorporated directly in the PLC.
(b) Certain control functions such as control relays, timers, and counters are already implemented in software internally in the PLC, so the user does not have to develop this software as would be the case with a microprocessor.
(c) Analog-to-digital and digital-to-analog converters are already incorporated into the PLC.
(d) Other input/output amenities such as handling of binary coded decimal data, DC and AC input/output, and remote I/O are built in.

We see that the programmable controller has been designed so that the applications engineer can use it as a black box. This is always a desirable characteristic of components being synthesized into a mechanical system. A schematic diagram of the black box is presented in Fig. 14-14.

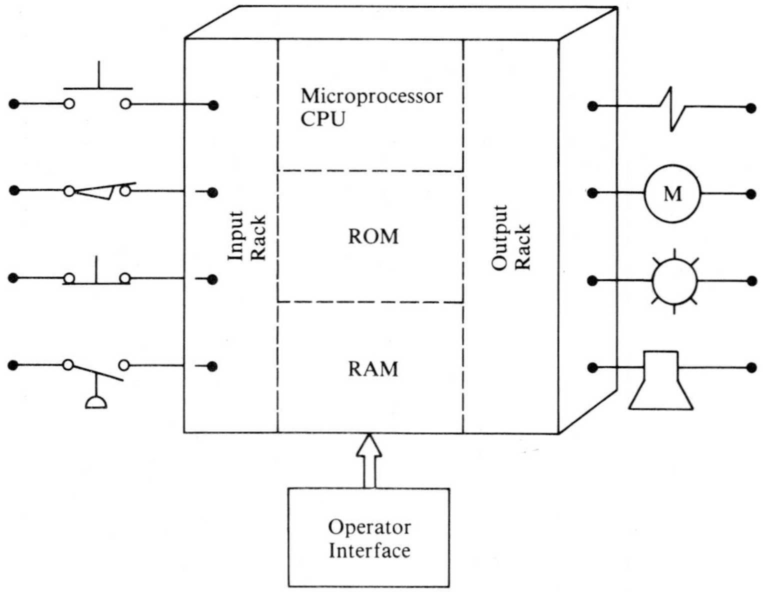

Figure 14-14. Programmable controller as a black box.

Included in the PLC are input and output racks which are simply electrical chasses for directly connecting the hardware inputs and outputs; a microprocessor which performs all mathematical operations, data handling, and diagnostic routines; an operator interface (keyboard, CRT, printer, etc.); a power supply; and associated memories for storing data, instructions, and the "executive".

The principal function of the microprocessor is to control the entire system. This is done by executing a program or programs called the executive. The programs are permanently stored in a ROM and are considered to be an integral part of the PLC. The executive performs all control, processing, communication, and housekeeping functions required in the execution of an applications program.

The applications program is the user's program and data that have been designed to implement the control of some particular mechanical system. The applications program is stored in an applications memory (either as a RAM or ROM, depending on how important the potential volatility of the RAM memory is in this application). When non-volatility is required, the applications program may be stored in an EPROM (Erasable Programmable Read Only Memory).

14-2-1 Scanning

One important function of the PLC's microprocessor is scanning. Scanning is the process of reading (polling) the inputs, executing the user's application program, and actuating the outputs. The time required to make a single scan varies from about 1 millisecond to 100 milliseconds. This process occurs continuously so all outputs are continuously updated in terms of current input values. There is a potential problem if the inputs are changing very rapidly in that the scan may not proceed fast enough to resolve the input behavior. In this event, the conventional method of scanning may have to be modified to utilize an interrupt scheme similar to the one we discussed earlier with regard to the microprocessor.

The scan time is an important specification to be made in selecting a PLC. If the PLC has a scan time of 15 milliseconds and the inputs are changing state every 5 milliseconds the PLC will not be able to implement all input signals. This will compromise the whole control system.

14-2-2 Addressing Real and Internal Input/Output

We stated earlier that a microprocessor is not a substitute for hardware devices such as limit switches, transducers, solenoids, and valves which are required to either actuate, or be actuated by, components of a mechanical system. The same thing is true for a programmable controller. We will call such devices real input/output.

The status of each real input and output can be defined by the logic level of a single bit. If a real input (or output) is ON, then its corresponding bit is at logic level 1 (in a positive logic system), while if it is OFF, then its bit is at logic level 0. Thus, if a PLC has 16 real inputs, 16 bits reserved in a memory will be required to define the status of these 16 inputs. Note that we can address these 16 bits with 4 binary digits in the range 0000 to 1111 (or 0 to F in hexadecimal). Similarly, 16 real output devices will require 16 status bits of a memory which are addressable with 4 address bits. Of course, a PLC having the capability of handling only 16 inputs and 16 outputs would be a small PLC.

Components such as control relays (and their contacts), timing relays, timers, and counters (which are implemented in hardware in hardwired electrical relay ladder circuits) are implemented internally by the executive program of the PLC. We call these internal inputs and outputs (or simply internals) because they are not real in the sense of being hardware. In fact, it is the ease by which these internals can be "rewired" in an applications program that is one of the major attributes of the PLC. The internals also are placed in one-to-one correspondence with bits of a memory, because a control relay, for example, needs to have a bit to define its ON or OFF status in terms of logic level 1 or 0 (regardless of whether it is real or internal). Internal bits are also readily addressable. For example, 176 internal bits can be addressed hexadecimally in the range 40 to F0.

Figure 14-15 is a block diagram of a PLC applications memory indicating where these various bits may be located. In addition to the memory bits required for real input and output and the internals, programmable controllers have the capability of addressing internal registers which are utilized in arithmetic, logical, and data transfer operations. These operations are more complex than those represented simply by internally implemented control relays. Registers may require bytes of memory rather than just bits. Finally, a portion of the applications memory is reserved for the storage of the user's program. In a small PLC, the user's program storage capability would typically be 984 locations (addressable, say, in the hexadecimal range 028 to 3FF).

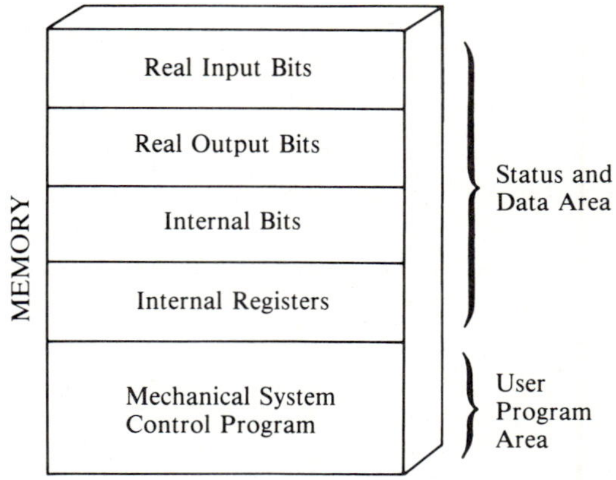

Figure 14-15. Applications memory map.

Addresses must be defined for all real inputs and outputs, internals, registers, and instructions in order to write a user's program (as we shall see). The addressing scheme used depends to a great extent on the PLC manufacturer. Some manufacturers use an octal number system for addressing real input/output in terms of the terminal, group, and rack of the input/output hardware chassis. Reference 14-2 discusses this type of I/O addressing system.

For the illustrations presented in this book, we will use the hexadecimal addressing scheme defined in Table 14-6. This is the system used by one PLC manufacturer to define I/O addresses.

TABLE 14-6. Memory Map for Defining I/O Addresses. (Entertron Industries SK-1600 PLC.)

Hex Address	Function	Remarks
00-0F	Real inputs	Provides 16 bits for storage of real input status.
10-1F	Real outputs	Provides 16 bits for storage of real output status.
20-2F	Real inputs	Provides 16 additional bits for 16 additional inputs.
30-37	Timers	Provides 8 bits for storage of the status of 8 timers.
38-3F	Counters	Provides 8 bits for storage of the status of 8 counters.
40-EF	Internals	Provides 176 bits for storage of the status of 176 internals.
F0-FF	Real outputs	Provides 16 additional bits for 16 additional outputs.

A schematic diagram showing the electrical connections for some typical real outputs and inputs (and their consecutive addresses) is presented in Fig. 14-16 for one manufacturer's PLC. Note the similarity of this figure to Fig. 14-14. Here the PLC is just a black box between real inputs and real outputs. In Fig. 14-16, no control relays, contacts, timers, or counters are shown as either input or output hardware. The reason is that the PLC istelf implements these items.

14-2-3 Programming the PLC

In Chapter 13, it was shown that a discrete-state control system of the kind used in machine sequence control can be represented in four equivalent ways: (a) with a relay ladder diagram, (b) with truth tables, (c) with logic gates, or (d) with Boolean algebra. These formats also apply to programming a programmable controller. The choice of format for programming a PLC could well depend on the background and training of the person undertaking the task.

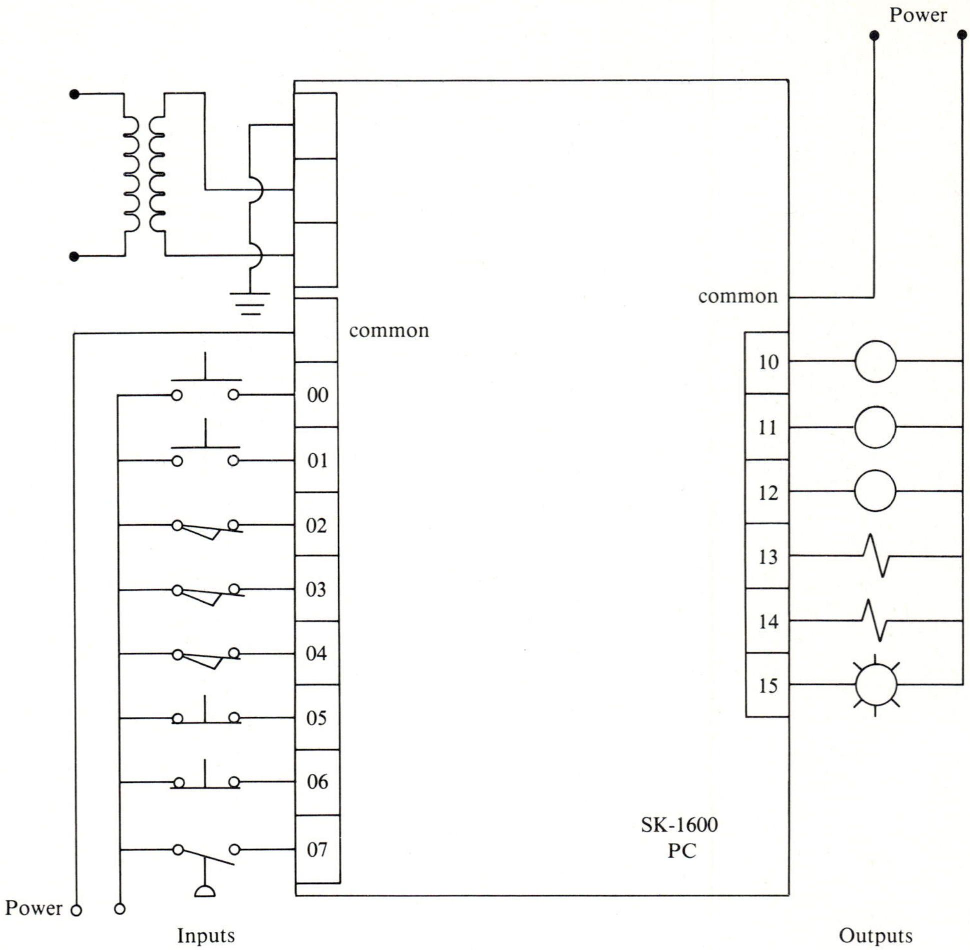

Figure 14-16. Typical connections and addresses of real input and output (Entertron Industries SK-1600 PLC).

For the engineering student or mechanical systems designer, only two of these programming formats will be of interest: the relay ladder diagram language format and the Boolean language format. Although it is certainly possible to use the relay ladder programming format when the control system has been defined in terms of logic gate circuits or truth tables, this will not be the most natural way of proceeding for most engineers or technicians involved in this work. Relay ladder logic is familiar to most systems designers, plant engineers, electricians, and maintenance personnel. It is desirable to use programming formats that these personnel will readily comprehend. Therefore, the two we will discuss are based on the control system having been defined, a priori, in terms of a relay ladder diagram.

In both of the programming formats to be discussed, programming of the PLC is actually carried out by inserting commands into the PLC via keyboard and video displays or by using simple switches on the operator panel. The executive program of the PLC will convert these key inputs into instructions necessary to implement the control scheme.

14-2-4 Ladder Diagram Language

Strictly speaking, the method of programming using ladder diagram "language" is not aptly named. It would be clearer, perhaps, if we termed this method ladder diagram "synthesis," because the programming technique is to actually synthesize the ladder diagram graphically on a CRT by keying in symbols from the PLC keyboard that represent wires, control relays, contacts, timers, and so forth. At the same time, addresses are assigned to these internals and to real inputs and outputs. In many PLC's, this programming process uses prompts and a cursor to assist the programmer in synthesizing the ladder diagram from the symbols on the keyboard which are to be used in representing various elements of the relay ladder diagram. The instruction set is composed of contact symbols which are referred to as contact symbology.

The attributes of this programming method are that it is graphical (on the CRT) and closely resembles the actual electrical relay ladder diagram. The contact symbology is slightly different from the original electrical relay ladder symbology discussed in Chapter 12. In effect, this programming method disguises the PLC's intrinsic logical operations with a familiar method of defining control circuitry. This approach produces ease of programming, understandability, straightforward editing capability, and ease of debugging and troubleshooting.

The method of programming via ladder diagram language is widely used and popular. In fact, floppy disks have been developed which makes it possible for this programming technique to be done remotely on, say, an IBM personal computer. Once the program has been synthesized on the personal computer CRT, it can be transported to the actual programmable controller via the floppy disk.

The ladder language method of programming is comprehensively discussed in Reference 14-2 and in each PLC manufacturer's user's manual.

EXAMPLE 14-4

Reconsider the relay ladder diagram of Fig. 14-17 for the control of two hydraulic cylinders discussed in Chapter 12 and defined in Fig. 12-12. The first step in programming this ladder diagram is to assign addresses for real inputs and outputs and for the internals. We have done this in Table 14-7 using the memory map of Table 14-6.

The synthesized ladder program for the circuit of Fig. 14-17 is shown in Fig. 14-18. This is the program as it would appear on the CRT of the PLC programmer as a result of the synthesis procedure we discussed earlier that involves pressing various keys to patch together the various symbols.

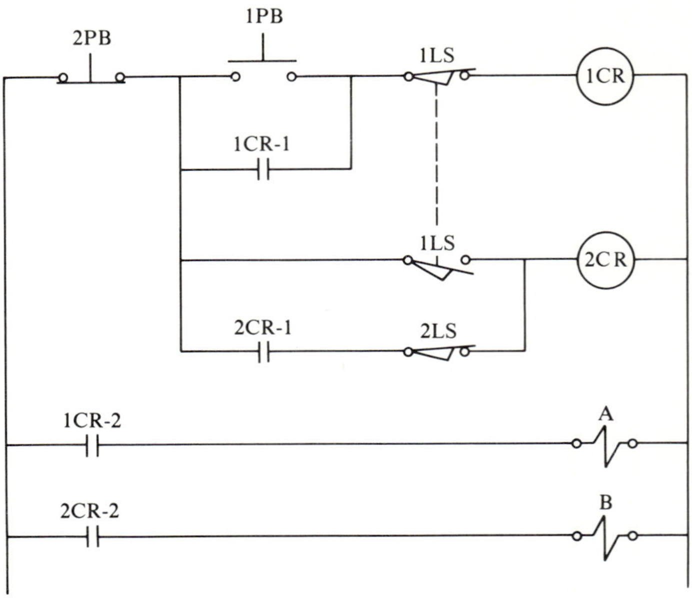

Figure 14-17. Ladder diagram for control of two cylinders.

Note that only two symbols are used. Identification of the devices that these symbols represent is done by the addresses (see Table 14-7) given above each

TABLE 14-7. I/O Address Assignment for Fig. 14-17.

Device	Type	Hex Address	Description
1PB	Real input	00	START pushbutton
2PB	Real input	01	STOP pushbutton
1LS	Real input	02	NO/NC limit switch
2LS	Real input	03	NC limit switch
Sol. A	Real output	10	Controls cylinder 1
Sol. B	Real output	11	Controls cylinder 2
1CR	Internal	40	Control relay
2CR	Internal	41	Control relay

symbol. The symbol for an input is ┤├, while that for an output is ─( )─. An internal, such as a control relay, will have the same address for both the relay itself (an internal output) and its contact(s) (an internal input); no distinction is made between the two. ■

The above example provides just a glimpse of how a relay ladder diagram is programmed for a PLC using the so-called ladder language. Obviously, there are many additional considerations, about which entire books have been written. One of these considerations involves an analysis of the control system to determine if,

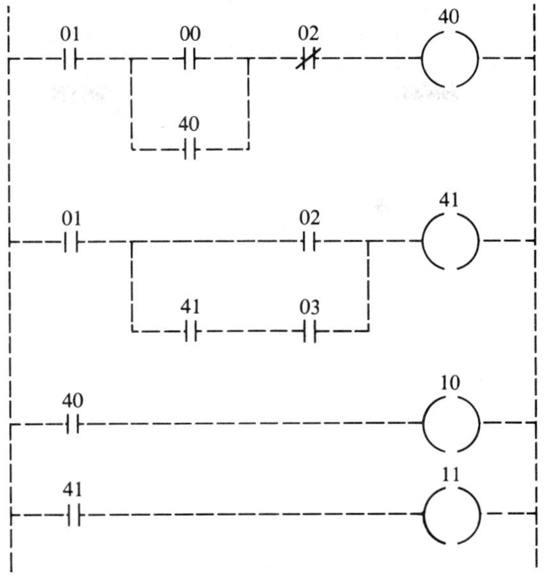

Figure 14-18. Ladder language program for Fig. 14-17.

and what, portions of the original ladder diagram should be left hardwired. For example, emergency stop conditions should remain hardwired independent of the PLC. There are many other safety and unique characteristics of particular control systems that will dictate how much of the electrical relay ladder diagram should be converted to PLC software.

Functional Blocks. Very few control problems can be implemented using only control relays and contacts as we discussed in Example 14-4. Most electrical relay ladder circuits will include timing relays, timers, counters, and other similar devices.

The question then arises as to whether more complex operations such as arithmetic, decision-making, data transfer, and sequencing can be made to fit the structure of the PLC ladder language method of programming. Indeed, this has been accomplished as PLC's have evolved. This implementation of complex operations makes it possible to program more complex functions in a graphical format that plant operating personnel have already learned and find easy to understand and interpret. The key concept facilitating this accommodation is the functional block.

The functional block can be considered to be a higher-level instruction that permits the PLC programmer to include complex functions within the ladder language format. The functional block itself may be equivalent to many more primitive control relay circuits which the programmer would otherwise have to implement in a lower-level language. It is associated with an enabling (input) contact and an output which is to be energized (see Fig. 14-19). Note that the functional block may include addressable registers internally.

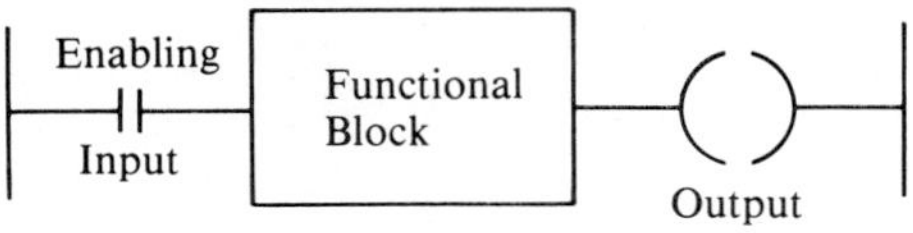

Figure 14-19. Basic concept of the functional block.

To illustrate the concept of a functional block, suppose the block in Fig. 14-19 is a timer. Then it will contain two registers (say, X and Y) which contain the pre-set time (X) and the accumulated time (Y) while timing. The block will begin to time when it is enabled by the contact shown at the left. The output, shown at the right, will be turned ON (energized) when the content of register Y equals that of register X (the timer has "timed out"). Of course, the enabling input, the output, and the registers must all be assigned addresses.

Logical operations can also be introduced into the ladder diagram using functional blocks. For example, the compare (CMP) functional block contains registers X and Y, whose contents are to be compared. If the compare is (>), then the output will be energized if $Y > X$.

Additional high-level operations such as register shifting, moving of data files, and arithmetic can be implemented using functional blocks. This format can even be made to include analog proportional, integral, and derivative (PID) control. For more information, consult Reference 14-2.

EXAMPLE 14-5

Consider a pumping load to be supplied by two pumps labeled 1 and 2. The motor of pump 1 is started by pressing pushbutton 1PB (see Fig. 14-20). If pump 1 cannot maintain a minimum pressure, as dictated by pressure switch 1PS, then the second pump motor, 2, will be started after a time delay dictated by timing relay 1TR.

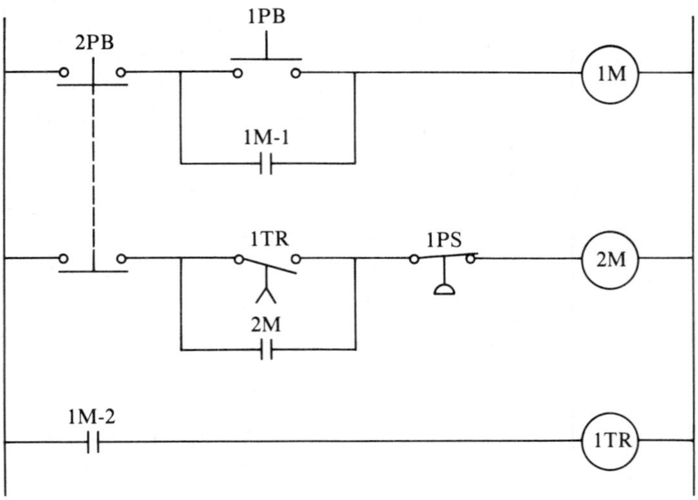

Figure 14-20. Ladder diagram for control of two pumps.

The ladder language program for implementing this control scheme is shown in Fig. 14-21. Addresses have been assigned as specified in Table 14-8. Again, the

TABLE 14-8. I/O Address Assignment for Fig. 14-20.

Device	Type	Hex Address	Description
1PB	Real input	00	START pushbutton
2PB	Real input	01	STOP pushbutton
1PS	Real input	02	Pressure switch
1M	Real output	10	Motor 1 relay
2M	Real output	11	Motor 2 relay
1TR	Internal	30	Timer functional block

memory map of Table 14-6 has been used. The addresses inside the timer functional block refer to registers where the pre-set time delay is to be stored. ■

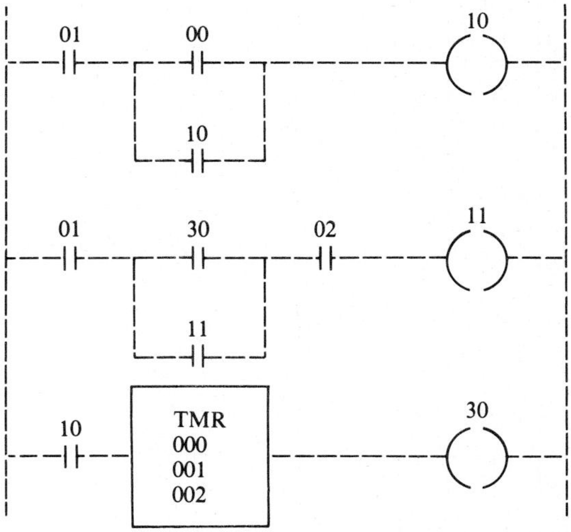

Figure 14-21. Ladder language program for Fig. 14-20.

14-2-5 Boolean Language

The Boolean language method of programming is another technique in which the PLC is programmed directly from the electrical relay ladder diagram. In some respects, the PLC Boolean language is similar to the assembly language used in programming microprocessors in that the instruction set consists of mnemonic instructions. However, the Boolean language PLC instruction set is a small number of high-level instructions in which the programming complexities of bringing in the inputs and sending out the outputs have been reduced to only a few instructions by incorporating the details into the PLC executive program. The other instructions implement the AND, OR, and NOT logical operations needed to represent the ladder devices which are in series, parallel, or are normally closed.

The mnemonics used vary from PLC manufacturer to manufacturer. Here we will use the simple instruction set defined in Table 14-9.

TABLE 14-9. Hypothetical Instruction Set for a PLC. (Entertron Industries SK-1600 PLC.)

Hex OP Code	Mnemonic Code	Meaning	Symbol
01	LOAD	Place a NO real or internal input in a ladder branch.	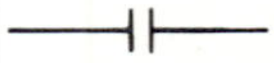
02	LOADC	Place a NC real or internal input in a ladder branch.	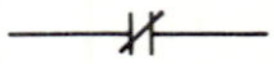
03	AND	Place a NO real or internal device in series.	
04	ANDC	Place a NC real or internal device in series.	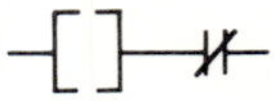
05	OR	Place a NO real or internal device in parallel.	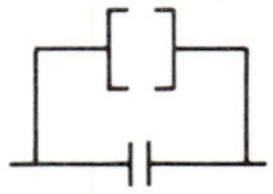
06	ORC	Place a NC real or internal device in parallel.	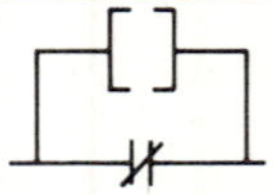
01 07	LOAD XNOR	Compare two contacts.	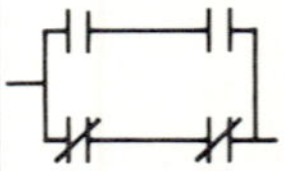
08	STORE	Place a ladder branch in a NO real or internal output.	
09	STOC	Place a ladder branch in a NC real or internal output.	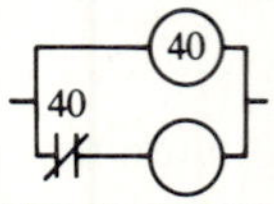
OC	TIME	Place a ladder branch in series with a timer.	

The technique of programming is very simple. Real or internal inputs are loaded into a ladder branch (actually, they are loaded into an accumulator). Then the next ladder device is either ANDed or ORed with the first input depending on whether it is in series or parallel with it. The process is continued until all components, except the branch output, have been synthesized in the ladder

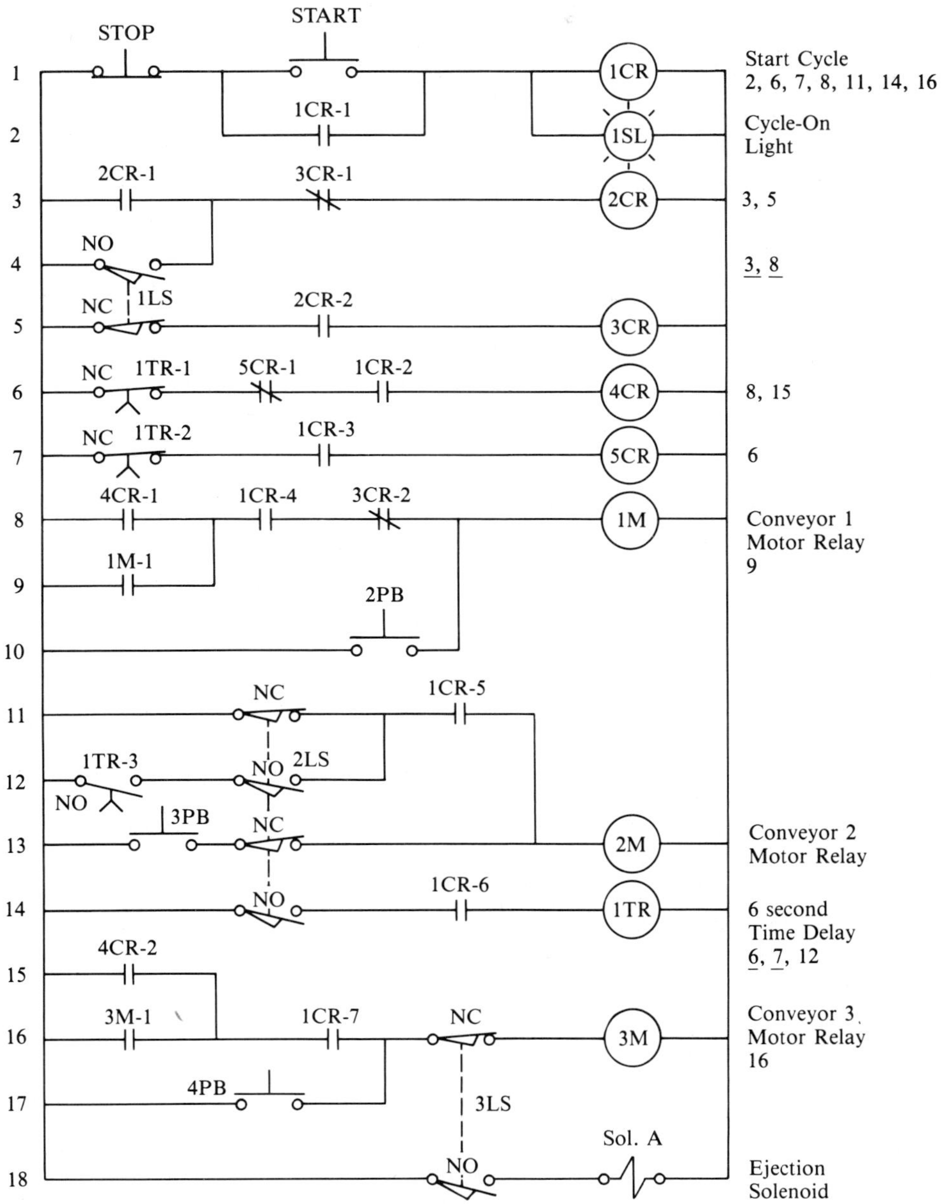

Figure 14-22. Ladder circuit for programmable controller implementation.

branch. Then the synthesized ladder branch is stored in the output (real or internal) by using a STORE instruction. The actual format for carrying out this programming process varies somewhat from manufacturer to manufacturer. In Example 14-6, we will use the format associated with the instruction set of Table 14-9, which is representative of one manufacturer.

EXAMPLE 14-6

As an illustration of PLC Boolean language programming, we will program the first ten branches of the electrical ladder diagram discussed in Example 12-3 and illustrated in Fig. 12-18. The circuit is represented in Fig. 14-22. The addresses are defined in Table 14-10.

TABLE 14-10. I/O Address Assignment for Fig. 14-22.

Device	Type	Hex address	Description
STOP PB	Real input	00	STOP pushbutton
START PB	Real input	01	START pushbutton
1LS	Real input	02	Limit switch
2PB	Real input	03	Pushbutton
1M	Real output	10	Conveyor motor relay
1SL	Real output	11	Cycle ON lamp
1CR	Internal	40	Control relay
2CR	Internal	41	Control relay
3CR	Internal	42	Control relay
4CR	Internal	43	Control relay
5CR	Internal	44	Control relay
1TR	Internal	30	Timing relay

The written program is tabulated in Table 14-11. Note that the instructions are stored in a memory addressable with 12 bits (three hexadecimal digits), with the first address at location 028. The addresses 000 through 027 are reserved for timer and counter functional blocks. Instructions can be written consecutively up to location 3FF (or a total of 984 instruction locations).

The program consists of instructions and I/O addresses stored alternately in consecutive memory locations. This is why the program code consists of alternate instruction OP codes and I/O addresses. The program is implemented by pressing the appropriate keys on the PLC keyboard. ■

14-2-6 Summary

A programmable controller has been defined by the National Electrical Manufacturers Association (NEMA) to be a digital electronic apparatus with a programmable memory for storing instructions to implement specific functions, such as logic, sequencing, timing, counting, and arithmetic, in order to control machines and processes. Physically, the programmable controller consists of two major components: its microprocessor CPU and its input/output interface.

TABLE 14-11. Boolean Language Program for Example 14-6.

Memory Address	Program Code	Mnemonic Instruction	I/O Address	Remarks
028	01	LOAD	01	Load (NO) start PB
029	01			
02A	05	OR	40	Parallel with (NO) 1CR-1
02B	40			
02C	04	ANDC	00	Series with (NC) STOP PB
02D	00			
02E	08	STORE	11	Energize indicator lamp
02F	11			
030	08	STORE	40	Energize relay 1CR
031	40			
032	01	LOAD	02	Load (NO) limit switch 1LS
033	02			
034	05	OR	41	Parallel with (NO) 2CR-1
035	41			
036	04	ANDC	42	Series with (NC) 3CR-1
037	42			
038	08	STORE	41	Energize relay 2CR
039	41			
03A	02	LOADC	02	Load (NC) limit switch 1LS
03B	02			
03C	03	AND	41	Series with (NO) 2CR-2
03D	41			
03E	08	STORE	42	Energize relay 3CR
03F	42			
040	02	LOADC	30	Load (NC) contact 1TR-1
041	30			
042	04	ANDC	44	Series with (NC) 5CR-1
043	44			
044	03	AND	40	Series with (NO) 1CR-2
045	40			
046	08	STORE	43	Energize relay 4CR
047	43			
048	02	LOADC	30	Load (NC) contact 1TR-2
049	30			
04A	03	AND	40	Series with (NO) 1CR-3
04B	40			
04C	08	STORE	44	Energize relay 5CR
04D	44			
04E	01	LOAD	43	Load (NO) 4CR-1
04F	43			
050	05	OR	10	Parallel with (NO) 1M-1
051	10			
052	03	AND	40	Series with (NO) 1CR-4
053	40			
054	04	ANDC	42	Series with (NC) 3CR-1
055	42			
056	05	OR	03	Parallel with (NO) 2PB
057	03			
058	08	STORE	10	Energize conveyor motor relay 1M
059	10			
.	.	.	.	. .
.	.	.	.	. .
.	.	.	.	. .

Insofar as the engineer is concerned, a PLC is a black box to be incorporated in a mechanical system just as are other components such as gear reducers, pumps, and so forth. The PLC is the mechanism by which mechanical system outputs are made responsive to the inputs it supplies in the form of switches and transducers.

Compared to the microprocessor, the PLC is very simple to program using either ladder language or Boolean language, both of which are capable of easily synthesizing an electrical relay ladder diagram. Although the PLC was originally developed to replace relay ladder logic hardware, it has evolved into a full-blown controller capable of more high-level control functions using functional blocks.

A photograph of a typical small PLC is shown in Fig. 14-23.

Figure 14-23. A typical small PLC. (Courtesy of Entertron Industries.)

14-3 REFERENCES

14-1. D. D. Givone and R. P. Roesser, *Microprocessors/Microcomputers: An Introduction*, McGraw-Hill, New York, 1980.

14-2. C. T. Jones and L. A. Bryan, *Programmable Controllers*, published by International Programmable Controls, Inc., 35 Glenlake Parkway, Suite 445, Atlanta, GA, 1983.

14-3. *Microprocessor Basics*, Penton/IPC, Cleveland. 1980. (Reprints of articles in *Machine Design* available from Penton/IPC.)

14-4. *Microprocessor Software*, Penton/IPC, Cleveland, 1979. (Reprints of articles in *Machine Design* available from Penton/IPC.)

14-5. R. J. Bibbero, *Microprocessors in Industrial Control*, published by Instrument Society of America, Box 12277, Research Triangle Park, NC, 1982.

14-6. A. K. Kochhar and N. D. Burns, *Microprocessors and Their Manufacturing Applications*, Edward Arnold, London, 1983.

PROBLEMS

14-1. Draw a flowchart and indicate how the ladder circuit of Example 13-12 can be implemented using the look-up table method.

14-2. Show that the resolution of an 8-bit ADC is approximately 0.4 percent if the method of successive approximations is used.

14-3. How would you turn on actuators A_4, A_5, and A_6 while keeping the others off using the scheme of Section 14-1-5?

14-4. How many instruction OP codes are possible with a 4-bit microprocessor?

In the following problems, assign I/O addresses in accordance with the memory map of Table 14-6.

14-5. Sketch the ladder language program that synthesizes the first six branches of the relay ladder diagram of Fig. P12-7(c).

14-6. Sketch the ladder language program that synthesizes the relay ladder diagram of Fig. 12-8.

14-7. Write a Boolean language program for implementing the relay ladder diagram of Fig. 13-17. Use the instruction set of Table 14-9.

14-8. Write a Boolean language program for implementing the relay ladder diagram of Fig. 12-8. Use the instruction set of Table 14-9.

14-9. Write a Boolean language program for implementing the relay ladder diagram of Fig. 14-17. Use the instruction set of Table 14-9.

14.10. Write a Boolean language program for implementing the first six branches of the relay ladder diagram of Fig. P12-7(c). Use the instruction set of Table 14-9.

14-11. Develop a microprocessor-based fuel gage system for a fuel tank of complicated geometry. Only the depth of fuel in the tank is to be sensed using a pressure transducer. This pressure measurement is signal conditioned and provided to an input port of the microprocessor. Use the look-up table method and sketch an elementary flowchart indicating how the system is to work. Assume the fuel gage display is driven by the pulse width of a signal that is stored in the look-up table. (See *Machine Design*, September 12, 1985, pp. 82–85.)

Appendix

Tensioning Systems

A common problem in the design of mechanical systems is that of controlling tension or speed in a continuous processing system. The material being processed could be a continuous web of paper, film, foil, or cloth as in printing, laminating, coating, or dyeing operations. This same type of problem arises in connection with mechanical designs requiring precise control of tension in wire-covering operations or in the wrapping of wire around cylinders.

Designs for the control of tension in a mechanical system can be open-loop or closed-loop. But almost all designs rest upon the principle of having consecutive rolls in the processing system operate at slightly different speeds.

A-1 OPEN-LOOP TENSIONING SYSTEMS

Consider the simple continuous-web configuration shown in Fig. A-1. Here, it is assumed that there is no slip between the rollers and the web, at each end of the configuration. Assume that the velocity V_2 of the material leaving the system slightly exceeds the velocity V_1 of the material entering the system. Then a tension T will be induced in the web between the two sets of rollers. We want to determine the relationship between tension T and the angular speeds ω_1 and ω_2 of the rollers.

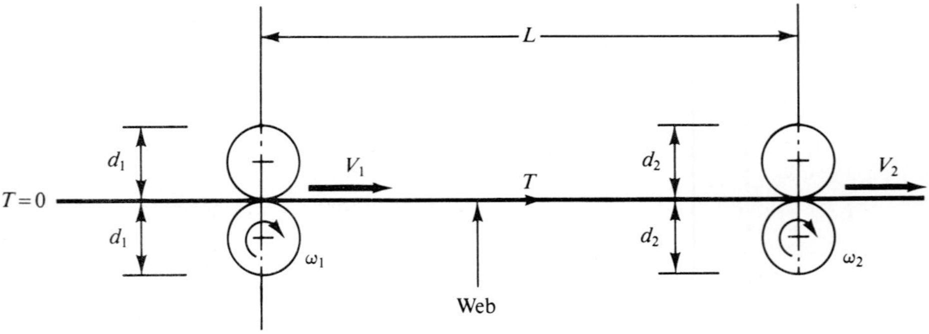

Figure A-1. Continuous-web processing configuration.

Suppose the material just to the left of the left set of rollers is unstressed. We identify time Δt as the time required for length L of unstressed web material to pass through the left set of rollers. In this same period of time, a slightly longer length $L + \delta L$ must pass through the right set of rollers because $V_2 > V_1$. Thus,

$$\Delta t = \frac{L}{V_1} = \frac{L + \delta L}{V_2} \tag{A-1}$$

or

$$\frac{V_2}{V_1} = \frac{L + \delta L}{L} = 1 + \varepsilon \tag{A-2}$$

where ε is the strain induced in the web material by the tension T existing between the left and right sets of rollers.

The relationship between stress and strain in the web material is

$$\frac{\delta L}{L} = \varepsilon = \frac{T}{AE} \tag{A-3}$$

where A is the cross-sectional area and E is the modulus of elasticity of the web material. Combining Eqs. (A-2) and (A-3) gives

$$\frac{V_2}{V_1} - 1 = \frac{T}{AE} \tag{A-4}$$

If d_1 and d_2 are the respective diameters of the left and right sets of rollers, then

$$V_1 = \frac{d_1}{2}\,\omega_1 \qquad V_2 = \frac{d_2}{2}\,\omega_2 \tag{A-5}$$

Thus, Eq. (A-4) becomes

$$\frac{\omega_2 d_2}{\omega_1 d_1} = 1 + \frac{T}{AE} \tag{A-6}$$

Equation (A-6) is the fundamental equation for open-loop tension control.

In an open-loop system, a fixed relationship is established between V_2 and V_1 by some kind of mechanical drive. One possible configuration is shown in Fig. A-2, where it is assumed that the sets of rollers have the same diameter ($d_1 = d_2$).

In Fig. A-2, a chain drive is used in which the drive sprocket is slightly larger than the driven sprocket. The configuration makes ω_2 slightly larger than ω_1. This arrangement is positive but may prove to be impractical for small tensions (because of limitations on available numbers of sprocket teeth). A V-belt drive could be used, but the potential creep and slippage inherent in such a drive could cause problems.

If the numbers of teeth on the sprockets do not give the desired ratio ω_2/ω_1, a clutch can be introduced between sprocket D_2 and roller d_2. By controlling the slip of the clutch, the ratio ω_2/ω_1 can be fine-tuned to give the nominal value

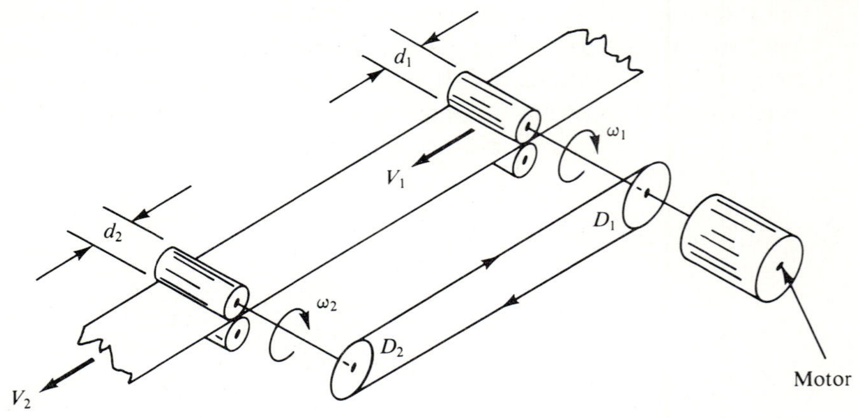

$$d_1 = d_2$$
$$D_1 > D_2$$

Figure A-2. Simple mechanical drive arrangement.

required. But in an open-loop system there is no feedback to the clutch and there could be a drift, or wandering, about the nominal speed ratio desired.

Another possibility, in Fig. A-2, is to make the sprockets of the same diameter but to make rollers d_2 slightly larger than rollers d_1. This choice, of course, builds some rather restrictive conditions into the design in that, to change the tension, the rollers would have to be changed. Changing the rollers would be more unwieldy than simply changing the sprockets or changing the slip of a variable-speed clutch.

The differential PIV drive, discussed in Chapter 3, is well suited for tension control but is expensive. One advantage of this drive is that it is both positive and

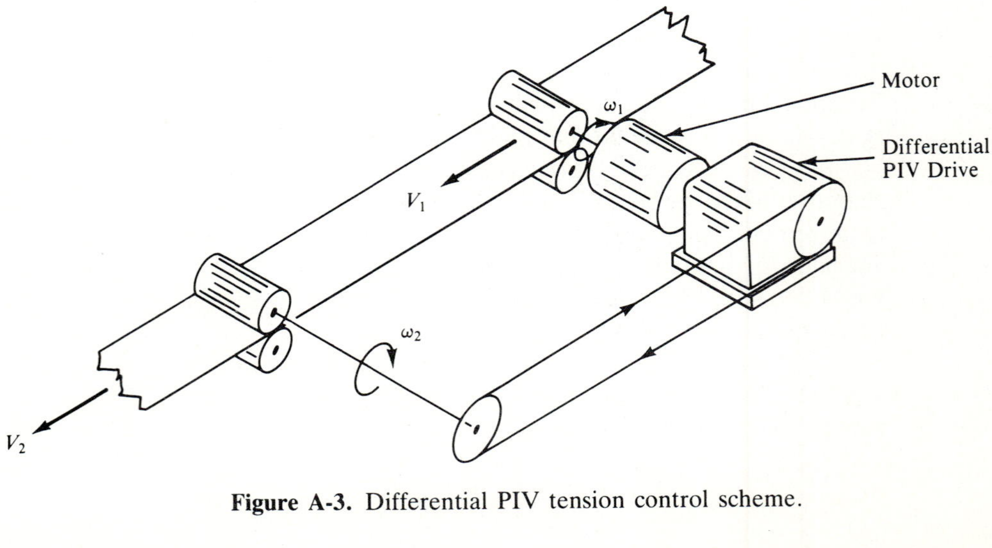

Figure A-3. Differential PIV tension control scheme.

variable in speed, so rather precise open-loop control can be achieved. It is also amenable to closed-loop control by having feedback to the PIV control mechanism. An open-loop differential PIV drive for tension control is shown in Fig. A-3. In the configuration shown, the narrow-speed-range drive would be used to make speed ω_2 slightly larger than, but synchronized with, speed ω_1.

A-2 CLOSED-LOOP TENSIONING SYSTEMS

In an open-loop tensioning system, tension in the system is not sensed to determine if it is the precise value intended by the design hardware. The tension is designed into the hardware, so to speak, and whether or not the nominal tension desired is actually produced by the design will depend on how good the design is.

Additional components must be added to an open-loop system if precise control is to be obtained for the tension. These additional components are in (a) a transducer for sensing tension in the web, (b) a command or setpoint device, and (c) a comparator for generating an error signal. An amplifier may also be needed if the various signals in this (now) closed-loop system are not properly matched in magnitude.

A-2-1 The Dancer Transducer

In a closed-loop tension control system, the transducer must sense changes in tension and convert them into a usable signal. One simple mechanical device for achieving this function is the dancer. A sketch of a dancer is shown in Fig. A-4.

A dancer is simply a weighted roller suspended in a loop of the web as shown in Fig. A-4. The nominal value of tension in the web is just half of the

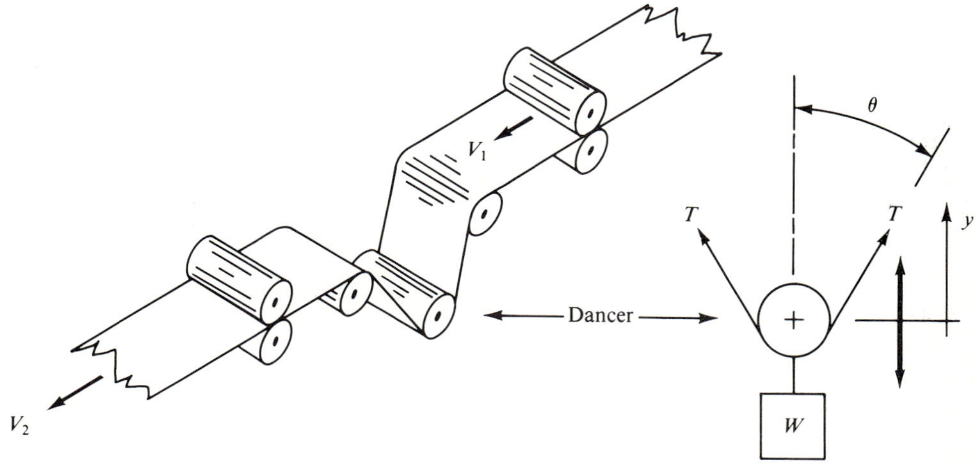

Figure A-4. Dancer transducer.

dancer weight value $W/\cos\theta$. This device is called a dancer because it will "dance" in response to changes in web tension that are slightly greater than or less than the nominal value. So, vertical position of the dancer is the direct mechanical output of the transducer while tension is the input. The basic datum value of tension is determined by the weight W, since θ is generally small. But note that as the dancer rises and falls, there are changes in θ, which means that some variation in the base level of tension can be obtained by raising or lowering the datum position of the dancer.

The dancer position signal can be changed into other signals quite easily. The dancer position can be changed into voltage in a variety of ways, the simplest of which would be to have the dancer actuate the wiper of a potentiometer.

A closed-loop system using the dancer arrangement shown in Fig. A-4 will operate in the following way. Suppose the tension in the web drops below the nominal value as dictated by weight W and reference angle θ_o.* Then the dancer will experience a downward motion, which will generate an error signal at a comparator. Then the error signal will cause an actuator (perhaps a motor) to speed up rollers d_2, and the resulting increase in tension will cause the dancer to return to its original position. The error signal will vanish when the dancer has returned to its datum position; the nominal tension desired will also have been restored.†

A-3 REWIND SYSTEMS

In many continuous-web processing machines, the web material is unwound from a supply reel (payoff), passes through the processing operation, and then is rewound on another reel (rewind). Printing operations epitomize this kind of continuous processing.

The rewind reel grows in diameter as the web material is wound on it. Therefore, the angular speed of the rewinder must diminish as the diameter becomes larger if the linear speed of the web is to be maintained constant. Constant linear speed is generally mandated by the web operation occurring. As usual, there are both open-loop and closed-loop schemes for attempting to achieve constancy of web speed (feed).

A-3-1 Open-Loop Rewinding

In an open-loop rewind operation, it is necessary to know how the angular speed ω of the rewind reel changes with time. Once this relationship is known, it can be programmed directly into the controller to directly control the rewinder.

*This would be caused by an inadvertent speeding up of velocity V_1 or an inadvertent slowing down of velocity V_2.

†In the arrangement shown in Fig. A-2, the error could actuate a variable-speed clutch located between roller d_2 and sprocket D_2 in such a way as to reduce clutch slip and hence increase speed ω_2.

Suppose the thickness of the web material is b. Then if dt is the time corresponding to one revolution of the rewinder, the increase in rewind radius in time dt will be $dr = b$. Since $dt = 2\pi/\omega$, we have the relationship

$$\frac{dr}{dt} = \frac{b}{2\pi/\omega} = \frac{b\omega}{2\pi} \qquad \textbf{(A-7)}$$

But the linear and angular speeds are related by

$$V = r\omega \qquad \textbf{(A-8)}$$

Differentiating Eq. (A-8) will show that (if $V = \text{constant}$)

$$dr = -\frac{V}{\omega^2}\,d\omega \qquad \textbf{(A-9)}$$

Then combining Eqs. (A-7) and (A-9) produces the equation

$$\frac{d\omega}{dt} = -\frac{b\omega^3}{2\pi V}\,dt \qquad \textbf{(A-10)}$$

Integration of Eq. (A-10) yields

$$\left(\frac{\omega}{\omega_i}\right)^2 = \frac{1}{1 + \dfrac{b\omega_i^2 t}{\pi V}} \qquad \textbf{(A-11)}$$

where ω_i is the reel angular speed when $t = 0$.

In an open-loop control system for a rewinder, a controller would be designed to produce the relationship of Eq. (A-11) in real time. A rewinder designed in this fashion would require a clutch to get the reel started properly before the controller takes over.

Open-loop control of a rewinder in the manner suggested is not really practical because any small deviations of the actual system from the controller program will accumulate with time in such a way as to eventually cause slack in the web or break the web. The solution to this problem is to use a closed-loop control system.

A-3-2 Closed-Loop Rewinding

In an open-loop rewind system, the tension in the web is not sensed. Instead, the relationship between rewinder speed and time is determined, and then a programmer is used to control the rewinder in accordance with this speed-time relationship. Note that although rewinder speed is controlled, tension in the web is not. This can result in an accumulated slack (or overtension) in the web which would be very detrimental to performance.

One way around this problem would be to add a slip clutch (e.g., a magnetic clutch) to the system so that a small slip exists between the actual web speed and the speed the controller would otherwise require.

A better alternative to rewinder control is to use a closed-loop control system. A schematic diagram of such a scheme is shown in Fig. A-5. Here, a

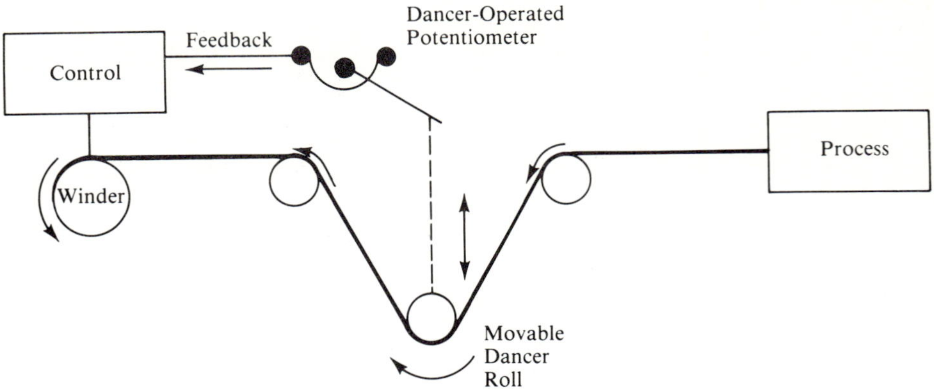

Figure A-5. Closed-loop rewinder control.

dancer is used to set the basic tension level and sense changes in tension. The dancer actuates a potentiometer whose signal is compared to a setpoint value established for some reference position of the dancer. If the dancer is out of position, an error signal is generated that actuates the rewinder control in such a way as to return the dancer to its reference position.

Note the two alternative modes in which such a closed-loop control system might work. One mode would be to program Eq. (A-11) into the rewind motor controller and then let the dancer correct only the tension departures from the basic program. The error signal would be null if the system was perfectly designed (i.e., if it was a perfect open-loop control system).

The alternative mode is simply to sense the web tension in the absence of any subsidiary speed control program. An error signal will always be present. It is this error signal itself, rather than a programmed control, that is used to slow down the drive motor as rewind radius increases.

In the final analysis, some actual hardware (electrical, hydraulic, pneumatic, or mechanical) will be required to slow down the rewinder reel as it accumulates rewound web. One ingenious scheme for doing this is shown in Fig. A-6. In the machine illustrated, the velocity V_1 of the web upstream of the dancer is held constant by a machine line shaft which operates several successive sets of rolls between print stations. The line shaft speed ω_1 is constant and is fed into one of the side gears of a differential. The rewinder reel is driven by the spider (arm) of the differential. The other side gear of the differential is driven at a variable speed ω_2 dictated by the error signal generated by the dancer. This will cause ω_2 to slow down continuously, pass through zero speed, reverse direction, and speed up continuously as the reel goes from minimum diameter to its full maximum diameter.

A-3-3 Variable-Speed Rewinders

The rewind systems discussed above are constant-horsepower systems, because both tension and linear speed of the web are maintained constant by the control

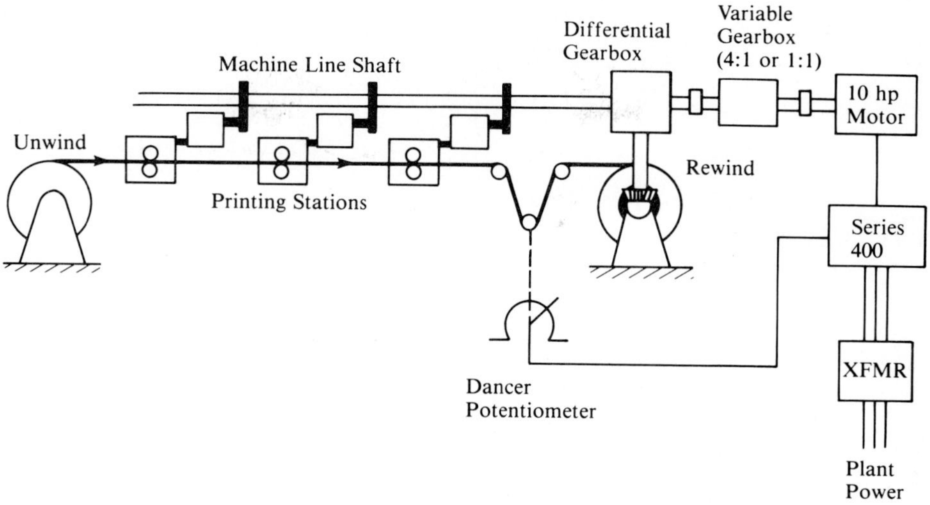

Figure A-6. Synchronized differential-drive rewinder.

arrangement. If there is no necessity to maintain the web linear velocity constant (as perhaps in a simple unwind-rewind operation), then a simple brake can be used.

Now, though, the drive at the rewind end must supply increasing horsepower if it is a simple constant-speed (rpm) drive, because the torque requirement increases as the radius of the rewind reel increases.

Index

W

X